VETERINARY FEE REFERENCE

Fifth Edition

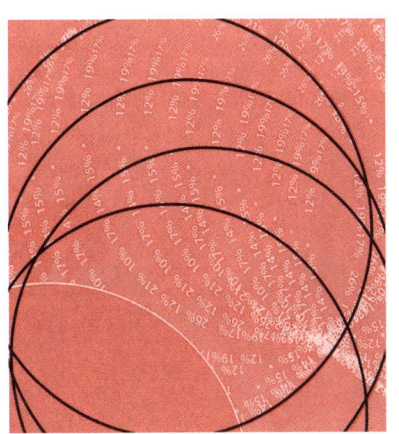
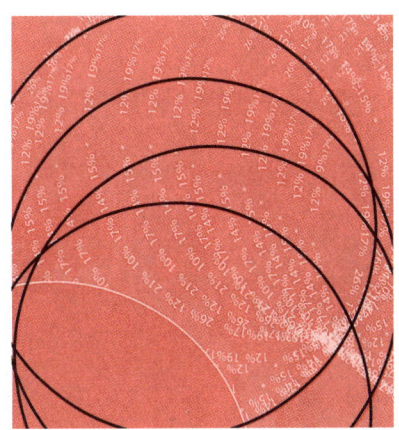
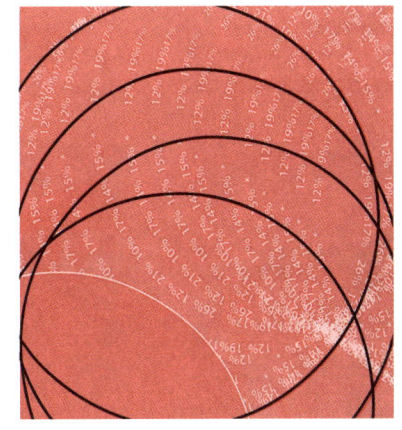

VETERINARY FEE REFERENCE

Fifth Edition

Vital Statistics for Your Veterinary Practice

Edited by Erin Landeck

American Animal Hospital Association Press
Lakewood, Colorado

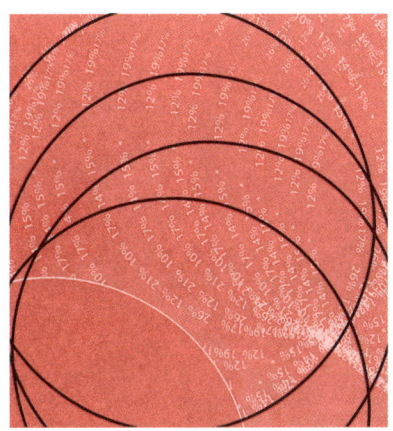
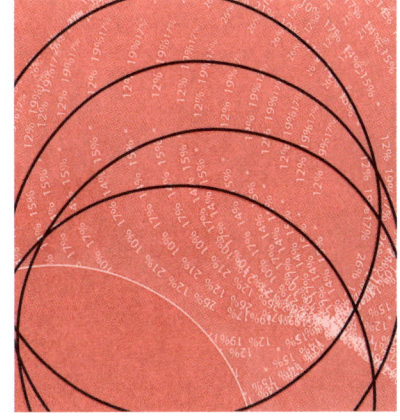
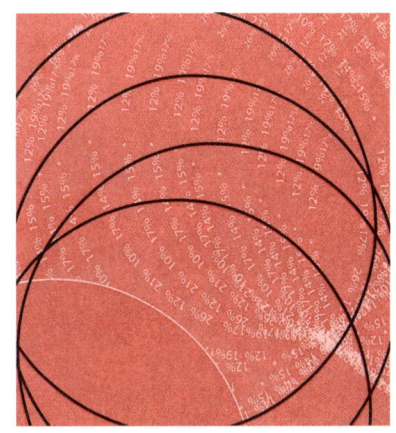

American Animal Hospital Association Press
12575 West Bayaud Avenue
Lakewood, Colorado 80228
800/252-2242 or 303/986-2800
AAHAPress@aahanet.org
www.aahanet.org

© 2007 by American Animal Hospital Association Press
All rights reserved. No part of this publication may be reproduced or transmitted in any form or by any means, electronic or mechanical, including photocopying, recording, or in an information storage and retrieval system, without permission in writing from the publisher.

Printed in the United States of America

ISBN 978-1-58326-076-0

Editor-in-Chief: Erin Landeck, MS, CPA
Graphic Designer: Elizabeth Lahey
Statistical Analyst: Donald Bacon, PhD

This information is intended to help you make good management decisions, but it is not a replacement for appropriate financial, legal, or other advice. AAHA does not assume responsibility for and makes no representation about the suitability or accuracy of the information contained in this work for any purpose, and makes no warranties, either express or implied, including the warranties of merchantability and fitness for a particular purpose. AAHA shall not be held liable for adverse reactions to or damage resulting from the application of this information or any misstatement or error contained in this work. AAHA shall be held harmless from any and all claims that may arise as a result of any reliance on the information provided.

Library of Congress Cataloging-in-Publication Data

The veterinary fee reference : vital statistics for your veterinary
practice. – 5th ed. / edited by Erin Landeck.
    p. ; cm.
 Includes bibliographical references.
 ISBN 978-1-58326-076-0
 1. Veterinary medicine–Economic aspects. 2. Veterinarians—Fees. I. Landeck, Erin. II. American Animal Hospital Association.
 [DNLM: 1. Veterinary Medicine—economics—United States-Statistics.
2. Fee Schedules—statistics & numerical data—United States. 3. Fees and Charges—statistics & numerical data—United States. SF 756.4 V5848 2007]

SF756.4.V42 2007
636.089'0681—dc22
                        2007023264

Antitrust Compliance

This fee reference has been prepared to assist veterinarians in managing their practices. One very strong word of caution about the use of these materials: Federal and state antitrust laws are designed to promote competition by prohibiting certain restraints of trade. Price fixing (or agreements among competitors concerning price or any part of price, such as discounts or credit terms) can result in criminal fines, imprisonment, and triple damages for offenders.

Agreements about price do not have to be in writing. Oral agreements or "knowing winks" are enough to prove a violation. Competitors should never discuss their prices with one another because they may create the appearance of an agreement as to prices. We remind you that you should not discuss your prices with your competitors—avoid violating antitrust laws.

The data in this publication has been collected in complete compliance with the law, and there is nothing illegal or improper about using this information to make your own independent decisions about your practice. Don't decide or discuss your prices with other veterinarians in your community.

Recommended Reading

The resources listed below are a valuable addition to your management library and are available through the American Animal Hospital Association. For more information, to order a Practice Resource Catalog, or to order, call the Member Service Center at 800/883-6301 or visit www.aahanet.org and click "AAHA Store."

AAHA Chart of Accounts, Second Edition
　　A veterinary-specific method of organizing and categorizing income, expenses, assets, and liabilities

AAHA Guide to Creating an Employee Handbook, Second Edition
　　A complete guide to developing an employee handbook that works

Compensation and Benefits, Fourth Edition
　　Detailed compensation and benefits statistics for 15 positions

Financial and Productivity Pulsepoints, Fourth Edition
　　Helpful benchmarks for your practice's revenue, expenses, and productivity levels

Legal Consent Forms for Veterinary Practices, Fourth Edition
　　Helps you make your medical records and client communication bullet proof

A Practical Guide to Performance Appraisals
　　Helps you use effectively implement and manage performance appraisals

Practice Made Perfect
Your complete guide to all facets of practice management

Valuation of Veterinary Practices
　　Learn why your practice is worth what it is

TABLE OF CONTENTS

List of Tables — x

Introduction — xxix

Survey Methodology — xxxi

Chapter 1: Overview — 1
How to Use the Data — 1
Which Factors Affect Fees? — 3
Significant Increases in Fees from 2004 to 2006 — 6
Fee Tables: Summary of Fees — 9
Fee Tables: Fee Adjustments — 28
Fee Tables: Demographics of Respondents — 30

Chapter 2: Examination and Wellness Services — 33
Training Staff for the Efficient Exam — 33
Creating Continuity of Care — 34
The Importance of Communication — 34
Choosing Appropriate Appointment Times and Lengths — 37
Fee Tables: Examinations and Consultations — 43
Fee Tables: Wellness and Preventive Care Services — 56

Chapter 3: Discounts on Services — 77
The Dangers of Discounting — 77
Discounting Dos and Don'ts — 78
Attracting High-Quality Clients without Discounts — 80
Tables: Discounts — 85

Chapter 4: Vaccination Services and Protocols — 91
Is There a Vaccine Compliance Problem? — 91
Choosing a Protocol — 92
Improving Compliance — 92
Fee Tables: Vaccination Services and Protocols — 95

Chapter 5: Grooming and Boarding Services — 107

   Trends in Grooming and Boarding Services — 107

   Does Adding Services Make Sense? — 108

   Planning Ahead — 109

   Fee Tables: Grooming Services — 115

   Fee Tables: Boarding Services — 120

Chapter 6: Dental Services — 127

   The Importance of Dental Health — 127

   Educating Clients About Preventive Dentistry — 128

   Creating Bundled Services to Increase Compliance — 129

   Fee Tables: Dental Services — 133

Chapter 7: Laboratory Services — 147

   The Lab as a Profit Center — 147

   Creating Lab Testing Protocols — 148

   Educating Clients About the Need for Diagnostics — 148

   Getting Clients to Say "Yes" to Diagnostic Tests — 149

   Fee Tables: General Lab Services — 155

   Fee Tables: Laboratory Tests Performed In-house — 160

   Fee Tables: Laboratory Tests Performed by Outside Lab — 180

Chapter 8: Diagnostic and Imaging Services — 225

   Getting Clients to Accept Diagnostic Services and Imaging — 225

   Developing Effective Referral Relationships with Specialists — 227

   Fee Tables: Diagnostic Services — 231

   Fee Tables: Imaging Services — 240

   Fee Tables: Diagnostic Imaging Cases — 255

Chapter 9: Prescription Medications — 263

   Educating Clients Pays Off — 263

   Buying Strategies — 264

   Value Your Expertise — 265

   Fee Tables: Medications and Markups — 267

Chapter 10: Fluid Therapy Services — 273

- The Importance of Fluid Therapy — 273
- Incorporating Fluid Therapy into Your Protocols — 274
- Establishing Fluid Therapy Fees — 275
- Fee Tables: Fluid Therapy Services — 277

Chapter 11: Hospitalization Services — 291

- Using Estimates to Increase Compliance — 291
- Offering Optimal Care for Hospitalized Patients — 292
- Charging Appropriately for Hospitalization — 292
- Fee Tables: Hospitalization Services — 295

Chapter 12: Anesthesia Services — 315

- Educating Clients About Anesthesia — 316
- Calming Nervous Clients — 316
- Fee Tables: Anesthesia Services — 321

Chapter 13: Treatment Procedures — 335

- Add Value to Your Treatment Procedures — 336
- Ensure Continuity of Care with Exceptional Home-Care Instructions — 336
- Fee Tables: Treatment Procedures and Cases — 341

Chapter 14: Surgical Procedures — 389

- Building Trust — 389
- Easing the Pain — 390
- Fee Tables: Surgical Setup — 395
- Fee Tables: Elective Procedures — 398
- Fee Tables: Nonelective Procedures — 416
- Fee Tables: Surgical Cases — 448

Chapter 15: End-of-Life Services — 471

- The Strength of the Human-Animal Bond — 471
- Womb-to-Tomb Care — 472
- How to Handle Billing — 473

Helping Clients Grieve for Their Pets	474
Fee Tables: Euthanasia	479
Fee Tables: Necropsy	497
Fee Tables: Cremation and Disposal	504

Surveys 509

List of Figures

2.1 Client-Education Checklist for New Clients	35
2.2 Checklist of Discussion Topics for New Pet Owners	36
3.1 Exam Report Card	40
3.1 Sample Letter of Agreement with Rescue Group	81
3.2 Sample Referral Rewards Letter	82
6.1 Educational Handout for Dental Cleaning	129
6.2 Sample Dental Reminder	131
7.1 Sample Preanesthesia Client Questionnaire	151
11.1 Checklist for Medical-Record Audit	294
12.1 Sample Postanesthesia-Care Client-Education Letter	318
13.1 Home-Care Instruction Form	339

Tables

1.A Summary of Analysis of Factors that Affect Fees	4
1.1 Average Fees for All Practices	9
1.2 Changes in Fees from 2004 to 2006	20
1.3 Frequency of Fee Adjustments	28
1.4 Method Used for Adjusting Fees	28
1.5 Fee Adjustment Percentage	29
1.6 Number of Full-Time Equivalent Veterinarians in Practice	30
1.7 Gender of Practice Ownership	30
1.8 Practice Type	31
1.9 Practice Focus	31
1.10 AAHA Membership Status	32
1.11 Metropolitan Status	32
2.A Clients' Ranking of Veterinary Practices' Communication Tools	34

2.B Appointment-Time Guidelines	37
2.1 Examination: Wellness	43
2.2 Examination: Senior Pet	43
2.3 Examination: Sick Pet	44
2.4 Examination: Single Problem (Includes Recheck Exam Fee)	44
2.5 Examination: Single Problem (Excludes Recheck Exam Fee)	45
2.6 Recheck Exam Fee Included in Single-Problem Exam Fee	45
2.7 Examination: Recheck of Previous Problem	46
2.8 Examination: Health Certificate (Includes Certificate Preparation Fee)	46
2.9 Examination: Health Certificate (Excludes Certificate Preparation Fee)	47
2.10 Certificate-Preparation Fee Included in Health-Certificate Exam Fee	47
2.11 Certificate-Preparation Fee	48
2.12 Examination: Inpatient	48
2.13 Examination: Behavior Consultation	49
2.14 Examination: Second Opinion at Owner's Request	49
2.15 Examination: Referral from Another Veterinarian	50
2.16 Examination: Emergency During Routine Office Hours	50
2.17 Examination: Emergency After Hours (6 p.m.–12 a.m.)	51
2.18 Examination: Emergency After Hours (12 a.m.–7 a.m.)	51
2.19 Examination: Avian	52
2.20 Examination: Reptile	52
2.21 Examination: Rabbit	53
2.22 Examination: Small Mammal	53
2.23 Examination: Ferret	54
2.24 Office Consultation: Pet Not Present	54
2.25 Minutes Allotted for Examination and Consultation Appointments	55
2.26 Pediatric Canine Wellness Visit: Prophylactic Deworming Treatment	56
2.27 Pediatric Canine Wellness Visit: Total	57
2.28 Adult Canine Wellness Visit: Heartworm Test Most Commonly Used	58
2.29 Adult Canine Wellness Visit: Total	59
2.30 Senior Canine Wellness Visit: Total	60
2.31 Pediatric Feline Wellness Visit: Bartonella Test	61

2.32 Pediatric Feline Wellness Visit: Total — 62

2.33 Adult Feline Wellness Visit: Heartworm Test Most Commonly Used — 63

2.34 Adult Feline Wellness Visit: Total — 64

2.35 Senior Feline Wellness Visit: Total — 65

2.36 Avian Wellness Visit: Fecal Microscopic Examination — 66

2.37 Avian Wellness Visit: Choanal Slit Culture/Sensitivity — 66

2.38 Avian Wellness Visit: Cloacal Culture/Sensitivity — 67

2.39 Avian Wellness Visit: Complete Blood Count — 67

2.40 Avian Wellness Visit: Total — 68

2.41 Reptile Wellness Visit: Fecal Examination — 69

2.42 Reptile Wellness Visit: Fecal Culture — 69

2.43 Reptile Wellness Visit: Complete Blood Count — 70

2.44 Reptile Wellness Visit: Gender Determination — 70

2.45 Reptile Wellness Visit: Total — 71

2.46 Small Mammal Wellness Visit: Ear Mite Examination — 72

2.47 Small Mammal Wellness Visit: Total — 73

2.48 Ferret Wellness Visit: Heartworm Antigen Test — 74

2.49 Ferret Wellness Visit: Plasma Chemistry Profile — 74

2.50 Ferret Wellness Visit: Total — 75

2.51 Wellness Visits Offered — 75

3.1 Wellness Exam Fees Discounted if Three or More Pets in for Exam and Booster Vaccines at Same Time — 87

3.2 Multiple-Pet Wellness Exam Discount — 87

3.3 Senior-Citizen Discount Offered on Services — 88

3.4 Senior-Citizen Discount Offered on Prescription Medications — 88

3.5 Senior-Citizen Discount Offered on Over-the-Counter Products — 89

3.6 Senior-Citizen Discounts (% Off Regular Fee) — 89

3.7 Early Spay/Neuter Discount for Puppies/Kittens Offered — 90

4.1 One-Year Rabies Vaccine — 97

4.2 Three-Year Rabies Vaccine — 97

4.3 DHPP Vaccine — 98

4.4 Parvovirus Vaccine — 98

4.5 Leptospirosis Vaccine	99
4.6 Bordetella Vaccine	99
4.7 Lyme Disease Vaccine	100
4.8 FVRCP Vaccine	100
4.9 FeLV Vaccine	101
4.10 FIP Vaccine	101
4.11 Distemper Vaccine	102
4.12 Risk-Based Determination of Vaccination for Canine/Feline Patients	102
4.13 Use Core/Noncore Vaccination Protocols	103
4.14 Use AAHA Canine Vaccine Guidelines	103
4.15 Use AAFP Feline Vaccine Recommendations	104
4.16 Recommended Frequency of Administration of Core Vaccines (DHPP or FVRCP)	104
4.17 Use Leptospirosis Vaccine	105
4.18 Frequency of Leptospirosis Vaccinations	105
4.19 Checking Immunity Titers Rather Than Vaccinating	106
4.20 Immunity Titer Check	106
5.1 Nail Trim: Dog	115
5.2 Nail Trim: Cat	115
5.3 Nail Trim: Bird	116
5.4 Nail Trim: Reptile	116
5.5 Nail Trim: Small Mammal	117
5.6 Tooth Trim: Small Mammal	117
5.7 Anal Gland Expression	118
5.8 Bath and Brush	118
5.9 Wing Trim	119
5.10 Beak Trim	119
5.11 Boarding: Dog in Small Cage, < 30 Pounds	120
5.12 Boarding: Dog in Small Run, < 30 Pounds	121
5.13 Boarding: Dog in Medium Run, 30-60 Pounds	121
5.14 Boarding: Dog in Large Run, 61-90 Pounds	122
5.15 Boarding: Dog in Large Run, > 90 Pounds	122
5.16 Boarding: Cat	123

5.17 Boarding: Bird	123
5.18 Boarding: Reptile	124
5.19 Boarding: Small Mammal	124
5.20 Boarding: Ferret	125
6.1 Dental Case: Preanesthetic Exam	135
6.2 Dental Case: CBC with Differential	136
6.3 Dental Case: Chemistry Panel with Eight Chemistries	136
6.4 Dental Case: Anesthesia, 30 Minutes	137
6.5 Dental Case: IV Catheter and Placement	137
6.6 Dental Case: IV Fluids	138
6.7 Dental Case: Dental Scaling and Polishing	139
6.8 Dental Case: Subgingival Curettage	139
6.9 Dental Case: Fluoride Application	140
6.10 Dental Case: Electronic Monitoring	141
6.11 Dental Case: Post-procedure Pain Medication	141
6.12 Dental Case: Post-procedure Injectable Antibiotics	142
6.13 Dental Case: Hospitalization	142
6.14 Dental Case: Postprocedure Antibiotics, One-Week Supply	143
6.15 Dental Case: Total	144
6.16 Extraction of Moderately Loose Premolar Tooth	145
6.17 Extraction of Firmly Implanted Upper Fourth Premolar Tooth	145
6.18 Endodontic Treatment of Upper Fourth Premolar Tooth	146
6.19 Endodontic Treatment of Lower First Molar	146
6.20 Endodontic Treatment of Upper Canine Tooth	146
7.A Sample Wellness Laboratory Screening Protocols	148
7.1 Tests Performed In-house or by Outside Lab	155
7.2 Percentage of Respondents that Do Not Offer Test	157
7.3 Lab and Blood Collection Fee: Canine	158
7.4 Lab and Blood Collection Fee: Feline	159
7.5 Lab and Blood Collection Fee for Other Species	159
7.6 In-house Lab Test Fees Not Reported	160
7.7 Fees for Select In-house Lab Tests	160

7.8 Blood Parasite Test: Ehrlichia	161
7.9 CBC Automated	161
7.10 CBC with 8–12 Chemistries	162
7.11 CBC with 16–24 Chemistries and T4	162
7.12 CBC with Manual Differential	163
7.13 CBC with No Differential	163
7.14 1 Chemistry	164
7.15 2 Chemistries	164
7.16 3 Chemistries	165
7.17 4 Chemistries	165
7.18 5–7 Chemistries	166
7.19 8–12 Chemistries	166
7.20 Cytology: Fine-Needle Aspirate	167
7.21 Cytology: Ear Swab	167
7.22 Cytology: Skin Swab	168
7.23 Cytology: Vaginal	168
7.24 Electrolytes	169
7.25 Fecal Dif-Quik Stain	169
7.26 Fecal Examination: Direct Smear	170
7.27 Fecal Examination: Flotation (Gravitational)	170
7.28 Fecal Examination: Flotation (Centrifugation, Zinc Sulfate)	171
7.29 Fecal Examination: Giardia Wet Mount	171
7.30 Fecal Gram's Stain	172
7.31 Feline Leukemia (FeLV) Test	172
7.32 FeLV and FIV Test	173
7.33 Fungal Culture	173
7.34 Giardia Antigen Test	174
7.35 Glucose Curve (6)	174
7.36 Glucose Single	175
7.37 Heartworm Test, Canine: Occult/Antigen	175
7.38 Heartworm Test, Canine: Occult/Antigen plus Lyme and E. Canis	176
7.39 Parvovirus Test	176

7.40 T4	177
7.41 Urinalysis: Complete	177
7.42 Urinalysis: Specific Gravity	178
7.43 Urinalysis: Dip Stick	178
7.44 Urinalysis: Sediment	179
7.45 Urinalysis: Microalbuminaria	179
7.46 Urinalysis: Urine Protein:Creatinine (UP:C) Ratio	180
7.47 Outside Lab Test Fees Not Reported	180
7.48 Fees for Select Outside Lab Tests	181
7.49 ACTH Stimulation	182
7.50 ACTH Stimulation (Fee Paid to Outside Lab)	182
7.51 Avian Chlamydia	183
7.52 Avian Chlamydia (Fee Paid to Outside Lab)	183
7.53 Avian Chromosomal Sexing	184
7.54 Avian Chromosomal Sexing (Fee Paid to Outside Lab)	184
7.55 Bacterial Culture and Sensitivity	185
7.56 Bacterial Culture and Sensitivity (Fee Paid to Outside Lab)	185
7.57 Bladder Stone Analysis	186
7.58 Bladder Stone Analysis (Fee Paid to Outside Lab)	186
7.59 Blood Parasite Test: Babesia	187
7.60 Blood Parasite Test: Babesia (Fee Paid to Outside Lab)	187
7.61 Blood Parasite Test: Ehrlichia	188
7.62 Blood Parasite Test: Ehrlichia (Fee Paid to Outside Lab)	188
7.63 Blood Parasite Test: Haemobartonella	189
7.64 Blood Parasite Test: Haemobartonella (Fee Paid to Outside Lab)	189
7.65 CBC Automated	190
7.66 CBC Automated (Fee Paid to Outside Lab)	190
7.67 CBC with 8-12 Chemistries	191
7.68 CBC with 8-12 Chemistries (Fee Paid to Outside Lab)	191
7.69 CBC with 16-24 Chemistries and T4	192
7.70 CBC with 16-24 Chemistries and T4 (Fee Paid to Outside Lab)	192
7.71 CBC with Manual Differential	193

7.72 CBC with Manual Differential (Fee Paid to Outside Lab)	193
7.73 8-12 Chemistries	194
7.74 8-12 Chemistries (Fee Paid to Outside Lab)	194
7.75 16-24 Chemistries	195
7.76 16-24 Chemistries (Fee Paid to Outside Lab)	195
7.77 Cytology: Fine-Needle Aspirate	196
7.78 Cytology: Fine-Needle Aspirate (Fee Paid to Outside Lab)	196
7.79 Dex Suppression	197
7.80 Dex Suppression (Fee Paid to Outside Lab)	197
7.81 Electrolytes	198
7.82 Electrolytes (Fee Paid to Outside Lab)	198
7.83 Fecal Examination: Flotation (Centrifugation, Zinc Sulfate)	199
7.84 Fecal Examination: Flotation (Centrifugation, Zinc Sulfate) (Fee Paid to Outside Lab)	199
7.85 FIV Test	200
7.86 FIV Test (Fee Paid to Outside Lab)	200
7.87 Fructosamine Test	201
7.88 Fructosamine Test (Fee Paid to Outside Lab)	201
7.89 Fungal Culture	202
7.90 Fungal Culture (Fee Paid to Outside Lab)	202
7.91 Giardia Antigen Test	203
7.92 Giardia Antigen Test (Fee Paid to Outside Lab)	203
7.93 Heartworm Test, Canine: Occult/Antigen	204
7.94 Heartworm Test, Canine: Occult/Antigen (Fee Paid to Outside Lab)	204
7.95 Heartworm Test, Feline: Occult/Antibody	205
7.96 Heartworm Test, Feline: Occult/Antibody (Fee Paid to Outside Lab)	205
7.97 Heartworm Test, Feline: Occult/Antigen	206
7.98 Heartworm Test, Feline: Occult/Antigen (Fee Paid to Outside Lab)	206
7.99 Heartworm Test, Feline: Occult/Antibody and Occult/Antigen	207
7.100 Heartworm Test, Feline: Occult/Antibody and Occult/Antigen (Fee Paid to Outside Lab)	207
7.101 Histopathology: Multiple Tissues	208
7.102 Histopathology: Multiple Tissues (Fee Paid to Outside Lab)	208
7.103 Histopathology: Single Tissue	209

7.104 Histopathology: Single Tissue (Fee Paid to Outside Lab)	209
7.105 Lyme Testing	210
7.106 Lyme Testing (Fee Paid to Outside Lab)	210
7.107 Pancreatic Evaluation	211
7.108 Pancreatic Evaluation (Fee Paid to Outside Lab)	211
7.109 Trypsin-like Immunoreactivity (TLI)	212
7.110 Trypsin-like Immunoreactivity (TLI) (Fee Paid to Outside Lab)	212
7.111 Pancreatic Lipase Immunoreactivity (PLI)	213
7.112 Pancreatic Lipase Immunoreactivity (PLI) (Fee Paid to Outside Lab)	213
7.113 Canine Pancreatic Lipase (CPL)	214
7.114 Canine Pancreatic Lipase (CPL) (Fee Paid to Outside Lab)	214
7.115 Specific Canine Pancreatic Lipase (Spec CPL)	215
7.116 Specific Canine Pancreatic Lipase (Spec CPL) (Fee Paid to Outside Lab)	215
7.117 Parvovirus Test	216
7.118 Parvovirus Test (Fee Paid to Outside Lab)	216
7.119 PTH Assay	217
7.120 PTH Assay (Fee Paid to Outside Lab)	217
7.121 Reticulocyte Count	218
7.122 Reticulocyte Count (Fee Paid to Outside Lab)	218
7.123 Serum Testing for Allergen-specific IgE	219
7.124 Serum Testing for Allergen-specific IgE (Fee Paid to Outside Lab)	219
7.125 T4	220
7.126 T4 (Fee Paid to Outside Lab)	220
7.127 T4, T3, Free T4, and Free T4ED	221
7.128 T4, T3, Free T4, and Free T4ED (Fee Paid to Outside Lab)	221
7.129 TSH Level	222
7.130 TSH Level (Fee Paid to Outside Lab)	222
7.131 Urinalysis: Urine Protein:Creatinine (UP:C) Ratio	223
7.132 Urinalysis: Urine Protein:Creatinine (UP:C) Ratio (Fee Paid to Outside Lab)	223
7.133 Increases in Markups on Outside Lab Services from 2004 to 2006	224
8.1 Routine ECG: In-house, Six-Lead	231
8.2 Routine ECG: In-house, Lead II Only	232

8.3 Routine ECG: Outside Service	233
8.4 Schirmer Tear Test	234
8.5 Corneal Stain	235
8.6 Tonometry	236
8.7 Blood Pressure Evaluation	237
8.8 Ear Swab Exam/Stain	238
8.9 Wood's Lamp Examination	239
8.10 Radiographic Setup Fee	240
8.11 Basis of Radiograph Charges	241
8.12 Routine Radiograph: One View, 8 x 10 Cassette, Dog or Cat	242
8.13 Routine Radiograph: One View, 8 x 10 Additional Cassette	242
8.14 Routine Radiograph: One View, 14 x 17 Cassette, Dog or Cat	243
8.15 Routine Radiograph: One View, 14 x 17 Additional Cassette	243
8.16 Routine Radiographs: Two Views, Chest, 60-Pound Dog	244
8.17 Routine Radiographs: Three Views, Chest, 60-Pound Dog	244
8.18 Routine Radiographs: Two Views, Pelvis, 60-Pound Dog	245
8.19 Routine Radiographs: Two Views, Abdomen, 60-Pound Dog	245
8.20 Routine Radiographs: Two Views, Spine, Dachshund	246
8.21 Routine Radiographs: Two Views, Abdomen, Cat	246
8.22 Routine Radiographs: Two Views, Chest, Cat	247
8.23 Routine Radiographs: Two Views, Forearm, Cat	247
8.24 Routine Radiograph: One View, 8 x 10 Cassette, Bird/Reptile/Small Mammal	248
8.25 Routine Radiographs: One View, 14 x 17 Cassette, Bird/Reptile/Small Mammal	248
8.26 Routine Radiographs: Two Views, 8 x 10 Cassette, Bird/Reptile/Small Mammal	249
8.27 Routine Radiographs: Two Views, 14 x 17 Cassette, Bird/Reptile/Small Mammal	249
8.28 Routine Radiography: Dental	250
8.29 Myelograms: Two Views, Cervical Spine	250
8.30 Ultrasound Unit In-house	251
8.31 Ultrasound: Chest and Abdomen	251
8.32 Ultrasound: Chest Only	252
8.33 Ultrasound: Abdomen Only	253
8.34 Ultrasound: Guided Biopsy Collection, Liver	254

8.35 Diagnostic Imaging Case One: Two Films and Your Interpretation	255
8.36 Diagnostic Imaging Case One: Two Films and Specialist Interpretation Fee	256
8.37 Diagnostic Imaging Case Two: Cystogram Procedure	257
8.38 Diagnostic Imaging Case Two: Contrast Materials	258
8.39 Diagnostic Imaging Case Two: Bladder Catheterization	259
8.40 Diagnostic Imaging Case Two: Double Contrast Cystogram Procedure	260
8.41 Diagnostic Imaging Case Two: All Films and Your Interpretation	261
8.42 Diagnostic Imaging Case Two: Total	262
9.A Heartworm and Flea Preventive Combined Sales	265
9.1 Prescription Fee for Medication Dispensed from Your Hospital	269
9.2 Fee Assessed if Client Has Prescription Filled Elsewhere	269
9.3 Prescription Fee if Client Has Prescription Filled Elsewhere	270
9.4 Average Markup: Prescription Medications	270
9.5 Average Markup: Heartworm Preventives	271
9.6 Average Markup: Food	271
9.7 Average Markup: Flea and Tick Products	272
9.8 Average Markup: Over-the-Counter Products	272
10.1 Butterfly Catheter	279
10.2 IV Indwelling Catheter	280
10.3 Jugular Catheter	281
10.4 Catheter Placement: Butterfly Catheter	282
10.5 Catheter Placement: IV Indwelling Catheter	283
10.6 Catheter Placement: Jugular Catheter	284
10.7 Initial Bag of Fluids: 1,000 ml Lactated Ringer's	285
10.8 Fluid Infusion Pump Setup	285
10.9 Fluid Infusion Pump Use (Eight Hours)	286
10.10 Lactated Ringer's Solution	286
10.11 Ringer's Solution (1,000 ml)	287
10.12 Normosol-R (1,000 ml)	287
10.13 0.9% NaCl (1,000 ml)	288
10.14 D5W (5% Dextrose) (1,000 ml)	288
10.15 Dextran (1,000 ml)	289

10.16 Hetastarch (1,000 ml)	289
11.1 Hospitalization with IV: No Overnight Stay, 10-Pound Cat	297
11.2 Hospitalization without IV: No Overnight Stay, 10-Pound Cat	298
11.3 Hospitalization with IV: No Overnight Stay, 25-Pound Dog	298
11.4 Hospitalization without IV: No Overnight Stay, 25-Pound Dog	299
11.5 Hospitalization with IV: No Overnight Stay, 60-Pound Dog	299
11.6 Hospitalization without IV: No Overnight Stay, 60-Pound Dog	300
11.7 Hospitalization with IV: No Overnight Stay, 100-Pound Dog	300
11.8 Hospitalization without IV: No Overnight Stay, 100-Pound Dog	301
11.9 Hospitalization with IV: No Overnight Stay, Bird	301
11.10 Hospitalization without IV: No Overnight Stay, Bird	302
11.11 Hospitalization with IV: No Overnight Stay, Reptile	302
11.12 Hospitalization without IV: No Overnight Stay, Reptile	303
11.13 Hospitalization with IV: No Overnight Stay, Small Mammal	303
11.14 Hospitalization without IV: No Overnight Stay, Small Mammal	304
11.15 Hospitalization with IV: No Overnight Stay, Ferret	304
11.16 Hospitalization without IV: No Overnight Stay, Ferret	305
11.17 Hospitalization with IV: Overnight Stay, 10-Pound Cat	305
11.18 Hospitalization without IV: Overnight Stay, 10-Pound Cat	306
11.19 Hospitalization with IV: Overnight Stay, 25-Pound Dog	306
11.20 Hospitalization without IV: Overnight Stay, 25-Pound Dog	307
11.21 Hospitalization with IV: Overnight Stay, 60-Pound Dog	307
11.22 Hospitalization without IV: Overnight Stay, 60-Pound Dog	308
11.23 Hospitalization with IV: Overnight Stay, 100-Pound Dog	308
11.24 Hospitalization without IV: Overnight Stay, 100-Pound Dog	309
11.25 Hospitalization with IV and Tube Feeding: Overnight Stay, Bird	309
11.26 Hospitalization without IV and Tube Feeding: Overnight Stay, Bird	310
11.27 Hospitalization with IV: Overnight Stay, Reptile	310
11.28 Hospitalization without IV: Overnight Stay, Reptile	311
11.29 Hospitalization with IV: Overnight Stay, Small Mammal	311
11.30 Hospitalization without IV: Overnight Stay, Small Mammal	312
11.31 Hospitalization with IV: Overnight Stay, Ferret	312

11.32 Hospitalization without IV: Overnight Stay, Ferret	313
11.33 Staff Physically Present in Hospital 24 Hours a Day	313
12.1 IV Sedative	323
12.2 IM Sedative	323
12.3 Client Charged for Preanesthetic Exam	324
12.4 Preanesthetic Exam	324
12.5 Preanesthetic Sedation	325
12.6 IV Induction	326
12.7 Intubation	326
12.8 Inhalant: 30 Minutes Isoflurane	327
12.9 Inhalant: 60 Minutes Isoflurane	328
12.10 Inhalant: Additional 60 Minutes Isoflurane	329
12.11 Anesthetic Monitoring: Electronic	330
12.12 Anesthetic Monitoring: Manual	330
12.13 Mask Inhalation Anesthesia: Small Mammal	331
12.14 Mask Inhalation Anesthesia: Bird	331
12.15 Pain Management Included in Elective Procedure Fees	332
12.16 Pain Management Included in Nonelective Procedure Fees	332
12.17 Fentanyl Patch Use	333
12.18 Fentanyl Patch	333
13.1 Estimate Routinely Provided for Treatment Plans	343
13.2 Deposit Routinely Required for Treatment	343
13.3 Deposit Amount Required for Treatment	344
13.4 Treatment Case One: Admitting Examination	344
13.5 Treatment Case One: Catheterization	345
13.6 Treatment Case One: 1,500 ml Lactated Ringer's Solution	346
13.7 Treatment Case One: Fluid Infusion Pump	347
13.8 Treatment Case One: Hospitalization	348
13.9 Treatment Case One: Veterinarian/Technician Supervision	349
13.10 Treatment Case One: Medical Waste Disposal	350
13.11 Treatment Case One: Total	351
13.12 Treatment Case Two: Admitting Examination	352

13.13 Treatment Case Two: Light Sedation — 352
13.14 Treatment Case Two: Abscess Curettage, Debridement, and Flush — 353
13.15 Treatment Case Two: Antibiotic Injection — 353
13.16 Treatment Case Two: Pain Medication — 354
13.17 Treatment Case Two: Hospitalization — 355
13.18 Treatment Case Two: Medical Progress Examination — 356
13.19 Treatment Case Two: Antibiotic, Oral or Injection — 357
13.20 Treatment Case Two: Antibiotics, One-Week Supply — 358
13.21 Treatment Case Two: Medical Waste Disposal — 359
13.22 Treatment Case Two: Total — 360
13.23 Treatment Case Three: Admitting Examination — 361
13.24 Treatment Case Three: IV Catheter and Placement — 362
13.25 Treatment Case Three: Liter of IV Fluids — 362
13.26 Treatment Case Three: IV Pump Use — 363
13.27 Treatment Case Three: Two Antibiotic Injections — 364
13.28 Treatment Case Three: Antiemetics — 365
13.29 Treatment Case Three: Hospitalization — 366
13.30 Treatment Case Three: Two Abdominal Radiographs — 366
13.31 Treatment Case Three: Pain Medication, Administered Twice — 367
13.32 Treatment Case Three: Medical Waste Disposal — 368
13.33 Treatment Case Three: Total, Day One — 369
13.34 Treatment Case Three: Inpatient Examination (Veterinarian Supervision) — 370
13.35 Treatment Case Three: Liter of Fluids — 371
13.36 Treatment Case Three: CBC with Six Chemistries — 372
13.37 Treatment Case Three: Total, Day Two — 373
13.38 Treatment Case Three: CBC with Two Chemistries — 374
13.39 Treatment Case Three: Special Diet — 375
13.40 Treatment Case Three: Total, Day Three — 376
13.41 Treatment Case Three: Total — 377
13.42 Treatment Case Four: Examination — 378
13.43 Treatment Case Four: Preanesthetic Lab Tests — 378
13.44 Treatment Case Four: Anesthesia, 30 Minutes — 379

13.45 Treatment Case Four: Ear Swab/Cytology	380
13.46 Treatment Case Four: Ear Cleaning	381
13.47 Treatment Case Four: Antibiotic Injection	382
13.48 Treatment Case Four: Pain Medication	383
13.49 Treatment Case Four: Hospitalization, Day Charge	384
13.50 Treatment Case Four: Antibiotic, One-Week Supply	385
13.51 Treatment Case Four: Analgesics, Three-Day Supply	386
13.52 Treatment Case Four: Medical Waste Disposal	387
13.53 Treatment Case Four: Total	388
14.1 Surgery Setup Fees	395
14.2 Electronic Monitoring Fee Charged	395
14.3 Electronic Monitoring	396
14.4 Additional Fee Charged for Emergency Surgery	396
14.5 Additional Fee for Emergency Surgery	397
14.6 Minutes Allotted for Elective Surgical Procedures	398
14.7 Canine Spay: < 25 Pounds	398
14.8 Canine Spay: 25-50 Pounds	399
14.9 Canine Spay: 51-75 Pounds	400
14.10 Canine Spay: > 75 Pounds	401
14.11 Canine Neuter: < 25 Pounds	402
14.12 Canine Neuter: 25-50 Pounds	403
14.13 Canine Neuter: 51-75 Pounds	404
14.14 Canine Neuter: > 75 Pounds	405
14.15 Feline Spay	406
14.16 Feline Neuter	407
14.17 Rabbit Spay	408
14.18 Rabbit Neuter	409
14.19 Ferret Spay	410
14.20 Ferret Neuter	411
14.21 Declaw Two Paws	412
14.22 Declaw Four Paws	413
14.23 Ear Cropping	414

14.24 Tail Docking (Neonatal)	415
14.25 Minutes Allotted for Nonelective Surgical Procedures	416
14.26 Abdominal Exploratory: Cat	417
14.27 Abdominal Exploratory: 30-Pound Dog	418
14.28 Abdominal Exploratory: 75-Pound Dog	418
14.29 Adrenalectomy: Ferret	419
14.30 Amputation: Pelvic Limb, Cat	419
14.31 Amputation: Tail, Cat	420
14.32 Amputation: Thoracic Limb, Cat	420
14.33 Amputation: Pelvic Limb, 30-Pound Dog	421
14.34 Amputation: Pelvic Limb, 75-Pound Dog	421
14.35 Amputation: Tail, 30-Pound Dog	422
14.36 Amputation: Tail, 75-Pound Dog	422
14.37 Amputation: Thoracic Limb, 30-Pound Dog	423
14.38 Amputation: Thoracic Limb, 75-Pound Dog	423
14.39 Anal Gland Resection: Bilateral, 30-Pound Dog	424
14.40 Aural Hematoma Repair	424
14.41 Broken Blood Feather Quill Removal	425
14.42 Cesarean Section: Cat with Three Kittens	425
14.43 Cesarean Section: 30-Pound Dog with Four Puppies	426
14.44 Cesarean Section: 75-Pound Dog with Four Puppies	426
14.45 Colonoscopy with Biopsy	427
14.46 Cruciate Repair: Anterior, 60-Pound Dog	427
14.47 Dewclaw Removal: Non-neonate Canine, Two Dewclaws, Front Limbs	428
14.48 Diaphragmatic Hernia: Cat	428
14.49 Diaphragmatic Hernia: 30-Pound Dog	429
14.50 Diaphragmatic Hernia: 75-Pound Dog	430
14.51 Ear Resection: Unilateral, 30-Pound Dog	431
14.52 Endoscopy: Upper GI, with Biopsy	432
14.53 Enucleation: One Eye, 30-Pound Dog	432
14.54 Femoral Head Removal: Cat	433
14.55 Gastrotomy: Foreign Body Removal, Cat	433

14.56 Gastrotomy: Foreign Body Removal, 30-Pound Dog	434
14.57 Gastrotomy: Foreign Body Removal, 75-Pound Dog	434
14.58 Incisor Extraction: Rabbit	435
14.59 Intestinal Resection/Anastomosis: Cat	435
14.60 Intestinal Resection/Anastomosis: 30-Pound Dog	436
14.61 Intestinal Resection/Anastomosis: 75-Pound Dog	436
14.62 Laser Surgery Unit In-house	437
14.63 Access to Laser Surgery Unit via Referral Center	437
14.64 Laser Surgery Fees	438
14.65 Luxating Patella Repair: Miniature Poodle	438
14.66 Mastectomy, Unilateral: Cat	439
14.67 Mastectomy, Unilateral: 30-Pound Dog	439
14.68 Mastectomy, Unilateral: 75-Pound Dog	440
14.69 Skin Tumor Removal (4 cm)	440
14.70 Splenectomy: 30-Pound Dog	441
14.71 Suture Laceration (4 cm)	441
14.72 Tibial Fracture Repair: Midshaft, Amazon Parrot	442
14.73 Tonsillectomy	442
14.74 TPLO	442
14.75 Urethrostomy: Cat	443
14.76 Wing Fracture Repair: Budgie, Simple	444
14.77 Unclassified Surgery: Per-Minute Fee, Surgical Suite	444
14.78 Minimum Fee for Unclassified Surgery Performed in Surgical Suite	445
14.79 Minimum Fee for Unclassified Surgery Performed in Surgical Suite	445
14.80 Unclassified Surgery: Per-Minute Fee, Nonsurgical Suite	446
14.81 Minimum Fee for Unclassified Surgery Performed in Nonsurgical Suite	446
14.82 Minimum Fee for Unclassified Surgery Performed in Nonsurgical Suite	447
14.83 Surgical Case One: Presurgical Examination	448
14.84 Surgical Case One: CBC	449
14.85 Surgical Case One: Chemistry Panel with Six Chemistries	450
14.86 Surgical Case One: Sedation	450
14.87 Surgical Case One: Fluids	451

14.88 Surgical Case One: Monitoring	451
14.89 Surgical Case One: Total Surgery Setup Fee	452
14.90 Surgical Case One: Surgical Suite Use	452
14.91 Surgical Case One: Surgical Pack	452
14.92 Surgical Case One: Disposables	453
14.93 Surgical Case One: Medical Waste Disposal	453
14.94 Surgical Case One: One Hour Assistant Time	454
14.95 Surgical Case One: Surgical Removal of Bladder Calculus	454
14.96 Surgical Case One: Anesthesia	455
14.97 Surgical Case One: Suture Material	456
14.98 Surgical Case One: Staples	456
14.99 Surgical Case One: Post-operative Pain Medication	457
14.100 Surgical Case One: Hospitalization	457
14.101 Surgical Case One: Total	458
14.102 Surgical Case Two: Presurgical Examination	459
14.103 Surgical Case Two: CBC	460
14.104 Surgical Case Two: Chemistry Panel with Six Chemistries	461
14.105 Surgical Case Two: Sedation	462
14.106 Surgical Case Two: Fluids	463
14.107 Surgical Case Two: Monitoring	464
14.108 Surgical Case Two: Total Surgery Setup Fee	465
14.109 Surgical Case Two: Surgical Suite Use	465
14.110 Surgical Case Two: Surgical Pack	465
14.111 Surgical Case Two: Disposables	466
14.112 Surgical Case Two: Medical Waste Disposal	466
14.113 Surgical Case Two: One Hour Assistant Time	467
14.114 Surgical Case Two: Surgical Fracture Repair (with IM Pin)	467
14.115 Surgical Case Two: Suture Material	468
14.116 Surgical Case Two: Staples	468
14.117 Surgical Case Two: Post-operative Pain Medication	468
14.118 Surgical Case Two: Hospitalization	469
14.119 Surgical Case Two: Total	470

15.1 Euthanasia Case One: IV Catheter and Placement	479
15.2 Euthanasia Case One: Preanesthetic Sedative	480
15.3 Euthanasia Case One: Euthanasia	481
15.4 Euthanasia Case One: Total	482
15.5 Euthanasia Case Two: IV Catheter and Placement	483
15.6 Euthanasia Case Two: Preanesthetic Sedative	484
15.7 Euthanasia Case Two: Euthanasia	485
15.8 Euthanasia Case Two: Total	486
15.9 Euthanasia Case Three: IV Catheter and Placement	487
15.10 Euthanasia Case Three: Preanesthetic Sedative	488
15.11 Euthanasia Case Three: Euthanasia	489
15.12 Euthanasia Case Three: Total	490
15.13 Euthanasia Case Four: IV Catheter and Placement	491
15.14 Euthanasia Case Four: Preanesthetic Sedative	492
15.15 Euthanasia Case Four: Euthanasia	493
15.16 Euthanasia Case Four: Total	494
15.17 Euthanasia Case Five: Preanesthetic Sedative	494
15.18 Euthanasia Case Five: Euthanasia	495
15.19 Euthanasia Case Five: Total	496
15.20 Necropsy: 30-Pound Dog	497
15.21 Necropsy: 75-Pound Dog	498
15.22 Necropsy: Cat	499
15.23 Necropsy: Small Mammal	500
15.24 Necropsy: Ferret	501
15.25 Necropsy: Bird	502
15.26 Necropsy: Reptile	503
15.27 Communal Cremation: 30-Pound Dog	504
15.28 Communal Cremation: Cat	505
15.29 Individual Cremation and Return of Ashes to Owner: 30-Pound Dog	506
15.30 Individual Cremation and Return of Ashes to Owner: Cat	506
15.31 Disposal of Small Mammal	507
15.32 Disposal of Reptile or Bird	507

INTRODUCTION

From a practice management perspective, the decisions you make about the fees you charge are critically important. From an economic standpoint, the fees must cover the cost of providing the product or service and provide a sufficient return to ensure that the practice can provide high-quality care. As my former partner in practice famously observed about spays (tongue firmly in cheek), "We lose money on every one we do, but we make it up in volume."

But there is an important strategic component of decision-making about fees. They can be used to position the practice relative to the other practices in the community. They can be used to grow a new practice or even control growth in an overly busy practice.

Most importantly, however, fees should reflect the value of the service provided. Clients will judge the fee for a service based on their perceived value of that service and not on the fee alone. This edition of *The Veterinary Fee Reference* will help you in your decision-making process by providing the fees charged by similar practices but also by providing information about vaccine protocols, wellness packages, senior-citizen and multiple-pet discounts, and more.

Our commitment is to help you achieve your goals so that you, in turn, can provide the highest possible quality of care to your patients.

This publication is just one of the many products and services available to you through AAHA. We hope you find it beneficial and continue to explore how we can work together to make your practice more successful than ever.

John W. Albers, DVM

Executive Director
American Animal Hospital Association

SURVEY METHODOLOGY

In mid-October, 2006, a total of 10,000 surveys were mailed to a random sample of small-animal-exclusive veterinary practices in the United States. Both AAHA members and nonmembers were included in the sample. The results are representative of all small-animal-exclusive practices, not just practices that have one or more AAHA members.

To reduce the number of questions any given respondent would have to answer, we mailed two different surveys—each containing 50% of the total survey questions plus demographic questions. One survey was sent to 5,000 respondents, and another survey was sent to the remaining 5,000 respondents.

A total of 1,043 completed surveys were returned and postmarked by the deadline of November 8, 2006, which yielded a response rate of 10.4%. As with all mail surveys, there is a possibility of nonresponse bias if those who chose to respond to the survey differ in some systematic way from those who chose not to respond. However, since the completed surveys are geographically representative of the universe of small-animal-exclusive veterinary practices in the United States, we are confident that the findings will provide a benchmark for comparison of your practice to others with similar demographics.

CHAPTER 1

OVERVIEW

In this chapter, you'll learn how to use the data in this book, see how fees have changed over time, and understand which demographic variables affect fees.

HOW TO USE THE DATA

The data tables in this book include information segmented by number of full-time equivalent (FTE) veterinarians working in the practice, AAHA member status, metropolitan status, and median household income in the area surrounding the practice. This segmentation allows you to analyze the results based on your practice's demographics.

For example, let's assume your practice has two full-time veterinarians, is AAHA accredited, and is located in an urban area. In addition, the median household income in the area surrounding the practice is $45,000. You should pay the closest attention to results for practices in the 1.1 to 2.0 FTE veterinarian range, accredited practice members, urban practices, and practices with an area household income of $35,000 to $49,999. By looking at the data in this way, you can make educated decisions based on a range of values that make sense for your practice.

Results by Region

In the past, we divided practices into four geographic regions and reported data for each of those regions. We received feedback from users that the geographic regions were too large to yield any meaningful data, and we agreed. In addition, users have suggested that even if data were reported for a relatively small geographic region (northern California, for example), that the differences among practices in that region would be too great to accurately compare to the benchmarks.

Because we do not receive enough survey responses to provide statistically valid data for geographic regions small enough to be meaningful, we did not report results by geographic region at all. Instead, we reported data for a number of other demographic segments that we believe are more meaningful in the benchmarking process.

Results by Number of Full-Time Equivalent Veterinarians

Large practices differ from small practices in many ways. For this reason, we reported results segmented by the number of full-time equivalent (FTE) veterinarians in the practice. The categories are 1.0 or less, 1.1 to 2.0, 2.1 to 3.0, and 3.1 or more. We did not receive a sufficient number of responses to break the respondents down into more categories.

A full-time equivalent is defined as a person who works 40 or more hours per week. Persons who work less than 40 hours per week were reported as fractions of a 40-hour workweek. For example, a veterinarian who works an average of 30 hours per week was reported as a .75 FTE veterinarian.

Results by AAHA Member Status

In our previous research, we have found significant differences between accredited members, nonaccredited members, and nonmembers. For that reason, we detailed the results by member status. Be sure to compare the fees reported for your member status in order to obtain the most accurate benchmarks.

Results by Metropolitan Status

We have found that reporting survey results by metropolitan status has allowed practices to obtain more accurate benchmarks than reporting data for large geographic regions (see discussion of "Results by Geographic Region" above). We divided respondents into five metropolitan status categories: urban, suburban, second city, town, and rural. These classifications were based on data from Claritas, a company that provides demographic data for research and marketing.

Urban areas have the largest population density, with several thousand people per square mile. Suburban, second city, and town areas have lower population densities but also differ in their proximity to other areas of larger population. Suburban areas are residential areas adjacent to urban areas. Second cities are satellite cities near larger cities, and towns are small cities that are not close to other metropolitan areas. Rural areas have the smallest population densities, often less than 100 people per square mile. Rural areas may or may not be close to a town, but certainly are not close to a major metropolitan area.

Results by Median Household Income

Practices in wealthier areas sometimes charge higher fees than practices in poorer areas, perhaps in part because the cost of doing business is higher in more affluent areas. U.S. census data, projected to 2006, was mapped to the ZIP code of each respondent so that we could estimate the median household income in the area around the practice.

We categorized median household income in four groups in order to approximate the commonly used socioeconomic categories of lower class (below $35,000), lower-middle class ($35,000 to $49,999), upper-middle class ($50,000 to $69,999), and upper class ($70,000 or more). Most practices in this study, and indeed most Americans, are located in middle-class neighborhoods (lower- or upper-middle class).

Averages and Medians

We reported averages and medians for each fee. The average was calculated by adding all of the values in a data set and dividing that number by the number of responses in the list. Zeros were not included in the calculation of the average. For example, the average for the sample data set below is $39.02.

$44.54, $30.85, $32.00, $58.75, $42.50,
$29.95, $47.49, $35.25, $30.00, $40.00, $37.90

The median, also called the 50th percentile, is the value of the middle number in a data set that has been arranged from smallest to largest. Fifty percent of the responses fall above the median, and 50% fall below it. In the following list of fees, which is the same data set as above rearranged from smallest to largest, the median is $37.90. Half of the values in the list are above $37.90, and half of the values are below $37.90.

$29.95, $30.00, $30.85, $32.00, $35.25,
$37.90, $40.00, $42.50, $44.54, $47.49, $58.75

Note that the median is not the same as the average for this list of numbers: $37.90 versus $39.02. A median is not skewed by extremely high or extremely low values like the average is. The median is still an imperfect measure, however, which is why you must consider all of the data when making decisions.

Percentiles

In addition to the averages and medians, we also report the 25th and 75th percentiles for each fee. The 25th percentile column indicates the point at which 25% of the responses are lower and 75% of the responses are higher than that number. For example, if the 25th percentile for a procedure fee is $20, you can conclude that 25% of respondents charge a fee that is equal to or lower than $20. Conversely, you can also conclude that 75% of respondents charge a fee higher than $20 for that procedure.

The 75th percentile column indicates the point at which 75% of the responses are lower and 25% of the responses are higher than that number. For example, if the 75th percentile for a procedure fee is $100, you can conclude that 75% of respondents charge a fee that is equal to or lower than $100 for that procedure. Conversely, you can also conclude that 25% of respondents charge a fee higher than $100 for that procedure.

The percentile data demonstrates the range of fees for each demographic segment and gives you a good idea where you stand in comparison to other practices—well above average or well below average.

Number of Respondents

Use caution when interpreting data based on a small number of respondents. It is impossible to give guidelines for how many respondents is "enough" to generate meaningful results. The standard error for any mean is different depending on the number of respondents and the variability in the data reported. A national mean based on several hundred respondents is highly reliable, but if there are few respondents from a particular metropolitan status or other demographic segment, the mean for that group of respondents is less reliable.

We did not report data when there were fewer than 25 respondents in any particular demographic segment. In the tables, an asterisk denotes that data was not reported due to an insufficient number of responses (less than 25). Be sure to take the number of respondents into consideration before you make decisions based on the data.

Excluded Data

With regard to "aberrant" responses—those that fall outside the expected range of values—some variation in fees across practices is natural and expected. However, some extreme responses likely reflect typographical errors or a misunderstanding on the part of the respondents. These extreme responses can distort the averages if they are not culled from the data set.

We used a systematic approach to identifying aberrant responses, wherein responses that were more than three standard deviations from the mean were dropped from the data set. Such extreme responses are expected about 0.3% of the time. Therefore, eliminating these responses resulted in very little loss of valid data and resulted in data that was substantially more reliable.

WHICH FACTORS AFFECT FEES?

All veterinary practices are different. Rural practices are different than urban practices. Multiveterinarian practices differ greatly from single-veterinarian practices. With these factors in mind, we analyzed the data to determine whether there were factors common to those respondents who charged higher fees.

The tables in this report show how pricing can vary by each of several demographics when considering one demographic at a time. All of these variables may have a combined effect on pricing, however. To explore the combined impact of these factors, we conducted a multiple-regression analysis. Multiple regression estimates the impact of each of several variables (in this case the demographics of the respondents' practices) on some variable of interest (in this case, a fee for a service).

Of course, the impact of these variables might differ depending on the fee we're analyzing. To explore the relationship between demographic variables and pricing, we conducted a regression analysis on several fees, including the fees for:

TABLE 1.A
SUMMARY OF ANALYSIS OF FACTORS THAT AFFECT FEES

	Wellness Examination	DHPP Vaccine	CBC	Fecal Examination	Radiograph	ECG	Hospitalization	Anesthesia	Treatment Case	Feline Spay	Gastrotomy	Dentistry Case
Number of FTE Veterinarians					+							
Member Status												
Accredited Practice Member					+			+		+		
Nonaccredited Member									-			
Nonmember	-		-									-
Metropolitan Status												
Urban	+	+			+			+				
Suburban	+	+		+								
Second City												
Town							-	-	-	-	-	-
Rural	-		-	-			-	-	-	-	-	-
Median Area Household Income												
Less than $35,000	-			-				-	-	-	-	-
$35,000 to $49,999	-			-				-	-	-	-	
$50,000 to $69,999					+					-		
$70,000 or More					+		+		+			+
Frequency of Fee Adjustments		+								+		
100% Female-Owned Practice				-			-					
Total Variance Explained	25%	3%	9%	16%	11%	4%	5%	16%	24%	23%	14%	14%

- Wellness examination
- DHPP vaccine
- CBC with manual differential
- Fecal exam with flotation (centrifugation)
- Single view 8 x 10 cassette radiograph for a dog or cat
- Routine ECG (lead II only)
- Day hospitalization with IV for a 25-pound dog
- Anesthesia (30 minutes of isoflurane for a 40-pound dog)
- Treatment case two (total)
- Feline spay
- Gastrotomy (foreign body removal for a 30-pound dog)
- Dentistry case (total)

The demographic variables we included in this analysis were:

- Number of FTE veterinarians in the practice
- Metropolitan status
- Median household income in the area containing the practice's ZIP code
- Whether the practice's owners were all female or not
- How often the practice adjusted its fees (monthly, every six months, or annually)

We did not include type of practice (general, emergency only, feline only, etc.) in the analysis because nearly all practices in this study were general practices.

Table 1.A shows the variables that were significant in the analysis of each fee. Plus signs indicate a positive association, meaning respondents within that demographic segment charged significantly higher fees than the others. For example, the plus sign for number of FTE veterinarians in the radiograph column indicates that practices with more veterinarians charge higher fees for that service than practices with fewer veterinarians. Similarly, the plus sign in the anesthesia column for urban practices means that urban practices charge more for that service than practices with other metropolitan statuses.

Negative signs indicate a negative association, meaning respondents within that demographic segment charged significantly lower fees than the others. For example, practices in the two lowest household income categories charged significantly less for a wellness examination than practices in the two highest household income categories.

You can get a general sense of which demographic variables had the most consistent effect on prices by glancing across the entire row for each demographic variable. For example, there are several positive signs for the practices in the $70,000 or more household income category, indicating that practices in areas with the highest household income charge more than other practices. Similarly, there are several negative signs for practices in the town and rural metropolitan status categories, indicating that those practices often charge lower prices than second city, urban, and suburban practices, other things being equal.

The gender of practice owners was significant only for two fees, and even then it may be an overstatement to say that some practices charge less for these services because they are 100% owned by women. Practices owned by all women tend to be smaller than mixed or 100% male-owned practices, and practice size is associated with fees.

These results should be interpreted with caution. There was still much more variation in prices than could be explained by this model. In fact, these models explain between 3 and 24% of the variation in prices (see the last row in Table 1.A). Thus, the results serve only as a general descriptor of how various demographic variables may affect pricing. Many other variables likely play a role, including the personal preferences of practice owners and managers, the cost of providing the service, and whether the practice is seeking to grow or limit growth.

SIGNIFICANT INCREASES IN FEES FROM 2004 TO 2006

In order to determine which fees increased significantly since the last time we collected this data, we compared this year's results with data from *The Veterinary Fee Reference*, Fourth Edition (published in 2005 based on fees charged in the fall of 2004). We adjusted the 2004 means for inflation before comparing them to the 2006 data, using a compound rate of inflation equal to the national Consumer Price Index.

Certain fees were not included in this table because the questions were not asked the same way in the 2004 and 2006 surveys. Also, some fees were not tested due to an insufficient number of responses (<25).

As you can see in Table 1.2 (p. 20), several of the fees increased significantly, even after inflation was taken into account. Of the 284 fees we tested, 116 (41%) were significantly higher than the inflation-adjusted fee charged two years ago. Six percent of the fees were significantly lower than in 2004. The remaining 53% of the fees we tested had not changed by a statistically significant amount.

In terms of which *types* of fees are increasing faster than others, relatively more fees had increased in these categories:

- Vaccine services
- Treatment cases
- Anesthesia services

Relatively fewer fees had increased in these categories:

- Examination and wellness services
- Grooming and boarding services
- Laboratory services
- Diagnostic and imaging services

■ ■ ■

In the tables that follow, you'll find the average fee for each of the services we asked about (all in one place for your convenience), the comparison of fees from 2004 to 2006, and the demographics of the survey respondents.

CHAPTER 1

DATA TABLES

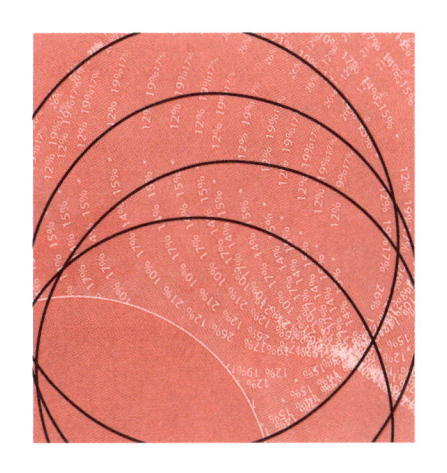

SUMMARY OF FEES

TABLE 1.1
AVERAGE FEES FOR ALL PRACTICES

Service	Average Fee	Number of Respondents
Examination and Consultation Services		
Examination: Wellness	$37.35	556
Examination: Senior Pet	$39.26	531
Examination: Sick Pet	$40.95	560
Examination: Single Problem	$38.78	546
Examination: Recheck of Previous Problem	$28.42	471
Examination: Health Certificate (Excludes Certificate Preparation Fee)	$37.32	234
Certificate-Preparation Fee	$18.31	231
Examination: Inpatient	$28.48	378
Examination: Behavior Consultation	$44.17	458
Examination: Second Opinion at Owner's Request	$40.97	536
Examination: Referral from Another Veterinarian	$40.56	498
Examination: Emergency During Routine Office Hours	$52.86	551
Examination: Emergency After Hours (6 p.m. – 12 a.m.)	$81.87	338
Examination: Emergency After Hours (12 a.m. – 7 a.m.)	$90.12	252
Examination: Avian	$43.01	234
Examination: Reptile	$42.10	214
Examination: Rabbit	$39.67	370
Examination: Small Mammal	$39.75	369
Examination: Ferret	$39.61	372
Office Consultation: Pet Not Present	$37.12	336
Wellness Services		
Pediatric Canine Wellness Visit: Prophylactic Deworming Treatment	$11.50	437
Pediatric Canine Wellness Visit: Total	$133.01	554
Adult Canine Wellness Visit: Total	$132.63	555
Senior Canine Wellness Visit: Total	$539.00	555
Pediatric Feline Wellness Visit: Bartonella Test	$59.82	141
Pediatric Feline Wellness Visit: Total	$168.05	564
Adult Feline Wellness Visit: Total	$193.05	543
Senior Feline Wellness Visit: Total	$584.22	543
Avian Wellness Visit: Fecal Microscopic Examination	$22.56	126
Avian Wellness Visit: Choanal Slit Culture/Sensitivity	$75.05	66
Avian Wellness Visit: Cloacal Culture/Sensitivity	$75.42	67
Avian Wellness Visit: Complete Blood Count	$44.34	101
Avian Wellness Visit: Total	$344.82	135
Reptile Wellness Visit: Fecal Examination	$20.71	118
Reptile Wellness Visit: Fecal Culture	$66.39	69
Reptile Wellness Visit: Complete Blood Count	$42.69	98
Reptile Wellness Visit: Gender Determination	$46.38	42
Reptile Wellness Visit: Total	$229.81	138
Small Mammal Wellness Visit: Ear Mite Examination	$19.54	248
Small Mammal Wellness Visit: Total	$132.82	316
Ferret Wellness Visit: Heartworm Antigen Test	$31.76	108
Ferret Wellness Visit: Plasma Chemistry Profile	$70.46	155
Ferret Wellness Visit: Total	$161.44	350
Discounts on Services		
Multiple-Pet Wellness Exam Discount ($)	$13.18	101
Senior Citizen Discount on Services	10%	227
Senior Citizen Discount on Prescription Medications	10%	149
Senior Citizen Discount on Over-the-Counter Products	9%	130
Vaccination Services		
One-Year Rabies Vaccine	$16.35	524
Three-Year Rabies Vaccine	$20.13	509

TABLE 1.1 (CONTINUED)
AVERAGE FEES FOR ALL PRACTICES

Service	Average Fee	Number of Respondents
Vaccination Services (continued)		
DHPP Vaccine	$21.36	524
Parvovirus Vaccine	$16.17	218
Leptospirosis Vaccine	$16.80	270
Bordetella Vaccine	$16.61	522
Lyme Disease Vaccine	$23.59	452
FVRCP Vaccine	$19.47	534
FeLV Vaccine	$21.22	535
FIP Vaccine	$21.58	192
Distemper Vaccine	$20.40	327
Immunity Titer Check	$77.34	91
Grooming and Boarding Services		
Nail Trim: Dog	$12.40	542
Nail Trim: Cat	$12.03	549
Nail Trim: Bird	$14.12	297
Nail Trim: Reptile	$12.97	116
Nail Trim: Small Mammal	$12.01	373
Tooth Trim: Small Mammal	$16.99	254
Anal Gland Expression	$16.91	543
Bath and Brush	$30.74	347
Wing Trim	$17.15	269
Beak Trim	$17.49	241
Boarding: Dog in Small Cage, < 30 Pounds	$15.30	366
Boarding: Dog in Small Run, < 30 Pounds	$15.99	323
Boarding: Dog in Medium Run, 30–60 Pounds	$17.05	355
Boarding: Dog in Large Run, 61–90 Pounds	$18.43	358
Boarding: Dog in Large Run, > 90 Pounds	$19.17	346
Boarding: Cat	$13.25	407
Boarding: Bird	$12.15	181
Boarding: Reptile	$12.63	123
Boarding: Small Mammal	$12.32	209
Boarding: Ferret	$12.84	221
Dental Services		
Dental Case: Preanesthetic Exam	$36.07	147
Dental Case: CBC with Differential	$47.46	344
Dental Case: Chemistry Panel with Eight Chemistries	$61.86	311
Dental Case: Anesthesia, 30 Minutes	$76.47	330
Dental Case: IV Catheter and Placement	$40.96	322
Dental Case: IV Fluids (1,000 ml Lactated Ringer's Solution)	$31.97	248
Dental Case: Dental Scaling and Polishing	$86.40	406
Dental Case: Subgingival Curettage	$54.00	38
Dental Case: Fluoride Application	$12.04	88
Dental Case: Electronic Monitoring	$20.99	114
Dental Case: Post-procedure Pain Medication	$21.96	318
Dental Case: Post-procedure Injectable Antibiotics	$21.01	310
Dental Case: Hospitalization	$26.76	181
Dental Case: Antibiotics, One-Week Supply	$23.23	409
Dental Case: Total	$356.97	436
Extraction of Moderately Loose Premolar Tooth	$21.53	399
Extraction of Firmly Implanted Upper Fourth Premolar Tooth	$54.26	415
Endodontic Treatment of Upper Fourth Premolar Tooth	$117.46	74
Endodontic Treatment of Lower First Molar	$116.06	69
Endodontic Treatment of Upper Canine Tooth	$109.59	71
Laboratory Services		
Lab and Blood Collection Fee: Canine	$27.96	124

TABLE 1.1 (CONTINUED)
AVERAGE FEES FOR ALL PRACTICES

Service	Average Fee	Number of Respondents
Laboratory Services (continued)		
Lab and Blood Collection Fee: Feline	$27.31	124
Lab and Blood Collection Fee: Avian	$15.62	30
Lab and Blood Collection Fee: Small Mammal (Excludes Ferret)	$17.56	42
Lab and Blood Collection Fee: Ferret	$26.28	59
Laboratory Tests Completed In-house		
ACTH Stimulation	$116.40	45
Arterial/Venous Blood Gases	$41.93	47
Bacterial Culture and Sensitivity	$51.34	39
Blood Parasite Test: Ehrlichia	$40.07	114
Blood Parasite Test: Haemobartonella	$27.32	32
CBC Automated	$36.97	275
CBC with 8–12 Chemistries	$89.35	188
CBC with 16–24 Chemistries and T4	$115.52	82
CBC with Manual Differential	$38.12	161
CBC with No Differential	$33.90	147
1 Chemistry	$20.25	340
2 Chemistries	$33.59	246
3 Chemistries	$44.12	207
4 Chemistries	$51.98	182
5–7 Chemistries	$50.33	303
8–12 Chemistries	$71.07	298
16–24 Chemistries	$78.80	56
Cytology: Fine-Needle Aspirate	$29.81	387
Cytology: Ear Swab	$20.68	484
Cytology: Skin Swab	$22.76	436
Cytology: Vaginal	$27.86	420
Dex Suppression	$100.09	49
Electrolytes	$30.90	277
Fecal Dif-Quik Stain	$21.70	244
Fecal Examination: Direct Smear	$16.62	446
Fecal Examination: Flotation (Gravitational)	$17.99	394
Fecal Examination: Flotation (Centrifugation, Zinc Sulfate)	$19.78	190
Fecal Examination: Sedimentation (Baermann)	$22.63	29
Fecal Examination: Giardia Wet Mount	$18.34	252
Fecal Gram's Stain	$24.19	140
Feline Leukemia (FeLV) Test	$33.71	207
FeLV and FIV Test	$44.85	485
FIV Test	$39.85	61
Fructosamine Test	$42.37	29
Fungal Culture	$31.93	361
Giardia Antigen Test	$28.03	232
Glucose Curve (6)	$85.03	432
Glucose Single	$17.50	474
Heartworm Test, Feline: Occult/Antibody	$31.40	47
Heartworm Test, Feline: Occult/Antigen	$28.46	58
Heartworm Test, Canine: Occult/Antigen	$29.71	255
Heartworm Test, Canine: Occult/Antigen plus Lyme and E. Canis	$38.75	317
Lyme Testing	$44.07	80
Pancreatic Evaluation	$57.24	52
Canine Pancreatic Lipase (CPL)	$24.33	50
Parvovirus Test	$41.92	381
Reticulocyte Count	$28.94	53
T4	$40.74	126
Urinalysis: Complete	$31.05	447
Urinalysis: Urine Specific Gravity	$12.73	275

TABLE 1.1 (CONTINUED)
AVERAGE FEES FOR ALL PRACTICES

Service	Average Fee	Number of Respondents
Laboratory Tests Completed In-house (continued)		
Urinalysis: Dipstick	$17.70	335
Urinalysis: Sediment	$20.40	183U
rinalysis: Microalbuminaria	$25.66	85
Urinalysis: Urine Protein:Creatinine (UP:C) Ratio	$42.47	84
Laboratory Tests Completed by Outside Lab		
ACTH Stimulation	$128.31	423
Avian Chlamydia	$69.93	81
Avian Chromosomal Sexing	$71.02	85
Avian PBFDV	$80.62	49
Avian Polyomavirus Test	$73.64	59
Bacterial Culture and Sensitivity	$77.95	432
Bladder Stone Analysis	$74.95	415
Blood Parasite Test: Babesia	$90.45	163
Blood Parasite Test: Ehrlichia	$83.38	206
Blood Parasite Test: Haemobartonella	$48.21	188
CBC Automated	$38.19	118
CBC with 8–12 Chemistries	$80.40	193
CBC with 16–24 Chemistries and T4	$106.36	347
CBC with Manual Differential	$38.92	166
CBC with No Differential	$35.62	44
1 Chemistry	$28.03	27
5–7 Chemistries	$50.08	29
8–12 Chemistries	$61.38	71
16–24 Chemistries	$78.09	242
Cytology: Fine-Needle Aspirate	$74.76	102
Dex Suppression	$115.58	350
Electrolytes	$44.69	115
Fecal Examination: Flotation (Centrifugation, Zinc Sulfate)	$25.58	85
Fecal Examination: Sedimentation (Baermann)	$68.51	39
Fecal Gram's Stain	$41.71	30
Feline Leukemia (FeLV) Test	$35.92	75
FeLV and FIV Test	$46.98	31
FIV Test	$51.50	133
Fructosamine Test	$54.99	376
Fungal Culture	$68.01	121
Giardia Antigen Test	$55.86	130
Heartworm Test, Canine: Occult/Antigen	$28.79	126
Heartworm Test, Feline: Occult/Antibody	$49.77	179
Heartworm Test, Feline: Occult/Antigen	$29.50	157
Heartworm Test, Feline: Occult/Antibody and Occult/Antigen	$64.46	189
Histopathology: Multiple Tissues	$104.29	382
Histopathology: Single Tissue	$90.51	447
Lyme Testing	$74.10	253
Pancreatic Evaluation	$98.12	141
Trypsin-like Immunoreactivity (TLI)	$98.48	315
Pancreatic Lipase Immunoreactivity (PLI)	$89.41	167
Canine Pancreatic Lipase (CPL)	$59.18	95
Specific Canine Pancreatic Lipase (Spec CPL)	$73.07	112
Parvovirus Test	$61.14	90
PTH Assay	$122.78	155
Reticulocyte Count	$29.50	252
Serum Testing for Allergen-Specific IgE	$207.24	193
T4	$43.16	345
T4, T3, Free T4, and Free T4ED	$101.73	315
TSH Level	$65.23	232
Uric Acid: Avian	$34.91	27

TABLE 1.1 (CONTINUED)
AVERAGE FEES FOR ALL PRACTICES

Service	Average Fee	Number of Respondents
Laboratory Tests Completed by Outside Lab (continued)		
Urinalysis: Complete	$33.78	62
Urinalysis: Microalbuminaria	$28.32	62
Urinalysis: Urine Protein:Creatinine (UP:C) Ratio	$57.56	283
Diagnostic and Imaging Services		
Routine ECG: In-house, Six-Lead	$63.07	221
Routine ECG: In-house, Lead II Only	$39.72	244
Routine ECG: Outside Service	$95.06	223
Schirmer Tear Test	$18.38	529
Corneal Stain	$17.94	535
Tonometry	$26.02	366
Blood Pressure Evaluation	$26.78	363
Ear Swab Exam/Stain	$20.87	517
Wood's Lamp Examination	$12.25	204
Radiographic Setup Fee	$66.81	125
Routine Radiograph: One View, 8 x 10 Cassette, Dog or Cat	$66.61	451
Routine Radiograph: One View, 8 x 10 Additional Cassette	$42.86	450
Routine Radiograph: One View, 14 x 17 Cassette, Dog or Cat	$69.19	477
Routine Radiograph: One View, 14 x 17 Additional Cassette	$44.91	480
Routine Radiographs: Two Views, Chest, 60-Pound Dog	$103.52	486
Routine Radiographs: Three Views, Chest, 60-Pound Dog	$140.02	477
Routine Radiographs: Two Views, Pelvis, 60-Pound Dog	$103.30	484
Routine Radiographs: Two Views, Abdomen, 60-Pound Dog	$103.50	485
Routine Radiographs: Two Views, Spine, Dachshund	$99.46	486
Routine Radiographs: Two Views, Abdomen, Cat	$97.25	491
Routine Radiographs: Two Views, Chest, Cat	$97.14	491
Routine Radiographs: Two Views, Forearm, Cat	$94.83	492
Routine Radiograph: One View, 8 x 10 Cassette, Bird/Reptile/Small Mammal	$66.66	298
Routine Radiograph: One View, 14 x 17 Cassette, Bird/Reptile/Small Mammal	$69.72	302
Routine Radiographs: Two Views, 8 x 10 Cassette, Bird-Reptile-Small Mammal	$94.65	306
Routine Radiographs: Two Views, 14 x 17 Cassette, Bird/Reptile/Small Mammal	$98.22	317
Routine Radiographs: Dental	$68.48	237
Myelogram: Dye (Omnipaque)	$100.06	26
Myelograms: Two Views, Cervical Spine	$109.44	60
Myelogram: CSF Examination	$81.64	33
Ultrasound: Chest and Abdomen	$271.33	197
Ultrasound: Chest Only	$201.30	216
Ultrasound: Abdomen Only	$192.13	244
Ultrasound: Guided Biopsy Collection, Liver	$150.32	158
Diagnostic Imaging Case One: Two Films and Your Interpretation	$108.95	510
Diagnostic Imaging Case One: Two Films and Specialist Interpretation Fee	$160.79	237
Diagnostic Imaging Case Two: Cystogram Procedure	$94.54	183
Diagnostic Imaging Case Two: Contrast Materials	$31.69	163
Diagnostic Imaging Case Two: Bladder Catheterization	$32.90	234
Diagnostic Imaging Case Two: Double Contrast Cystogram Procedure	$102.64	167
Diagnostic Imaging Case Two: All Films and Your Interpretation	$166.48	241
Diagnostic Imaging Case Two: Total	$283.85	308
Prescription Medications		
Prescription Fee for Medication Dispensed from Your Hospital	$7.95	476
Prescription Fee if Client Has Prescription Filled Elsewhere	$8.91	96
Average Markup Applied to Prescription Medications	127%	540
Average Markup: Heartworm Preventives	94%	506
Average Markup: Food	42%	483
Average Markup: Flea and Tick Products	87%	507
Average Markup: Over-the-Counter Products	102%	497

TABLE 1.1 (CONTINUED)
AVERAGE FEES FOR ALL PRACTICES

Service	Average Fee	Number of Respondents
Fluid Therapy Services		
Butterfly Catheter	$26.63	216
IV Indwelling Catheter	$35.73	343
Catheter Placement: Butterfly Catheter	$29.73	132
Jugular Catheter	$46.48	158
Catheter Placement: IV Indwelling Catheter	$36.78	215
Catheter Placement: Jugular Catheter	$49.19	101
Initial Bag of Fluids (1,000 ml Lactated Ringer's)	$31.87	401
Fluid Infusion Pump Setup	$20.60	99
Fluid Infusion Pump Use (Eight Hours)	$19.40	152
Lactated Ringer's Solution (1,000 ml)	$32.47	411
Ringer's Solution (1,000 ml)	$32.53	194
Normosol-R (1,000 ml)	$33.12	243
0.9% NaCl (1,000 ml)	$32.43	397
D5W (5% Dextrose) (1,000 ml)	$33.98	328
Dextran (1,000 ml)	$44.84	69
Hetastarch (1,000 ml)	$71.37	198
Hospitalization Services		
Hospitalization with IV: No Overnight Stay, 10-Pound Cat	$76.00	406
Hospitalization without IV: No Overnight Stay, 10-Pound Cat	$29.29	412
Hospitalization with IV: No Overnight Stay, 25-Pound Dog	$79.56	399
Hospitalization without IV: No Overnight Stay, 25-Pound Dog	$30.25	402
Hospitalization with IV: No Overnight Stay, 60-Pound Dog	$80.85	394
Hospitalization without IV: No Overnight Stay, 60-Pound Dog	$31.85	401
Hospitalization with IV: No Overnight Stay, 100-Pound Dog	$83.63	391
Hospitalization without IV: No Overnight Stay, 100-Pound Dog	$34.59	402
Hospitalization with IV: No Overnight Stay, Bird	$66.37	68
Hospitalization without IV: No Overnight Stay, Bird	$29.91	107
Hospitalization with IV: No Overnight Stay, Reptile	$63.64	76
Hospitalization without IV: No Overnight Stay, Reptile	$29.52	114
Hospitalization with IV: No Overnight Stay, Small Mammal	$66.23	121
Hospitalization without IV: No Overnight Stay, Small Mammal	$28.19	162
Hospitalization with IV: No Overnight Stay, Ferret	$65.39	141
Hospitalization without IV: No Overnight Stay, Ferret	$28.58	169
Hospitalization with IV: Overnight Stay, 10-Pound Cat	$87.98	399
Hospitalization without IV: Overnight Stay, 10-Pound Cat	$38.53	404
Hospitalization with IV: Overnight Stay, 25-Pound Dog	$91.02	392
Hospitalization without IV: Overnight Stay, 25-Pound Dog	$40.58	396
Hospitalization with IV: Overnight Stay, 60-Pound Dog	$94.58	391
Hospitalization without IV: Overnight Stay, 60-Pound Dog	$44.03	399
Hospitalization with IV: Overnight Stay, 100-Pound Dog	$96.73	382
Hospitalization without IV: Overnight Stay, 100-Pound Dog	$44.81	394
Hospitalization with IV and Tube Feeding: Overnight Stay, Bird	$79.73	71
Hospitalization without IV and Tube Feeding: Overnight Stay, Bird	$39.77	90
Hospitalization with IV: Overnight Stay, Reptile	$72.07	82
Hospitalization without IV: Overnight Stay, Reptile	$40.52	92
Hospitalization with IV: Overnight Stay, Small Mammal	$72.88	116
Hospitalization without IV: Overnight Stay, Small Mammal	$38.04	137
Hospitalization with IV: Overnight Stay, Ferret	$73.03	129
Hospitalization without IV: Overnight Stay, Ferret	$37.19	142
Anesthesia Services		
IV Sedative: 30-Minute Radiology Procedure, No Intubation/Inhalant, 30-Pound Dog	$47.93	421
IM Sedative: Abscess Treatment, Fractious Cat	$42.13	417
Preanesthetic Exam	$30.14	67
Preanesthetic Sedation: 40-Pound Dog	$30.59	221

TABLE 1.1 (CONTINUED)
AVERAGE FEES FOR ALL PRACTICES

Service	Average Fee	Number of Respondents
Anesthesia Services (continued)		
IV Induction: 40-Pound Dog	$44.20	205
Intubation: 40-Pound Dog	$44.46	29
Inhalant: 30 Minutes, Isoflurane, 40-Pound Dog	$72.40	387
Inhalant: 60 Minutes, Isoflurane, 40-Pound Dog	$102.00	398
Inhalant: Additional Hour, Isoflurane, 40-Pound Dog	$89.60	368
Anesthetic Monitoring: Electronic	$26.19	173
Anesthetic Monitoring: Manual	$26.76	48
Mask Inhalation Anesthesia: Small Mammal	$56.76	275
Mask Inhalation Anesthesia: Bird	$56.53	125
Fentanyl Patch: 30-Pound Dog	$56.80	151
Treatment Procedures and Services		
Treatment Case One: Admitting Examination	$40.76	437
Treatment Case One: Catheterization	$42.76	428
Treatment Case One: 1,500 ml Lactated Ringer's Solution Over 24 Hours	$45.44	408
Treatment Case One: Fluid Infusion Pump	$20.39	165
Treatment Case One: Hospitalization	$42.73	435
Treatment Case One: Veterinarian/Technician Supervision	$29.64	121
Treatment Case One: Medical Waste Disposal	$3.48	121
Treatment Case One: Total	$177.95	453
Treatment Case Two: Admitting Examination	$41.56	457
Treatment Case Two: Light Sedation	$42.77	445
Treatment Case Two: Abscess Curettage, Debridement, and Flush	$53.14	449
Treatment Case Two: Antibiotic Injection	$20.91	442
Treatment Case Two: Pain Medication	$21.04	397
Treatment Case Two: Hospitalization	$34.55	422
Treatment Case Two: Medical Progress Examination	$27.25	160
Treatment Case Two: Antibiotic, Oral or Injection	$18.80	314
Treatment Case Two: Antibiotics, One-Week Supply	$19.35	444
Treatment Case Two: Medical Waste Disposal	$3.63	122
Treatment Case Two: Total	$246.43	460
Treatment Case Three: Admitting Examination	$40.78	436
Treatment Case Three: IV Catheter and Placement	$42.78	426
Treatment Case Three: First Liter IV Fluids	$32.50	379
Treatment Case Three: IV Pump Use	$19.96	159
Treatment Case Three: 2 Antibiotic Injections	$39.08	438
Treatment Case Three: Antiemetics	$24.42	392
Treatment Case Three: Hospitalization	$42.88	434
Treatment Case Three: Two Abdominal Radiographs	$105.83	437
Treatment Case Three: Pain Medication, Administered Twice	$36.08	413
Treatment Case Three: Medical Waste Disposal	$3.75	121
Treatment Case Three: Total, Day One	$353.58	448
Treatment Case Three: Inpatient Examination (Veterinarian Supervision)	$28.41	192
Treatment Case Three: Liter of Fluids	$27.66	432
Treatment Case Three: IV Pump	$19.39	134
Treatment Case Three: Two Subcutaneous Antibiotic Injections	$38.14	427
Treatment Case Three: CBC with Six Chemistries	$77.13	439
Treatment Case Three: Antiemetics	$24.49	382
Treatment Case Three: Hospitalization	$41.76	430
Treatment Case Three: Pain Medication, Administered Twice	$34.84	402
Treatment Case Three: Medical Waste Disposal	$3.65	99
Treatment Case Three: Total, Day Two	$251.31	445
Treatment Case Three: Inpatient Examination (Veterinarian Supervision)	$28.27	189
Treatment Case Three: Liter of Fluids	$27.59	430
Treatment Case Three: Fluid Pump Use	$19.42	133
Treatment Case Three: Two Subcutaneous Antibiotic Injections	$38.06	423
Treatment Case Three: CBC with Two Chemistries	$65.16	417

TABLE 1.1 (CONTINUED)
AVERAGE FEES FOR ALL PRACTICES

Service	Average Fee	Number of Respondents
Treatment Procedures and Services (continued)		
Treatment Case Three: Hospitalization	$41.05	419
Treatment Case Three: Pain Medication, Administered Twice	$34.61	398
Treatment Case Three: Medical Waste Disposal	$3.85	103
Treatment Case Three: Special Diet	$5.75	179
Treatment Case Three: Total, Day Three	$214.52	445
Treatment Case Three: Total	$816.29	448
Treatment Case Four: Examination	$40.35	437
Treatment Case Four: Preanesthetic Lab Tests	$65.63	421
Treatment Case Four: Anesthesia, 30 Minutes	$73.33	433
Treatment Case Four: Ear Swab/Cytology	$25.21	430
Treatment Case Four: Ear Cleaning	$25.55	405
Treatment Case Four: Antibiotic Injection	$21.29	432
Treatment Case Four: Post-treatment Pain Medication	$21.24	396
Treatment Case Four: Hospitalization, Day Charge	$24.32	324
Treatment Case Four: Antibiotic, One-Week Supply	$21.29	432
Treatment Case Four: Analgesics, Three-Day Supply	$14.93	376
Treatment Case Four: Medical Waste Disposal	$3.48	121
Treatment Case Four: Total	$308.70	451
Surgical Services		
Total Surgery Setup Fee	$51.47	70
Surgical Suite Use	$41.05	48
Surgical Pack	$35.56	97
Disposables	$21.13	36
Medical Waste Disposal	$3.84	65
Assistant Time (One Hour)	$57.21	69
Suture Material	$15.07	67
Staples	$19.67	68
Use of Cold Tray Instruments	$23.38	45
Electronic Monitoring	$25.94	169
Additional Fee for Emergency Surgery	$70.74	143
Canine Spay: < 25 Pounds	$189.40	425
Canine Spay: 25–50 Pounds	$204.13	410
Canine Spay: 51–75 Pounds	$226.28	411
Canine Spay: > 75 Pounds	$250.74	415
Canine Neuter: < 25 Pounds	$164.92	425
Canine Neuter: 25–50 Pounds	$175.46	414
Canine Neuter: 51–75 Pounds	$189.19	415
Canine Neuter: > 75 Pounds	$206.81	417
Feline Spay	$156.59	428
Feline Neuter	$102.77	431
Rabbit Spay	$166.86	223
Rabbit Neuter	$126.80	243
Ferret Spay	$162.28	121
Ferret Neuter	$125.66	124
Declaw Two Paws	$207.62	428
Declaw Four Paws	$275.56	295
Ear Cropping	$337.58	90
Tail Docking (Neonatal)	$31.09	322

Note: Fees for nonelective procedures below are for procedure only.

Service	Average Fee	Number of Respondents
Abdominal Exploratory: Cat	$275.61	419
Abdominal Exploratory: 30-Pound Dog	$303.21	406
Abdominal Exploratory: 75-Pound Dog	$346.54	407
Adrenalectomy: Ferret	$308.03	76
Amputation: Pelvic Limb, Cat	$362.15	369
Amputation: Tail, Cat	$143.38	396

TABLE 1.1 (CONTINUED)
AVERAGE FEES FOR ALL PRACTICES

Service	Average Fee	Number of Respondents
Surgical Services (continued)		
Amputation: Thoracic Limb, Cat	$353.08	370
Amputation: Pelvic Limb, 30-Pound Dog	$404.30	357
Amputation: Pelvic Limb, 75-Pound Dog	$448.93	354
Amputation: Tail, 30-Pound Dog	$164.44	396
Amputation: Tail, 75-Pound Dog	$178.89	393
Amputation: Thoracic Limb, 30-Pound Dog	$391.16	356
Amputation: Thoracic Limb, 75-Pound Dog	$447.03	359
Anal Gland Resection, Bilateral: 30-Pound Dog	$303.23	307
Aural Hematoma Repair	$144.42	421
Broken Blood Feather Quill Removal	$48.24	90
Cesarean Section: Cat with Three Kittens	$304.08	413
Cesarean Section: 30-Pound Dog with Four Puppies	$368.53	411
Cesarean Section: 75-Pound Dog with Four Puppies	$418.03	408
Colonoscopy with Biopsy	$285.15	71
Cruciate Repair: Anterior, 60-Pound Dog	$610.99	266
Dewclaw Removal: Non-neonate Canine, Two Dewclaws, Front Limbs	$106.14	393
Diaphragmatic Hernia: Cat	$410.27	303
Diaphragmatic Hernia: 30-Pound Dog	$442.30	299
Diaphragmatic Hernia: 75-Pound Dog	$496.34	297
Ear Resection: Unilateral, 30-Pound Dog	$332.04	218
Endoscopy: Upper GI, with Biopsy	$274.96	93
Enucleation: One Eye, 30-Pound Dog	$262.48	385
Femoral Head Removal: Cat	$370.41	323
Gastrotomy: Foreign Body Removal, Cat	$338.27	404
Gastrotomy: Foreign Body Removal, 30-Pound Dog	$368.75	399
Gastrotomy: Foreign Body Removal, 75-Pound Dog	$408.54	399
Incisor Extraction: Rabbit	$66.79	106
Intestinal Resection/Anastamosis: Cat	$419.69	368
Intestinal Resection/Anastamosis: 30-Pound Dog	$454.57	359
Intestinal Resection/Anastamosis: 75-Pound Dog	$493.88	355
Laser Surgery: Ear Surgery for Chronic Otitis	$288.60	43
Laser Surgery: Feline Declaw, Two Paws	$166.30	76
Laser Surgery: Oral Surgery	$122.38	57
Laser Surgery: Small Tumor Removal	$109.41	78
Luxating Patella Repair: Miniature Poodle	$444.97	199
Mastectomy: Unilateral, Cat	$258.54	303
Mastectomy: Unilateral, 30-Pound Dog	$310.21	303
Mastectomy: Unilateral, 75-Pound Dog	$361.86	301
Skin Tumor Removal: 4 cm	$128.00	343
Splenectomy: 30-Pound Dog	$427.68	302
Suture Laceration: 4 cm	$98.42	339
Tibial Fracture Repair: Midshaft, Amazon Parrot	$262.30	29
Tonsillectomy	$187.95	76
TPLO	$1,457.69	32
Urethrostomy: Cat	$386.53	241
Wing Fracture Repair: Budgie, Simple	$82.00	40
Unclassified Surgery: Per-Minute Fee, Surgical Suite	$5.18	161
Minimum Surgical Fee for Surgery Performed in Surgical Suite	$83.63	95
Unclassified Surgery: Per-Minute Fee, Nonsurgical Suite	$4.77	148
Minimum Surgical Fee for Surgery Performed in Clean Surgery Area	$66.68	72
Surgical Case One: Presurgical Examination	$39.24	278
Surgical Case One: CBC	$44.13	374
Surgical Case One: Chemistry Panel with Six Chemistries	$53.57	325
Surgical Case One: Sedation	$43.41	230
Surgical Case One: Fluids	$47.50	393
Surgical Case One: Monitoring	$25.25	158

TABLE 1.1 (CONTINUED)
AVERAGE FEES FOR ALL PRACTICES

Service	Average Fee	Number of Respondents
Surgical Services (continued)		
Surgical Case One: Total Surgery Setup Fee	$62.30	62
Surgical Case One: Surgical Suite Use	$42.94	50
Surgical Case One: Surgical Pack	$35.67	100
Surgical Case One: Disposables	$21.48	46
Surgical Case One: Medical Waste Disposal	$4.16	110
Surgical Case One: One Hour Assistant Time	$54.98	83
Surgical Case One: Surgical Removal of Bladder Calculus	$297.79	403
Surgical Case One: Suture Material	$20.86	81
Surgical Case One: Anesthesia, 45 Minutes	$98.41	393
Surgical Case One: Staples	$21.56	70
Surgical Case One: Post-operative Pain Medication	$25.27	376
Surgical Case One: Hospitalization, Day	$34.13	294
Surgical Case One: Total	$626.64	438
Surgical Case Two: Presurgical Examination	$38.81	226
Surgical Case Two: CBC	$43.35	280
Surgical Case Two: Chemistry Panel with Six Chemistries	$54.29	246
Surgical Case Two: Sedation	$57.46	193
Surgical Case Two: Fluids	$46.59	296
Surgical Case Two: Monitoring	$27.11	113
Surgical Case Two: Total Surgery Setup Fee	$80.07	50
Surgical Case Two: Surgical Suite Use	$48.06	44
Surgical Case Two: Surgical Pack	$41.20	77
Surgical Case Two: Disposables	$23.18	36
Surgical Case Two: Medical Waste Disposal	$4.17	81
Surgical Case Two: One Hour Assistant Time	$65.75	67
Surgical Case Two: Surgical Fracture Repair with IM Pin	$432.42	294
Surgical Case Two: Suture Material	$21.64	62
Surgical Case Two: Staples	$23.35	56
Surgical Case Two: Post-operative Pain Medication	$30.97	278
Surgical Case Two: Hospitalization, Day	$34.52	225
Surgical Case Two: Total	$676.46	330
End-of-Life Services		
Euthanasia Case One: IV Catheter and Placement, 30-Pound Dog, Dropped Off	$33.30	167
Euthanasia Case One: Preanesthetic Sedative	$25.21	151
Euthanasia Case One: Euthanasia	$48.00	512
Euthanasia Case One: Total	$64.53	526
Euthanasia Case Two: IV Catheter and Placement, 30-Pound Dog, Owner Present	$33.25	200
Euthanasia Case Two: Preanesthetic Sedative	$26.95	175
Euthanasia Case Two: Euthanasia	$50.12	515
Euthanasia Case Two: Total	$69.76	533
Euthanasia Case Three: IV Catheter and Placement, Cat	$33.10	197
Euthanasia Case Three: Preanesthetic Sedative	$25.71	170
Euthanasia Case Three: Euthanasia	$45.46	523
Euthanasia Case Three: Total	$64.32	539
Euthanasia Case Four: IV Catheter and Placement, Small Mammal	$32.21	78
Euthanasia Case Four: Preanesthetic Sedative	$26.49	88
Euthanasia Case Four: Euthanasia	$34.67	383
Euthanasia Case Four: Total	$45.88	395
Euthanasia Case Five: Preanesthetic Sedative, Bird/Reptile	$29.22	51
Euthanasia Case Five: Euthanasia	$32.35	254
Euthanasia Case Five: Total	$37.34	260
Necropsy: 30-Pound Dog	$138.74	377
Necropsy: 75-Pound Dog	$155.74	381
Necropsy: Cat	$134.27	384
Necropsy: Small Mammal	$122.28	210
Necropsy: Ferret	$123.58	213

TABLE 1.1 (CONTINUED)
AVERAGE FEES FOR ALL PRACTICES

Service	Average Fee	Number of Respondents
End-of-Life Services (continued)		
Necropsy: Bird	$108.82	149
Necropsy: Reptile	$117.71	131
Communal Cremation: 30-Pound Dog	$54.18	482
Communal Cremation: Cat	$47.16	483
Individual Cremation and Return of Ashes to Owner: 30-Pound Dog	$143.23	472
Individual Cremation and Return of Ashes to Owner: Cat	$130.00	478
Disposal of Small Mammal	$32.14	268
Disposal of Reptile or Bird	$31.84	218

Note 1: Cases and wellness visits are described in detail in the notes for the first table for that case or wellness visit.
Note 2: Fee data was not reported if there were fewer than 25 responses for that service.

TABLE 1.2
CHANGES IN FEES FROM 2004 TO 2006

Service	2004 Inflation-Adjusted Average	2006 Average	Percentage Increase	Statistically Significant Change	Minimum Number of Respondents
Examination and Consultation Services					
Examination: Single Problem (Includes Recheck Exam Fee)	$38.01	$39.02	3%	No	403
Examination: Single Problem (Excludes Recheck Exam Fee)	$36.48	$38.13	5%	No	48
Examination: Recheck of Previous Problem	$27.00	$28.42	5%	Yes	402
Examination: Health Certificate (Includes Certificate Preparation Fee)	$43.94	$39.23	-11%	No	46
Examination: Inpatient	$28.26	$28.48	1%	No	378
Examination: Behavior Consultation	$41.22	$44.17	7%	Yes	458
Examination: Second Opinion at Owner's Request	$40.27	$40.97	2%	No	536
Examination: Referral from Another Veterinarian	$39.55	$40.56	3%	No	498
Examination: Avian	$40.52	$43.01	6%	Yes	234
Examination: Reptile	$39.68	$42.10	6%	Yes	214
Examination: Rabbit	$38.23	$39.67	4%	Yes	370
Examination: Small Mammal	$37.93	$39.75	5%	Yes	369
Office Consultation: Pet Not Present	$36.33	$37.12	2%	No	336
Wellness Services					
Pediatric Canine Wellness Visit: Prophylactic Deworming Treatment	$12.31	$11.50	-7%	Yes	438
Avian Wellness Visit: Fecal Microscopic Examination	$18.60	$22.56	21%	Yes	126
Avian Wellness Visit: Choanal Slit Culture/Sensitivity	$71.70	$75.05	5%	No	66
Avian Wellness Visit: Cloacal Culture/Sensitivity	$71.65	$75.42	5%	No	67
Reptile Wellness Visit: Fecal Examination	$19.12	$20.71	8%	Yes	118
Reptile Wellness Visit: Fecal Culture	$69.66	$66.39	-5%	No	69
Discounts on Services					
Multiple-Pet Wellness Exam Discount ($)	$14.65	$13.18	-10%	No	101
Vaccination Services and Protocols					
One-Year Rabies Vaccine	$15.11	$16.35	8%	Yes	524
Parvovirus Vaccine	$15.54	$16.17	4%	No	218
Leptospirosis Vaccine	$16.28	$16.80	3%	No	199
Bordetella Vaccine	$15.57	$16.61	7%	Yes	523
Lyme Disease Vaccine	$21.94	$23.59	7%	Yes	454
FVRCP Vaccine	$17.33	$19.47	12%	Yes	534
FeLV Vaccine	$19.09	$21.22	11%	Yes	535
FIP Vaccine	$19.72	$21.58	9%	Yes	192
Immunity Titer Check	$74.15	$77.34	4%	No	91
Grooming and Boarding Services					
Nail Trim: Dog	$11.88	$12.40	4%	Yes	542
Nail Trim: Cat	$11.53	$12.03	4%	Yes	549
Nail Trim: Bird	$13.62	$14.12	4%	No	297
Nail Trim: Small Mammal	$11.58	$12.01	4%	No	373
Anal Gland Expression	$16.47	$16.91	3%	No	543
Bath and Brush	$28.59	$30.74	8%	Yes	347
Wing Trim	$16.94	$17.15	1%	No	269
Beak Trim	$17.01	$17.49	3%	No	241
Boarding: Dog in Small Cage, < 30 Pounds	$14.94	$15.30	2%	No	366
Boarding: Dog in Small Run, < 30 Pounds	$15.70	$15.99	2%	No	323
Boarding: Dog in Medium Run, 30–60 Pounds	$16.84	$17.05	1%	No	355

TABLE 1.2 (CONTINUED)
CHANGES IN FEES FROM 2004 TO 2006

Service	2004 Inflation-Adjusted Average	2006 Average	Percentage Increase	Statistically Significant Change	Minimum Number of Respondents
Grooming and Boarding Services (continued)					
Boarding: Dog in Large Run, 61–90 Pounds	$18.16	$18.43	1%	No	358
Boarding: Cat	$12.91	$13.25	3%	No	407
Boarding: Bird	$11.61	$12.15	5%	No	181
Boarding: Reptile	$11.89	$12.63	6%	No	123
Boarding: Small Mammal	$11.94	$12.32	3%	No	209
Dental Services					
Dental Case: Preanesthetic Exam	$32.18	$36.07	12%	Yes	147
Dental Case: CBC with Differential	$36.72	$47.46	29%	Yes	344
Dental Case: Chemistry Panel with Eight Chemistries	$61.82	$61.86	0%	No	311
Dental Case: Anesthesia, 30 Minutes	$62.69	$76.47	22%	Yes	330
Dental Case: IV Catheter and Placement	$34.52	$40.96	19%	Yes	322
Dental Case: IV Fluids (1,000 ml Bag of Lactated Ringer's Solution)	$32.06	$31.97	0%	No	248
Dental Case: Dental Scaling and Polishing	$77.59	$86.40	11%	Yes	406
Dental Case: Subgingival Curettage	$35.24	$54.00	53%	Yes	38
Dental Case: Fluoride Application	$12.42	$12.04	-3%	No	88
Dental Case: Electronic Monitoring	$18.86	$20.99	11%	No	114
Dental Case: Post-procedure Pain Medication	$19.09	$21.96	15%	Yes	318
Dental Case: Post-procedure Injectable Antibiotics	$21.47	$21.01	-2%	No	310
Dental Case: Hospitalization	$23.86	$26.76	12%	Yes	181
Extraction of Moderately Loose Premolar Tooth	$20.31	$21.53	6%	No	399
Extraction of Firmly Implanted Upper Fourth Premolar Tooth	$50.44	$54.26	8%	No	415
Endodontic Treatment of Lower First Molar	$136.76	$116.06	-15%	No	69
Laboratory Services					
Lab and Blood Collection Fee: Canine	$21.54	$27.96	30%	Yes	124
Lab and Blood Collection Fee: Feline	$21.64	$27.31	26%	Yes	124
Lab and Blood Collection Fee: Avian	$17.28	$15.62	-10%	No	30
Lab and Blood Collection Fee: Small Mammal	$16.07	$17.56	9%	No	42
Laboratory Tests Completed In-house					
ACTH Stimulation	$125.75	$116.40	-7%	No	45
Arterial/Venous Blood Gases	$53.35	$41.93	-21%	Yes	27
CBC Automated	$34.87	$36.97	6%	Yes	275
CBC with 8–12 Chemistries	$83.03	$89.35	8%	Yes	188
CBC with 16–24 Chemistries and T4	$114.32	$115.52	1%	No	82
CBC with Manual Differential	$38.03	$38.12	0%	No	161
CBC with No Differential	$32.78	$33.90	3%	No	147
1 Chemistry	$20.16	$20.25	0%	No	340
2 Chemistries	$33.55	$33.59	0%	No	246
3 Chemistries	$44.35	$44.12	-1%	No	207
4 Chemistries	$52.45	$51.98	-1%	No	182
Cytology: Fine-Needle Aspirate	$27.63	$29.81	8%	Yes	387
Cytology: Vaginal	$28.87	$27.86	-4%	No	420
Dex Suppression	$100.23	$100.09	0%	No	42
Fecal Dif-Quik Stain	$23.57	$21.70	-8%	Yes	244
Fecal Examination: Giardia Wet Mount	$19.08	$18.34	-4%	No	252
Fecal Gram's Stain	$23.61	$24.19	2%	No	140
Feline Leukemia (FeLV) Test	$35.20	$33.71	-4%	No	207
FIV Test	$38.82	$39.85	3%	No	61

TABLE 1.2 (CONTINUED)
CHANGES IN FEES FROM 2004 TO 2006

Service	2004 Inflation-Adjusted Average	2006 Average	Percentage Increase	Statistically Significant Change	Minimum Number of Respondents
Laboratory Tests Completed In-house (continued)					
Fructosamine Test	$51.91	$42.37	-18%	No	29
Fungal Culture	$31.49	$31.93	1%	No	361
Glucose Curve (6)	$91.77	$85.03	-7%	Yes	432
Glucose Single	$18.36	$17.50	-5%	No	474
T4	$38.32	$40.74	6%	Yes	126
Urinalysis: Complete	$29.46	$31.05	5%	Yes	447
Laboratory Tests Completed by Outside Lab					
ACTH Stimulation	$126.11	$128.31	2%	No	423
Avian Chlamydia	$75.44	$69.93	-7%	No	81
Avian Chromosomal Sexing	$78.35	$71.02	-9%	No	85
Avian PBFDV	$94.04	$80.62	-14%	No	49
Avian Polyomavirus Test	$72.10	$73.64	2%	No	51
Bladder Stone Analysis	$77.52	$74.95	-3%	No	415
CBC Automated	$39.80	$38.19	-4%	No	118
CBC with 8–12 Chemistries	$75.43	$80.40	7%	Yes	193
CBC with 16–24 Chemistries and T4	$101.93	$106.36	4%	Yes	347
CBC with Manual Differential	$39.12	$38.92	-1%	No	166
CBC with No Differential	$35.85	$35.62	-1%	No	44
1 Chemistry	$27.54	$28.03	2%	No	27
Cytology: Fine-Needle Aspirate	$69.60	$74.76	7%	No	102
Dex Suppression	$116.44	$115.58	-1%	No	350
Fecal Gram's Stain	$40.45	$41.71	3%	No	30
Feline Leukemia (FeLV) Test	$35.59	$35.92	1%	No	75
FIV Test	$48.85	$51.50	5%	No	133
Fructosamine Test	$56.76	$54.99	-3%	No	376
Fungal Culture	$66.38	$68.01	2%	No	121
Histopathology: Multiple Tissues	$103.71	$104.29	1%	No	382
Histopathology: Single Tissue	$84.36	$90.51	7%	Yes	447
Trypsin-like Immunoreactivity (TLI)	$98.69	$98.48	0%	No	315
PTH Assay	$123.31	$122.78	0%	No	155
Serum Testing for Allergen-Specific IgE	$204.39	$207.24	1%	No	193
T4	$42.81	$43.16	1%	No	345
TSH Level	$63.91	$65.23	2%	No	232
Urinalysis: Complete	$34.92	$33.78	-3%	No	62
Diagnostic and Imaging Services					
Routine ECG: In-house, Six-Lead	$62.74	$63.07	1%	No	221
Routine ECG: In-house, Lead II Only	$43.78	$39.72	-9%	Yes	244
Routine ECG: Outside Service	$91.24	$95.06	4%	No	223
Schirmer Tear Test	$18.65	$18.38	-1%	No	529
Corneal Stain	$17.86	$17.94	0%	No	535
Tonometry	$26.62	$26.02	-2%	No	366
Blood Pressure Evaluation	$29.03	$26.78	-8%	Yes	363
Ear Swab Exam/Stain	$22.59	$20.87	-8%	Yes	517
Wood's Lamp Examination	$12.74	$12.25	-4%	No	204
Routine Radiograph: One View, 8 x 10 Cassette, Dog or Cat	$66.45	$66.61	0%	No	451
Routine Radiograph: One View, 8 x 10 Additional Cassette	$41.29	$42.86	4%	No	450
Routine Radiograph: One View, 14 x 17 Cassette, Dog or Cat	$70.01	$69.19	-1%	No	477
Routine Radiograph: One View, 14 x 17 Additional Cassette	$44.04	$44.91	2%	No	480
Routine Radiographs: Two Views, Chest, 60-Pound Dog	$101.59	$103.52	2%	No	486

TABLE 1.2 (CONTINUED)
CHANGES IN FEES FROM 2004 TO 2006

Service	2004 Inflation-Adjusted Average	2006 Average	Percentage Increase	Statistically Significant Change	Minimum Number of Respondents
Diagnostic and Imaging Services (continued)					
Routine Radiographs: Three Views, Chest, 60-Pound Dog	$135.84	$140.02	3%	No	477
Routine Radiographs: Two Views, Pelvis, 60-Pound Dog	$102.25	$103.30	1%	No	484
Routine Radiographs: Two Views, Spine, Dachshund	$96.48	$99.46	3%	No	486
Routine Radiographs: Two Views, Abdomen, Cat	$95.31	$97.25	2%	No	491
Routine Radiographs: Two Views, Forearm, Cat	$92.01	$94.83	3%	No	492
Routine Radiograph: One View, 8 x 10 Cassette, Bird/Reptile/Small Mammal	$65.59	$66.66	2%	No	298
Routine Radiograph: One View, 14 x 17 Cassette, Bird/Reptile/Small Mammal	$69.16	$69.72	1%	No	302
Routine Radiographs: Two Views, 8 x 10 Cassette, Bird/Reptile/Small Mammal	$91.30	$94.65	4%	No	306
Routine Radiographs: Two Views, 14 x 17 Cassette, Bird/Reptile/Small Mammal	$95.62	$98.22	3%	No	317
Routine Radiographs: Dental	$79.23	$68.48	-14%	Yes	237
Myelograms: Two Views, Cervical Spine	$108.27	$109.44	1%	No	60
Myelogram: CSF Examination	$88.93	$81.64	-8%	No	33
Ultrasound: Chest and Abdomen	$294.46	$271.33	-8%	Yes	197
Ultrasound: Chest Only	$220.27	$201.30	-9%	Yes	216
Ultrasound: Abdomen Only	$208.55	$192.13	-8%	Yes	244
Ultrasound: Guided Biopsy Collection, Liver	$163.49	$150.32	-8%	No	158
Diagnostic Imaging Case One: Two Films and Your Interpretation	$104.86	$108.95	4%	Yes	510
Diagnostic Imaging Case One: Two Films and Specialist Interpretation	$145.93	$173.22	19%	Yes	193
Prescription Medications					
Prescription Fee for Medication Dispensed from Your Hospital	$8.12	$7.95	-2%	No	476
Prescription Fee if Client Has Prescription Filled Elsewhere	$10.77	$8.91	-17%	No	96
Hospitalization Services					
Hospitalization with IV: No Overnight Stay, 10-Pound Cat	$71.52	$76.00	6%	No	406
Hospitalization without IV: No Overnight Stay, 10-Pound Cat	$25.60	$29.29	14%	Yes	412
Hospitalization with IV: No Overnight Stay, 25-Pound Dog	$73.35	$79.56	8%	Yes	399
Hospitalization without IV: No Overnight Stay, 25-Pound Dog	$26.90	$30.25	12%	Yes	402
Hospitalization with IV: Overnight Stay, 10-Pound Cat	$82.84	$87.98	6%	No	399
Hospitalization without IV: Overnight Stay, 10-Pound Cat	$34.40	$38.53	12%	Yes	404
Hospitalization with IV: Overnight Stay, 25-Pound Dog	$85.21	$91.02	7%	Yes	392
Hospitalization without IV: Overnight Stay, 25-Pound Dog	$36.39	$40.58	12%	Yes	396
Hospitalization with IV: Overnight Stay, 60-Pound Dog	$89.11	$94.58	6%	No	391
Hospitalization without IV: Overnight Stay, 60-Pound Dog	$38.18	$44.03	15%	Yes	399

TABLE 1.2 (CONTINUED)
CHANGES IN FEES FROM 2004 TO 2006

Service	2004 Inflation-Adjusted Average	2006 Average	Percentage Increase	Statistically Significant Change	Minimum Number of Respondents
Hospitalization Services (continued)					
Hospitalization with IV: Overnight Stay, 100-Pound Dog	$91.94	$96.73	5%	No	382
Hospitalization without IV: Overnight Stay, 100-Pound Dog	$39.70	$44.81	13%	Yes	394
Hospitalization with IV and Tube Feeding: Overnight Stay, Bird	$68.33	$79.73	17%	No	71
Hospitalization without IV and Tube Feeding: Overnight Stay, Bird	$34.52	$39.77	15%	No	90
Hospitalization with IV: Overnight Stay, Reptile	$66.07	$72.07	9%	No	82
Hospitalization without IV: Overnight Stay, Reptile	$34.88	$40.52	16%	No	92
Hospitalization with IV: Overnight Stay, Small Mammal	$69.35	$72.88	5%	No	116
Hospitalization without IV: Overnight Stay, Small Mammal	$32.88	$38.04	16%	Yes	137
Anesthesia Services					
IV Sedative: 30-Minute Radiology Procedure, No Intubation/Inhalant, 30-Pound Dog	$41.63	$47.93	15%	Yes	421
IM Sedative: Abscess Treatment, Fractious Cat	$36.55	$42.13	15%	Yes	417
Preanesthetic Exam	$31.51	$30.14	-4%	No	67
Preanesthetic Sedation: 40-Pound Dog	$27.42	$30.59	12%	Yes	221
IV Induction: 40-Pound Dog	$37.66	$44.20	17%	Yes	205
Intubation: 40-Pound Dog	$34.30	$44.46	30%	No	29
Inhalant: 30 Minutes, Isoflurane, 40-Pound Dog	$62.64	$72.40	16%	Yes	387
Inhalant: 60 Minutes, Isoflurane, 40-Pound Dog	$90.07	$102.00	13%	Yes	398
Inhalant: Additional Hour, Isoflurane, 40-Pound Dog	$80.60	$89.60	11%	Yes	368
Anesthetic Monitoring: Electronic	$21.05	$26.19	24%	Yes	173
Anesthetic Monitoring: Manual	$22.35	$26.76	20%	No	48
Fentanyl Patch: 30-Pound Dog	$57.13	$56.80	-1%	No	151
Treatment Procedures and Services					
Treatment Case One: Catheterization	$36.24	$42.76	18%	Yes	428
Treatment Case One: 1,500 ml Lactated Ringer's Solution Over 24 Hours	$43.16	$45.44	5%	No	408
Treatment Case One: Fluid Infusion Pump	$19.88	$20.39	3%	No	165
Treatment Case One: Hospitalization	$37.59	$42.73	14%	Yes	435
Treatment Case One: Veterinarian/Technician Supervision	$26.42	$29.64	12%	Yes	121
Treatment Case One: Medical Waste Disposal	$3.20	$3.48	9%	No	121
Treatment Case Two: Admitting Examination	$37.90	$41.56	10%	Yes	457
Treatment Case Two: Light Sedation	$35.33	$42.77	21%	Yes	445
Treatment Case Two: Abscess Curettage, Debridement, and Flush	$44.98	$53.14	18%	Yes	449
Treatment Case Two: Antibiotic Injection	$18.84	$20.91	11%	Yes	442
Treatment Case Two: Pain Medication	$18.98	$21.04	11%	Yes	397
Treatment Case Two: Hospitalization	$28.21	$34.55	22%	Yes	422
Treatment Case Two: Medical Progress Examination	$24.38	$27.25	12%	Yes	160
Treatment Case Two: Antibiotic, Oral or Injection	$17.09	$18.80	10%	Yes	314

TABLE 1.2 (CONTINUED)
CHANGES IN FEES FROM 2004 TO 2006

Service	2004 Inflation-Adjusted Average	2006 Average	Percentage Increase	Statistically Significant Change	Minimum Number of Respondents
Treatment Procedures and Services (continued)					
Treatment Case Two: Antibiotics, One-Week Supply	$17.17	$19.35	13%	Yes	444
Treatment Case Two: Medical Waste Disposal	$3.29	$3.63	10%	No	122
Treatment Case Two: Total	$212.89	$246.43	16%	Yes	460
Treatment Case Three: Admitting Examination, Day 1	$38.29	$40.78	7%	Yes	436
Treatment Case Three: IV Catheter and Placement, Day 1	$37.57	$42.78	14%	Yes	426
Treatment Case Three: First Liter IV Fluids	$30.56	$32.50	6%	Yes	379
Treatment Case Three: IV Pump Use	$19.42	$19.96	3%	No	159
Treatment Case Three: Two Antibiotic Injections	$33.71	$39.08	16%	Yes	438
Treatment Case Three: Hospitalization	$36.92	$42.88	16%	Yes	434
Treatment Case Three: Two Abdominal Radiographs	$94.79	$105.83	12%	Yes	437
Treatment Case Three: Pain Medication, Administered Twice	$35.40	$36.08	2%	No	413
Treatment Case Three: Medical Waste Disposal	$3.35	$3.75	12%	No	121
Treatment Case Three: Inpatient Examination (Veterinarian Supervision), Day Two	$24.80	$28.41	15%	Yes	192
Treatment Case Three: Liter of Fluids, Day Two	$25.84	$27.66	7%	Yes	432
Treatment Case Three: CBC with Two Chemistries	$60.74	$65.16	7%	Yes	417
Treatment Case Three: Special Diet	$14.55	$5.75	-60%	Yes	179
Treatment Case Four: Examination	$37.72	$40.35	7%	Yes	437
Treatment Case Four: Preanesthetic Lab Tests	$64.12	$65.63	2%	No	421
Treatment Case Four: Anesthesia, 30 Minutes	$61.25	$73.33	20%	Yes	433
Treatment Case Four: Ear Swab/Cytology	$20.73	$25.21	22%	Yes	430
Treatment Case Four: Ear Cleaning	$23.70	$25.55	8%	Yes	405
Treatment Case Four: Antibiotic Injection	$19.12	$21.29	11%	Yes	432
Treatment Case Four: Hospitalization, Day Charge	$23.19	$24.32	5%	No	324
Treatment Case Four: Antibiotic, One-Week Supply	$22.15	$21.29	0%	No	430
Treatment Case Four: Medical Waste Disposal	$3.07	$3.48	13%	No	121
Surgical Services					
Total Surgery Setup Fee	$62.14	$51.47	-17%	No	70
Surgical Suite Use	$37.83	$41.05	9%	No	48
Surgical Pack	$28.47	$35.56	25%	Yes	97
Disposables	$13.40	$21.13	58%	Yes	36
Medical Waste Disposal	$3.79	$3.84	1%	No	65
Assistant Time (One Hour)	$42.58	$57.21	34%	Yes	69
Suture Material	$13.11	$15.07	15%	No	67
Staples	$20.15	$19.67	-2%	No	68
Use of Cold Tray Instruments	$19.99	$23.38	17%	No	45
Electronic Monitoring	$20.89	$25.94	24%	Yes	169
Canine Spay: < 25 Pounds	$156.30	$189.40	21%	Yes	425
Canine Spay: 25–50 Pounds	$168.33	$204.13	21%	Yes	410
Canine Spay: 51–75 Pounds	$186.19	$226.28	22%	Yes	411
Canine Spay: > 75 Pounds	$205.74	$250.74	22%	Yes	415
Canine Neuter: < 25 Pounds	$134.94	$164.92	22%	Yes	425
Canine Neuter: 25–50 Pounds	$143.46	$175.46	22%	Yes	414
Canine Neuter: 51–75 Pounds	$154.82	$189.19	22%	Yes	415
Canine Neuter: > 75 Pounds	$165.02	$206.81	25%	Yes	417
Feline Spay	$132.11	$156.59	19%	Yes	428

TABLE 1.2 (CONTINUED)
CHANGES IN FEES FROM 2004 TO 2006

Service	2004 Inflation-Adjusted Average	2006 Average	Percentage Increase	Statistically Significant Change	Minimum Number of Respondents
Surgical Services (continued)					
Feline Neuter	$87.89	$102.77	17%	Yes	431
Rabbit Spay	$138.69	$166.86	20%	Yes	223
Rabbit Neuter	$107.98	$126.80	17%	Yes	243
Ferret Spay	$134.28	$162.28	21%	Yes	121
Ferret Neuter	$102.67	$125.66	22%	Yes	124
Adrenalectomy, Ferret	$303.66	$308.03	1%	No	76
Anal Gland Resection, Bilateral: 30-Pound Dog	$289.19	$303.23	5%	No	307
Aural Hematoma Repair	$139.37	$144.42	4%	No	421
Broken Blood Feather Quill Removal	$31.37	$48.24	54%	Yes	90
Cesarean Section: 30-Pound Dog with Four Puppies	$339.13	$368.53	9%	Yes	411
Colonoscopy with Biopsy	$266.94	$285.15	7%	No	71
Cruciate Repair: Anterior, 60-Pound Dog	$508.57	$610.99	20%	Yes	266
Diaphragmatic Hernia: 30-Pound Dog	$433.52	$442.30	2%	No	299
Ear Resection: Unilateral, 30-Pound Dog	$325.60	$332.04	2%	No	218
Endoscopy: Upper GI, with Biopsy	$273.59	$274.96	0%	No	93
Enucleation: One Eye, 30-Pound Dog	$254.86	$262.48	3%	No	385
Femoral Head Removal: Cat	$345.98	$370.41	7%	Yes	323
Gastrotomy: Foreign Body Removal, Cat	$301.48	$338.27	12%	Yes	404
Gastrotomy: Foreign-Body Removal, 30-Pound Dog	$331.52	$368.75	11%	Yes	399
Incisor Extraction: Rabbit	$79.18	$66.79	-16%	Yes	106
Intestinal Resection/Anastamosis: Cat	$423.09	$419.69	-1%	No	336
Intestinal Resection/Anastamosis: 30-Pound Dog	$390.37	$454.57	16%	Yes	359
Laser Surgery: Small Tumor Removal	$100.79	$109.41	9%	No	78
Luxating Patella Repair: Miniature Poodle	$392.72	$444.97	13%	Yes	199
Mastectomy: Unilateral, 30-Pound Dog	$295.87	$310.21	5%	No	303
Skin Tumor Removal: 4 cm	$119.54	$128.00	7%	No	343
Splenectomy: 30-Pound Dog	$384.40	$427.68	11%	Yes	302
Suture Laceration: 4 cm	$83.76	$98.42	18%	Yes	339
Tibial Fracture Repair: Midshaft, Amazon Parrot	$266.91	$262.30	-2%	No	29
Tonsillectomy	$185.30	$187.95	1%	No	76
Urethrostomy: Cat	$357.13	$386.53	8%	Yes	241
Wing Fracture Repair: Budgie, Simple	$86.21	$82.00	-5%	No	40
Unclassified Surgery: Per-Minute Fee, Surgical Suite	$8.05	$5.18	-36%	Yes	161
Unclassified Surgery: Per-Minute Fee, Nonsurgical Suite	$7.25	$4.77	-34%	Yes	148
Surgical Case One: Surgical Suite Use	$47.74	$42.94	-10%	No	50
Surgical Case One: Surgical Pack	$31.04	$35.67	15%	Yes	100
Surgical Case One: Surgical Removal of Bladder Calculus	$274.08	$297.79	9%	Yes	403
Surgical Case Two: Surgical Suite Use	$88.35	$48.06	-46%	Yes	44
Surgical Case Two: Surgical Pack	$39.51	$41.20	4%	No	77
Surgical Case Two: One Hour Assistant Time	$47.99	$65.75	37%	Yes	67
Surgical Case Two: Surgical Fracture Repair with IM Pin	$412.75	$432.42	5%	No	294
End-of-Life Services					
Communal Cremation: 30-Pound Dog	$51.31	$54.18	6%	Yes	482
Communal Cremation: Cat	$46.31	$47.16	2%	No	483
Individual Cremation and Return of Ashes to Owner: 30-Pound Dog	$135.85	$143.23	5%	Yes	472

TABLE 1.2 (CONTINUED)
CHANGES IN FEES FROM 2004 TO 2006

Service	2004 Inflation-Adjusted Average	2006 Average	Percentage Increase	Statistically Significant Change	Minimum Number of Respondents
End-of-Life Services (continued)					
Individual Cremation and Return of Ashes to Owner: Cat	$124.61	$130.00	4%	Yes	478
Disposal of Small Mammal	$31.78	$32.14	1%	No	268
Disposal of Reptile or Bird	$31.25	$31.84	2%	No	218

Note 1: Yes = 2004 and 2006 fees are statistically significantly different at the .05 level.
Note 2: Each t-test tests the difference between the means of two groups. "Minimum Number of Respondents" is the size of the smaller of the two groups.
Note 3: Data was not reported if there were less than 25 observations in either group.
Note 4: 2004 fees were adjusted with a compound rate of inflation equal to the increase in the national Consumer Price Index for the same period.

FEE ADJUSTMENTS

TABLE 1.3
FREQUENCY OF FEE ADJUSTMENTS

	Monthly	Every Six Months	Annually	Other	Number of Respondents
All Practices	**2%**	**27%**	**54%**	**17%**	**1,027**
Number of FTE Veterinarians					
1.0 or Less	1%	30%	51%	18%	294
1.1 to 2.0	2%	28%	54%	17%	310
2.1 to 3.0	4%	22%	60%	15%	203
3.1 or More	2%	28%	53%	17%	196
Member Status					
Accredited Practice Member	4%	29%	49%	18%	275
Nonaccredited Member	2%	31%	48%	20%	117
Nonmember	1%	25%	57%	16%	371
Metropolitan Status					
Urban	2%	27%	51%	20%	86
Suburban	3%	27%	51%	19%	292
Second City	2%	25%	56%	17%	210
Town	2%	29%	55%	14%	298
Rural	1%	28%	56%	16%	122
Median Area Household Income					
$35,000 or Less	1%	28%	52%	19%	124
$35,000 to $49,999	3%	25%	56%	16%	372
$50,000 to $69,999	2%	28%	51%	19%	336
$70,000 or More	2%	27%	58%	13%	172

TABLE 1.4
METHOD USED FOR ADJUSTING FEES

	All at the Same Time	Targeted Fees	Number of Respondents
All Practices	**38%**	**62%**	**1,015**
Number of FTE Veterinarians			
1.0 or Less	32%	68%	292
1.1 to 2.0	37%	63%	310
2.1 to 3.0	46%	54%	201
3.1 or More	42%	59%	191
Member Status			
Accredited Practice Member	42%	58%	268
Nonaccredited Member	37%	63%	115
Nonmember	35%	65%	370
Metropolitan Status			
Urban	35%	65%	84
Suburban	40%	60%	288
Second City	43%	58%	205
Town	38%	62%	297
Rural	31%	69%	122
Median Area Household Income			
$35,000 or Less	35%	65%	119
$35,000 to $49,999	41%	59%	371
$50,000 to $69,999	39%	61%	333
$70,000 or More	33%	67%	170

TABLE 1.5
FEE ADJUSTMENT PERCENTAGE

	25th Percentile	Median	Average	75th Percentile	Number of Respondents
All Practices	**3.0%**	**5.0%**	**5.3%**	**7.0%**	**262**
Number of FTE Veterinarians					
1.0 or Less	3.0%	5.0%	6.1%	9.0%	65
1.1 to 2.0	3.1%	5.0%	5.3%	6.0%	76
2.1 to 3.0	3.0%	5.0%	5.2%	7.0%	63
3.1 or More	2.0%	5.0%	4.6%	5.0%	55
Member Status					
Accredited Practice Member	3.0%	5.0%	4.5%	5.0%	80
Nonaccredited Member	3.0%	4.5%	5.3%	7.3%	26
Nonmember	4.0%	5.0%	6.1%	8.0%	91
Metropolitan Status					
Urban	3.0%	4.0%	4.4%	5.0%	19
Suburban	3.3%	5.0%	5.4%	7.9%	72
Second City	3.0%	5.0%	5.2%	7.0%	66
Town	3.0%	5.0%	5.4%	7.3%	81
Rural	3.0%	5.0%	6.7%	8.5%	17
Median Area Household Income					
$35,000 or Less	3.0%	5.0%	5.4%	8.0%	33
$35,000 to $49,999	3.0%	5.0%	5.2%	6.0%	99
$50,000 to $69,999	3.0%	5.0%	5.5%	7.4%	88
$70,000 or More	4.0%	5.0%	5.7%	7.0%	35

Note: Only those respondents who adjust their fees for all services at one time were asked to report the percentage by which they adjust their fees.

DEMOGRAPHICS OF RESPONDENTS

TABLE 1.6
NUMBER OF FULL-TIME EQUIVALENT VETERINARIANS IN PRACTICE

	25th Percentile	Median	Average	75th Percentile	Number of Respondents
All Practices	**1.0**	**2.0**	**2.5**	**3.0**	**1,013**
Number of FTE Veterinarians					
1.0 or Less	1.0	1.0	1.0	1.0	298
1.1 to 2.0	1.5	2.0	1.8	2.0	313
2.1 to 3.0	2.5	3.0	2.8	3.0	204
3.1 or More	4.0	4.0	5.8	5.0	198
Member Status					
Accredited Practice Member	2.0	2.8	2.9	4.0	264
Nonaccredited Member	1.5	2.0	2.5	3.0	114
Nonmember	1.0	1.8	2.0	2.5	370
Metropolitan Status					
Urban	1.0	2.0	2.5	3.0	84
Suburban	1.0	2.0	2.4	3.0	284
Second City	1.0	2.0	3.3	3.0	209
Town	1.1	2.0	2.5	3.0	296
Rural	1.0	1.5	1.9	2.1	122
Median Area Household Income					
$35,000 or Less	1.0	1.8	1.9	2.0	120
$35,000 to $49,999	1.0	2.0	2.7	3.0	372
$50,000 to $69,999	1.0	2.0	2.5	3.0	332
$70,000 or More	1.0	2.0	2.3	3.0	166

TABLE 1.7
GENDER OF PRACTICE OWNERS

	Females Only	Males Only	Females and Males	Number of Respondents
All Practices	**27%**	**55%**	**18%**	**1,026**
Number of FTE Veterinarians				
1.0 or Less	30%	62%	8%	297
1.1 to 2.0	30%	53%	17%	309
2.1 to 3.0	24%	55%	20%	201
3.1 or More	18%	52%	30%	196
Member Status				
Accredited Practice Member	22%	57%	21%	273
Nonaccredited Member	32%	50%	18%	117
Nonmember	29%	55%	16%	369
Metropolitan Status				
Urban	22%	64%	14%	83
Suburban	30%	54%	16%	294
Second City	22%	58%	20%	208
Town	24%	57%	19%	298
Rural	38%	46%	16%	124
Median Area Household Income				
$35,000 or Less	27%	61%	12%	121
$35,000 to $49,999	27%	53%	20%	374
$50,000 to $69,999	26%	57%	16%	335
$70,000 or More	30%	55%	16%	172

TABLE 1.8
PRACTICE TYPE

	Small Animal	Mixed Animal	Other	Number of Respondents
All Practices	**94%**	**4%**	**3%**	**1,037**
Number of FTE Veterinarians				
1.0 or Less	94%	4%	4%	298
1.1 to 2.0	95%	4%	4%	313
2.1 to 3.0	95%	3%	2%	204
3.1 or More	92%	6%	3%	198
Member Status				
Accredited Practice Member	95%	4%	3%	276
Nonaccredited Member	98%	2%	3%	117
Nonmember	94%	4%	3%	373
Metropolitan Status				
Urban	94%	2%	3%	86
Suburban	96%	2%	4%	294
Second City	94%	5%	2%	212
Town	93%	4%	4%	301
Rural	89%	10%	1%	125
Median Area Household Income				
$35,000 or Less	88%	12%	1%	124
$35,000 to $49,999	93%	4%	4%	377
$50,000 to $69,999	96%	3%	3%	338
$70,000 or More	95%	2%	4%	174

TABLE 1.9
PRACTICE FOCUS

	General Practice	Emergency Only	Referral/ Specialty	Emergency and Referral/ Specialty	Feline Only	Exotic/ Avian Only	Exotic Only	Avian Only	Other	Number of Respondents
All Practices	**94%**	**3%**	**4%**	**3%**	**2%**	**2%**	**1%**	**1%**	**6%**	**1,028**
Number of FTE Veterinarians										
1.0 or Less	91%	2%	3%	0%	3%	1%	1%	1%	9%	295
1.1 to 2.0	96%	2%	3%	1%	3%	3%	1%	1%	4%	308
2.1 to 3.0	94%	4%	4%	3%	1%	1%	0%	0%	5%	204
3.1 or More	95%	6%	7%	7%	2%	2%	1%	1%	6%	197
Member Status										
Accredited Practice Member	98%	4%	7%	4%	3%	1%	1%	1%	3%	272
Nonaccredited Member	95%	1%	4%	2%	2%	3%	0%	0%	6%	116
Nonmember	91%	3%	4%	3%	2%	2%	1%	1%	8%	370
Metropolitan Status										
Urban	92%	1%	6%	2%	4%	0%	0%	0%	6%	85
Suburban	92%	4%	6%	3%	4%	1%	1%	0%	7%	289
Second City	91%	6%	6%	5%	2%	2%	0%	1%	5%	211
Town	97%	2%	3%	2%	1%	2%	1%	1%	4%	299
Rural	94%	2%	2%	1%	1%	0%	0%	0%	9%	125
Median Area Household Income										
$35,000 or Less	93%	3%	2%	4%	0%	2%	0%	0%	8%	121
$35,000 to $49,999	94%	4%	5%	3%	2%	1%	0%	1%	5%	374
$50,000 to $69,999	96%	2%	4%	2%	4%	2%	1%	1%	5%	337
$70,000 or More	90%	5%	4%	2%	3%	2%	1%	1%	8%	172

Note: Respondents were asked to check all categories that applied. Responses for individual rows may not add up to 100%.

TABLE 1.10
AAHA MEMBERSHIP STATUS

	Accredited Member	Nonaccredited Member	Nonmember	Number of Respondents
All Practices	**28%**	**24%**	**48%**	**991**
Number of FTE Veterinarians				
1.0 or Less	11%	22%	67%	278
1.1 to 2.0	25%	23%	52%	303
2.1 to 3.0	37%	26%	37%	198
3.1 or More	46%	25%	29%	191
Member Status				
Accredited Practice Member	100%	0%	0%	276
Nonaccredited Member	0%	100%	0%	117
Nonmember	0%	0%	100%	368
Metropolitan Status				
Urban	22%	30%	48%	83
Suburban	32%	27%	41%	278
Second City	31%	25%	44%	202
Town	27%	21%	52%	296
Rural	14%	22%	64%	116
Median Area Household Income				
$35,000 or Less	18%	22%	60%	118
$35,000 to $49,999	26%	24%	50%	361
$50,000 to $69,999	29%	26%	45%	321
$70,000 or More	36%	19%	45%	170

TABLE 1.11
METROPOLITAN STATUS

	Urban	Suburban	Second City	Town	Rural	Number of Respondents
All Practices	**8%**	**29%**	**21%**	**30%**	**12%**	**1,019**
Number of FTE Veterinarians						
1.0 or Less	8%	28%	21%	25%	18%	291
1.1 to 2.0	10%	25%	18%	35%	13%	310
2.1 to 3.0	8%	32%	24%	28%	8%	199
3.1 or More	7%	32%	24%	30%	7%	195
Member Status						
Accredited Practice Member	9%	34%	22%	29%	7%	271
Nonaccredited Member	4%	33%	21%	31%	11%	116
Nonmember	10%	26%	20%	30%	14%	366
Metropolitan Status						
Urban	100%	0%	0%	0%	0%	86
Suburban	0%	100%	0%	0%	0%	295
Second City	0%	0%	100%	0%	0%	212
Town	0%	0%	0%	100%	0%	301
Rural	0%	0%	0%	0%	100%	125
Median Area Household Income						
$35,000 or Less	12%	2%	30%	32%	24%	124
$35,000 to $49,999	11%	17%	27%	31%	14%	374
$50,000 to $69,999	7%	39%	16%	27%	11%	334
$70,000 or More	5%	55%	11%	27%	2%	173

Note: See page 2 for the definitions of the metropolitan status categories.

CHAPTER 2

EXAMINATION AND WELLNESS SERVICES

This chapter contains advice on communication tools, client preferences for appointment times, and setting exam and consultation fees. You will see how important it is to take the time to train your staff to interact with and educate clients while the veterinarian prepares for the examination.

TRAINING STAFF FOR THE EFFICIENT EXAM

At Best Friends Veterinary Center in Grafton, Wisconsin, an assistant and a technician support each doctor in an exam room. The assistant greets the client, asks questions about the patient's history, reviews a checklist on preventive care, and prepares vaccines before the doctor enters the room. Technicians draw blood, run lab tests, and see their own appointments in between. This system lets Dr. Nan Boss and her associates leverage a well-trained team, increase efficiency, and generate more revenue per doctor.

To realize this level of efficiency requires staff training, along with some patience. "I gradually built up to this system because it takes an amazing amount of time—sometimes up to a year—to train support staff," says Dr. Boss, the owner of Best Friends Veterinary Center and author of *Educating Clients From A to Z* (AAHA Press, 1999).

She dedicates up to seven hours a week for group staff training, including a staff meeting each Friday from noon to 2:00 p.m. and doctors' rounds from 2:00 p.m. to 3:00 p.m. New team members spend two hours a week in their own training class. Team members take nine months to complete all of Dr. Boss's basic training modules, which include periodic tests. A typical weekly training schedule includes a general staff meeting, department meetings, and training sessions on client service and new equipment. "Every time you get a new piece of equipment, you have to learn again," Dr. Boss says.

For example, when Dr. Boss purchased a dental radiograph unit, she scheduled a two-hour staff meeting to explain why pets need dental radiographs, how it changes treatment plans, and what team members should tell clients. At a follow-up meeting with doctors, technicians, and assistants, Dr. Boss demonstrated how to use the equipment and conducted problem-solving exercises. "You have to invest time in staff training," Dr. Boss advises. "Pick the most common scenarios you get, decide what your protocols are, and then do it one disease at a time. We didn't start with seven hours a week. We did a little at a time, working from monthly to weekly staff meetings."

Implementing a protocol like Dr. Boss's in your practice will help you train your staff to conduct the most efficient and client-friendly exams. It also helps practice teams keep up with the rapid changes in medicine today. Fifty percent of medical protocols will be out of date in five years, which means we have to turn over 10%–15% of what we do every year just to stay current. New drugs, new tests, new diagnostic tools—all require training.

CREATING CONTINUITY OF CARE

Empowering your team to educate clients is the most effective way to improve your efficiency in exam rooms. At the Colorado State University Veterinary Teaching Hospital in Fort Collins, Colorado, staff members, veterinary students, and doctors follow a checklist of discussion topics for new pet owners (see Figure 2.1).

Doctors also need to follow standards for which topics they will discuss with clients during pediatric, adult, senior, and geriatric wellness visits. Consistent messages and continuity of care is even more critical in multidoctor hospitals. At Westwood Animal Hospital in Westwood, Kansas, Dr. Wayne Hunthausen schedules 30-minute appointments for clients with puppies and kittens. He uses a checklist to spread information over a series of visits (see Figure 2.2).

Training staff on which topics to discuss with clients, when to discuss them, and what to say promotes good client service and a consistent message. Create your own checklist based on the ones we've provided to ensure that team members remember to discuss all of these important issues with clients.

THE IMPORTANCE OF COMMUNICATION

You and your team should employ a variety of communication techniques. Why? Different clients learn in different ways. At Westwood Animal Hospital, Dr. Hunthausen uses a series of behavior pamphlets that he coauthored for AAHA. "We also sell books and videos," he says. "People remember what they hear, even more what they see, and even more of what they do. If I send clients home with information, they're more likely to get the idea."

In the 2002 Pfizer Veterinary Practice Evaluation Survey of 17,769 veterinary clients, survey participants were asked to rank the importance of in-clinic communication tools on a scale of 0 to 5, with 5 being the most important. A ranking of 12 communication tools is detailed in Table 2.A below.

TABLE 2.A
CLIENTS' RANKING OF VETERINARY PRACTICES' COMMUNICATION TOOLS

Communication Tool	Rank
Exam/Vaccination reminder cards	4.4
Exam/Vaccination reminder phone calls	3.2
Brochures	2.9
Anatomical drawings	2.7
In-clinic posters	2.4
Clinic newsletters	2.3
Anatomical models	2.2
Other clinic direct mail	2.1
Clinic library	2.0
Magazines in waiting room	1.8
Videos for in-clinic viewing	1.8
Videos for home viewing	1.4

FIGURE 2.1
CLIENT EDUCATION CHECKLIST FOR NEW CLIENTS

Owner: _____ Pet's Name: _____
Date of birth: _____ Species: _____ Breed: _____

	Visit 1 Date: Age:	Visit 2 Date: Age:	Visit 3 Date: Age:	Visit 4 Date: Age:
1. Behavior				
Socialization				
Safety / Pet-proofing				
Crate training				
House- / Litter-training				
Reward training				
Control / Handling				
Destructive chewing and scratching				
Play biting / Nipping				
Puppy / Kitten classes				
Neutering				
2. Health Care				
Vaccines				
Deworming / Fecals				
Heartworms				
Fleas				
Grooming (ears and skin)				
Feeding / Nutrition				
Microchipping / Licensing				
Dental care				
3. General Advice				
Clinic Services				
After-hours care				
Insurance				
4. Handouts and Samples				
Puppy / Kitten kit				
Reading list				
Rabies tag				
Clinic brochure / Handout				
5. Products				
Food / Treats				
Heartworm products				
Flea control products				
Head halter				
Grooming products				
Chew toys				
Dental care products				
Microchips				
Books, videos, and pamphlets				
6. Follow-up				
Welcome letter / Package				
Magnet / Business card				

FIGURE 2.2
CHECKLIST OF DISCUSSION TOPICS FOR NEW PET OWNERS

First Veterinary Visit

Puppy
- ❏ Housetraining
- ❏ Puppy-proofing the home
- ❏ Introduction to other pets at home
- ❏ Socialization and training classes
- ❏ Desensitizing to all-over touching
- ❏ Biting and mouthing issues
- ❏ Vaccination protocol, parasite control, and zoonoses
- ❏ Nutrition

Kitten
- ❏ Litter-box use
- ❏ Kitten-proofing the home
- ❏ Introduction to other pets at home
- ❏ Socialization
- ❏ Scratching posts
- ❏ Desensitizing to all-over touching
- ❏ Biting and scratching issues
- ❏ Vaccination protocol, parasite control, and zoonoses
- ❏ Nutrition

Second Veterinary Visit

Puppy
- ❏ Grooming (fur, nails, teeth, ears)
- ❏ Pet ID tag, collar, microchip
- ❏ Spay or neuter
- ❏ Toys, exercise, playtime
- ❏ Pet insurance
- ❏ Socialization follow-up discussion

Kitten
- ❏ Grooming (fur, nails, teeth, ears)
- ❏ Pet ID tag, collar, microchip
- ❏ Spay or neuter
- ❏ Toys, playtime
- ❏ Pet insurance
- ❏ Socialization follow-up discussion

Using checklists for puppy and kitten visits as well as wellness visits for adult and senior pets helps to ensure that doctors or exam-room assistants cover every topic with every client. At Best Friends Veterinary Center, assistants also give clients handouts before they leave the exam room to notify the doctor that the client and patient are ready to be seen. To polish their client communication skills, assistants use scripts and cue cards during training exercises. "They learn how to talk with people effectively and develop an underlying knowledge of medicine," Dr. Boss says.

When Dr. Boss enters the exam room, she examines the pet, discusses appropriate diagnostic tests, and makes recommendations. "I listen to clients and ask them questions," she says. "How do you feel about this? Where are you in your decision-making process? Are you prepared to proceed with treatment? Then I take a breath, step back, and wait for their feedback. This technique lets clients feel it's their decision."

Using communication tools such as handouts, brochures, videos, and interactive discussions or demonstrations helps increase client retention of information. In *101 Secrets of a High-Performance Veterinary Practice* (Veterinary Medicine Publishing Group, 1996), author and management consultant Bob Levoy quotes research that reveals how comprehension rates improve based on the method of learning:

- **Reading** training materials generates a 10% comprehension rate.

- **Hearing** yields a 20% comprehension rate.
- **Seeing** increases comprehension to 30%.
- **Watching** someone perform the task brings about a 50% comprehension rate.
- **Participating** in the task produces a 70% comprehension rate.
- **Doing** the task or performing a simulation results in a 90% comprehension rate.

CHOOSING APPROPRIATE APPOINTMENT TIMES AND LENGTHS

The purpose of the veterinary visit often determines the appointment length, so make sure your front-desk team knows how to create a flexible appointment book. Your team should follow guidelines for wellness visits, sick-patient appointments, and same-day emergencies. Dr. Nan Boss recommends the appointment and scheduling guidelines as shown in Table 2.B.

You may also need to keep an open block of time for same-day emergency appointments. The number of emergency slots depends on the number of doctors, staff, and exam rooms in your practice. Try one 20-minute emergency slot in the morning, such as 10:00 a.m. to 10:20 a.m., and another in the afternoon, such as 2:00 p.m. to 2:20 p.m. Planning for emergencies can help you manage the appointment book, keep the entire day on schedule, and provide same-day service for emergencies and worried clients.

Clients still prefer 15-minute appointments for routine wellness exams, according to the Pfizer 2002 Veterinary Practice Evaluation Survey. Survey respondents prefer spending an average of 14 minutes to meet with the veterinarian for physical exams and vaccinations. The study showed 59% of clients prefer 11–20

TABLE 2.B
APPOINTMENT-TIME GUIDELINES

10-minute Appointments	20-minute Appointments	30-minute Appointments	40-minute Appointments
• Suture removal • Lyme, rabies, FIP, or FeLV vaccinations with no other services • Nail trim • Express anal glands • Bandage or Fentanyl patch removal • Blood-pressure recheck • Reweigh medical case	• Illness other than vomiting/diarrhea in adult animals • Heartworm test • Recheck blood draw • Recheck illness • Third puppy visit • Fourth puppy or kitten visit • Release of a complicated medical/surgical case • Routine surgery discharge • Routine surgery admission	• Annual vaccine for adult animals • Illness in geriatric pets • Vomiting/diarrhea for any age pet • Second puppy visit • Second kitten visit • Third kitten visit (combination FeLV/FIV test) • Chest, abdomen, or bladder radiographs (chest and abdomen radiographs may be taken by technicians, but the doctor needs 10–20 minutes to interpret them and discuss results with owners)	• Annual vaccine for geriatric pets • First puppy visit • First kitten visit • New-pet exam • Second opinion • Canine-rehabilitation exam • Routine exam for a new doctor on staff

minutes for appointments, 27% want to spend 1–10 minutes with the veterinarian, 12% chose 21–30 minutes, and 2% prefer more than 30 minutes.

What Are Clients' Preferred Times for Veterinary Care?

Clients ranked Saturday mornings as their first choice of appointment time, followed by weekday mornings and weekday evenings. How do your hours and availability match up to what clients want?

Clients' Time and Day Preference	Rank
Saturdays, 8 a.m. to 12 p.m.	34%
Weekdays, 8 a.m. to 11 a.m.	28%
Weekdays, 5 p.m. to 8 p.m.	26%
Weekdays, 1 p.m. to 5 p.m.	18%
Saturdays, 12 p.m. to 5 p.m.	13%
Weekdays, 11 a.m. to 1 p.m.	11%
Weekdays, 6 a.m. to 8 a.m.	9%
Sundays	7%
Weekdays, 8 p.m. to 10 p.m.	7%

Source: Pfizer 2002 Veterinary Practice Evaluation Survey. Totals equal more than 100% because respondents could choose more than one appointment time.

Certain appointments may require extended visits and merit an extended exam fee. Puppy and kitten visits, geriatric-pet exams, and behavior consultations often take more time and expertise. For example, some practices schedule a 45-minute appointment for the first puppy or kitten visit so that technicians and veterinarians can discuss life-stage wellness, housetraining, nutrition, vaccination, socialization, infectious diseases, and other topics.

At Westwood Animal Hospital, Dr. Hunthausen varies the appointment time based on the behavior problem. He schedules one hour for cat house-soiling cases; one-and-a-half hours for separation anxiety or thunderstorm phobia; one to one-and-a-half hours for aggressive cats, out-of-control puppies, or impulsive disorders; and two hours for aggression. He charges $140 per hour for behavior consultations. "Some behaviorists charge by the case, but the complexity and time required to work up individual cases varies so much that an hourly rate works the best for me," Dr. Hunthausen says.

When offering any specialized service, you need to determine your level of expertise and charge appropriately. Dr. Hunthausen recommends this step-by-step approach to behavior consultation services:

1. Provide behavior information in puppy and kitten kits and sell training books in your reception area.

2. Offer training tools such as KONG© Toys, Gentle Leader™, motion alarms for countertops, and citronella bark collars.

3. Begin interactive behavior discussions in the exam room, and provide helpful resources. You can host puppy or kitten parties that let new pet owners socialize and ask questions about behavior issues.

4. Offer on-site training with a staff member or a trainer who works as an independent contractor. Puppy classes are a great way to get started.

5. Consult with owners on specific behavior problems, beginning with common behavior problems of puppies and kittens and working up to problems in adolescent pets.

6. Join the American Veterinary Society of Animal Behavior (www.avsabonline.org), attend continuing-education lectures on behavior, and read all the behavior literature you can get your hands on. Once you have a solid understanding of pet behavior, learning theory, and behavior modification, you can address more serious problems.

"When you first offered dentistry, you didn't jump into a root canal," Dr. Hunthausen says. "The same is true with behavior consultations."

Case Study

Why I Chose 30-minute Appointments

Ten years ago, when Dr. Patricia Kay opened a practice in Texas, she wanted to create lasting relationships with pet owners who saw themselves as caregivers as well as guardians of the human-animal bond. She knew the doctors in her practice would need more than 15-minute appointments to achieve her vision. The practice offers 30-minute appointments so that doctors can establish rapport with clients, discuss a total life plan, explain preventive services, and describe the extras that differentiate the hospital from others, including visiting board-certified specialists, luxury boarding, and gentle grooming.

"We take a phased approach to preventive medicine and want to have a good relationship with caregivers from the get-go," Dr. Kay says. "You need more than 15 minutes to achieve that. I want clients to have a sense of our practice and what we can offer their pets."

When setting the exam fee, Dr. Kay chose a higher rate to reflect longer appointments. Doctors take a proactive approach and emphasize preventive services such as routine dental cleanings and blood tests. "Longer appointments give you a chance to elaborate and talk about issues," she says. "Clients admire that trait about our hospital."

Dr. Kay cautions that you must use doctors' time sensibly for 30-minute appointments. To improve their efficiency, doctors are adding outpatient-exam assistants who will open and close each appointment and handle client education on preventive issues. "You must manage your time wisely when you have 30-minute appointments," Dr. Kay warns. "It's easy to get sidetracked and not use the time to the advantage that you intended."

Note: The name of the veterinarian who was interviewed for this case study has been changed.

Implementation Idea
Create an Exam Report Card

Exam report cards reiterate your exam findings, make recommendations stronger because they're in writing, and help clients explain your diagnosis to other family members at home (see Figure 2.3). Use this example as the basis of your own customized form, and have the forms printed in a carbonless, two-part format so that you can give one to the client and keep one for the medical record.

If you are able to insert digital photos into the document, add a picture of the pet. Personalization will have clients bragging about your level of care and attention to detail.

HELPFUL RESOURCES

Lifelearn Client Handouts on CD: Small Animal Series (800+ handouts), multiple authors. Lifelearn, Inc., 2004/2005. Available through the American Animal Hospital Association.

Educating Your Clients From A to Z, Nan Boss, DVM. American Animal Hospital Association Press, 1999.

Pet behavior pamphlets (14 titles), Wayne L. Hunthausen, DVM and Gary Landsberg, DVM, Dipl. ACVB. American Animal Hospital Association Press, 1998 and 2002.

Pet Behavior Protocols: What to Say, What to Do, When to Refer, Suzanne Hetts, PhD. American Animal Hospital Association Press, 1999.

Pet health brochures (16 titles), AAHA. American Animal Hospital Association Press, 2007.

Physical Exam Sticker for medical records, AAHA. American Animal Hospital Association Press.

FIGURE 2.3
EXAM REPORT CARD

Generic Animal Hospital
222 Anywhere
Any Town, State Zip
555-222-2222
www.genericah.com

Pet's Name _____ Species _____ Breed _____
Sex _____ Age _____ Weight _____ Temp. _____ Owner's Name _____

COAT & SKIN
- ○ Appear Normal
- ○ Dull
- ○ Scaly
- ○ Dry
- ○ Oily
- ○ Shedding
- ○ Matted
- ○ Tumors
- ○ Itchy
- ○ Parasites
- ○ Other _____

Recommendation: _____

EYES
- ○ Appear normal
- ○ Discharge
- ○ Inflamed
- ○ Eyelid Deformities
- ○ Infection
- ○ Cataract: L R
- ○ Other _____

Recommendation: _____

EARS
- ○ Appear Normal
- ○ Inflamed
- ○ Itchy
- ○ Mites
- ○ Tumors: L R
- ○ Excessive Hair
- ○ Other _____

Recommendation: _____

NOSE & THROAT
- ○ Appear Normal
- ○ Nasal Discharge
- ○ Inflamed Throat
- ○ Inflamed Tonsils
- ○ Enlarged Lymph Glands
- ○ Other _____

Recommendation: _____

MOUTH, TEETH, GUMS
- ○ Appear Normal
- ○ Broken Teeth
- ○ Tartar Build-up
- ○ Tumors
- ○ Ulcers
- ○ Inflamed Lips
- ○ Loose Teeth
- ○ Pyorrhea
- ○ Other _____

Recommendation: _____

ANAL SACS
- ○ Appear Normal
- ○ Excessively Full
- ○ Infected
- ○ Abcessed
- ○ Other _____

Recommendation: _____

LEGS & PAWS
- ○ Appear Normal
- ○ Lameness
- ○ Damaged Ligaments
- ○ Joint Problems
- ○ Nail Problems
- ○ Other _____

Recommendation: _____

HEART
- ○ Appears Normal
- ○ Murmur
- ○ Slow
- ○ Fast
- ○ Other _____

Recommendation: _____

ABDOMEN
- ○ Appears Normal
- ○ Enlarged Organs
- ○ Fluid
- ○ Abnormal Mass
- ○ Tense / Painful
- ○ Other _____

Recommendation: _____

LUNGS
- ○ Appear Normal
- ○ Abnormal Sound
- ○ Coughing
- ○ Congestion
- ○ Breathing Difficulty
- ○ Rapid Respiration
- ○ Other _____

Recommendation: _____

GASTROINTESTINAL SYSTEM
- ○ Appears Normal
- ○ Excessive Gas
- ○ Vomiting Problem
- ○ Anorexia
- ○ Abnormal Feces
- ○ Parasites
- ○ Other _____

Recommendation: _____

UROGENITAL SYSTEM
- ○ Appears Normal
- ○ Abnormal Urination
- ○ Genital Discharge
- ○ Abnormal Testicles
- ○ Enlarged Prostate
- ○ Mammary Tumors
- ○ Other _____

Recommendation: _____

VACCINATIONS
Vac Due _____ Parvo _____ DHPP _____ Rabies _____ FIV _____ FeLV _____ Other _____
Vac Given _____ Parvo _____ DHPP _____ Rabies _____ FIV _____ FeLV _____ Other _____

CHAPTER 2

DATA TABLES

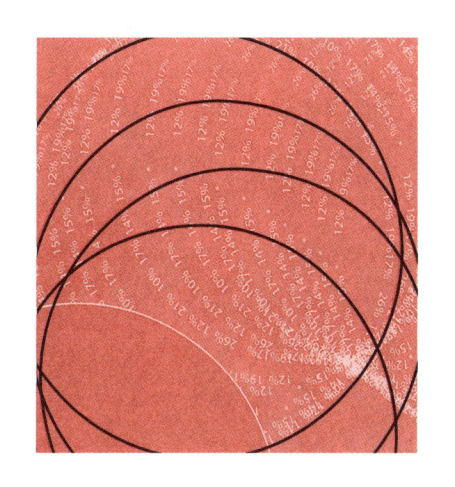

EXAMINATIONS AND CONSULTATIONS

TABLE 2.1
EXAMINATION: WELLNESS

	25th Percentile	Median	Your Data	Average	75th Percentile	Number of Respondents
All Practices	$32.00	$38.00		$37.35	$43.00	556
Number of FTE Veterinarians						
1.0 or Less	$30.00	$35.15		$35.97	$41.75	148
1.1 to 2.0	$30.00	$36.24		$35.93	$42.00	171
2.1 to 3.0	$34.63	$39.00		$39.07	$44.87	112
3.1 or More	$35.00	$39.50		$39.23	$44.00	107
Member Status						
Accredited Practice Member	$35.71	$40.00		$40.71	$46.00	146
Nonaccredited Member	$32.50	$39.00		$38.08	$43.38	116
Nonmember	$29.00	$35.40		$34.66	$40.00	243
Metropolitan Status						
Urban	$33.50	$40.00		$39.74	$46.54	41
Suburban	$37.00	$40.25		$41.26	$46.00	160
Second City	$32.00	$38.50		$37.74	$42.25	110
Town	$31.00	$35.50		$35.78	$41.00	171
Rural	$23.00	$30.00		$29.76	$36.00	65
Median Area Household Income						
$35,000 or Less	$26.00	$34.50		$32.99	$39.95	67
$35,000 to $49,999	$29.75	$35.00		$34.54	$40.00	201
$50,000 to $69,999	$35.00	$39.50		$39.43	$44.00	181
$70,000 or More	$38.00	$42.80		$42.37	$47.00	94

TABLE 2.2
EXAMINATION: SENIOR PET

	25th Percentile	Median	Your Data	Average	75th Percentile	Number of Respondents
All Practices	$35.00	$39.00		$39.26	$44.85	531
Number of FTE Veterinarians						
1.0 or Less	$32.39	$37.52		$37.68	$43.72	144
1.1 to 2.0	$34.00	$38.00		$37.88	$42.53	160
2.1 to 3.0	$35.89	$40.00		$40.33	$45.00	108
3.1 or More	$37.63	$40.00		$41.86	$45.25	101
Member Status						
Accredited Practice Member	$37.83	$40.50		$42.66	$47.43	141
Nonaccredited Member	$35.00	$40.00		$40.23	$45.00	111
Nonmember	$31.50	$37.50		$36.31	$41.87	233
Metropolitan Status						
Urban	$35.75	$41.50		$42.26	$49.71	40
Suburban	$38.00	$42.00		$42.52	$47.05	153
Second City	$35.24	$39.60		$40.06	$45.00	105
Town	$33.00	$37.50		$37.50	$41.00	163
Rural	$28.50	$34.30		$32.76	$38.00	61
Median Area Household Income						
$35,000 or Less	$27.19	$35.00		$34.47	$40.00	65
$35,000 to $49,999	$32.00	$37.25		$36.47	$40.00	194
$50,000 to $69,999	$36.25	$40.50		$41.66	$45.00	169
$70,000 or More	$39.95	$45.00		$44.03	$48.00	91

Note: 4% of respondents reported that they do not offer this service.

TABLE 2.3
EXAMINATION: SICK PET

	25th Percentile	Median	Your Data	Average	75th Percentile	Number of Respondents
All Practices	**$35.00**	**$40.00**		**$40.95**	**$45.00**	**560**
Number of FTE Veterinarians						
1.0 or Less	$34.40	$39.00		$39.05	$45.00	153
1.1 to 2.0	$35.00	$39.50		$39.51	$43.30	168
2.1 to 3.0	$37.52	$41.00		$42.45	$47.00	114
3.1 or More	$38.00	$42.00		$43.69	$47.88	108
Member Status						
Accredited Practice Member	$38.50	$42.00		$43.92	$48.00	147
Nonaccredited Member	$35.00	$41.15		$41.51	$46.00	115
Nonmember	$34.38	$38.50		$38.65	$42.93	246
Metropolitan Status						
Urban	$36.50	$42.00		$42.07	$47.75	41
Suburban	$39.00	$43.00		$43.94	$48.00	159
Second City	$37.24	$40.00		$42.32	$45.00	113
Town	$35.00	$39.00		$39.38	$43.00	172
Rural	$30.01	$35.00		$35.22	$40.00	65
Median Area Household Income						
$35,000 or Less	$29.75	$35.00		$36.15	$41.17	69
$35,000 to $49,999	$34.00	$38.00		$38.62	$42.20	205
$50,000 to $69,999	$38.00	$41.08		$42.97	$45.25	178
$70,000 or More	$40.37	$45.25		$45.75	$48.00	94

TABLE 2.4
EXAMINATION: SINGLE PROBLEM (INCLUDES RECHECK EXAM FEE)

	25th Percentile	Median	Your Data	Average	75th Percentile	Number of Respondents
All Practices	**$33.56**	**$37.50**		**$38.13**	**$42.96**	**116**
Number of FTE Veterinarians						
1.0 or Less	$30.00	$35.00		$34.93	$39.75	36
1.1 to 2.0	$34.00	$37.50		$38.21	$42.65	35
2.1 to 3.0	$34.00	$38.00		$37.68	$43.00	27
3.1 or More	*	*		*	*	14
Member Status						
Accredited Practice Member	$37.00	$41.00		$44.29	$51.50	29
Nonaccredited Member	*	*		*	*	21
Nonmember	$30.25	$35.00		$34.68	$40.00	48
Metropolitan Status						
Urban	*	*		*	*	7
Suburban	$35.00	$40.57		$41.74	$46.53	30
Second City	$30.50	$37.09		$37.80	$42.40	25
Town	$33.00	$35.00		$36.12	$39.00	35
Rural	*	*		*	*	14
Median Area Household Income						
$35,000 or Less	*	*		*	*	20
$35,000 to $49,999	$31.85	$35.43		$35.71	$39.00	40
$50,000 to $69,999	$36.00	$41.75		$43.06	$48.65	37
$70,000 or More	*	*		*	*	15

Note: An asterisk indicates that data was not reported due to an insufficient number of responses.

TABLE 2.5
EXAMINATION: SINGLE PROBLEM (EXCLUDES RECHECK EXAM FEE)

	25th Percentile	Median	Your Data	Average	75th Percentile	Number of Respondents
All Practices	**$34.00**	**$39.40**		**$39.02**	**$44.00**	**403**
Number of FTE Veterinarians						
1.0 or Less	$30.00	$37.54		$37.51	$44.00	103
1.1 to 2.0	$34.00	$39.48		$37.75	$42.08	120
2.1 to 3.0	$34.50	$40.00		$40.59	$45.00	82
3.1 or More	$35.00	$39.00		$39.75	$44.44	87
Member Status						
Accredited Practice Member	$38.00	$40.50		$42.08	$46.20	106
Nonaccredited Member	$34.00	$39.50		$39.28	$45.00	87
Nonmember	$31.13	$38.00		$36.78	$42.00	184
Metropolitan Status						
Urban	$35.00	$42.00		$40.50	$45.00	31
Suburban	$36.50	$41.15		$41.88	$47.00	119
Second City	$34.00	$39.00		$39.34	$44.00	83
Town	$32.19	$38.00		$37.93	$42.53	121
Rural	$29.25	$34.00		$33.42	$39.73	45
Median Area Household Income						
$35,000 or Less	$28.13	$35.62		$34.85	$41.00	46
$35,000 to $49,999	$32.00	$37.00		$36.43	$40.00	153
$50,000 to $69,999	$35.23	$40.13		$40.88	$45.00	126
$70,000 or More	$40.00	$45.00		$43.75	$48.00	71

Note: 3% of respondents reported that they do not offer this service.

TABLE 2.6
RECHECK EXAM FEE INCLUDED IN SINGLE-PROBLEM EXAM FEE

	Yes	No	Number of Respondents
All Practices	**22%**	**78%**	**537**
Number of FTE Veterinarians			
1.0 or Less	26%	74%	144
1.1 to 2.0	22%	78%	161
2.1 to 3.0	24%	76%	112
3.1 or More	14%	86%	103
Member Status			
Accredited Practice Member	22%	78%	140
Nonaccredited Member	19%	81%	110
Nonmember	20%	80%	240
Metropolitan Status			
Urban	18%	82%	38
Suburban	20%	80%	157
Second City	23%	77%	111
Town	22%	78%	161
Rural	23%	77%	61
Median Area Household Income			
$35,000 or Less	29%	71%	68
$35,000 to $49,999	21%	79%	196
$50,000 to $69,999	21%	79%	173
$70,000 or More	18%	82%	89

TABLE 2.7
EXAMINATION: RECHECK OF PREVIOUS PROBLEM

	25th Percentile	Median	Your Data	Average	75th Percentile	Number of Respondents
All Practices	**$23.00**	**$29.00**		**$28.42**	**$33.99**	**471**
Number of FTE Veterinarians						
1.0 or Less	$20.00	$27.48		$27.44	$33.25	123
1.1 to 2.0	$21.84	$27.00		$27.07	$33.00	146
2.1 to 3.0	$27.00	$30.00		$30.48	$35.15	95
3.1 or More	$25.00	$29.03		$29.48	$34.00	95
Member Status						
Accredited Practice Member	$26.63	$30.70		$30.62	$35.00	128
Nonaccredited Member	$24.91	$29.50		$29.19	$33.85	102
Nonmember	$20.00	$26.00		$26.47	$31.95	208
Metropolitan Status						
Urban	$26.00	$29.50		$30.14	$36.00	35
Suburban	$26.13	$30.20		$31.02	$36.25	137
Second City	$24.00	$30.00		$29.28	$34.00	93
Town	$20.62	$27.00		$27.00	$32.50	145
Rural	$19.96	$24.50		$23.33	$28.00	56
Median Area Household Income						
$35,000 or Less	$17.00	$23.00		$23.75	$30.50	53
$35,000 to $49,999	$21.08	$26.00		$26.54	$31.81	172
$50,000 to $69,999	$26.00	$30.00		$29.93	$35.00	159
$70,000 or More	$29.00	$33.00		$32.58	$38.00	76

Note: 3% of respondents reported that they do not offer this service.

TABLE 2.8
EXAMINATION: HEALTH CERTIFICATE (INCLUDES CERTIFICATE PREPARATION FEE)

	25th Percentile	Median	Your Data	Average	75th Percentile	Number of Respondents
All Practices	**$32.50**	**$39.00**		**$39.57**	**$46.51**	**546**
Number of FTE Veterinarians						
1.0 or Less	$31.00	$37.00		$38.89	$45.00	147
1.1 to 2.0	$32.00	$39.00		$38.52	$45.25	169
2.1 to 3.0	$35.09	$40.90		$42.39	$49.19	112
3.1 or More	$33.56	$39.00		$39.08	$45.00	100
Member Status						
Accredited Practice Member	$37.00	$41.00		$43.26	$49.14	143
Nonaccredited Member	$34.00	$40.00		$39.66	$45.50	113
Nonmember	$30.00	$37.00		$37.21	$44.80	241
Metropolitan Status						
Urban	$37.00	$45.00		$43.56	$49.44	38
Suburban	$37.00	$42.00		$43.12	$48.05	155
Second City	$31.50	$40.00		$40.25	$48.23	108
Town	$31.85	$37.00		$37.53	$42.98	172
Rural	$25.00	$34.65		$33.35	$40.00	64
Median Area Household Income						
$35,000 or Less	$25.00	$35.00		$33.49	$41.80	69
$35,000 to $49,999	$30.90	$37.54		$37.46	$44.00	199
$50,000 to $69,999	$35.35	$41.00		$42.00	$48.18	173
$70,000 or More	$38.00	$45.00		$45.16	$51.00	91

TABLE 2.9
EXAMINATION: HEALTH CERTIFICATE (EXCLUDES CERTIFICATE PREPARATION FEE)

	25th Percentile	Median	Your Data	Average	75th Percentile	Number of Respondents
All Practices	**$30.00**	**$36.75**		**$37.32**	**$42.53**	**234**
Number of FTE Veterinarians						
1.0 or Less	$30.00	$35.15		$37.22	$42.00	66
1.1 to 2.0	$30.00	$35.00		$35.50	$40.58	69
2.1 to 3.0	$29.93	$37.29		$38.25	$43.43	45
3.1 or More	$34.00	$38.80		$38.65	$44.00	47
Member Status						
Accredited Practice Member	$35.23	$39.65		$41.84	$48.20	62
Nonaccredited Member	$29.88	$39.00		$38.46	$45.00	45
Nonmember	$28.50	$35.00		$33.99	$38.50	107
Metropolitan Status						
Urban	*	*		*	*	13
Suburban	$35.00	$39.00		$41.80	$47.50	61
Second City	$30.00	$37.60		$37.76	$43.80	51
Town	$30.38	$35.90		$35.92	$40.00	77
Rural	$24.00	$29.85		$30.06	$37.00	29
Median Area Household Income						
$35,000 or Less	$24.23	$32.00		$31.32	$38.50	34
$35,000 to $49,999	$29.35	$35.00		$34.84	$39.65	89
$50,000 to $69,999	$35.00	$38.90		$40.62	$44.00	66
$70,000 or More	$34.00	$45.00		$44.38	$50.23	38

Note 1: An asterisk indicates that data was not reported due to an insufficient number of responses.
Note 2: 2% of respondents reported that they do not offer this service.

TABLE 2.10
CERTIFICATE-PREPARATION FEE INCLUDED IN HEALTH-CERTIFICATE EXAM FEE

	Yes	No	Number of Respondents
All Practices	**55%**	**45%**	**545**
Number of FTE Veterinarians			
1.0 or Less	53%	47%	148
1.1 to 2.0	58%	42%	165
2.1 to 3.0	59%	41%	114
3.1 or More	50%	50%	101
Member Status			
Accredited Practice Member	56%	44%	142
Nonaccredited Member	57%	43%	108
Nonmember	53%	47%	245
Metropolitan Status			
Urban	65%	35%	40
Suburban	57%	43%	152
Second City	52%	48%	113
Town	53%	47%	166
Rural	55%	45%	64
Median Area Household Income			
$35,000 or Less	49%	51%	69
$35,000 to $49,999	54%	46%	199
$50,000 to $69,999	59%	41%	173
$70,000 or More	57%	43%	90

TABLE 2.11
CERTIFICATE-PREPARATION FEE

	25th Percentile	Median	Your Data	Average	75th Percentile	Number of Respondents
All Practices	**$12.00**	**$16.00**		**$18.31**	**$23.10**	**231**
Number of FTE Veterinarians						
1.0 or Less	$10.00	$15.00		$17.05	$23.73	66
1.1 to 2.0	$11.00	$17.00		$18.62	$23.52	69
2.1 to 3.0	$14.42	$17.25		$19.52	$25.00	43
3.1 or More	$13.00	$15.93		$18.30	$21.00	47
Member Status						
Accredited Practice Member	$15.00	$18.00		$19.84	$25.00	59
Nonaccredited Member	$14.00	$15.73		$17.45	$20.00	52
Nonmember	$10.19	$15.81		$18.48	$25.00	104
Metropolitan Status						
Urban	*	*		*	*	11
Suburban	$13.21	$17.25		$18.58	$20.19	64
Second City	$11.93	$18.00		$18.54	$25.00	53
Town	$13.89	$15.95		$19.00	$25.00	77
Rural	*	*		*	*	23
Median Area Household Income						
$35,000 or Less	$10.00	$15.00		$17.22	$25.00	35
$35,000 to $49,999	$11.98	$16.29		$18.37	$22.65	82
$50,000 to $69,999	$12.03	$15.15		$17.20	$20.00	68
$70,000 or More	$15.00	$20.00		$20.95	$25.00	41

Note 1: An asterisk indicates that data was not reported due to an insufficient number of responses.
Note 2: 18% of respondents reported that they do not offer this service.

TABLE 2.12
EXAMINATION: INPATIENT

	25th Percentile	Median	Your Data	Average	75th Percentile	Number of Respondents
All Practices	**$20.50**	**$27.00**		**$28.48**	**$35.00**	**378**
Number of FTE Veterinarians						
1.0 or Less	$22.04	$27.00		$28.66	$33.58	104
1.1 to 2.0	$19.98	$25.00		$26.09	$33.03	110
2.1 to 3.0	$23.00	$30.00		$30.50	$37.00	79
3.1 or More	$20.00	$26.50		$28.56	$35.93	70
Member Status						
Accredited Practice Member	$22.82	$28.88		$30.34	$36.76	108
Nonaccredited Member	$20.50	$27.24		$28.36	$34.39	85
Nonmember	$20.00	$25.50		$27.44	$33.90	148
Metropolitan Status						
Urban	$23.00	$27.50		$29.63	$35.50	26
Suburban	$23.50	$30.00		$31.25	$38.88	118
Second City	$20.75	$26.78		$28.28	$34.01	86
Town	$20.00	$26.00		$27.07	$32.10	107
Rural	$15.00	$22.37		$23.12	$29.00	35
Median Area Household Income						
$35,000 or Less	$21.00	$27.50		$26.64	$31.00	35
$35,000 to $49,999	$19.88	$25.00		$26.68	$32.13	134
$50,000 to $69,999	$20.09	$27.12		$28.43	$35.00	132
$70,000 or More	$25.00	$30.00		$33.39	$45.00	67

Note 1: The inpatient exam includes veterinarian supervision of a hospitalized patient. The fee excludes a hospitalization fee.
Note 2: 13% of respondents reported that they do not offer this service.

TABLE 2.13
EXAMINATION: BEHAVIOR CONSULTATION

	25th Percentile	Median	Your Data	Average	75th Percentile	Number of Respondents
All Practices	**$35.00**	**$41.00**		**$44.17**	**$49.00**	**458**
Number of FTE Veterinarians						
1.0 or Less	$34.63	$40.00		$42.00	$47.00	120
1.1 to 2.0	$35.00	$40.00		$41.21	$46.60	136
2.1 to 3.0	$37.37	$42.00		$45.67	$51.19	100
3.1 or More	$38.28	$44.55		$49.42	$55.00	88
Member Status						
Accredited Practice Member	$39.00	$46.00		$48.32	$54.25	125
Nonaccredited Member	$36.68	$42.00		$45.47	$49.38	92
Nonmember	$34.38	$39.00		$41.02	$45.00	202
Metropolitan Status						
Urban	$38.00	$43.25		$48.32	$53.38	36
Suburban	$38.95	$43.75		$47.00	$50.25	130
Second City	$37.20	$42.58		$44.84	$50.00	88
Town	$34.38	$39.75		$42.51	$48.00	146
Rural	$31.01	$38.00		$38.38	$42.18	49
Median Area Household Income						
$35,000 or Less	$28.50	$35.00		$36.02	$41.34	55
$35,000 to $49,999	$34.13	$39.60		$43.04	$48.38	160
$50,000 to $69,999	$38.00	$42.00		$47.01	$50.94	152
$70,000 or More	$40.00	$47.00		$47.01	$51.00	79

Note: 13% of respondents reported that they do not offer this service.

TABLE 2.14
EXAMINATION: SECOND OPINION AT OWNER'S REQUEST

	25th Percentile	Median	Your Data	Average	75th Percentile	Number of Respondents
All Practices	**$35.00**	**$40.00**		**$40.97**	**$45.00**	**536**
Number of FTE Veterinarians						
1.0 or Less	$34.00	$38.80		$39.50	$45.00	144
1.1 to 2.0	$34.98	$39.00		$38.86	$42.95	161
2.1 to 3.0	$36.50	$41.00		$42.66	$47.10	111
3.1 or More	$38.00	$41.88		$43.72	$48.36	104
Member Status						
Accredited Practice Member	$38.50	$42.08		$44.44	$48.89	140
Nonaccredited Member	$36.00	$41.40		$42.08	$47.25	113
Nonmember	$33.50	$38.00		$38.16	$42.08	236
Metropolitan Status						
Urban	$35.00	$41.50		$42.48	$48.25	38
Suburban	$38.50	$42.00		$43.30	$48.00	149
Second City	$36.25	$40.00		$42.45	$46.70	110
Town	$34.88	$39.00		$40.07	$45.00	166
Rural	$29.50	$35.00		$34.97	$40.00	63
Median Area Household Income						
$35,000 or Less	$29.50	$35.00		$35.58	$41.34	67
$35,000 to $49,999	$34.00	$38.00		$39.03	$43.00	199
$50,000 to $69,999	$38.00	$41.08		$43.19	$47.50	170
$70,000 or More	$40.00	$45.00		$45.23	$48.00	86

Note: 2% of respondents reported that they do not offer this service.

TABLE 2.15
EXAMINATION: REFERRAL FROM ANOTHER VETERINARIAN

	25th Percentile	Median	Your Data	Average	75th Percentile	Number of Respondents
All Practices	**$35.00**	**$40.00**		**$40.56**	**$45.00**	**498**
Number of FTE Veterinarians						
1.0 or Less	$34.45	$38.38		$38.73	$45.00	134
1.1 to 2.0	$34.00	$39.00		$38.71	$42.98	152
2.1 to 3.0	$36.90	$41.99		$42.47	$47.05	105
3.1 or More	$38.00	$40.00		$43.43	$48.34	93
Member Status						
Accredited Practice Member	$38.23	$42.00		$44.08	$48.00	129
Nonaccredited Member	$35.00	$41.00		$41.64	$46.93	104
Nonmember	$33.61	$38.00		$38.01	$42.51	222
Metropolitan Status						
Urban	$35.00	$42.00		$41.89	$48.59	36
Suburban	$38.73	$42.25		$43.87	$48.00	138
Second City	$35.50	$40.00		$41.12	$45.00	102
Town	$35.00	$39.00		$39.66	$43.25	153
Rural	$29.00	$35.00		$33.95	$40.00	59
Median Area Household Income						
$35,000 or Less	$29.75	$35.00		$35.17	$41.67	65
$35,000 to $49,999	$33.00	$38.00		$38.36	$42.15	179
$50,000 to $69,999	$37.55	$41.00		$42.82	$45.25	161
$70,000 or More	$40.00	$45.00		$45.33	$48.00	79

Note: 8% of respondents reported that they do not offer this service.

TABLE 2.16
EXAMINATION: EMERGENCY DURING ROUTINE OFFICE HOURS

	25th Percentile	Median	Your Data	Average	75th Percentile	Number of Respondents
All Practices	**$39.00**	**$50.00**		**$52.86**	**$65.00**	**551**
Number of FTE Veterinarians						
1.0 or Less	$35.15	$46.00		$49.71	$63.72	149
1.1 to 2.0	$39.50	$50.00		$51.44	$63.98	168
2.1 to 3.0	$42.51	$50.00		$54.71	$67.96	112
3.1 or More	$39.70	$51.50		$55.67	$69.00	105
Member Status						
Accredited Practice Member	$45.00	$57.49		$58.97	$70.00	142
Nonaccredited Member	$40.00	$49.85		$52.62	$65.00	115
Nonmember	$36.83	$45.00		$49.15	$60.00	245
Metropolitan Status						
Urban	$41.00	$55.00		$53.73	$60.00	41
Suburban	$44.00	$55.00		$57.77	$70.00	160
Second City	$40.00	$49.00		$54.72	$65.00	109
Town	$37.50	$48.00		$51.14	$65.00	168
Rural	$32.00	$39.50		$42.17	$53.50	65
Median Area Household Income						
$35,000 or Less	$31.00	$38.00		$44.63	$54.50	65
$35,000 to $49,999	$36.16	$48.00		$50.13	$63.34	205
$50,000 to $69,999	$41.94	$50.00		$54.99	$66.04	178
$70,000 or More	$48.00	$60.00		$60.63	$71.75	92

Note: 2% of respondents reported that they do not offer this service.

TABLE 2.17
EXAMINATION: EMERGENCY AFTER HOURS (6 P.M. – 12 A.M.)

	25th Percentile	Median	Your Data	Average	75th Percentile	Number of Respondents
All Practices	$69.57	$79.70		$81.87	$95.00	338
Number of FTE Veterinarians						
1.0 or Less	$60.30	$77.50		$78.93	$95.00	88
1.1 to 2.0	$65.00	$75.00		$78.29	$87.53	96
2.1 to 3.0	$75.00	$85.41		$85.92	$100.00	66
3.1 or More	$72.86	$84.38		$85.06	$96.38	74
Member Status						
Accredited Practice Member	$70.55	$84.46		$84.53	$99.17	96
Nonaccredited Member	$75.00	$81.62		$83.36	$94.75	64
Nonmember	$63.00	$77.25		$80.21	$93.63	154
Metropolitan Status						
Urban	*	*		*	*	18
Suburban	$65.00	$77.75		$78.63	$89.75	84
Second City	$71.60	$80.38		$82.45	$95.01	62
Town	$75.00	$85.00		$85.56	$99.50	118
Rural	$58.20	$75.00		$77.71	$89.65	49
Median Area Household Income						
$35,000 or Less	$51.00	$76.50		$80.35	$99.50	44
$35,000 to $49,999	$69.00	$78.00		$79.38	$90.00	131
$50,000 to $69,999	$65.75	$80.00		$83.79	$99.25	98
$70,000 or More	$75.00	$85.00		$83.78	$95.00	55

Note 1: An asterisk indicates that data was not reported due to an insufficient number of responses.
Note 2: 37% of respondents reported that they do not offer this service.

TABLE 2.18
EXAMINATION: EMERGENCY AFTER HOURS (12 A.M. – 7 A.M.)

	25th Percentile	Median	Your Data	Average	75th Percentile	Number of Respondents
All Practices	$75.00	$86.00		$90.12	$104.82	252
Number of FTE Veterinarians						
1.0 or Less	$74.25	$82.50		$86.83	$100.00	70
1.1 to 2.0	$71.60	$80.00		$86.99	$101.23	74
2.1 to 3.0	$78.00	$95.05		$95.52	$108.00	47
3.1 or More	$75.00	$91.15		$93.40	$110.00	52
Member Status						
Accredited Practice Member	$75.00	$90.00		$93.21	$107.70	70
Nonaccredited Member	$75.00	$91.50		$91.81	$108.50	50
Nonmember	$75.00	$85.00		$88.05	$101.31	120
Metropolitan Status						
Urban	*	*		*	*	10
Suburban	$70.00	$85.00		$84.47	$95.07	50
Second City	$75.00	$85.00		$89.63	$101.75	51
Town	$77.00	$95.00		$96.93	$111.50	95
Rural	$60.00	$79.00		$80.86	$100.00	39
Median Area Household Income						
$35,000 or Less	$75.00	$85.00		$93.59	$113.00	35
$35,000 to $49,999	$75.00	$87.00		$89.65	$102.99	110
$50,000 to $69,999	$74.50	$85.00		$88.55	$101.00	65
$70,000 or More	$75.00	$85.82		$87.38	$100.00	33

Note 1: An asterisk indicates that data was not reported due to an insufficient number of responses.
Note 2: 51% of respondents reported that they do not offer this service.

TABLE 2.19
EXAMINATION: AVIAN

	25th Percentile	Median	Your Data	Average	75th Percentile	Number of Respondents
All Practices	**$36.00**	**$41.00**		**$43.01**	**$48.00**	**234**
Number of FTE Veterinarians						
1.0 or Less	$30.26	$37.77		$38.58	$45.00	40
1.1 to 2.0	$35.00	$40.00		$40.11	$45.00	67
2.1 to 3.0	$37.00	$45.00		$44.55	$51.00	55
3.1 or More	$38.00	$42.50		$45.34	$48.00	63
Member Status						
Accredited Practice Member	$39.00	$45.00		$48.22	$52.00	83
Nonaccredited Member	$35.00	$41.50		$41.76	$47.60	48
Nonmember	$33.00	$39.00		$38.68	$45.00	91
Metropolitan Status						
Urban	*	*		*	*	15
Suburban	$37.38	$42.88		$45.73	$49.25	54
Second City	$36.00	$40.00		$43.29	$51.00	47
Town	$35.55	$41.70		$42.70	$48.00	88
Rural	$30.01	$36.00		$35.30	$40.50	25
Median Area Household Income						
$35,000 or Less	$29.50	$35.00		$35.09	$42.25	29
$35,000 to $49,999	$35.00	$39.45		$40.59	$46.75	88
$50,000 to $69,999	$39.05	$43.00		$46.66	$52.00	69
$70,000 or More	$40.00	$46.20		$47.69	$51.00	39

Note 1: An asterisk indicates that data was not reported due to an insufficient number of responses.
Note 2: 55% of respondents reported that they do not offer this service.

TABLE 2.20
EXAMINATION: REPTILE

	25th Percentile	Median	Your Data	Average	75th Percentile	Number of Respondents
All Practices	**$35.00**	**$40.25**		**$42.10**	**$48.00**	**214**
Number of FTE Veterinarians						
1.0 or Less	$31.03	$40.00		$39.51	$47.25	36
1.1 to 2.0	$35.00	$40.00		$39.51	$45.00	62
2.1 to 3.0	$37.00	$43.00		$43.99	$50.00	51
3.1 or More	$37.50	$41.00		$43.76	$48.00	59
Member Status						
Accredited Practice Member	$39.00	$45.00		$46.49	$51.00	73
Nonaccredited Member	$35.00	$40.00		$40.85	$46.65	43
Nonmember	$33.00	$39.00		$38.91	$45.00	87
Metropolitan Status						
Urban	*	*		*	*	12
Suburban	$36.73	$41.50		$41.80	$48.00	46
Second City	$35.88	$40.00		$43.55	$51.00	42
Town	$35.00	$41.50		$42.57	$48.00	86
Rural	*	*		*	*	23
Median Area Household Income						
$35,000 or Less	$29.63	$35.00		$35.68	$43.50	26
$35,000 to $49,999	$34.56	$39.20		$40.10	$45.00	80
$50,000 to $69,999	$38.30	$42.87		$45.01	$49.61	62
$70,000 or More	$40.00	$45.42		$45.74	$49.25	38

Note 1: An asterisk indicates that data was not reported due to an insufficient number of responses.
Note 2: 58% of respondents reported that they do not offer this service.

EXAMINATION AND WELLNESS SERVICES / 53

TABLE 2.21
EXAMINATION: RABBIT

	25th Percentile	Median	Your Data	Average	75th Percentile	Number of Respondents
All Practices	$35.00	$39.55		$39.67	$45.00	370
Number of FTE Veterinarians						
1.0 or Less	$31.00	$37.77		$37.91	$43.12	82
1.1 to 2.0	$32.88	$38.45		$38.01	$42.86	118
2.1 to 3.0	$36.50	$41.00		$41.81	$47.00	81
3.1 or More	$37.71	$40.00		$41.41	$45.88	78
Member Status						
Accredited Practice Member	$37.95	$42.00		$42.99	$47.15	97
Nonaccredited Member	$35.00	$40.00		$40.38	$45.00	85
Nonmember	$31.40	$38.00		$37.04	$42.13	162
Metropolitan Status						
Urban	*	*		*	*	22
Suburban	$36.00	$40.00		$40.69	$45.00	108
Second City	$35.96	$40.00		$41.39	$48.23	66
Town	$34.00	$38.50		$39.07	$45.00	125
Rural	$29.50	$35.00		$34.33	$40.00	43
Median Area Household Income						
$35,000 or Less	$29.25	$35.00		$34.71	$40.75	40
$35,000 to $49,999	$32.00	$38.00		$37.41	$42.00	139
$50,000 to $69,999	$36.00	$40.71		$41.72	$46.10	119
$70,000 or More	$39.71	$43.00		$43.65	$48.00	62

Note 1: An asterisk indicates that data was not reported due to an insufficient number of responses.
Note 2: 31% of respondents reported that they do not offer this service.

TABLE 2.22
EXAMINATION: SMALL MAMMAL

	25th Percentile	Median	Your Data	Average	75th Percentile	Number of Respondents
All Practices	$34.85	$40.00		$39.75	$45.00	369
Number of FTE Veterinarians						
1.0 or Less	$30.75	$37.77		$37.73	$43.12	82
1.1 to 2.0	$32.00	$38.50		$37.89	$42.90	114
2.1 to 3.0	$37.15	$41.00		$41.37	$46.75	81
3.1 or More	$37.89	$40.00		$42.28	$47.75	81
Member Status						
Accredited Practice Member	$37.66	$42.00		$43.26	$47.63	102
Nonaccredited Member	$35.00	$40.00		$40.43	$46.00	83
Nonmember	$30.25	$38.00		$36.92	$42.08	160
Metropolitan Status						
Urban	*	*		*	*	20
Suburban	$35.98	$40.00		$40.54	$45.25	110
Second City	$35.96	$40.00		$41.44	$47.51	66
Town	$32.09	$39.00		$38.97	$45.00	128
Rural	$29.50	$34.30		$34.05	$40.00	39
Median Area Household Income						
$35,000 or Less	$28.88	$35.00		$35.05	$42.13	38
$35,000 to $49,999	$32.00	$38.25		$37.67	$42.78	134
$50,000 to $69,999	$36.00	$41.00		$42.03	$48.00	122
$70,000 or More	$37.75	$42.60		$42.15	$47.35	65

Note 1: An asterisk indicates that data was not reported due to an insufficient number of responses.
Note 2: 30% of respondents reported that they do not offer this service.

TABLE 2.23
EXAMINATION: FERRET

	25th Percentile	Median	Your Data	Average	75th Percentile	Number of Respondents
All Practices	**$35.00**	**$39.55**		**$39.61**	**$45.00**	**372**
Number of FTE Veterinarians						
1.0 or Less	$31.10	$38.00		$38.11	$44.00	83
1.1 to 2.0	$34.00	$38.75		$38.07	$42.89	112
2.1 to 3.0	$36.00	$40.00		$41.00	$46.63	82
3.1 or More	$36.38	$40.00		$41.28	$46.63	84
Member Status						
Accredited Practice Member	$37.58	$42.00		$42.61	$47.18	104
Nonaccredited Member	$36.00	$40.00		$40.94	$46.00	79
Nonmember	$31.65	$37.78		$37.09	$42.87	161
Metropolitan Status						
Urban	*	*		*	*	22
Suburban	$37.00	$40.00		$41.07	$46.00	107
Second City	$36.00	$40.00		$41.30	$47.84	68
Town	$34.00	$38.50		$38.88	$45.00	127
Rural	$29.13	$35.00		$33.98	$40.00	42
Median Area Household Income						
$35,000 or Less	$28.88	$35.00		$34.41	$40.00	42
$35,000 to $49,999	$33.00	$38.00		$37.47	$42.00	137
$50,000 to $69,999	$35.70	$40.71		$41.32	$46.30	117
$70,000 or More	$39.48	$44.00		$44.11	$48.75	65

Note 1: An asterisk indicates that data was not reported due to an insufficient number of responses.
Note 2: 29% of respondents reported that they do not offer this service.

TABLE 2.24
OFFICE CONSULTATION: PET NOT PRESENT

	25th Percentile	Median	Your Data	Average	75th Percentile	Number of Respondents
All Practices	**$30.13**	**$37.60**		**$37.12**	**$45.00**	**336**
Number of FTE Veterinarians						
1.0 or Less	$28.00	$35.30		$36.14	$44.00	87
1.1 to 2.0	$31.37	$36.80		$36.25	$42.70	100
2.1 to 3.0	$30.00	$38.00		$37.73	$45.78	69
3.1 or More	$32.00	$39.70		$38.32	$46.00	71
Member Status						
Accredited Practice Member	$34.00	$39.65		$39.60	$47.15	96
Nonaccredited Member	$29.75	$38.00		$36.42	$45.00	71
Nonmember	$30.00	$37.00		$36.30	$42.00	135
Metropolitan Status						
Urban	*	*		*	*	20
Suburban	$30.75	$39.45		$38.17	$46.05	109
Second City	$30.00	$37.54		$36.67	$45.00	65
Town	$31.00	$37.00		$35.99	$43.18	101
Rural	$28.00	$34.30		$34.41	$40.00	35
Median Area Household Income						
$35,000 or Less	*	*		*	*	23
$35,000 to $49,999	$30.00	$35.95		$35.18	$42.00	118
$50,000 to $69,999	$30.00	$38.25		$37.41	$45.00	123
$70,000 or More	$36.25	$43.68		$41.68	$48.75	64

Note 1: An asterisk indicates that data was not reported due to an insufficient number of responses.
Note 2: 20% of respondents reported that they do not offer this service.

TABLE 2.25
MINUTES ALLOTTED FOR EXAMINATION AND CONSULTATION APPOINTMENTS

	25th Percentile	Median	Your Data	Average	75th Percentile	Number of Respondents
Examination: Wellness	15	20		20	30	536
Examination: Senior Pet	15	20		22	30	507
Examination: Sick Pet	20	30		25	30	528
Examination: Single Problem	15	20		22	30	514
Examination: Recheck of Previous Problem	15	15		18	20	439
Examination: Health Certificate	15	20		20	24	493
Examination: Inpatient	10	15		15	20	247
Examination: Behavior Consultation	20	30		28	30	415
Examination: Second Opinion at Owner's Request	20	28		25	30	470
Examination: Referral from Another Veterinarian	15	30		25	30	446
Examination: Emergency During Routine Office Hours	20	30		26	30	342
Examination: Emergency After Hours (6 p.m. – 12 a.m.)	20	30		30	30	146
Examination: Emergency After Hours (12 a.m. – 7 a.m.)	20	30		31	30	106
Examination: Avian	20	30		25	30	214
Examination: Reptile	15	30		25	30	198
Examination: Rabbit	15	20		23	30	332
Examination: Small Mammal	15	20		22	30	347
Examination: Ferret	15	20		22	30	347
Office Consultation: Pet Not Present	15	20		20	30	332

Note: Data was reported only for all practices because there was no significant variation in the data based on practice size, member status, metropolitan status, or median area household income.

WELLNESS AND PREVENTIVE CARE SERVICES

TABLE 2.26
PEDIATRIC CANINE WELLNESS VISIT: PROPHYLACTIC DEWORMING TREATMENT

	25th Percentile	Median	Your Data	Average	75th Percentile	Number of Respondents
All Practices	$7.00	$10.00		$11.50	$15.00	437
Number of FTE Veterinarians						
1.0 or Less	$6.75	$10.00		$11.94	$16.00	115
1.1 to 2.0	$6.34	$10.00		$10.73	$15.00	129
2.1 to 3.0	$7.40	$10.00		$11.54	$15.00	97
3.1 or More	$8.00	$10.50		$12.07	$15.28	87
Member Status						
Accredited Practice Member	$8.00	$12.00		$12.80	$16.91	111
Nonaccredited Member	$6.88	$10.00		$10.93	$15.00	90
Nonmember	$6.00	$9.00		$10.42	$14.25	193
Metropolitan Status						
Urban	$8.00	$13.00		$13.14	$19.00	31
Suburban	$8.00	$11.00		$12.59	$16.00	114
Second City	$8.00	$11.09		$12.70	$16.00	90
Town	$6.00	$10.00		$10.71	$15.00	134
Rural	$5.00	$8.00		$8.28	$10.00	59
Median Area Household Income						
$35,000 or Less	$6.00	$10.00		$10.28	$14.25	54
$35,000 to $49,999	$5.92	$8.50		$10.08	$14.32	164
$50,000 to $69,999	$8.00	$11.50		$12.89	$17.38	142
$70,000 or More	$9.50	$12.00		$13.03	$15.00	63

Note 1: 3% of respondents reported that they do not charge a separate fee for this service, though the service is included as part of the wellness visit.
Note 2: 6% of respondents reported that they do not offer this service.

TABLE 2.27
PEDIATRIC CANINE WELLNESS VISIT: TOTAL

	25th Percentile	Median	Your Data	Average	75th Percentile	Number of Respondents
All Practices	$104.00	$127.23		$133.01	$159.74	554
Number of FTE Veterinarians						
1.0 or Less	$94.00	$116.00		$121.06	$143.15	149
1.1 to 2.0	$99.85	$125.81		$128.97	$154.68	168
2.1 to 3.0	$106.90	$137.25		$139.46	$165.86	114
3.1 or More	$118.50	$138.00		$145.77	$164.63	105
Member Status						
Accredited Practice Member	$116.50	$141.50		$146.89	$177.21	143
Nonaccredited Member	$107.64	$135.00		$136.93	$157.94	114
Nonmember	$95.38	$117.29		$122.57	$149.68	245
Metropolitan Status						
Urban	$114.83	$138.50		$144.57	$177.00	38
Suburban	$116.58	$143.75		$150.22	$182.67	154
Second City	$102.31	$127.13		$129.33	$150.86	112
Town	$100.00	$120.75		$126.51	$151.43	174
Rural	$76.00	$103.31		$108.91	$138.69	66
Median Area Household Income						
$35,000 or Less	$81.50	$106.80		$111.74	$140.70	70
$35,000 to $49,999	$94.88	$117.75		$120.88	$143.69	202
$50,000 to $69,999	$114.00	$135.70		$143.80	$171.41	176
$70,000 or More	$123.00	$150.08		$153.57	$181.38	92

Note: The total fee for the puppy wellness visit is the sum of the fees for a wellness physical examination, DHPP vaccine, one-year rabies vaccine, bordetella vaccine, prophylactic deworming treatment (one treatment), fecal examination (centrifugation), and giardia antigen test. Some respondents do not charge for some of the individual services included in this package, though the services are provided as part of the wellness visit. In addition, some respondents reported that they do not offer some of the individual services. Therefore, the average total fee for the wellness visit may be significantly lower than the sum of the average fees for the individual services.

TABLE 2.28
ADULT CANINE WELLNESS VISIT: HEARTWORM TEST MOST COMMONLY USED

	Occult/ Antigen Only	Occult/Antigen Plus Lyme and E. Canis	Not Routinely Performed	Other	Number of Respondents
All Practices	50%	43%	3%	8%	554
Number of FTE Veterinarians					
1.0 or Less	51%	41%	6%	7%	148
1.1 to 2.0	48%	43%	2%	10%	165
2.1 to 3.0	50%	45%	2%	10%	114
3.1 or More	50%	44%	1%	6%	108
Member Status					
Accredited Practice Member	53%	42%	0%	7%	143
Nonaccredited Member	52%	43%	0%	9%	113
Nonmember	48%	42%	6%	9%	246
Metropolitan Status					
Urban	62%	28%	3%	10%	39
Suburban	56%	41%	2%	5%	155
Second City	53%	41%	3%	10%	113
Town	46%	46%	2%	11%	171
Rural	32%	55%	8%	9%	66
Median Area Household Income					
$35,000 or Less	52%	42%	4%	9%	69
$35,000 to $49,999	50%	40%	4%	8%	206
$50,000 to $69,999	47%	47%	2%	7%	174
$70,000 or More	54%	42%	1%	10%	91

Note: Respondents were asked to check all categories that applied. Responses for individual rows may not add up to 100%.

TABLE 2.29
ADULT CANINE WELLNESS VISIT: TOTAL

	25th Percentile	Median	Your Data	Average	75th Percentile	Number of Respondents
All Practices	$101.01	$129.15		$132.63	$163.00	555
Number of FTE Veterinarians						
1.0 or Less	$92.44	$111.13		$119.84	$147.42	148
1.1 to 2.0	$97.20	$127.00		$128.82	$150.50	165
2.1 to 3.0	$103.50	$132.00		$138.58	$171.12	115
3.1 or More	$121.13	$138.73		$146.09	$167.91	108
Member Status						
Accredited Practice Member	$119.20	$143.53		$147.71	$179.38	144
Nonaccredited Member	$101.71	$137.00		$137.37	$166.50	113
Nonmember	$93.69	$117.58		$121.56	$146.00	246
Metropolitan Status						
Urban	$116.50	$146.50		$148.92	$183.79	39
Suburban	$117.00	$141.85		$149.50	$174.90	155
Second City	$101.11	$125.00		$127.29	$150.00	113
Town	$99.00	$125.50		$127.97	$150.88	172
Rural	$73.57	$93.00		$103.97	$123.88	66
Median Area Household Income						
$35,000 or Less	$78.55	$105.00		$114.70	$150.00	69
$35,000 to $49,999	$94.98	$120.50		$121.03	$143.00	206
$50,000 to $69,999	$107.75	$137.08		$140.80	$173.00	174
$70,000 or More	$124.09	$151.00		$154.85	$183.25	92

Note: The total fee for the adult canine wellness visit is the sum of the fees for a wellness physical examination, DHPP vaccine, three-year rabies vaccine, fecal examination (centrifugation), giardia antigen test, and heartworm test (occult/antigen). Some respondents do not charge for some of the individual services included in this package, though the services are provided as part of the wellness visit. In addition, some respondents reported that they do not offer some of the individual services. Therefore, the average total fee for the wellness visit may be significantly lower than the sum of the average fees for the individual services.

TABLE 2.30
SENIOR CANINE WELLNESS VISIT: TOTAL

	25th Percentile	Median	Your Data	Average	75th Percentile	Number of Respondents
All Practices	$374.00	$518.00		$539.00	$662.46	555
Number of FTE Veterinarians						
1.0 or Less	$283.93	$448.00		$458.82	$574.20	148
1.1 to 2.0	$374.85	$516.25		$528.50	$631.80	165
2.1 to 3.0	$405.00	$541.00		$570.44	$700.00	115
3.1 or More	$478.00	$599.16		$619.11	$755.08	108
Member Status						
Accredited Practice Member	$494.28	$600.88		$632.53	$795.25	144
Nonaccredited Member	$413.03	$586.50		$585.99	$694.83	113
Nonmember	$338.88	$440.50		$470.43	$572.43	246
Metropolitan Status						
Urban	$420.00	$576.59		$606.71	$749.00	39
Suburban	$431.80	$577.60		$595.84	$742.65	155
Second City	$410.68	$537.00		$576.35	$747.08	113
Town	$357.10	$483.50		$506.86	$618.29	172
Rural	$260.81	$405.15		$408.37	$552.31	66
Median Area Household Income						
$35,000 or Less	$283.38	$365.75		$412.07	$553.35	69
$35,000 to $49,999	$352.45	$492.72		$523.63	$651.66	206
$50,000 to $69,999	$450.58	$563.20		$584.05	$704.06	174
$70,000 or More	$464.23	$571.62		$600.99	$724.73	92

Note: The total fee for the senior canine wellness visit is the sum of the fees for a senior pet physical examination, DHPP vaccine, three-year rabies vaccine, CBC with manual differential, chemistry panel (16–24), fecal examination (centrifugation), heartworm test (occult/antigen), urinalysis, microalbuminaria, UP:C ratio, routine ECG: lead II only, radiographs: two thoracic views and two abdomen views, and ultrasound: chest and abdomen. Some respondents do not charge for some of the individual services included in this package, though the services are provided as part of the wellness visit. In addition, some respondents reported that they do not offer some of the individual services. Therefore, the average total fee for the wellness visit may be significantly lower than the sum of the average fees for the individual services.

TABLE 2.31
PEDIATRIC FELINE WELLNESS VISIT: BARTONELLA TEST

	25th Percentile	Median	Your Data	Average	75th Percentile	Number of Respondents
All Practices	$32.38	$50.00		$59.82	$81.76	141
Number of FTE Veterinarians						
1.0 or Less	$35.00	$54.00		$60.13	$70.00	39
1.1 to 2.0	$28.50	$48.00		$56.73	$91.90	35
2.1 to 3.0	$41.15	$50.32		$60.50	$73.63	28
3.1 or More	$29.05	$48.00		$58.79	$90.00	31
Member Status						
Accredited Practice Member	$31.81	$51.00		$61.00	$92.44	38
Nonaccredited Member	$35.00	$50.13		$64.68	$100.00	27
Nonmember	$32.00	$50.00		$58.31	$70.00	67
Metropolitan Status						
Urban	*	*		*	*	13
Suburban	$45.00	$59.32		$67.73	$92.00	47
Second City	$32.75	$47.75		$58.51	$90.45	28
Town	$26.50	$45.00		$53.36	$69.53	37
Rural	*	*		*	*	13
Median Area Household Income						
$35,000 or Less	*	*		*	*	15
$35,000 to $49,999	$32.88	$50.00		$58.79	$74.98	44
$50,000 to $69,999	$35.00	$50.00		$60.96	$76.50	52
$70,000 or More	$31.00	$59.32		$64.65	$92.00	27

Note 1: An asterisk indicates that data was not reported due to an insufficient number of responses.
Note 2: 70% of respondents reported that they do not offer this service.

TABLE 2.32
PEDIATRIC FELINE WELLNESS VISIT: TOTAL

	25th Percentile	Median	Your Data	Average	75th Percentile	Number of Respondents
All Practices	$133.25	$162.00		$168.05	$197.13	564
Number of FTE Veterinarians						
1.0 or Less	$124.93	$155.23		$160.87	$191.75	152
1.1 to 2.0	$131.47	$153.63		$159.94	$177.39	172
2.1 to 3.0	$139.50	$165.70		$173.91	$200.75	115
3.1 or More	$146.55	$174.75		$181.99	$204.75	106
Member Status						
Accredited Practice Member	$143.21	$167.25		$179.48	$206.25	148
Nonaccredited Member	$132.50	$166.00		$172.53	$202.40	115
Nonmember	$125.68	$155.08		$159.61	$185.67	248
Metropolitan Status						
Urban	$146.60	$185.54		$187.55	$209.75	40
Suburban	$152.08	$175.00		$185.98	$211.25	161
Second City	$135.50	$158.20		$168.28	$201.13	112
Town	$126.70	$153.00		$158.13	$178.25	175
Rural	$110.88	$137.25		$137.77	$167.84	66
Median Area Household Income						
$35,000 or Less	$116.33	$144.48		$146.03	$172.15	70
$35,000 to $49,999	$127.45	$148.68		$155.25	$175.13	206
$50,000 to $69,999	$142.13	$175.25		$179.50	$204.30	180
$70,000 or More	$154.00	$176.93		$188.92	$208.59	94

Note: The total fee for the kitten wellness visit is the sum of the fees for a wellness physical examination, one-year rabies vaccine, prophylactic deworming treatment (one treatment), FVRCP vaccine, FeLV vaccine, Bartonella test, fecal examination (centrifugation), and FeLV and FIV tests. Some respondents do not charge for some of the individual services included in this package, though the services are provided as part of the wellness visit. In addition, some respondents reported that they do not offer some of the individual services. Therefore, the average total fee for the wellness visit may be significantly lower than the sum of the average fees for the individual services.

TABLE 2.33
ADULT FELINE WELLNESS VISIT: HEARTWORM TEST MOST COMMONLY USED

	Occult/ Antibody	Occult/ Antigen	Combined Occult/ Antibody and Occult/ Antigen	Number of Respondents
All Practices	36%	31%	33%	211
Number of FTE Veterinarians				
1.0 or Less	27%	49%	24%	55
1.1 to 2.0	29%	24%	47%	66
2.1 to 3.0	49%	21%	30%	43
3.1 or More	41%	32%	27%	41
Member Status				
Accredited Practice Member	43%	20%	38%	56
Nonaccredited Member	31%	36%	33%	45
Nonmember	31%	34%	34%	90
Metropolitan Status				
Urban	*	*	*	22
Suburban	30%	30%	41%	61
Second City	44%	23%	33%	43
Town	38%	35%	26%	65
Rural	*	*	*	16
Median Area Household Income				
$35,000 or Less	15%	58%	27%	26
$35,000 to $49,999	39%	30%	30%	79
$50,000 to $69,999	38%	25%	37%	68
$70,000 or More	39%	22%	39%	36

Note: 1: An asterisk indicates that data was not reported due to an insufficient number of responses.
Note 2: Some row totals do not equal 100% due to rounding.

TABLE 2.34
ADULT FELINE WELLNESS VISIT: TOTAL

	25th Percentile	Median	Your Data	Average	75th Percentile	Number of Respondents
All Practices	$153.50	$192.20		$193.05	$231.05	543
Number of FTE Veterinarians						
1.0 or Less	$147.13	$182.13		$183.46	$222.38	144
1.1 to 2.0	$153.96	$185.04		$189.68	$225.05	168
2.1 to 3.0	$152.03	$196.20		$196.39	$237.00	113
3.1 or More	$165.25	$211.37		$206.47	$241.08	101
Member Status						
Accredited Practice Member	$171.76	$208.00		$206.00	$240.15	143
Nonaccredited Member	$163.83	$199.50		$199.35	$236.06	110
Nonmember	$145.95	$174.30		$180.98	$222.00	239
Metropolitan Status						
Urban	$183.43	$211.50		$211.31	$235.88	40
Suburban	$171.85	$211.00		$208.76	$243.10	157
Second City	$146.50	$185.54		$189.10	$231.77	109
Town	$153.89	$184.50		$189.01	$223.63	166
Rural	$121.50	$149.38		$155.85	$185.23	62
Median Area Household Income						
$35,000 or Less	$127.73	$173.92		$172.64	$211.25	68
$35,000 to $49,999	$150.75	$176.70		$180.41	$214.50	195
$50,000 to $69,999	$156.94	$201.29		$203.20	$241.75	176
$70,000 or More	$178.36	$215.75		$211.49	$245.63	93

Note: The total fee for the adult feline wellness visit is the sum of the fees for a wellness physical examination, one-year rabies vaccine, FVRCP vaccine, FeLV vaccine, FIP vaccine, fecal examination (centrifugation), FeLV and FIV tests, and heartworm test (occult/antibody). Some respondents do not charge for some of the individual services included in this package, though the services are provided as part of the wellness visit. In addition, some respondents reported that they do not offer some of the individual services. Therefore, the average total fee for the wellness visit may be significantly lower than the sum of the average fees for the individual services.

TABLE 2.35
SENIOR FELINE WELLNESS VISIT: TOTAL

	25th Percentile	Median	Your Data	Average	75th Percentile	Number of Respondents
All Practices	$409.00	$568.00		$584.22	$719.75	543
Number of FTE Veterinarians						
1.0 or Less	$347.63	$491.17		$510.40	$630.04	144
1.1 to 2.0	$413.19	$559.13		$562.44	$678.13	168
2.1 to 3.0	$473.62	$593.00		$617.92	$728.30	113
3.1 or More	$512.83	$643.00		$671.42	$829.58	101
Member Status						
Accredited Practice Member	$529.00	$662.00		$682.41	$837.08	143
Nonaccredited Member	$417.98	$603.00		$612.03	$732.17	110
Nonmember	$367.00	$512.65		$520.36	$610.50	239
Metropolitan Status						
Urban	$490.75	$612.00		$656.70	$810.25	40
Suburban	$487.00	$621.43		$639.70	$756.81	157
Second City	$395.84	$590.25		$613.41	$813.00	109
Town	$385.51	$526.75		$551.94	$672.00	166
Rural	$276.63	$470.33		$455.41	$572.45	62
Median Area Household Income						
$35,000 or Less	$302.20	$381.75		$441.11	$599.63	68
$35,000 to $49,999	$394.40	$559.50		$567.71	$687.20	195
$50,000 to $69,999	$483.75	$602.83		$626.63	$752.70	176
$70,000 or More	$482.65	$617.70		$649.96	$792.90	93

Note: The total fee for the senior feline wellness visit is the sum of the fees for a senior pet physical examination, one-year rabies vaccine, FVRCP vaccine, fecal examination (centrifugation), FeLV and FIV tests, heartworm test (occult/antibody), CBC with 16–24 chemistries and T4, urinalysis, microalbuminaria, UP:C ratio, routine ECG: lead II only, radiographs: two thoracic views and two abdomen views, and ultrasound: chest and abdomen. Some respondents do not charge for some of the individual services included in this package, though the services are provided as part of the wellness visit. In addition, some respondents reported that they do not offer some of the individual services. Therefore, the average total fee for the wellness visit may be significantly lower than the sum of the average fees for the individual services.

TABLE 2.36
AVIAN WELLNESS VISIT: FECAL MICROSCOPIC EXAMINATION

	25th Percentile	Median	Your Data	Average	75th Percentile	Number of Respondents
All Practices	$16.50	$20.76		$22.56	$25.51	126
Number of FTE Veterinarians						
1.0 or Less	*	*		*	*	20
1.1 to 2.0	$16.00	$18.50		$20.91	$24.30	32
2.1 to 3.0	$17.99	$22.00		$21.85	$24.00	26
3.1 or More	$17.28	$21.35		$24.14	$28.89	42
Member Status						
Accredited Practice Member	$18.00	$23.30		$24.13	$28.85	51
Nonaccredited Member	*	*		*	*	22
Nonmember	$16.38	$20.00		$21.07	$23.91	46
Metropolitan Status						
Urban	*	*		*	*	8
Suburban	$20.00	$24.50		$25.63	$31.25	28
Second City	$15.79	$20.00		$21.14	$24.00	30
Town	$16.00	$19.00		$21.08	$23.95	46
Rural	*	*		*	*	10
Median Area Household Income						
$35,000 or Less	*	*		*	*	20
$35,000 to $49,999	$16.00	$20.00		$20.68	$24.05	37
$50,000 to $69,999	$18.50	$23.60		$25.33	$29.00	39
$70,000 or More	*	*		*	*	23

Note 1: An asterisk indicates that data was not reported due to an insufficient number of responses.
Note 2: 4% of respondents reported that they do not offer this service.

TABLE 2.37
AVIAN WELLNESS VISIT: CHOANAL SLIT CULTURE/SENSITIVITY

	25th Percentile	Median	Your Data	Average	75th Percentile	Number of Respondents
All Practices	$57.88	$74.76		$75.05	$90.29	66

Note 1: Results were reported only for all practices due to an insufficient number of responses.
Note 2: 43% of respondents reported that they do not offer this service.

TABLE 2.38
AVIAN WELLNESS VISIT: CLOACAL CULTURE/SENSITIVITY

	25th Percentile	Median	Your Data	Average	75th Percentile	Number of Respondents
All Practices	$58.50	$74.45		$75.42	$90.17	67

Note 1: Results were reported only for all practices due to an insufficient number of responses.
Note 2: 3% of respondents reported that they do not charge a separate fee for this service, though the service is included as part of the wellness visit.
Note 3: 40% of respondents reported that they do not offer this service.

TABLE 2.39
AVIAN WELLNESS VISIT: COMPLETE BLOOD COUNT

	25th Percentile	Median	Your Data	Average	75th Percentile	Number of Respondents
All Practices	$35.00	$41.06		$44.34	$48.00	101
Number of FTE Veterinarians						
1.0 or Less	*	*		*	*	17
1.1 to 2.0	$34.50	$40.00		$42.61	$46.13	29
2.1 to 3.0	*	*		*	*	23
3.1 or More	$31.50	$44.20		$47.78	$56.28	27
Member Status						
Accredited Practice Member	$36.13	$44.25		$45.33	$50.53	40
Nonaccredited Member	*	*		*	*	20
Nonmember	$35.00	$40.00		$42.33	$46.26	36
Metropolitan Status						
Urban	*	*		*	*	6
Suburban	*	*		*	*	21
Second City	$38.38	$44.30		$49.52	$58.00	29
Town	$33.00	$41.84		$42.79	$46.38	33
Rural	*	*		*	*	9
Median Area Household Income						
$35,000 or Less	*	*		*	*	15
$35,000 to $49,999	$35.00	$42.00		$44.31	$48.00	33
$50,000 to $69,999	$36.38	$42.28		$46.29	$50.18	30
$70,000 or More	*	*		*	*	18

Note 1: An asterisk indicates that data was not reported due to an insufficient number of responses.
Note 2: 15% of respondents reported that they do not offer this service.

TABLE 2.40
AVIAN WELLNESS VISIT: TOTAL

	25th Percentile	Median	Your Data	Average	75th Percentile	Number of Respondents
All Practices	$218.80	$330.00		$344.82	$443.00	135
Number of FTE Veterinarians						
1.0 or Less	*	*		*	*	23
1.1 to 2.0	$218.80	$295.00		$325.42	$410.00	35
2.1 to 3.0	$303.26	$361.89		$378.89	$507.69	27
3.1 or More	$218.15	$314.05		$334.38	$444.31	44
Member Status						
Accredited Practice Member	$240.43	$333.08		$357.95	$463.86	54
Nonaccredited Member	$218.75	$361.89		$341.41	$434.00	25
Nonmember	$196.00	$313.00		$324.67	$434.40	49
Metropolitan Status						
Urban	*	*		*	*	8
Suburban	$219.63	$360.13		$361.50	$463.50	28
Second City	$288.12	$359.00		$376.52	$490.92	33
Town	$217.75	$307.70		$329.58	$428.40	51
Rural	*	*		*	*	11
Median Area Household Income						
$35,000 or Less	*	*		*	*	21
$35,000 to $49,999	$219.00	$303.26		$323.55	$410.00	43
$50,000 to $69,999	$260.85	$333.08		$354.31	$452.02	40
$70,000 or More	*	*		*	*	24

Note 1: The total fee for the avian wellness visit is the sum of the fees for an avian physical examination, fecal microscopic examination, choanal slit culture/sensitivity, cloacal culture/sensitivity, polyoma vaccine, Pacheco's vaccine, CBC, four chemistries, chlamydia test, fecal Gram's stain, nail trim, beak trim, and wing trim. Some respondents do not charge for some of the individual services included in this package, though the services are provided as part of the wellness visit. In addition, some respondents reported that they do not offer some of the individual services. Therefore, the average total fee for the wellness visit may be significantly lower than the sum of the average fees for the individual services.

Note 2: An asterisk indicates that data was not reported due to an insufficient number of responses.

EXAMINATION AND WELLNESS SERVICES / 69

TABLE 2.41
REPTILE WELLNESS VISIT: FECAL EXAMINATION

	25th Percentile	Median	Your Data	Average	75th Percentile	Number of Respondents
All Practices	$16.08	$19.95		$20.71	$23.62	118
Number of FTE Veterinarians						
1.0 or Less	*	*		*	*	17
1.1 to 2.0	$15.25	$17.75		$18.91	$21.75	32
2.1 to 3.0	$15.08	$20.00		$19.39	$22.85	25
3.1 or More	$17.86	$20.00		$22.31	$26.38	37
Member Status						
Accredited Practice Member	$16.40	$21.00		$22.04	$24.25	46
Nonaccredited Member	*	*		*	*	19
Nonmember	$16.00	$18.00		$18.75	$20.94	48
Metropolitan Status						
Urban	*	*		*	*	5
Suburban	$20.00	$24.00		$25.16	$30.50	25
Second City	$15.15	$19.50		$19.71	$22.00	27
Town	$16.00	$18.00		$18.92	$21.50	49
Rural	*	*		*	*	9
Median Area Household Income						
$35,000 or Less	*	*		*	*	14
$35,000 to $49,999	$15.00	$17.50		$18.91	$22.00	35
$50,000 to $69,999	$18.13	$21.41		$23.55	$25.26	40
$70,000 or More	*	*		*	*	23

Note 1: An asterisk indicates that data was not reported due to an insufficient number of responses.
Note 2: 8% of respondents reported that they do not offer this service.

TABLE 2.42
REPTILE WELLNESS VISIT: FECAL CULTURE

	25th Percentile	Median	Your Data	Average	75th Percentile	Number of Respondents
All Practices	$43.50	$65.00		$66.39	$81.66	69

Note 1: Results were reported only for all practices due to an insufficient number of responses.
Note 2: 38% of respondents reported that they do not offer this service.

TABLE 2.43
REPTILE WELLNESS VISIT: COMPLETE BLOOD COUNT

	25th Percentile	Median	Your Data	Average	75th Percentile	Number of Respondents
All Practices	$34.40	$40.00		$42.69	$46.26	98
Number of FTE Veterinarians						
1.0 or Less	*	*		*	*	15
1.1 to 2.0	$34.13	$38.25		$39.51	$42.60	28
2.1 to 3.0	*	*		*	*	20
3.1 or More	$33.00	$42.00		$46.40	$55.88	29
Member Status						
Accredited Practice Member	$34.50	$41.84		$43.45	$50.00	39
Nonaccredited Member	*	*		*	*	16
Nonmember	$34.00	$38.00		$41.20	$45.00	39
Metropolitan Status						
Urban	*	*		*	*	5
Suburban	*	*		*	*	20
Second City	$35.05	$42.55		$46.02	$53.88	25
Town	$34.00	$38.25		$41.85	$44.86	38
Rural	*	*		*	*	7
Median Area Household Income						
$35,000 or Less	*	*		*	*	13
$35,000 to $49,999	$34.75	$39.30		$42.18	$46.40	29
$50,000 to $69,999	$36.00	$41.40		$44.80	$47.00	31
$70,000 or More	*	*		*	*	20

Note 1: An asterisk indicates that data was not reported due to an insufficient number of responses.
Note 2: 16% of respondents reported that they do not offer this service.

TABLE 2.44
REPTILE WELLNESS VISIT: GENDER DETERMINATION

	25th Percentile	Median	Your Data	Average	75th Percentile	Number of Respondents
All Practices	$30.00	$49.00		$46.38	$59.04	42

Note 1: Results were reported only for all practices due to an insufficient number of responses.
Note 2: 45% of respondents reported that they do not offer this service.

TABLE 2.45
REPTILE WELLNESS VISIT: TOTAL

	25th Percentile	Median	Your Data	Average	75th Percentile	Number of Respondents
All Practices	$164.22	$224.63		$229.81	$286.30	138
Number of FTE Veterinarians						
1.0 or Less	*	*		*	*	22
1.1 to 2.0	$160.38	$204.63		$217.42	$272.87	36
2.1 to 3.0	$141.55	$229.71		$225.27	$293.03	29
3.1 or More	$198.33	$225.86		$248.13	$293.01	44
Member Status						
Accredited Practice Member	$209.50	$247.40		$256.99	$297.50	51
Nonaccredited Member	*	*		*	*	23
Nonmember	$136.63	$188.80		$198.40	$258.03	58
Metropolitan Status						
Urban	*	*		*	*	6
Suburban	$156.35	$226.50		$222.18	$280.25	29
Second City	$164.04	$245.20		$247.54	$302.63	32
Town	$168.06	$225.01		$231.68	$286.26	54
Rural	*	*		*	*	13
Median Area Household Income						
$35,000 or Less	*	*		*	*	15
$35,000 to $49,999	$142.68	$210.85		$217.20	$285.05	45
$50,000 to $69,999	$165.00	$237.43		$234.70	$289.27	47
$70,000 or More	*	*		*	*	24

Note 1: The total fee for the reptile wellness visit is the sum of the fees for a reptile physical examination, fecal examination, fecal culture, complete blood count, four chemistries, gender determination, nail trim, and one radiograph: 8 x 10 cassette. Some respondents do not charge for some of the individual services included in this package, though the services are provided as part of the wellness visit. In addition, some respondents reported that they do not offer some of the individual services. Therefore, the average total fee for the wellness visit may be significantly lower than the sum of the average fees for the individual services.
Note 2: An asterisk indicates that data was not reported due to an insufficient number of responses.

TABLE 2.46
SMALL MAMMAL WELLNESS VISIT: EAR MITE EXAMINATION

	25th Percentile	Median	Your Data	Average	75th Percentile	Number of Respondents
All Practices	$13.00	$18.00		$19.54	$24.18	248
Number of FTE Veterinarians						
1.0 or Less	$13.50	$16.00		$19.26	$24.00	53
1.1 to 2.0	$10.49	$17.00		$18.52	$24.40	78
2.1 to 3.0	$14.65	$19.50		$19.44	$23.65	48
3.1 or More	$13.00	$19.80		$20.47	$26.40	59
Member Status						
Accredited Practice Member	$13.00	$20.16		$20.69	$25.41	71
Nonaccredited Member	$14.17	$19.80		$20.62	$27.43	49
Nonmember	$12.45	$17.00		$18.61	$22.87	109
Metropolitan Status						
Urban	*	*		*	*	12
Suburban	$14.75	$19.65		$21.66	$28.14	74
Second City	$13.75	$18.00		$19.77	$23.33	46
Town	$10.71	$17.10		$18.15	$23.93	84
Rural	$11.00	$16.00		$16.70	$21.40	27
Median Area Household Income						
$35,000 or Less	*	*		*	*	24
$35,000 to $49,999	$12.60	$17.00		$18.96	$23.60	85
$50,000 to $69,999	$12.00	$17.50		$18.75	$24.00	87
$70,000 or More	$16.88	$22.25		$24.02	$30.14	42

Note 1: An asterisk indicates that data was not reported due to an insufficient number of responses.
Note 2: 14% of respondents reported that they do not charge a separate fee for this service, though the service is included as part of the wellness visit.
Note 3: 3% of respondents reported that they do not offer this service.

TABLE 2.47
SMALL MAMMAL WELLNESS VISIT: TOTAL

	25th Percentile	Median	Your Data	Average	75th Percentile	Number of Respondents
All Practices	$98.20	$128.50		$132.82	$160.69	316
Number of FTE Veterinarians						
1.0 or Less	$75.62	$109.75		$113.34	$139.23	74
1.1 to 2.0	$105.43	$127.00		$128.81	$147.55	93
2.1 to 3.0	$97.16	$140.60		$141.63	$180.25	67
3.1 or More	$114.79	$138.63		$147.27	$176.38	72
Member Status						
Accredited Practice Member	$116.00	$139.26		$149.72	$183.00	83
Nonaccredited Member	$102.38	$143.73		$143.17	$173.00	62
Nonmember	$90.00	$119.00		$124.30	$153.00	145
Metropolitan Status						
Urban	*	*		*	*	17
Suburban	$103.51	$134.77		$139.70	$183.12	96
Second City	$90.13	$136.75		$139.48	$167.59	60
Town	$104.25	$124.50		$129.35	$153.60	103
Rural	$68.69	$122.00		$114.30	$150.38	34
Median Area Household Income						
$35,000 or Less	$81.55	$105.10		$113.27	$141.56	32
$35,000 to $49,999	$90.88	$121.00		$122.94	$148.80	116
$50,000 to $69,999	$110.63	$137.00		$141.48	$183.30	106
$70,000 or More	$118.63	$139.44		$149.85	$180.78	52

Note 1: The total fee for the small mammal wellness visit is the sum of the fees for a small mammal physical examination, ear mite examination, tooth trim, CBC with manual differential, 4 chemistries, fecal examination (centrifugation), and nail trim. Some respondents do not charge for some of the individual services included in this package, though the services are provided as part of the wellness visit. In addition, some respondents reported that they do not offer some of the individual services. Therefore, the average total fee for the wellness visit may be significantly lower than the sum of the average fees for the individual services.
Note 2: An asterisk indicates that data was not reported due to an insufficient number of responses.

TABLE 2.48
FERRET WELLNESS VISIT: HEARTWORM ANTIGEN TEST

	25th Percentile	Median	Your Data	Average	75th Percentile	Number of Respondents
All Practices	**$26.00**	**$29.78**		**$31.76**	**$37.00**	**108**
Number of FTE Veterinarians						
1.0 or Less	$27.20	$30.00		$31.70	$35.60	34
1.1 to 2.0	$25.00	$28.50		$30.98	$38.64	33
2.1 to 3.0	*	*		*	*	16
3.1 or More	*	*		*	*	21
Member Status						
Accredited Practice Member	$28.43	$31.23		$34.12	$37.25	26
Nonaccredited Member	$25.00	$33.00		$32.88	$39.50	25
Nonmember	$25.00	$28.50		$30.12	$35.00	53
Metropolitan Status						
Urban	*	*		*	*	6
Suburban	$28.25	$34.00		$33.05	$36.00	33
Second City	*	*		*	*	23
Town	$25.00	$28.00		$30.97	$37.40	35
Rural	*	*		*	*	8
Median Area Household Income						
$35,000 or Less	*	*		*	*	18
$35,000 to $49,999	$25.55	$29.00		$30.78	$36.75	36
$50,000 to $69,999	$25.38	$29.50		$31.77	$35.00	30
$70,000 or More	*	*		*	*	20

Note 1: An asterisk indicates that data was not reported due to an insufficient number of responses.
Note 2: 62% of respondents reported that they do not offer this service.

TABLE 2.49
FERRET WELLNESS VISIT: PLASMA CHEMISTRY PROFILE

	25th Percentile	Median	Your Data	Average	75th Percentile	Number of Respondents
All Practices	**$50.00**	**$66.00**		**$70.46**	**$86.90**	**155**
Number of FTE Veterinarians						
1.0 or Less	$47.50	$59.00		$64.84	$74.20	37
1.1 to 2.0	$48.50	$70.50		$70.30	$86.90	47
2.1 to 3.0	$52.80	$67.49		$72.65	$87.13	25
3.1 or More	$51.25	$65.45		$71.41	$79.80	41
Member Status						
Accredited Practice Member	$54.00	$66.00		$75.97	$92.50	43
Nonaccredited Member	$54.05	$70.35		$73.53	$87.68	34
Nonmember	$46.00	$60.25		$65.60	$80.00	69
Metropolitan Status						
Urban	*	*		*	*	8
Suburban	$52.50	$71.25		$74.32	$87.94	50
Second City	$48.88	$66.00		$70.84	$84.26	32
Town	$49.80	$63.87		$67.54	$80.00	47
Rural	*	*		*	*	15
Median Area Household Income						
$35,000 or Less	*	*		*	*	18
$35,000 to $49,999	$49.38	$65.00		$66.26	$79.80	57
$50,000 to $69,999	$53.00	$79.00		$80.74	$98.00	47
$70,000 or More	$55.43	$67.49		$69.49	$82.45	29

Note 1: An asterisk indicates that data was not reported due to an insufficient number of responses.
Note 2: 45% of respondents reported that they do not offer this service.

EXAMINATION AND WELLNESS SERVICES / 75

TABLE 2.50
FERRET WELLNESS VISIT: TOTAL

	25th Percentile	Median	Your Data	Average	75th Percentile	Number of Respondents
All Practices	$121.99	$155.93		$161.44	$198.38	350
Number of FTE Veterinarians						
1.0 or Less	$112.60	$138.63		$149.65	$183.75	78
1.1 to 2.0	$131.25	$155.71		$160.22	$185.50	100
2.1 to 3.0	$123.35	$164.50		$165.55	$203.31	78
3.1 or More	$129.96	$162.00		$168.41	$206.50	83
Member Status						
Accredited Practice Member	$136.86	$167.55		$174.91	$201.73	94
Nonaccredited Member	$122.28	$167.40		$167.35	$203.88	76
Nonmember	$118.38	$146.14		$153.19	$184.00	154
Metropolitan Status						
Urban	*	*		*	*	17
Suburban	$121.75	$173.70		$171.33	$220.55	103
Second City	$123.00	$161.50		$163.44	$200.00	67
Town	$128.93	$151.58		$156.31	$179.35	120
Rural	$93.20	$144.00		$142.81	$178.91	37
Median Area Household Income						
$35,000 or Less	$117.00	$138.00		$145.59	$174.90	37
$35,000 to $49,999	$119.50	$151.00		$152.46	$178.85	125
$50,000 to $69,999	$122.70	$163.75		$164.49	$202.25	114
$70,000 or More	$135.15	$166.75		$179.55	$229.90	65

Note 1: The total fee for the ferret wellness visit is the sum of the fees for a ferret physical examination, ear mite examination, distemper vaccine, heartworm antigen test, one-year rabies vaccine, CBC with manual differential, four chemistries, fecal examination (centrifugation), and nail trim. Some respondents do not charge for some of the individual services included in this package, though the services are provided as part of the wellness visit. In addition, some respondents reported that they do not offer some of the individual services. Therefore, the average total fee for the wellness visit may be significantly lower than the sum of the average fees for the individual services.
Note 2: An asterisk indicates that data was not reported due to an insufficient number of responses.

TABLE 2.51
WELLNESS VISITS OFFERED

	Yes	No	Number of Respondents
Puppy Wellness Visit	97%	3%	570
Adult Canine Wellness Visit	97%	3%	570
Pediatric Feline Wellness Visit	99%	1%	572
Adult Feline Wellness Visit	96%	4%	567
Avian Wellness Visit	25%	75%	546
Reptile Wellness Visit	25%	75%	543
Small Mammal Wellness Visit	57%	43%	550
Ferret Wellness Visit	64%	36%	545

Note 1: Data was reported only for all practices because there was no significant variation in the data based on practice size, member status, metropolitan status, or median area household income.
Note 2: Some row totals do not equal 100% due to rounding.

CHAPTER 3

DISCOUNTS ON SERVICES

In this chapter, you'll find guidelines on discounting dos and don'ts, learn how to offer value-added services, and understand the value of carefully tracking your return on investment.

THE DANGERS OF DISCOUNTING

When you opened your practice 15 years ago, just a few colleagues owned practices nearby. Now it may seem like veterinary practices are springing up like dandelions, dotting the suburban landscape of your neighborhood. Your new-client numbers have slowly declined for the past year. Should you print a coupon on the back of the neighborhood grocery receipt? Try a bigger Yellow Pages ad? Give new clients the first exam free?

It may be tempting, but discounting is dangerous. Why? You'll attract a coupon-clipper clientele, devalue your professional services, lower client expectations, and force good clients to pick up the tab for discount divas. "The vast majority of discounting you're doing probably doesn't make sense," warns Dr. Steve Fisher, former owner of 10 veterinary practices, author, and executive with VCA Antech in Los Angeles, California. "With consistent discounting, you'll build a practice of coupon clippers who won't do anything except when coupons are available. You'll also have a difficult if not impossible time of converting them to good clients. Discounts are a marketing program—not a way of life—because most practices net only 25% to 35% of total revenue."

Most veterinarians discount due to 1) compassion or 2) competition. The compassion motivator causes you to feel sympathy toward apparently destitute pet owners. The competition motivator makes you concerned that you'll lose clients to your competitor because your fees are higher.

"Veterinarians feel a lot of compassion for people, and many feel guilty that their fees are higher than some clients think they should be," says Dr. Karen Felsted, CPA, MS, CVPM of Gatto McFerson CPAs in Santa Monica, California. "Discounts get used as a marketing tool, but guilt, compassion, and perceived competition are often the real reasons."

Understanding the Effects of Discounting

You need to understand that every discount means you must serve additional full-fare clients to make up the difference. For example, in a practice grossing $1,000,000, giving a 10% discount to only 10% of the clients equals a giveaway of $10,000 of services. If this practice started with a 20% profit margin, the net profit after discounting is $190,000 ($200,000 – $10,000), or 19%. Now the veterinarians at this practice must see 555 more patients (5% more than before!) to bring profits back to $200,000. "It doesn't seem fair to give discounts to whining clients, because good clients will bear the expense," Dr. Felsted says.

If your new client numbers are up, why give away professional services? Veterinarians in most areas of the country need to replace 25 to 30 clients every month just to keep up with the ones that move or leave. If your new client numbers are lower than that industry average, try a new marketing program, but be sure to track and measure the results so that you can determine your return on investment. For example, if you give a free first exam to people who adopt a pet from a local shelter or humane society, create a computer code for the exam so that you can monitor the redemption rate on a monthly basis. Also look at the additional visits that resulted from the free first exams. You can then calculate the return on your investment in the campaign.

If you're going to offer a discount, it should be for a specific time and a specific procedure, and the results should be measurable so that you can determine whether or not you're achieving your goal. For example, you might offer a vaccine clinic once a month on a day that is typically slower than others. Increase the number of staff for a busy day so that clients will enjoy good customer service. If you skimp on staffing, you risk delivering poor service and losing those clients forever.

Also be aware of freebies that don't make it onto the invoice, because that translates into a discount of 100%. For example, you might clean a kitten's ears without charging the client. "Do a medical records audit and compare what's in the medical record to what's on the invoice," Dr. Felsted says. "This will show you the holes in your fee-capture system."

DISCOUNTING DOS AND DON'TS

Common discounts include employee discounts, senior-citizen discounts, multiple-pet discounts, breeder discounts, bundled-services discounts, special promotional discounts, and rescue-group discounts. We'll discuss each of them here and give you some tips that will help you avoid discounting away the profit you could reinvest in staff training, new equipment, client education, and other products and programs that improve quality of care.

Employee Discounts

Many practice owners offer free or significantly reduced veterinary care for their staff as an employee benefit. The Internal Revenue Service's position is that any discount greater than the limits they set triggers taxable income to the employee. (As of 2007, these limits are 20% off of the retail price of services and, for products, the average gross profit percentage for all products multiplied by the retail price.) Consult with your CPA, and set written guidelines in your employee handbook.

Employers with a large number of employees will most likely find it necessary to limit the number of animals per staff member that qualify for pet-care discounts, or limit the annual discount dollar amount per staff member, advises Dr. James F. Wilson, JD, of Priority Veterinary Consultants in Yardley, Pennsylvania, in his book, *Contracts, Benefits, and Practice Management for the Veterinary Profession* (Priority Press, Ltd., 2000).

This is especially true considering the level of care available today, where specialty diagnostic and surgical procedures, oncology drugs, and radiation therapy for employees' pets can be exceptionally costly for employers. "Practice owners are encouraged to limit the number of animals that qualify for their employee pet-care discounts to prevent abuse by some staff members and subsequent ill will among the others," advises Dr. Wilson

Dr. Wilson recommends these guidelines for employee pet-care plans:

- Maintain individual medical records for each pet.
- Record sales of veterinary products and professional services in each pet's medical record contemporaneously with the provision of care.
- Require that employees pay charges within 30 days of the date that products are dispensed or services are rendered, or, if they cannot pay within that timeframe, you set up and they comply with a payment plan for the outstanding balance.

Another alternative is to provide pet health insurance in lieu of a discount on veterinary services. According to *Compensation and Benefits*, Fourth Edition by AAHA Press, only 1% of small-animal practices provided this benefit to employees at the end of 2005.

A quote from one company for an insurance policy for a mixed-breed, two-year-old dog was $255 per year, which covered more than $200 of basic vaccination and well care plus a portion of the fees charged to treat a whole host of medical problems such as ear infections and stomach disorders. Of course, the cost of the premiums increases depending upon the species, breed, age, and type of coverage. Consider the annual cost of a pet insurance policy versus the dollar amount of the discount you're giving.

Senior-Citizen Discounts

Offer this discount only after you've carefully analyzed your clients' demographics and decided that you want to attract senior citizens as clients. Seniors tend to be good clients but are used to receiving a 10%–15% discount on many products and services. Often, though, these discounts are offered only during certain days of the week or times of the day. You'll want to fill empty space in your appointment book by offering the senior discount during the particular times and days that are typically slower for you.

Multiple-Pet Discounts

A veterinarian may offer a multiple-pet discount out of guilt, due to compassion for people who share his passion for pets, or because the prior owners of the practice he purchased had always done it. "Don't do multi-pet discounts unless you're looking at a litter," Dr. Felsted advises. "It almost always takes two appointment times to see two pets, whether they come in separately or together." You must examine each pet, and you often have to repeat information, tailor recommendations, and review different educational materials for each pet. Charge based on your time and services delivered, not the number of bodies in the exam room.

Breeder Discounts

If you offer discounts to breeders, you must receive referrals in return. Select only breeders whom you trust and with whom you can develop long-term relationships. As with most business dealings, it's best to put the discount and terms in writing. Clearly explain the discount amount, applicable services, and number of pets qualified for discounted services.

Don't encourage do-it-yourself breeders who nurture do-it-yourself pet owners. Partner with breeders who value your veterinary expertise, purchase services and products, and passionately recommend you to new pet owners. Give breeders information about your clinic to include in their adoption kits, such as brochures, magnets, handouts, and food samples.

Bundled-Services Discounts

Use bundled plans that slightly discount the total amount only if you are upgrading clients to a higher level of care for their pets. For example, you might bundle a spay or neuter procedure with a preanesthetic test that you discount by 10%. The savings might entice a client who would otherwise decline preanesthetic testing for a young, healthy animal.

Promotional Discounts

When promoting a specific service, it is not necessary to discount that service. For example, during National Pet Dental Health Month each February, consider alternatives to discounts on dental cleanings. Tie a giveaway that markets your practice into the theme of the month, such as a branded toothbrush in February.

Ask vendors about promotional items they can supply, such as finger toothbrushes, diet samples, pet toothpaste samples, and dental chews. Give clients a goody bag rather than a discount. The goal of the promotion is to increase revenue and entice owners to bring pets in for the care that might otherwise get overlooked.

Rescue-Group Discounts

People who are passionate about certain breeds or causes are Good Samaritans you can reward with win-win relationships with your practice. If you offer discounts to rescue groups, select only one or two so you can deliver exceptional service, and set guidelines that clearly define each party's expectations and limitations.

Draft a letter of agreement that outlines your contributions and the rescue group's reciprocation. The written annual agreement can be a simple one-page letter that outlines each party's responsibilities, such as the one in Figure 3.1.

> **Implementation Idea**
> **Letter of Agreement with Rescue Group**
>
> It's important to delineate your respective obligations when working with a rescue group. Use the sample letter in Figure 3.1 of agreement as the basis of your own contract.

ATTRACTING HIGH-QUALITY CLIENTS WITHOUT DISCOUNTS

The 2002 Pfizer Veterinary Practice Evaluation Survey showed that a referral is the number-one reason clients choose a clinic, with location being second most important. Study after study shows that fees do not drive clients to or away from veterinary practices, so there is not much evidence to support discounting as a valid promotional strategy.

Factors that Influence Clients' Choice of Veterinary Practice

1.	Referrals	51%
2.	Location	45%
3.	Other	18%
4.	Yellow Pages	6%
5.	Road sign	2%
6.	Newspaper ad	0%
7.	TV ad	0%
8.	Radio ad	0%

Source: 2002 Pfizer Veterinary Practice Evaluation Survey. Note that answers total more than 100% because respondents could choose more than one factor.

Small-animal practices in stable populations should generate 20 new clients per month per doctor, while practices in growing or transient communities need 30 new clients per month per doctor to sustain adequate growth, according to Dr. Jim Wilson's *Contracts, Benefits, and Practice Management for the Veterinary Profession*. In *Financial and Productivity Pulsepoints*, Fourth Edition, by AAHA Press, the average number of new clients per full-time veterinarian per year is 303. How do your numbers compare?

Happy clients will refer like-minded friends, neighbors, and family, so providing an excellent service experience during every visit is key. Enhance service with

FIGURE 3.1
SAMPLE LETTER OF AGREEMENT WITH RESCUE GROUP

ABC Veterinary Hospital values your efforts to find loving homes for greyhounds and will support your program in 20__. This letter outlines the contributions and responsibilities of ABC Veterinary Hospital and the State Greyhound Association.

ABC Veterinary Hospital will provide the following services to State Greyhound Association in 20__:

- Provide a preadoption exam, dental cleaning, one dose of heartworm prevention, and one dose of flea control for up to 12 dogs (average of one dog per month)
- Display State Greyhound Association brochures in our reception area and insert them in new-client welcome kits
- Allow the State Greyhound Association to use our reception area/conference room for monthly volunteer meetings; the meeting schedule must be coordinated with our hospital manager at least one week in advance
- Post a link to State Greyhound Association on our hospital's website

The State Greyhound Association will provide the following services to ABC Veterinary Hospital in 20__:

- Include an ABC Veterinary Hospital brochure in each adoption kit
- Publish a half-page ad, which ABC Veterinary Hospital will provide, in our monthly/quarterly newsletter
- Invite ABC Veterinary Hospital veterinarians to speak at one of our events
- Post a link to ABC Veterinary Hospital on our website

This agreement covers the period of January 1 to December 31, 20__. Either party may terminate the agreement with 30 days' written notice. This agreement will be reviewed annually, and a new annual agreement will be drafted if both parties want to continue this mutually beneficial relationship. To accept this agreement, please sign and return a copy to Dr. John Myers, ABC Veterinary Hospital, 124 Main St., Goodvet, State 00000.

_____   _____
Dr. John Myers, ABC Veterinary Hospital   State Greyhound Association

_____   _____
Date   Date

value-added offerings rather than deep discounts. For example, you can partner with a trainer to offer a free introductory behavior class for puppy owners. The class meets for only one session and can lead to referrals for veterinary care as well as enrollments in training classes.

You can also partner with vendors for value-added offerings and giveaways. For example, a pet-food company can donate samples or small bags of dental diets for you to give clients after their pets' dental cleanings. Pharmaceutical companies might donate a trial dose of a long-term medication or provide samples of a new pain-management drug for your team to test. "I would rather see veterinarians give away a free bottle of shampoo or a dental diet with the purchase of a service than a free exam," Dr. Felsted says. "It may be the same dollar amount, but the perception is different when you give clients a gift."

Case Study

Send a Letter to Your Best Clients about Your Referral Program

To encourage referrals from his existing clientele, Dr. O. Barnea developed a promotional campaign for Cliffside Animal Hospital in Cliffside Park, New Jersey and Crossroads Animal Clinic in Fort Lee, New Jersey. Dr. Barnea sent a letter

FIGURE 3.2
SAMPLE REFERRAL REWARDS LETTER

Dear Client:

The doctors and staff at Cliffside Animal Hospital and Crossroads Animal Clinic are excited to introduce a **Referral Rewards Program**, with the goal of sharing our vision of advanced medicine and compassionate care with new clients. You've shown your confidence in us by telling your friends and neighbors about our hospitals, and we appreciate that.

I've enclosed three referral cards for you to pass along to your friends and neighbors. When your friends or neighbors visit Cliffside Animal Hospital or Crossroads Animal Clinic, they simply present the **Referral Rewards Card** with your name signed on the back to receive an introductory exam for $9.95.

To thank you for your referral, we will offer you your choice of one of these rewards:

- A $25 referral credit that will be applied to your next veterinary visit
- A $25 referral credit to purchase vitamins, flea medication, or shampoo for your pet
- A one-year subscription to *Catnip*, a newsletter for caring cat owners, or *Your Dog*, a newsletter for dog owners, published by Tufts University School of Veterinary Medicine

We take pride in the exceptional service and compassionate care we give clients and their pets. Our newly renovated hospital features state-of-the-art equipment, including ultrasound and laser surgery. Cliffside Animal Hospital and Crossroads Animal Clinic have earned a reputation for their innovative procedures, friendly staff, and exceptional client service.

We invite you to take advantage of this opportunity to tell your friends and neighbors about how their pets can get the same quality of care that your pets receive at Cliffside Animal Hospital and Crossroads Animal Clinic.

Again, thank you for joining us in a successful partnership in animal care.

Warm Regards,

Dr. O. Barnea

to 1,500 clients to gauge the response before mailing it to all active clients.

The promotion included a cover letter (see Figure 3.2) and three business cards. Within days, receptionists noticed an increase in calls from prospective clients who received the referral cards from friends and neighbors. "The feedback has been great because clients need to refer just one person to get rewarded," Dr. Barnea says.

Helpful Resources

Practice Made Perfect, Marsha L. Heinke, DVM, EA, CPA, CVPM and John B. McCarthy, DVM, MBA. American Animal Hospital Association Press, 2001.

Blackwell's Five-Minute Veterinary Practice Management Consult, Lowell Ackerman, DVM. Blackwell Publishing Limited, 2006.

CHAPTER 3

DATA TABLES

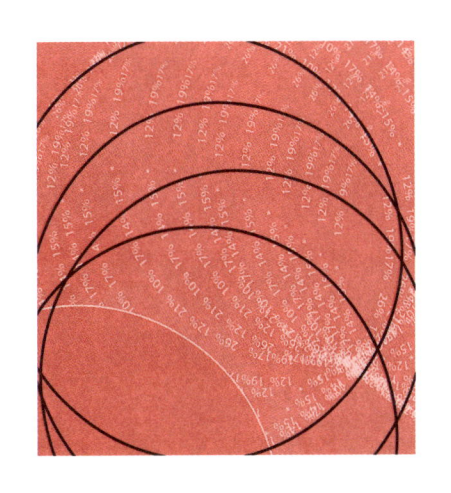

TABLE 3.1
WELLNESS EXAM FEES DISCOUNTED IF THREE OR MORE PETS IN FOR EXAM AND BOOSTER VACCINES AT SAME TIME

	Yes	No	Number of Respondents
All Practices	57%	43%	569
Number of FTE Veterinarians			
1.0 or Less	56%	44%	155
1.1 to 2.0	53%	47%	170
2.1 to 3.0	56%	44%	117
3.1 or More	59%	41%	108
Member Status			
Accredited Practice Member	57%	43%	149
Nonaccredited Member	59%	41%	117
Nonmember	55%	45%	251
Metropolitan Status			
Urban	62%	38%	42
Suburban	57%	43%	163
Second City	55%	45%	114
Town	58%	42%	175
Rural	52%	48%	65
Median Area Household Income			
$35,000 or Less	57%	43%	70
$35,000 to $49,999	57%	43%	207
$50,000 to $69,999	58%	42%	181
$70,000 or More	55%	45%	97

TABLE 3.2
MULTIPLE-PET WELLNESS EXAM DISCOUNT

	25th Percentile	Median	Average	75th Percentile	Number of Respondents
Flat Discount	$5.00	$10.00	$13.18	$18.00	101
Percent Discount	10%	10%	12%	15%	196

Note: Data was reported only for all practices due to an insufficient number of responses.

TABLE 3.3
SENIOR-CITIZEN DISCOUNT OFFERED ON SERVICES

	Yes	No	Number of Respondents
All Practices	**42%**	**58%**	**571**
Number of FTE Veterinarians			
1.0 or Less	34%	66%	154
1.1 to 2.0	40%	60%	172
2.1 to 3.0	46%	54%	117
3.1 or More	52%	48%	109
Member Status			
Accredited Practice Member	49%	51%	149
Nonaccredited Member	49%	51%	117
Nonmember	36%	64%	254
Metropolitan Status			
Urban	36%	64%	42
Suburban	48%	52%	164
Second City	46%	54%	114
Town	41%	59%	175
Rural	26%	74%	66
Median Area Household Income			
$35,000 or Less	27%	73%	70
$35,000 to $49,999	38%	63%	208
$50,000 to $69,999	46%	54%	182
$70,000 or More	53%	47%	97

Note: Some row totals do not equal 100% due to rounding.

TABLE 3.4
SENIOR-CITIZEN DISCOUNT OFFERED ON PRESCRIPTION MEDICATIONS

	Yes	No	Number of Respondents
All Practices	**27%**	**73%**	**570**
Number of FTE Veterinarians			
1.0 or Less	25%	75%	154
1.1 to 2.0	26%	74%	172
2.1 to 3.0	26%	74%	117
3.1 or More	33%	67%	108
Member Status			
Accredited Practice Member	32%	68%	149
Nonaccredited Member	29%	71%	117
Nonmember	24%	76%	254
Metropolitan Status			
Urban	26%	74%	42
Suburban	29%	71%	164
Second City	25%	75%	114
Town	30%	70%	174
Rural	17%	83%	66
Median Area Household Income			
$35,000 or Less	21%	79%	70
$35,000 to $49,999	25%	75%	208
$50,000 to $69,999	28%	72%	182
$70,000 or More	33%	67%	96

TABLE 3.5
SENIOR-CITIZEN DISCOUNT OFFERED ON OVER-THE-COUNTER PRODUCTS

	Yes	No	Number of Respondents
All Practices	**24%**	**76%**	**571**
Number of FTE Veterinarians			
1.0 or Less	21%	79%	154
1.1 to 2.0	25%	75%	172
2.1 to 3.0	21%	79%	117
3.1 or More	28%	72%	109
Member Status			
Accredited Practice Member	26%	74%	149
Nonaccredited Member	31%	69%	117
Nonmember	20%	80%	254
Metropolitan Status			
Urban	19%	81%	42
Suburban	24%	76%	164
Second City	25%	75%	114
Town	26%	74%	175
Rural	18%	82%	66
Median Area Household Income			
$35,000 or Less	20%	80%	70
$35,000 to $49,999	24%	76%	208
$50,000 to $69,999	23%	77%	182
$70,000 or More	28%	72%	97

TABLE 3.6
SENIOR-CITIZEN DISCOUNTS (% OFF REGULAR FEE)

	25th Percentile	Median	Average	75th Percentile	Number of Respondents
Discount on Services	10%	10%	10%	10%	227
Discount on Prescription Medications	10%	10%	10%	10%	149
Discount on Over-the-Counter Products	10%	10%	9%	10%	130

Note: Data was reported only for all practices because there was no significant variation in the data based on practice size, member status, metropolitan status, or median area household income.

TABLE 3.7
EARLY SPAY/NEUTER DISCOUNT FOR PUPPIES/KITTENS OFFERED

	Yes	No	Number of Respondents
All Practices	7%	93%	454
Number of FTE Veterinarians			
1.0 or Less	6%	94%	138
1.1 to 2.0	7%	93%	138
2.1 to 3.0	8%	92%	86
3.1 or More	10%	90%	84
Member Status			
Accredited Practice Member	11%	89%	126
Nonaccredited Member	6%	94%	78
Nonmember	4%	96%	217
Metropolitan Status			
Urban	2%	98%	43
Suburban	9%	91%	127
Second City	8%	92%	93
Town	11%	89%	123
Rural	2%	98%	57
Median Area Household Income			
$35,000 or Less	0%	100%	53
$35,000 to $49,999	7%	93%	163
$50,000 to $69,999	9%	91%	154
$70,000 or More	10%	90%	73

CHAPTER 4

VACCINATION SERVICES AND PROTOCOLS

In this chapter, we review and present helpful tips and advice about implementing vaccination protocols as well as how to improve client compliance for routine vaccinations.

IS THERE A VACCINE COMPLIANCE PROBLEM?

Vaccines have become the foundation of health-care management for cats and dogs. Not only are vaccines easy to administer, they provide a cost-effective method for controlling infectious diseases. The veterinary community has improved vaccine design and production, and now vaccines can provide up to several years of protection against deadly diseases.

But even with the improvements in vaccines, there is only 87% compliance for routine vaccinations. Why? Our research suggests that the problem is three-fold:

1. Lack of recommendation by the health-care team
2. Lack of client acceptance
3. Lack of follow-through by the health-care team

Routine vaccinations must be established as part of a regular health-care program, and it's the team's responsibility to do everything they can to make that happen. Clients must bring their pets in for vaccination on a regular basis, and you need to educate them to help them understand why it's so vitally important.

AAHA conducted an extensive compliance survey and studied six areas, including canine and feline core vaccines (*The Path to High-Quality Care*, American Animal Hospital Association Press, 2003). We determined that although compliance for core vaccines was significantly higher (at 87%) than in other areas we studied, there were still 12.4 million dogs and cats that were not protected against core diseases, including:

- Distemper
- Hepatitis
- Leptospirosis
- Parainfluenza virus
- Parvovirus
- Feline viral rhinotracheitis
- Calicivirus
- Panleukopenia

The AAHA compliance study results demonstrated that compliance is not something veterinarians or practice managers think about regularly. In addition, it revealed that many members of the veterinary healthcare team incorrectly assume that clients don't want to know about all of the treatment options. In the study, we proved otherwise, with 90% of clients reporting that they want their veterinarian to tell them about all of the recommended treatment options for their pet, even if they may not be able to afford them.

CHOOSING A PROTOCOL

The first step in ensuring 100% compliance with core vaccines is to choose the protocols you will follow and educate every team member about them.

Vaccine Protocols for Dogs

According to the *2006 AAHA Canine Vaccine Guidelines*, a vaccine protocol must take into account:

- Age
- Breed
- Health status
- Environment (potential exposure to harmful agents)
- Lifestyle (contact with other animals)
- Travel habits

Visit www.aahanet.org for the current recommended protocol for the core and noncore vaccines for dogs.

Vaccine Protocols for Cats

The American Association of Feline Practitioners (AAFP) provides the 2006 Feline Vaccine Advisory Panel Report on its website. In this report, the advisory panel recommends the following special considerations when developing a vaccine protocol for cats:

- Age, especially for kittens and senior cats
- Breed
- Vaccination of cats with preexisting illness
- Vaccination of retrovirus-infected cats
- Concurrent use of corticosteroids
- Vaccination of cats with prior vaccine-associated adverse events

Visit the AAFP website at www.aafponline.org for more information and the full report.

IMPROVING COMPLIANCE

With time and attention, you will be able to improve the rate of compliance with core vaccines in your practice. It's important to first measure your current rate of compliance, though, so that you can set goals for improvement. The CD that accompanies *The Path to High-Quality Care* has a compliance measurement tool that instructs you exactly how to measure compliance by auditing the medical records of 30 dogs and 30 cats that you've seen in the past year. (The compliance tool, along with the study results, are available for no charge to members at www.aahanet.org.)

Once you've measured your current compliance and set a goal for future compliance, lay out the steps you will take to achieve your goal. Make sure everyone is on the same page by establishing a vaccination protocol (see Implementation Idea). Educating every team member about your protocol and the importance of vaccination will boost compliance.

Another excellent way to increase client compliance is to send vaccine reminders. Does every pet have a vaccination reminder date set up in your electronic records? Are you sure? Many practice teams think they have been collecting historical vaccination data for every pet, but if you look, you may just find that you are missing reminder dates for more patients than you think.

The AAHA compliance study showed a direct correlation between clients' compliance with core vaccines and the practice of sending routine vaccine reminders. The study also showed that 65% of pet owners indicated that they would like to receive reminders via multiple delivery methods, including phone calls and emails. In the 2002 Pfizer Veterinary Practice Evaluation survey, the top two communication tools ranked by clients (on a scale of 0 to 5 with 5 being most important) were exam/vaccination reminder cards (4.4) and exam/vaccination reminder phone calls (3.2). Be sure to take full advantage of these communication avenues.

Take-home education pieces, such as the brochures offered by AAHA Press and free educational resources provided by manufacturers, also help to improve clients' awareness of the diseases that can be prevented with vaccines. In addition, you might consider putting vaccine-related articles in your client newsletter or posting bulletins or signs to keep clients updated on the latest research, success stories, or seasonal concerns.

Implementation Idea
Create a Standard of Care for Vaccination

When the entire team follows your standards of care, you will seem more credible to the client, and you and your clients will be more confident in the recommendations your staff makes. Consider adopting something along these lines:

1. The veterinarian adopts a vaccine protocol based on the AAHA and AAFP guidelines and other sources as appropriate.

2. The veterinarian educates the entire staff about the protocol and why it is important to follow the protocol, including a discussion of the diseases that vaccines prevent.

3. The receptionist provides information about the vaccines when a client calls to make an appointment.

4. The receptionist checks the client's record every time the client calls to see if the client's pet(s) is/are overdue for vaccines.

5. During appointments, technicians discuss with clients the diseases and infections the vaccines protect against.

6. Veterinarians make recommendations for specific vaccines based on the age, breed, and lifestyle of the pet.

7. The veterinarian or technician shows the client illustrations, videos, and/or animations that demonstrate the changes and damage caused by diseases and infections that could be prevented by vaccinating.

8. The veterinarian or technician gives the client one or more brochures about the diseases and infections the vaccines will protect against.

9. The receptionist asks to make a vaccination appointment as client checks out.

10. The receptionist sets a reminder date in the pet's electronic record to ensure the client receive notification in the future.

11. One week after the recommendation is made but has not yet been accepted, the receptionist makes a follow-up call to schedule the vaccination appointment.

13. The receptionist runs a weekly report that shows which clients are due for a reminder and sends reminders as follows: mails a card two months before the due date, sends an email one month before the due date (or mails another card if client does not have email), and makes a phone call one week before the due date.

14. If the client declines vaccination for the patient, the client must sign a release form that it is kept in the medical record.

HELPFUL RESOURCES

2006 AAHA Canine Vaccine Guidelines, American Animal Hospital Association, 2006. Available at www.aahanet.org/About_aaha/About_Guidelines_Canine06.html.

The 2006 American Association of Feline Practitioners Feline Vaccine Advisory Panel Report, American Association of Feline Practitioners, 2006. Available at www.aafponline.org/resources/practice_guidelines.htm.

Educating Your Clients from A to Z, Nan Boss, DVM. American Animal Hospital Association Press, 1999.

Legal Consent Forms for Veterinary Practices, Fourth Edition, James F. Wilson, DVM, JD. Priority Press, Ltd., 2006. Available through the American Animal Hospital Association.

The Path to High-Quality Care, AAHA. American Animal Hospital Association Press, 2003.

Pet Health Brochures (Canine Parvovirus, Feline Immune-Deficiency Viruses, Rabies, and Vaccinating Your Pet), AAHA. American Animal Hospital Association Press, 2007.

CHAPTER 4

DATA TABLES

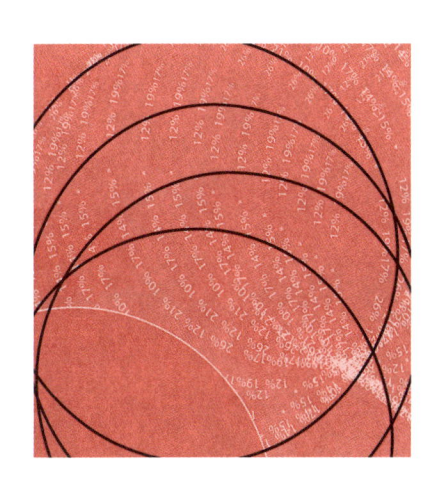

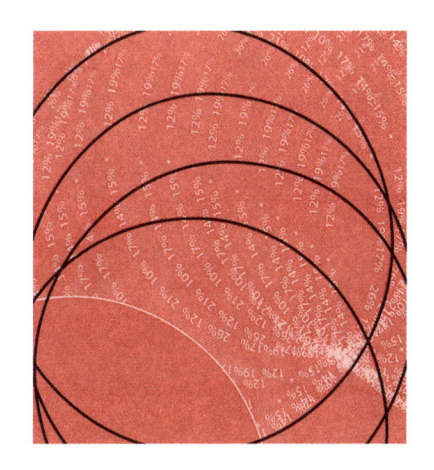

VACCINATION SERVICES AND PROTOCOLS / 97

TABLE 4.1
ONE-YEAR RABIES VACCINE

	25th Percentile	Median	Your Data	Average	75th Percentile	Number of Respondents
All Practices	$12.75	$15.00		$16.35	$18.75	523
Number of FTE Veterinarians						
1.0 or Less	$12.00	$15.00		$15.65	$18.00	140
1.1 to 2.0	$12.00	$15.00		$16.13	$18.00	161
2.1 to 3.0	$13.00	$15.00		$16.43	$19.26	105
3.1 or More	$13.13	$16.00		$17.21	$20.50	99
Member Status						
Accredited Practice Member	$14.00	$16.21		$17.35	$20.00	134
Nonaccredited Member	$13.00	$15.41		$16.81	$20.00	105
Nonmember	$12.00	$15.00		$15.48	$18.00	233
Metropolitan Status						
Urban	$14.13	$18.00		$19.16	$24.00	33
Suburban	$14.00	$16.75		$17.47	$20.00	149
Second City	$12.00	$14.58		$16.24	$18.85	105
Town	$12.00	$15.00		$15.83	$18.00	161
Rural	$10.00	$14.00		$13.88	$16.00	65
Median Area Household Income						
$35,000 or Less	$11.00	$14.00		$15.82	$18.00	61
$35,000 to $49,999	$12.00	$14.00		$14.86	$17.50	191
$50,000 to $69,999	$14.00	$16.00		$17.00	$20.00	170
$70,000 or More	$14.50	$17.50		$18.63	$22.00	87

TABLE 4.2
THREE-YEAR RABIES VACCINE

	25th Percentile	Median	Your Data	Average	75th Percentile	Number of Respondents
All Practices	$14.00	$18.00		$20.13	$25.00	507
Number of FTE Veterinarians						
1.0 or Less	$13.18	$17.00		$19.27	$23.67	128
1.1 to 2.0	$13.90	$17.00		$19.16	$24.00	151
2.1 to 3.0	$14.50	$18.00		$21.04	$25.00	108
3.1 or More	$14.58	$19.08		$21.06	$25.93	102
Member Status						
Accredited Practice Member	$15.00	$19.00		$21.06	$25.00	133
Nonaccredited Member	$14.35	$18.70		$20.80	$25.32	102
Nonmember	$13.00	$17.00		$18.77	$21.80	223
Metropolitan Status						
Urban	$18.14	$25.38		$27.82	$31.08	32
Suburban	$15.00	$19.55		$20.90	$24.81	144
Second City	$13.65	$17.60		$20.07	$25.80	104
Town	$14.00	$17.00		$19.59	$24.00	155
Rural	$12.00	$15.00		$16.07	$18.00	62
Median Area Household Income						
$35,000 or Less	$12.25	$16.50		$19.63	$24.93	60
$35,000 to $49,999	$13.00	$16.42		$18.38	$22.00	182
$50,000 to $69,999	$14.81	$18.68		$20.59	$25.00	164
$70,000 or More	$17.00	$21.42		$23.49	$29.50	87

Note: 4% of respondents reported that they do not offer this service.

TABLE 4.3
DHPP VACCINE

	25th Percentile	Median	Your Data	Average	75th Percentile	Number of Respondents
All Practices	**$15.00**	**$19.38**		**$21.36**	**$25.30**	**524**
Number of FTE Veterinarians						
1.0 or Less	$15.00	$19.95		$21.32	$25.00	141
1.1 to 2.0	$14.00	$19.13		$21.03	$25.50	160
2.1 to 3.0	$14.50	$19.00		$20.70	$24.75	108
3.1 or More	$16.20	$19.00		$22.48	$25.95	97
Member Status						
Accredited Practice Member	$15.00	$18.50		$21.25	$25.00	134
Nonaccredited Member	$15.00	$20.00		$21.45	$25.00	107
Nonmember	$15.00	$19.50		$21.42	$26.00	233
Metropolitan Status						
Urban	$18.30	$24.00		$25.94	$26.50	35
Suburban	$16.00	$20.25		$22.59	$26.75	149
Second City	$15.00	$18.98		$21.04	$25.60	106
Town	$14.00	$19.00		$20.47	$25.00	159
Rural	$12.75	$16.00		$18.27	$22.50	65
Median Area Household Income						
$35,000 or Less	$15.00	$20.00		$22.03	$26.00	67
$35,000 to $49,999	$14.00	$18.00		$19.94	$23.85	187
$50,000 to $69,999	$15.00	$19.95		$21.42	$26.00	169
$70,000 or More	$16.50	$21.00		$23.40	$28.90	87

TABLE 4.4
PARVOVIRUS VACCINE

	25th Percentile	Median	Your Data	Average	75th Percentile	Number of Respondents
All Practices	**$12.00**	**$15.11**		**$16.17**	**$19.80**	**218**
Number of FTE Veterinarians						
1.0 or Less	$12.25	$15.00		$15.66	$18.00	57
1.1 to 2.0	$10.74	$14.00		$15.41	$19.48	66
2.1 to 3.0	$11.75	$16.00		$16.13	$20.00	42
3.1 or More	$13.10	$16.70		$17.09	$20.13	45
Member Status						
Accredited Practice Member	$12.75	$16.94		$17.61	$22.34	60
Nonaccredited Member	$12.00	$16.85		$16.81	$20.13	38
Nonmember	$11.00	$15.00		$14.99	$18.00	97
Metropolitan Status						
Urban	*	*		*	*	18
Suburban	$13.75	$16.75		$17.37	$20.90	69
Second City	$10.51	$15.00		$14.76	$18.75	40
Town	$11.63	$14.00		$15.69	$19.15	61
Rural	*	*		*	*	24
Median Area Household Income						
$35,000 or Less	$9.50	$15.50		$16.24	$19.75	25
$35,000 to $49,999	$10.86	$14.00		$14.02	$17.25	78
$50,000 to $69,999	$12.88	$16.00		$16.61	$20.00	69
$70,000 or More	$15.81	$20.13		$19.47	$22.73	38

Note 1: An asterisk indicates that data was not reported due to an insufficient number of responses.
Note 2: 19% of respondents reported that they do not offer this service.

VACCINATION SERVICES AND PROTOCOLS / 99

TABLE 4.5
LEPTOSPIROSIS VACCINE

	25th Percentile	Median	Your Data	Average	75th Percentile	Number of Respondents
All Practices	$12.00	$16.00		$16.80	$20.00	270
Number of FTE Veterinarians						
1.0 or Less	$13.00	$16.00		$16.47	$20.00	69
1.1 to 2.0	$12.00	$15.56		$17.07	$21.00	84
2.1 to 3.0	$11.00	$15.00		$15.51	$19.15	55
3.1 or More	$12.24	$16.60		$17.31	$21.00	52
Member Status						
Accredited Practice Member	$12.00	$17.10		$17.52	$23.00	83
Nonaccredited Member	$13.00	$15.60		$16.99	$20.15	59
Nonmember	$11.00	$15.00		$15.55	$19.00	103
Metropolitan Status						
Urban	*	*		*	*	18
Suburban	$14.00	$18.00		$18.40	$22.50	93
Second City	$11.50	$15.00		$15.04	$19.00	43
Town	$12.00	$15.00		$15.89	$19.85	78
Rural	$10.00	$15.55		$15.62	$19.75	32
Median Area Household Income						
$35,000 or Less	$8.77	$14.00		$14.21	$17.50	25
$35,000 to $49,999	$10.98	$14.00		$14.92	$18.13	78
$50,000 to $69,999	$13.86	$17.50		$18.02	$22.00	102
$70,000 or More	$14.63	$17.75		$18.65	$23.00	56

Note 1: An asterisk indicates that data was not reported due to an insufficient number of responses.
Note 2: 23% of respondents reported that they do not offer this service.

TABLE 4.6
BORDETELLA VACCINE

	25th Percentile	Median	Your Data	Average	75th Percentile	Number of Respondents
All Practices	$12.75	$16.00		$16.61	$19.82	522
Number of FTE Veterinarians						
1.0 or Less	$12.03	$15.00		$15.68	$18.27	136
1.1 to 2.0	$12.00	$15.00		$16.59	$19.50	163
2.1 to 3.0	$13.03	$16.00		$16.76	$20.00	106
3.1 or More	$14.00	$16.75		$17.24	$20.00	99
Member Status						
Accredited Practice Member	$13.10	$16.39		$17.75	$21.00	136
Nonaccredited Member	$13.06	$16.00		$16.97	$20.00	105
Nonmember	$12.00	$15.00		$15.69	$19.00	231
Metropolitan Status						
Urban	$15.00	$18.00		$18.12	$20.00	35
Suburban	$14.75	$17.75		$18.25	$22.00	149
Second City	$12.56	$15.64		$16.78	$20.00	104
Town	$12.00	$14.85		$15.50	$18.00	159
Rural	$12.00	$15.00		$14.57	$18.18	65
Median Area Household Income						
$35,000 or Less	$11.00	$14.73		$14.98	$17.63	62
$35,000 to $49,999	$12.00	$14.54		$15.08	$18.00	190
$50,000 to $69,999	$14.00	$17.00		$17.36	$20.00	169
$70,000 or More	$16.00	$18.35		$19.47	$23.00	87

Note: 1% of respondents reported that they do not offer this service.

TABLE 4.7
LYME DISEASE VACCINE

	25th Percentile	Median	Your Data	Average	75th Percentile	Number of Respondents
All Practices	**$19.03**	**$22.50**		**$23.59**	**$27.00**	**452**
Number of FTE Veterinarians						
1.0 or Less	$18.25	$22.10		$22.53	$25.00	113
1.1 to 2.0	$19.00	$23.00		$23.78	$27.35	133
2.1 to 3.0	$19.88	$22.48		$23.74	$27.50	96
3.1 or More	$19.74	$22.50		$24.53	$29.16	94
Member Status						
Accredited Practice Member	$19.86	$23.00		$24.76	$28.25	119
Nonaccredited Member	$19.52	$22.38		$24.05	$27.88	96
Nonmember	$18.50	$22.00		$22.44	$25.19	196
Metropolitan Status						
Urban	$20.63	$25.00		$27.10	$30.00	28
Suburban	$20.00	$24.50		$24.88	$28.50	127
Second City	$19.97	$22.90		$24.06	$28.00	89
Town	$18.50	$22.00		$22.43	$25.00	139
Rural	$17.13	$20.68		$21.04	$24.88	60
Median Area Household Income						
$35,000 or Less	$17.50	$20.00		$21.83	$25.00	53
$35,000 to $49,999	$18.00	$21.90		$22.47	$25.00	155
$50,000 to $69,999	$20.00	$24.00		$24.11	$27.20	155
$70,000 or More	$21.00	$25.00		$26.10	$30.75	76

Note: 16% of respondents reported that they do not offer this service.

TABLE 4.8
FVRCP VACCINE

	25th Percentile	Median	Your Data	Average	75th Percentile	Number of Respondents
All Practices	**$14.00**	**$18.00**		**$19.47**	**$22.50**	**534**
Number of FTE Veterinarians						
1.0 or Less	$14.80	$18.00		$19.10	$21.93	146
1.1 to 2.0	$14.00	$18.00		$19.51	$24.00	163
2.1 to 3.0	$13.48	$18.00		$19.21	$22.75	109
3.1 or More	$14.25	$18.00		$20.04	$22.00	97
Member Status						
Accredited Practice Member	$14.00	$18.00		$19.14	$21.50	139
Nonaccredited Member	$14.00	$19.00		$20.01	$23.00	107
Nonmember	$14.00	$18.00		$19.29	$22.60	236
Metropolitan Status						
Urban	$18.00	$21.75		$23.56	$25.00	36
Suburban	$16.00	$19.38		$20.72	$24.00	156
Second City	$12.58	$16.75		$18.72	$21.96	106
Town	$13.88	$17.00		$18.71	$22.00	161
Rural	$12.63	$15.00		$16.82	$20.00	65
Median Area Household Income						
$35,000 or Less	$13.27	$18.00		$19.49	$22.00	67
$35,000 to $49,999	$13.00	$17.40		$18.14	$20.83	191
$50,000 to $69,999	$15.00	$18.75		$19.84	$23.75	172
$70,000 or More	$16.00	$20.00		$21.09	$24.05	90

TABLE 4.9
FeLV VACCINE

	25th Percentile	Median	Your Data	Average	75th Percentile	Number of Respondents
All Practices	$16.50	$20.00		$21.22	$25.00	535
Number of FTE Veterinarians						
1.0 or Less	$15.26	$19.00		$19.86	$23.00	145
1.1 to 2.0	$16.00	$20.00		$20.80	$24.28	162
2.1 to 3.0	$16.81	$21.50		$21.89	$26.65	108
3.1 or More	$17.75	$20.90		$22.56	$25.50	101
Member Status						
Accredited Practice Member	$17.30	$20.38		$22.90	$27.38	142
Nonaccredited Member	$17.25	$21.00		$21.68	$25.30	109
Nonmember	$16.00	$19.50		$19.93	$23.48	235
Metropolitan Status						
Urban	$20.00	$23.40		$25.34	$26.88	38
Suburban	$18.00	$21.50		$22.65	$26.53	154
Second City	$16.00	$20.00		$21.35	$25.00	108
Town	$16.00	$18.50		$20.05	$23.00	163
Rural	$15.00	$17.00		$18.16	$22.00	63
Median Area Household Income						
$35,000 or Less	$15.00	$19.00		$20.59	$25.00	63
$35,000 to $49,999	$16.00	$18.50		$19.79	$22.00	193
$50,000 to $69,999	$17.44	$21.50		$21.52	$25.00	174
$70,000 or More	$17.63	$22.25		$24.11	$29.04	92

TABLE 4.10
FIP VACCINE

	25th Percentile	Median	Your Data	Average	75th Percentile	Number of Respondents
All Practices	$17.07	$20.00		$21.58	$25.00	192
Number of FTE Veterinarians						
1.0 or Less	$17.00	$20.00		$20.19	$23.00	53
1.1 to 2.0	$17.19	$20.00		$21.59	$25.00	66
2.1 to 3.0	$18.50	$22.00		$23.21	$27.00	35
3.1 or More	$16.49	$20.00		$21.27	$24.50	33
Member Status						
Accredited Practice Member	$16.50	$20.20		$22.83	$25.75	41
Nonaccredited Member	$18.38	$21.00		$21.91	$24.00	38
Nonmember	$16.94	$19.20		$20.32	$24.00	86
Metropolitan Status						
Urban	*	*		*	*	15
Suburban	$17.13	$20.50		$22.24	$25.00	52
Second City	$15.93	$21.25		$22.26	$27.12	36
Town	$17.06	$19.00		$20.39	$23.40	63
Rural	*	*		*	*	22
Median Area Household Income						
$35,000 or Less	$17.00	$20.00		$21.41	$26.00	27
$35,000 to $49,999	$16.00	$19.93		$19.88	$22.00	71
$50,000 to $69,999	$19.24	$22.00		$22.98	$25.38	62
$70,000 or More	$14.75	$21.93		$23.18	$30.75	26

Note 1: An asterisk indicates that data was not reported due to an insufficient number of responses.
Note 2: 61% of respondents reported that they do not offer this service.

TABLE 4.11
DISTEMPER VACCINE

	25th Percentile	Median	Your Data	Average	75th Percentile	Number of Respondents
All Practices	**$15.00**	**$19.00**		**$20.40**	**$24.00**	**327**
Number of FTE Veterinarians						
1.0 or Less	$15.00	$19.05		$19.93	$24.00	74
1.1 to 2.0	$14.00	$18.70		$20.10	$25.00	93
2.1 to 3.0	$15.00	$19.50		$20.06	$25.00	75
3.1 or More	$15.90	$19.00		$21.10	$24.00	75
Member Status						
Accredited Practice Member	$15.98	$19.00		$21.23	$24.25	90
Nonaccredited Member	$16.00	$19.94		$20.69	$23.50	69
Nonmember	$15.00	$18.00		$19.44	$23.30	143
Metropolitan Status						
Urban	*	*		*	*	15
Suburban	$18.00	$21.00		$21.98	$25.00	97
Second City	$14.88	$19.00		$20.63	$25.00	62
Town	$14.00	$17.50		$18.87	$23.30	111
Rural	$14.25	$17.58		$18.20	$21.75	36
Median Area Household Income						
$35,000 or Less	$14.50	$19.70		$20.97	$25.00	35
$35,000 to $49,999	$15.00	$18.00		$18.81	$21.50	118
$50,000 to $69,999	$15.00	$19.23		$20.87	$25.00	106
$70,000 or More	$17.20	$22.00		$21.54	$25.00	59

Note 1: An asterisk indicates that data was not reported due to an insufficient number of responses.
Note 2: 3% of respondents reported that they do not offer this service.

TABLE 4.12
RISK-BASED DETERMINATION OF VACCINATION FOR CANINE/FELINE PATIENTS

	Yes	No	Number of Respondents
All Practices	**90%**	**10%**	**564**
Number of FTE Veterinarians			
1.0 or Less	90%	10%	154
1.1 to 2.0	89%	11%	168
2.1 to 3.0	90%	10%	116
3.1 or More	93%	7%	108
Member Status			
Accredited Practice Member	97%	3%	146
Nonaccredited Member	95%	5%	115
Nonmember	84%	16%	252
Metropolitan Status			
Urban	88%	12%	42
Suburban	94%	6%	162
Second City	85%	15%	112
Town	91%	9%	174
Rural	86%	14%	64
Median Area Household Income			
$35,000 or Less	81%	19%	69
$35,000 to $49,999	84%	16%	206
$50,000 to $69,999	96%	4%	180
$70,000 or More	98%	2%	95

TABLE 4.13
USE CORE/NON-CORE VACCINATION PROTOCOLS

	Yes	No	Number of Respondents
All Practices	86%	14%	543
Number of FTE Veterinarians			
1.0 or Less	85%	15%	147
1.1 to 2.0	86%	14%	159
2.1 to 3.0	87%	13%	115
3.1 or More	88%	12%	104
Member Status			
Accredited Practice Member	95%	5%	146
Nonaccredited Member	90%	10%	114
Nonmember	79%	21%	237
Metropolitan Status			
Urban	86%	14%	42
Suburban	91%	9%	158
Second City	83%	17%	109
Town	84%	16%	163
Rural	85%	15%	62
Median Area Household Income			
$35,000 or Less	77%	23%	64
$35,000 to $49,999	79%	21%	203
$50,000 to $69,999	95%	5%	173
$70,000 or More	91%	9%	90

TABLE 4.14
USE AAHA CANINE VACCINE GUIDELINES

	Yes	No	Don't Know	Number of Respondents
All Practices	57%	20%	23%	453
Number of FTE Veterinarians				
1.0 or Less	58%	16%	26%	121
1.1 to 2.0	48%	29%	23%	131
2.1 to 3.0	60%	13%	27%	96
3.1 or More	63%	21%	16%	91
Member Status				
Accredited Practice Member	73%	17%	10%	132
Nonaccredited Member	63%	20%	17%	101
Nonmember	43%	23%	34%	183
Metropolitan Status				
Urban	51%	14%	34%	35
Suburban	60%	21%	19%	132
Second City	56%	27%	17%	89
Town	54%	18%	28%	135
Rural	68%	13%	19%	53
Median Area Household Income				
$35,000 or Less	47%	20%	33%	51
$35,000 to $49,999	58%	21%	21%	160
$50,000 to $69,999	59%	19%	23%	155
$70,000 or More	62%	22%	16%	76

Note 1: Only those who responded that they use core/non-core vaccination protocols were asked this question.
Note 2: Some row totals do not equal 100% due to rounding.

TABLE 4.15
USE AAFP FELINE VACCINE RECOMMENDATIONS

	Yes	No	Don't Know	Number of Respondents
All Practices	**53%**	**21%**	**26%**	**451**
Number of FTE Veterinarians				
1.0 or Less	58%	17%	26%	121
1.1 to 2.0	44%	30%	27%	131
2.1 to 3.0	53%	17%	30%	96
3.1 or More	55%	23%	23%	88
Member Status				
Accredited Practice Member	65%	18%	17%	131
Nonaccredited Member	54%	23%	23%	100
Nonmember	43%	24%	33%	183
Metropolitan Status				
Urban	61%	14%	25%	36
Suburban	55%	19%	26%	133
Second City	59%	26%	15%	87
Town	43%	25%	32%	133
Rural	60%	17%	23%	53
Median Area Household Income				
$35,000 or Less	45%	24%	31%	51
$35,000 to $49,999	57%	20%	23%	160
$50,000 to $69,999	52%	22%	26%	154
$70,000 or More	52%	24%	24%	75

Note 1: Only those who responded that they use core/non-core vaccination protocols were asked this question.
Note 2: Some row totals do not equal 100% due to rounding.

TABLE 4.16
RECOMMENDED FREQUENCY OF ADMINISTRATION OF CORE VACCINES (DHPP OR FVRCP)

	Annually	Every Two Years	Every Three Years	Other	Number of Respondents
All Practices	**46%**	**5%**	**26%**	**24%**	**569**
Number of FTE Veterinarians					
1.0 or Less	55%	5%	26%	14%	152
1.1 to 2.0	47%	6%	20%	27%	172
2.1 to 3.0	38%	6%	29%	26%	117
3.1 or More	39%	4%	30%	28%	109
Member Status					
Accredited Practice Member	34%	4%	35%	27%	149
Nonaccredited Member	38%	5%	32%	25%	116
Nonmember	54%	6%	17%	23%	252
Metropolitan Status					
Urban	49%	10%	27%	15%	41
Suburban	42%	5%	33%	21%	163
Second City	48%	3%	20%	29%	115
Town	48%	5%	21%	26%	174
Rural	45%	5%	30%	20%	66
Median Area Household Income					
$35,000 or Less	59%	7%	19%	16%	70
$35,000 to $49,999	49%	5%	22%	25%	208
$50,000 to $69,999	43%	5%	29%	23%	181
$70,000 or More	38%	4%	32%	26%	96

Note: Some row totals do not equal 100% due to rounding.

TABLE 4.17
USE LEPTOSPIROSIS VACCINE

	Yes	No	Number of Respondents
All Practices	**78%**	**22%**	**570**
Number of FTE Veterinarians			
1.0 or Less	77%	23%	154
1.1 to 2.0	81%	19%	171
2.1 to 3.0	78%	22%	117
3.1 or More	75%	25%	109
Member Status			
Accredited Practice Member	77%	23%	149
Nonaccredited Member	84%	16%	116
Nonmember	78%	22%	253
Metropolitan Status			
Urban	67%	33%	42
Suburban	80%	20%	163
Second City	72%	28%	115
Town	82%	18%	174
Rural	80%	20%	66
Median Area Household Income			
$35,000 or Less	84%	16%	70
$35,000 to $49,999	74%	26%	209
$50,000 to $69,999	80%	20%	181
$70,000 or More	79%	21%	96

TABLE 4.18
FREQUENCY OF LEPTOSPIROSIS VACCINATIONS

	Annually	Biannually	Other	Number of Respondents
All Practices	**87%**	**4%**	**9%**	**445**
Number of FTE Veterinarians				
1.0 or Less	87%	5%	8%	119
1.1 to 2.0	89%	3%	8%	137
2.1 to 3.0	88%	2%	10%	91
3.1 or More	84%	5%	11%	81
Member Status				
Accredited Practice Member	90%	2%	8%	114
Nonaccredited Member	87%	2%	11%	97
Nonmember	86%	6%	8%	197
Metropolitan Status				
Urban	79%	7%	14%	28
Suburban	88%	2%	9%	129
Second City	90%	2%	7%	83
Town	85%	6%	9%	142
Rural	89%	2%	9%	53
Median Area Household Income				
$35,000 or Less	92%	2%	7%	59
$35,000 to $49,999	84%	6%	10%	154
$50,000 to $69,999	90%	3%	8%	144
$70,000 or More	86%	3%	12%	76

Note: Some row totals do not equal 100% due to rounding.

THE VETERINARY FEE REFERENCE / 106

TABLE 4.19
CHECKING IMMUNITY TITERS RATHER THAN VACCINATING

	Yes	No	Number of Respondents
All Practices	**19%**	**81%**	**556**
Number of FTE Veterinarians			
1.0 or Less	12%	88%	153
1.1 to 2.0	18%	82%	167
2.1 to 3.0	23%	77%	112
3.1 or More	22%	78%	105
Member Status			
Accredited Practice Member	21%	79%	143
Nonaccredited Member	24%	76%	113
Nonmember	15%	85%	248
Metropolitan Status			
Urban	31%	69%	42
Suburban	20%	80%	158
Second City	18%	82%	114
Town	19%	81%	169
Rural	9%	91%	64
Median Area Household Income			
$35,000 or Less	13%	87%	69
$35,000 to $49,999	15%	85%	206
$50,000 to $69,999	20%	80%	177
$70,000 or More	30%	70%	91

TABLE 4.20
IMMUNITY TITER CHECK

	25th Percentile	Median	Your Data	Average	75th Percentile	Number of Respondents
All Practices	**$61.60**	**$68.00**		**$77.34**	**$90.00**	**91**
Number of FTE Veterinarians						
1.0 or Less	*	*		*	*	18
1.1 to 2.0	$58.00	$65.50		$72.89	$76.00	27
2.1 to 3.0	*	*		*	*	18
3.1 or More	*	*		*	*	22
Member Status						
Accredited Practice Member	$62.74	$70.06		$81.77	$91.80	28
Nonaccredited Member	*	*		*	*	21
Nonmember	$52.25	$65.50		$76.88	$91.75	32
Metropolitan Status						
Urban	*	*		*	*	11
Suburban	$59.00	$67.00		$80.19	$97.30	29
Second City	*	*		*	*	18
Town	$62.53	$70.17		$76.51	$86.63	26
Rural	*	*		*	*	6
Median Area Household Income						
$35,000 or Less	*	*		*	*	8
$35,000 to $49,999	*	*		*	*	23
$50,000 to $69,999	$64.45	$76.00		$80.58	$98.50	33
$70,000 or More	$60.80	$67.00		$71.98	$80.75	25

Note: An asterisk indicates that data was not reported due to an insufficient number of responses.

CHAPTER 5

GROOMING AND BOARDING SERVICES

In this chapter, we will look at the trends in grooming and boarding services, tips for how to add grooming and boarding to your practice's list of services, and some ideas about setting fees for these services.

TRENDS IN GROOMING AND BOARDING SERVICES

As recently as ten years ago, often the only kennels or runs in veterinary hospitals were those used for patients hospitalized for treatments or procedures. As the demand for boarding grew, practice owners discovered they could generate additional revenue for their practices by using empty cages and runs for boarding. In addition, many technicians and assistants added grooming to their list of responsibilities as a way to expand the practice's service offerings and increase profit.

According to the pet industry spending forecast released by the American Pet Products Manufacturers Association (APPMA), projections of pet-owner spending on their furry friends will reach an all-time high of $40.8 billion in 2007. Of this amount, $9.8 billion will be spent on veterinary care, and an additional $2.9 billion will be used to purchase grooming and boarding services. These projections for veterinary care and grooming and boarding spending are up 6.7% and 6.5%, respectively, as compared to 2006.

Amazingly, the dollars spent annually on kennel boarding is comparable to the annual amount spent on routine veterinary appointments, according to the 2007 projections from the American Pet Products Manufacturers Association. In the United States, the average dog owner will spend $219 per year to take her dog to the veterinarian for routine exams and vaccinations, while the same dog owner will spend $225 per year on boarding her dog. For cats, the numbers are switched but still comparable. In 2007, the projections for cat owner spending are $175 for routine veterinary care compared to $149 for boarding.

As reported by the American Pet Products Manufacturers Association (APPMA), pet services are growing just as quickly as new-product introductions. Some of the fastest growing areas are:

- Pet hotels
- Doggie day care
- High-end grooming
- Training
- Dietician consultations
- Massage therapy

"People consider pets as part of the family, and as health becomes a more pressing issue in our country, people are putting extra thought and care into their pets' health as well," said Bob Vetere, President of the APPMA.

Implementation Idea
Make a Resort, not a Kennel

Building a "resort" for pets, rather than a traditional kennel, will keep you in step with trends in the industry and allow you to offer specialized care that older boarding facilities in your area may not have. According to the American Boarding Kennels Association's *Building, Buying, and Operating a Boarding Kennel*, there are ways to create a resort-like feel for your clients without spending a significant amount of money:

- Build a cat play room that includes a fish tank, large windows to let in a lot of sunlight, and lots of cat towers in front of the windows so that the cats can soak up the sun.
- Build luxury suites for the dogs, complete with little beds for the dogs to sleep in, televisions in the rooms, and large spaces for exercise.
- Keep a bakery case stocked with gourmet doggie goodies.
- Put name plates on the doors or outside each suite personalized with the guest's name.

DOES ADDING SERVICES MAKE SENSE?

But does the increase in the spending on grooming 2and boarding services mean that you should add those services to your offerings? One of the first factors to consider is how much revenue boarding and/or grooming will add to your revenue stream. As reported in the August 2005 issue of *Veterinary Economics*, Gary Glassman, CPA, commented that boarding is one of the least profitable areas in a veterinary practice. In addition, boarding generates only $100 in annual revenue per square foot in a two-doctor, 2,500-square-foot practice. Compare that to clinical services, which generate $1,045 per square foot annually, and the pharmacy, which generates an astounding $1,410 per square foot.

The *2005 Well-Managed Practice Study* conducted by Wutchiett Tumblin and Associates and *Veterinary Economics* reported that when veterinarians were asked what they would do with 200 extra square feet, adding more kennel space came up as eighth in a list of 12 answers. Only 5% of the respondents said they would add more kennel space if they had the opportunity. The highest-ranking answer, with a 33% response, was another exam room. Does this mean that veterinarians see kennel space as wasted potential? Not necessarily.

Meeting Clients' Needs

Veterinary practice today is about establishing a service-oriented business that meets the needs of clients and patients. As a result, many practice owners are designing grooming and boarding facilities into their new or remodeled practices. The last three winners from the *Veterinary Economics* Hospital Design Competition included significant space for grooming and boarding facilities in their winning designs.

The 2007 Hospital of the Year winner, Atascocita Animal Hospital in Humble, Texas, built two kennels with 14 runs and 10 cages for dogs, a cat playroom, and a spacious outdoor play yard. The 2006 Hospital of the Year winner, Yorba Regional Animal Hospital in Anaheim, California, boasts a cattery, 10 luxury suites, three dedicated boarding areas with 60 runs, and a large space dedicated to grooming.

And the 2005 Hospital of the Year winner, VCA Arroyo Animal Hospital in Lake Forest, California, specifically designed their clinic to function as two separate service areas—hospital services and grooming and boarding services—while maintaining the client convenience of having all of the services under one roof. Their facility includes a dedicated grooming area, a dedicated boarding area with 26 runs and 12 cages, and seven cat condos, all on the second floor.

Having an on-site boarding facility also improves client relationships by offering a full-service facility with medical personnel should the pets ever become ill while they are boarded. Denise L. Tumblin, CPA, and Jennette R. Lawson, CPA offered some thoughts about what clients want in a veterinary facility in the October 2005 issue of *Veterinary Economics*. Keep these in mind when you are evaluating fees and new services:

- Convenient location
- Sufficient parking with large spaces
- Clean hospital
- Friendly and professional staff
- Personal recognition by the staff
- Appointments that are kept on schedule

PLANNING AHEAD

A successful business must have a well-written and organized plan. You, too, should create a business plan, even if it's just for the addition of one or two services. The plan should include:

Which services will you offer:

- Your goals related to the services (number of transactions, revenue)
- Which markets will you target
- How you will market the services to the community
- How much financing you'll need
- How will you finance it initially
- How much your costs and net profit will be

Next, establish the grooming and boarding business as a separate profit center from your medical services. This will help you to track costs, track profits specifically related to that service, and help determine just the right amount of staffing needed to offer the services without going over budget.

Develop Your Staff

Adding grooming and boarding services can be a value-added option for clients and the practice. One benefit of offering this service is to provide the veterinary staff with an additional business to learn about and manage. A survey on VetMedTeam.com in July 2004 reported that 95% of a practice's team members felt that their skills were very important or important at work, and that they would leave their jobs or become very unhappy if they were not given new responsibilities. Adding a new service gives your staff the opportunity to learn, grow, and shine.

Keeping the books for these services separate from medical services also ensures that the overhead costs of the medical practice do not affect the grooming and boarding finances. A veterinary practice can have significant overhead that could lead you to incorrectly adjust your fees or targeted profit levels for the grooming and boarding business.

Keeping it as a separate profit center will also allow you to focus your marketing on a particular kind of customer. Not all veterinary hospitals offer boarding or grooming, so some clients may come to your facility just for these services and continue to use their own veterinarians for all medical services. This is not a bad thing—you want to keep up positive relations with other veterinary practices in your area. You do, however, want to track the customers separately from your medical practice so that you can evaluate their demographics.

Setting Fees

Setting the right fees for the grooming and boarding business is a key component to making the services profitable. Based on the findings reported in this edition of *The Veterinary Fee Reference*, there was not a significant increase in boarding fees from 2004 to 2006, but there was a significant 8% increase in the average grooming fee. The average daily rates for boarding dogs in 2006 ranged from $15.30 to $18.43, depending on size. The average daily rate for cats in 2006 was $13.25.

Before establishing a fee, conduct a competitive analysis to find out what similar facilities in your area are charging. If your price is higher, be sure to explain the services and the customer service you offer when potential customers call to inquire about price. Higher-than-average prices are acceptable if you can demonstrate to the customer the value that goes along with it.

Designing the Space

Plan a space and proper facilities for reducing noise and odor when designing grooming and boarding services for your facility. Most boarding facilities put the kennels at the back of the building to make the reception area calm and odor free. However, some clinics, such as the Pet Medical Center of San Antonio, Texas, have developed systems that allow them to put their boarding facilities toward the front of the building.

In the November 2005 issue of *Veterinary Economics*, Dr. Scott Weeks, the owner of Pet Medical Center, stated that he wanted to make the facility's layout easy on the staff and the clients by avoiding the need to walk through the entire hospital to see or pick up a pet. He also keeps the dogs separated from the cats to minimize stress on the cats.

Following Standards

Once the facility is open for business, follow the standards promulgated by the American Boarding Kennel Association (ABKA). Following these standards offers pet owners the assurance and peace of mind that your facility provides state-of-the-art animal care and management. The standards of care as established by ABKA address:

1. Grounds
2. Office and reception areas
3. Recordkeepin
4. Business practices
5. Personnel
6. Work areas
7. Boarding areas
8. Animal-care procedures
9. Environmental control
10. Sanitation
11. Trash and sewage disposal
12. Pest control
13. Fire safety
14. Boarding animals other than dogs and cats
15. Grooming roo
16. Facility vehicles
17. Community playtime

Continue to track the success of the business, including finances, customer satisfaction, and employee morale. Keep in mind that new services will open up opportunities to increase revenue, build your client base, bolster your reputation in the community, and continue to provide high-quality pet care.

HELPFUL RESOURCES

Building, Buying and Operating a Boarding Kennel, Jim Krack, CKO, CAE. American Boarding Kennel Association, 1990.

Boarding Kennel Starter Kit. American Boarding Kennel Association, 1990.

Design It Right: A Pre-architect Primer for Planning Your Veterinary Facility Flow, Thomas E. Catanzaro, DVM, MHA, FACHE. American Animal Hospital Association Press, 2006.

CHAPTER 5

DATA TABLES

GROOMING SERVICES

TABLE 5.1
NAIL TRIM: DOG

	25th Percentile	Median	Your Data	Average	75th Percentile	Number of Respondents
All Practices	**$10.00**	**$12.00**		**$12.40**	**$15.00**	**542**
Number of FTE Veterinarians						
1.0 or Less	$9.25	$11.00		$11.55	$14.26	143
1.1 to 2.0	$10.00	$11.70		$12.34	$15.00	164
2.1 to 3.0	$10.28	$12.50		$13.01	$15.49	110
3.1 or More	$10.50	$12.20		$12.65	$15.00	107
Member Status						
Accredited Practice Member	$10.08	$12.59		$12.92	$15.00	142
Nonaccredited Member	$10.03	$13.00		$13.21	$15.91	112
Nonmember	$9.50	$11.45		$11.80	$15.00	238
Metropolitan Status						
Urban	$11.40	$13.00		$13.64	$17.00	39
Suburban	$10.17	$13.00		$13.27	$15.69	150
Second City	$10.64	$12.38		$12.93	$14.93	108
Town	$9.00	$11.00		$11.27	$13.47	172
Rural	$9.00	$11.50		$11.79	$15.00	63
Median Area Household Income						
$35,000 or Less	$8.00	$10.00		$10.60	$13.00	67
$35,000 to $49,999	$10.00	$11.60		$12.07	$14.24	202
$50,000 to $69,999	$10.05	$12.20		$12.77	$15.00	173
$70,000 or More	$10.93	$14.50		$13.77	$16.06	86

Note: 2% of respondents reported that they do not offer this service.

TABLE 5.2
NAIL TRIM: CAT

	25th Percentile	Median	Your Data	Average	75th Percentile	Number of Respondents
All Practices	**$10.00**	**$11.79**		**$12.03**	**$14.59**	**549**
Number of FTE Veterinarians						
1.0 or Less	$9.00	$10.51		$11.39	$14.00	146
1.1 to 2.0	$9.65	$11.07		$11.85	$14.26	166
2.1 to 3.0	$10.00	$12.00		$12.41	$15.00	111
3.1 or More	$10.00	$12.00		$12.38	$15.00	107
Member Status						
Accredited Practice Member	$10.00	$12.20		$12.58	$15.00	147
Nonaccredited Member	$10.01	$12.00		$12.74	$15.00	113
Nonmember	$9.06	$11.00		$11.40	$14.00	238
Metropolitan Status						
Urban	$9.75	$12.75		$12.99	$16.58	41
Suburban	$10.00	$12.00		$12.71	$15.00	154
Second City	$10.54	$12.15		$12.69	$14.66	108
Town	$8.90	$10.50		$11.11	$13.38	173
Rural	$9.00	$10.50		$11.23	$14.00	63
Median Area Household Income						
$35,000 or Less	$8.00	$10.00		$10.27	$13.00	67
$35,000 to $49,999	$9.58	$11.28		$11.77	$14.00	204
$50,000 to $69,999	$10.00	$12.00		$12.23	$14.36	177
$70,000 or More	$10.30	$14.00		$13.54	$16.00	87

TABLE 5.3
NAIL TRIM: BIRD

	25th Percentile	Median	Your Data	Average	75th Percentile	Number of Respondents
All Practices	**$10.00**	**$13.00**		**$14.12**	**$16.50**	**297**
Number of FTE Veterinarians						
1.0 or Less	$9.31	$11.00		$12.89	$15.00	68
1.1 to 2.0	$10.00	$12.55		$13.95	$16.05	82
2.1 to 3.0	$11.03	$14.00		$15.14	$17.40	67
3.1 or More	$10.63	$13.70		$14.47	$16.94	72
Member Status						
Accredited Practice Member	$11.00	$13.95		$14.77	$17.25	90
Nonaccredited Member	$10.21	$14.00		$15.58	$19.00	61
Nonmember	$10.00	$12.00		$13.11	$15.00	123
Metropolitan Status						
Urban	*	*		*	*	17
Suburban	$10.70	$14.00		$15.15	$17.40	75
Second City	$10.51	$13.00		$14.67	$16.10	55
Town	$10.00	$12.40		$13.12	$15.25	107
Rural	$10.00	$11.40		$13.59	$15.33	37
Median Area Household Income						
$35,000 or Less	$9.63	$12.00		$14.13	$15.75	36
$35,000 to $49,999	$10.00	$11.50		$12.82	$15.00	103
$50,000 to $69,999	$10.31	$14.00		$14.84	$17.70	93
$70,000 or More	$11.03	$14.00		$15.20	$18.00	56

Note 1: An asterisk indicates that data was not reported due to an insufficient number of responses.
Note 2: 43% of respondents reported that they do not offer this service.

TABLE 5.4
NAIL TRIM: REPTILE

	25th Percentile	Median	Your Data	Average	75th Percentile	Number of Respondents
All Practices	**$10.00**	**$13.00**		**$12.97**	**$15.00**	**116**
Number of FTE Veterinarians						
1.0 or Less	*	*		*	*	19
1.1 to 2.0	$10.06	$12.00		$12.83	$15.00	29
2.1 to 3.0	$10.00	$12.11		$12.60	$15.00	26
3.1 or More	$10.50	$15.00		$13.92	$16.64	37
Member Status						
Accredited Practice Member	$10.58	$14.00		$13.89	$16.71	44
Nonaccredited Member	*	*		*	*	20
Nonmember	$9.98	$12.00		$11.97	$15.00	47
Metropolitan Status						
Urban	*	*		*	*	4
Suburban	$10.00	$12.61		$12.78	$15.21	26
Second City	$11.00	$13.16		$13.10	$16.08	28
Town	$9.04	$12.34		$12.50	$15.00	44
Rural	*	*		*	*	11
Median Area Household Income						
$35,000 or Less	*	*		*	*	13
$35,000 to $49,999	$10.08	$12.70		$12.96	$16.03	38
$50,000 to $69,999	$10.00	$13.40		$13.08	$15.00	37
$70,000 or More	*	*		*	*	22

Note 1: An asterisk indicates that data was not reported due to an insufficient number of responses.
Note 2: 64% of respondents reported that they do not offer this service.

TABLE 5.5
NAIL TRIM: SMALL MAMMAL

	25th Percentile	Median	Your Data	Average	75th Percentile	Number of Respondents
All Practices	$10.00	$11.50		$12.01	$14.33	373
Number of FTE Veterinarians						
1.0 or Less	$9.00	$10.50		$11.23	$14.00	80
1.1 to 2.0	$9.50	$11.35		$11.77	$14.00	119
2.1 to 3.0	$10.00	$12.00		$12.66	$15.13	78
3.1 or More	$10.00	$12.00		$12.34	$15.00	84
Member Status						
Accredited Practice Member	$10.00	$12.40		$12.68	$15.00	103
Nonaccredited Member	$10.00	$12.00		$12.06	$14.00	79
Nonmember	$9.51	$11.00		$11.64	$14.19	160
Metropolitan Status						
Urban	*	*		*	*	22
Suburban	$10.00	$12.00		$12.35	$15.00	110
Second City	$10.00	$11.70		$12.37	$14.00	64
Town	$9.00	$11.00		$11.23	$13.60	128
Rural	$9.82	$10.90		$11.69	$14.63	42
Median Area Household Income						
$35,000 or Less	$8.13	$10.00		$10.60	$12.75	36
$35,000 to $49,999	$9.00	$11.00		$11.58	$14.00	135
$50,000 to $69,999	$10.00	$12.00		$12.10	$14.23	125
$70,000 or More	$11.00	$13.00		$13.28	$15.69	65

Note 1: An asterisk indicates that data was not reported due to an insufficient number of responses.
Note 2: 30% of respondents reported that they do not offer this service.

TABLE 5.6
TOOTH TRIM: SMALL MAMMAL

	25th Percentile	Median	Your Data	Average	75th Percentile	Number of Respondents
All Practices	$10.00	$15.00		$16.99	$20.00	254
Number of FTE Veterinarians						
1.0 or Less	$10.08	$15.00		$16.98	$20.00	56
1.1 to 2.0	$10.00	$15.00		$16.15	$19.75	72
2.1 to 3.0	$10.00	$15.00		$16.20	$20.00	53
3.1 or More	$11.70	$15.00		$17.65	$21.68	64
Member Status						
Accredited Practice Member	$12.03	$15.43		$19.12	$23.00	72
Nonaccredited Member	$11.48	$15.50		$16.07	$20.00	49
Nonmember	$10.00	$15.00		$16.70	$20.00	115
Metropolitan Status						
Urban	*	*		*	*	14
Suburban	$10.00	$13.75		$15.99	$20.00	82
Second City	$10.00	$15.00		$17.35	$22.75	48
Town	$12.36	$15.00		$16.73	$20.00	81
Rural	$10.00	$13.40		$16.46	$20.00	25
Median Area Household Income						
$35,000 or Less	*	*		*	*	24
$35,000 to $49,999	$10.00	$13.35		$15.43	$20.00	86
$50,000 to $69,999	$11.00	$15.00		$17.68	$20.00	91
$70,000 or More	$10.00	$17.00		$18.64	$20.50	46

Note 1: An asterisk indicates that data was not reported due to an insufficient number of responses.
Note 2: 3% of respondents reported that they do not offer this service.

TABLE 5.7
ANAL GLAND EXPRESSION

	25th Percentile	Median	Your Data	Average	75th Percentile	Number of Respondents
All Practices	**$13.00**	**$16.00**		**$16.91**	**$20.00**	**543**
Number of FTE Veterinarians						
1.0 or Less	$12.00	$15.00		$16.22	$19.48	146
1.1 to 2.0	$12.41	$16.00		$16.16	$19.05	162
2.1 to 3.0	$14.00	$16.40		$17.61	$21.25	113
3.1 or More	$13.95	$18.00		$18.11	$20.75	105
Member Status						
Accredited Practice Member	$14.00	$17.55		$18.33	$21.50	144
Nonaccredited Member	$13.33	$16.00		$17.28	$20.00	109
Nonmember	$12.00	$15.50		$16.00	$19.00	241
Metropolitan Status						
Urban	$14.38	$18.38		$18.24	$22.10	38
Suburban	$15.00	$18.10		$18.84	$21.58	156
Second City	$13.87	$15.68		$17.26	$19.45	105
Town	$11.89	$15.00		$15.50	$18.00	170
Rural	$11.85	$14.25		$14.57	$17.77	64
Median Area Household Income						
$35,000 or Less	$12.00	$15.00		$14.76	$18.00	68
$35,000 to $49,999	$12.00	$16.00		$16.44	$19.49	200
$50,000 to $69,999	$13.79	$16.00		$17.16	$20.00	172
$70,000 or More	$15.00	$18.63		$19.33	$23.00	89

TABLE 5.8
BATH AND BRUSH

	25th Percentile	Median	Your Data	Average	75th Percentile	Number of Respondents
All Practices	**$25.00**	**$30.00**		**$30.74**	**$35.00**	**347**
Number of FTE Veterinarians						
1.0 or Less	$24.05	$27.00		$30.06	$35.00	93
1.1 to 2.0	$24.00	$29.00		$29.88	$35.00	103
2.1 to 3.0	$25.45	$30.00		$31.30	$37.23	73
3.1 or More	$25.00	$30.98		$32.40	$37.75	69
Member Status						
Accredited Practice Member	$25.00	$31.75		$31.87	$36.00	102
Nonaccredited Member	$25.00	$30.00		$32.01	$38.50	67
Nonmember	$24.00	$29.00		$29.35	$32.88	148
Metropolitan Status						
Urban	$23.30	$30.00		$31.09	$36.35	25
Suburban	$26.19	$32.50		$33.62	$40.00	93
Second City	$23.92	$28.50		$29.77	$34.96	74
Town	$23.00	$27.38		$29.12	$35.00	106
Rural	$25.00	$30.00		$30.01	$32.00	41
Median Area Household Income						
$35,000 or Less	$20.10	$25.00		$26.09	$30.00	53
$35,000 to $49,999	$23.67	$30.00		$30.33	$35.00	128
$50,000 to $69,999	$25.23	$30.00		$31.46	$36.00	106
$70,000 or More	$27.63	$34.65		$35.00	$40.00	53

Note: 33% of respondents reported that they do not offer this service.

TABLE 5.9
WING TRIM

	25th Percentile	Median	Your Data	Average	75th Percentile	Number of Respondents
All Practices	**$11.85**	**$15.00**		**$17.15**	**$20.75**	**269**
Number of FTE Veterinarians						
1.0 or Less	$10.00	$14.25		$15.63	$19.75	60
1.1 to 2.0	$10.64	$15.00		$17.53	$22.00	72
2.1 to 3.0	$11.84	$15.15		$16.79	$19.98	56
3.1 or More	$13.35	$16.05		$17.79	$20.75	72
Member Status						
Accredited Practice Member	$12.21	$16.50		$17.88	$21.00	87
Nonaccredited Member	$12.00	$18.00		$18.59	$24.75	52
Nonmember	$10.00	$15.00		$15.69	$18.00	111
Metropolitan Status						
Urban	*	*		*	*	16
Suburban	$12.00	$16.20		$17.64	$23.50	64
Second City	$11.97	$15.00		$18.32	$24.25	50
Town	$11.72	$15.00		$16.60	$19.54	98
Rural	$10.00	$15.00		$15.95	$16.83	34
Median Area Household Income						
$35,000 or Less	$10.00	$15.00		$17.36	$20.75	34
$35,000 to $49,999	$11.00	$15.00		$16.69	$20.63	98
$50,000 to $69,999	$11.86	$15.40		$16.91	$20.00	78
$70,000 or More	$12.75	$16.74		$17.87	$22.07	50

Note 1: An asterisk indicates that data was not reported due to an insufficient number of responses.
Note 2: 49% of respondents reported that they do not offer this service.

TABLE 5.10
BEAK TRIM

	25th Percentile	Median	Your Data	Average	75th Percentile	Number of Respondents
All Practices	**$12.00**	**$15.13**		**$17.49**	**$21.50**	**241**
Number of FTE Veterinarians						
1.0 or Less	$10.00	$14.62		$15.54	$19.25	51
1.1 to 2.0	$11.00	$15.00		$16.35	$19.75	65
2.1 to 3.0	$13.75	$18.00		$18.98	$23.98	49
3.1 or More	$13.00	$16.00		$18.58	$23.50	67
Member Status						
Accredited Practice Member	$13.75	$18.50		$19.02	$22.00	79
Nonaccredited Member	$13.00	$16.50		$18.44	$24.00	43
Nonmember	$10.00	$15.00		$15.90	$18.85	105
Metropolitan Status						
Urban	*	*		*	*	15
Suburban	$12.00	$16.00		$18.10	$25.00	55
Second City	$11.50	$15.00		$18.44	$24.25	46
Town	$12.00	$15.91		$17.03	$20.00	91
Rural	$10.63	$15.00		$16.14	$19.34	29
Median Area Household Income						
$35,000 or Less	$11.63	$15.00		$17.82	$25.50	33
$35,000 to $49,999	$11.00	$15.00		$16.38	$19.40	88
$50,000 to $69,999	$12.00	$15.91		$18.03	$24.00	67
$70,000 or More	$13.50	$18.25		$18.61	$22.07	46

Note 1: An asterisk indicates that data was not reported due to an insufficient number of responses.
Note 2: 52% of respondents reported that they do not offer this service.

BOARDING SERVICES

TABLE 5.11
BOARDING: DOG IN SMALL CAGE, < 30 POUNDS

	25th Percentile	Median	Your Data	Average	75th Percentile	Number of Respondents
All Practices	**$12.00**	**$15.00**		**$15.30**	**$18.00**	**366**
Number of FTE Veterinarians						
1.0 or Less	$11.00	$13.00		$13.62	$16.00	95
1.1 to 2.0	$11.91	$14.00		$14.81	$17.50	110
2.1 to 3.0	$13.45	$15.85		$16.43	$19.50	73
3.1 or More	$12.69	$15.75		$16.71	$20.00	76
Member Status						
Accredited Practice Member	$13.00	$15.40		$16.07	$18.90	95
Nonaccredited Member	$12.25	$15.00		$16.49	$19.75	77
Nonmember	$10.50	$14.00		$14.11	$16.08	158
Metropolitan Status						
Urban	$14.88	$16.00		$16.80	$18.50	26
Suburban	$15.00	$17.00		$17.63	$20.00	98
Second City	$12.00	$13.18		$14.42	$16.00	76
Town	$11.54	$13.75		$14.47	$17.09	114
Rural	$10.00	$12.00		$12.78	$15.00	44
Median Area Household Income						
$35,000 or Less	$10.00	$11.00		$12.16	$14.63	54
$35,000 to $49,999	$11.75	$13.00		$13.81	$15.89	135
$50,000 to $69,999	$13.69	$16.00		$16.71	$19.81	110
$70,000 or More	$16.00	$18.85		$19.06	$21.64	58

Note: 33% of respondents reported that they do not offer this service.

TABLE 5.12
BOARDING: DOG IN SMALL RUN, < 30 POUNDS

	25th Percentile	Median	Your Data	Average	75th Percentile	Number of Respondents
All Practices	$12.50	$15.05		$15.99	$19.00	323
Number of FTE Veterinarians						
1.0 or Less	$11.00	$13.00		$14.10	$16.00	83
1.1 to 2.0	$12.00	$15.00		$15.44	$18.00	105
2.1 to 3.0	$14.50	$16.75		$17.40	$20.00	59
3.1 or More	$14.10	$16.95		$17.84	$21.00	67
Member Status						
Accredited Practice Member	$14.00	$16.51		$16.96	$20.00	86
Nonaccredited Member	$13.00	$15.15		$16.78	$20.00	65
Nonmember	$11.55	$14.50		$14.85	$17.50	143
Metropolitan Status						
Urban	*	*		*	*	20
Suburban	$15.48	$18.00		$18.70	$22.00	86
Second City	$12.00	$14.25		$15.16	$18.00	62
Town	$12.00	$14.70		$15.05	$17.75	105
Rural	$10.00	$13.10		$13.45	$15.25	42
Median Area Household Income						
$35,000 or Less	$10.00	$12.00		$12.63	$15.00	49
$35,000 to $49,999	$12.00	$14.50		$14.60	$17.00	124
$50,000 to $69,999	$14.25	$16.00		$17.58	$20.00	93
$70,000 or More	$16.28	$19.25		$20.06	$23.60	48

Note 1: An asterisk indicates that data was not reported due to an insufficient number of responses.
Note 2: 39% of respondents reported that they do not offer this service.

TABLE 5.13
BOARDING: DOG IN MEDIUM RUN, 30–60 POUNDS

	25th Percentile	Median	Your Data	Average	75th Percentile	Number of Respondents
All Practices	$13.90	$16.50		$17.05	$20.00	355
Number of FTE Veterinarians						
1.0 or Less	$12.00	$15.00		$15.26	$18.00	93
1.1 to 2.0	$12.85	$16.00		$16.48	$19.78	110
2.1 to 3.0	$15.00	$17.90		$18.38	$22.00	67
3.1 or More	$15.48	$17.55		$18.67	$20.25	74
Member Status						
Accredited Practice Member	$15.10	$17.90		$17.98	$20.00	92
Nonaccredited Member	$14.08	$16.95		$18.41	$22.00	74
Nonmember	$12.00	$15.48		$15.86	$18.99	157
Metropolitan Status						
Urban	*	*		*	*	20
Suburban	$16.53	$19.00		$19.80	$22.90	96
Second City	$12.99	$15.00		$16.22	$18.12	74
Town	$12.55	$15.94		$16.05	$19.00	113
Rural	$11.00	$14.00		$14.24	$16.30	44
Median Area Household Income						
$35,000 or Less	$11.00	$12.75		$13.67	$15.79	52
$35,000 to $49,999	$12.97	$15.00		$15.72	$18.00	134
$50,000 to $69,999	$15.00	$18.00		$18.38	$20.63	106
$70,000 or More	$18.39	$20.00		$21.17	$24.00	54

Note 1: An asterisk indicates that data was not reported due to an insufficient number of responses.
Note 2: 34% of respondents reported that they do not offer this service.

TABLE 5.14
BOARDING: DOG IN LARGE RUN, 61–90 POUNDS

	25th Percentile	Median	Your Data	Average	75th Percentile	Number of Respondents
All Practices	**$15.00**	**$18.00**		**$18.43**	**$21.00**	**358**
Number of FTE Veterinarians						
1.0 or Less	$13.00	$15.55		$16.27	$18.63	90
1.1 to 2.0	$14.50	$18.00		$17.88	$20.43	114
2.1 to 3.0	$17.00	$20.00		$19.63	$22.25	69
3.1 or More	$17.00	$19.75		$20.61	$24.00	75
Member Status						
Accredited Practice Member	$16.00	$19.00		$19.61	$21.85	94
Nonaccredited Member	$15.15	$18.52		$19.55	$23.40	75
Nonmember	$13.85	$17.00		$17.03	$20.00	156
Metropolitan Status						
Urban	*	*		*	*	21
Suburban	$18.00	$20.00		$20.88	$23.95	96
Second City	$14.50	$16.00		$17.82	$20.00	75
Town	$13.90	$17.00		$17.18	$19.53	113
Rural	$12.50	$15.00		$15.50	$18.00	45
Median Area Household Income						
$35,000 or Less	$12.13	$14.70		$15.08	$17.38	52
$35,000 to $49,999	$14.00	$16.60		$17.05	$19.93	134
$50,000 to $69,999	$17.00	$19.40		$20.03	$22.00	110
$70,000 or More	$18.76	$21.00		$21.99	$24.75	53

Note 1: An asterisk indicates that data was not reported due to an insufficient number of responses.
Note 2: 34% of respondents reported that they do not offer this service.

TABLE 5.15
BOARDING: DOG IN LARGE RUN, > 90 POUNDS

	25th Percentile	Median	Your Data	Average	75th Percentile	Number of Respondents
All Practices	**$15.00**	**$18.50**		**$19.17**	**$22.00**	**346**
Number of FTE Veterinarians						
1.0 or Less	$14.00	$17.10		$17.41	$20.00	91
1.1 to 2.0	$15.00	$18.00		$18.59	$21.88	108
2.1 to 3.0	$17.73	$20.00		$20.43	$23.13	66
3.1 or More	$17.00	$20.00		$20.98	$25.00	71
Member Status						
Accredited Practice Member	$16.70	$20.00		$20.40	$23.00	91
Nonaccredited Member	$15.50	$20.00		$20.56	$25.00	71
Nonmember	$14.00	$18.00		$17.76	$20.51	150
Metropolitan Status						
Urban	*	*		*	*	23
Suburban	$18.73	$20.75		$21.74	$24.00	92
Second City	$15.00	$17.50		$18.34	$20.54	70
Town	$14.00	$17.80		$17.94	$20.00	111
Rural	$12.75	$15.00		$15.78	$18.00	42
Median Area Household Income						
$35,000 or Less	$13.00	$15.00		$16.10	$18.69	51
$35,000 to $49,999	$14.50	$17.10		$17.81	$20.85	129
$50,000 to $69,999	$17.75	$20.00		$20.62	$23.23	105
$70,000 or More	$19.00	$22.00		$22.64	$25.00	53

Note 1: An asterisk indicates that data was not reported due to an insufficient number of responses.
Note 2: 34% of respondents reported that they do not offer this service.

TABLE 5.16
BOARDING: CAT

	25th Percentile	Median	Your Data	Average	75th Percentile	Number of Respondents
All Practices	**$10.50**	**$12.60**		**$13.25**	**$15.00**	**407**
Number of FTE Veterinarians						
1.0 or Less	$9.50	$12.00		$12.12	$14.00	103
1.1 to 2.0	$10.00	$12.00		$12.90	$15.00	125
2.1 to 3.0	$12.00	$13.90		$14.07	$16.00	84
3.1 or More	$11.74	$14.00		$14.13	$16.00	82
Member Status						
Accredited Practice Member	$11.60	$14.00		$14.17	$16.00	108
Nonaccredited Member	$11.70	$13.00		$14.18	$16.21	87
Nonmember	$10.00	$12.00		$12.33	$14.20	175
Metropolitan Status						
Urban	$12.00	$14.00		$14.06	$16.00	29
Suburban	$12.00	$14.75		$15.07	$17.00	109
Second City	$10.98	$12.50		$12.80	$14.10	81
Town	$10.00	$12.00		$12.57	$14.88	129
Rural	$8.50	$10.00		$11.37	$13.00	51
Median Area Household Income						
$35,000 or Less	$9.00	$10.00		$10.69	$12.00	54
$35,000 to $49,999	$10.00	$12.00		$12.38	$14.08	150
$50,000 to $69,999	$12.00	$14.00		$14.27	$16.00	132
$70,000 or More	$13.00	$15.00		$15.35	$17.00	62

Note: 25% of respondents reported that they do not offer this service.

TABLE 5.17
BOARDING: BIRD

	25th Percentile	Median	Your Data	Average	75th Percentile	Number of Respondents
All Practices	**$9.00**	**$12.00**		**$12.15**	**$15.00**	**181**
Number of FTE Veterinarians						
1.0 or Less	$8.00	$10.00		$10.50	$14.00	46
1.1 to 2.0	$8.75	$12.00		$12.07	$14.50	45
2.1 to 3.0	$9.50	$13.20		$13.00	$16.05	33
3.1 or More	$10.00	$12.00		$12.90	$17.00	51
Member Status						
Accredited Practice Member	$11.00	$13.20		$14.05	$17.02	59
Nonaccredited Member	$9.75	$12.00		$12.65	$15.62	33
Nonmember	$8.00	$10.00		$10.49	$12.51	74
Metropolitan Status						
Urban	*	*		*	*	9
Suburban	$9.81	$12.00		$13.04	$16.75	44
Second City	$8.80	$10.50		$11.35	$14.00	35
Town	$9.00	$12.00		$12.09	$15.00	69
Rural	*	*		*	*	20
Median Area Household Income						
$35,000 or Less	$8.00	$9.00		$9.75	$10.50	27
$35,000 to $49,999	$9.19	$11.32		$11.82	$14.45	64
$50,000 to $69,999	$9.63	$12.25		$12.94	$16.20	57
$70,000 or More	$11.34	$13.28		$13.74	$17.75	28

Note 1: An asterisk indicates that data was not reported due to an insufficient number of responses.
Note 2: 65% of respondents reported that they do not offer this service.

TABLE 5.18
BOARDING: REPTILE

	25th Percentile	Median	Your Data	Average	75th Percentile	Number of Respondents
All Practices	**$10.00**	**$12.00**		**$12.63**	**$15.00**	**123**
Number of FTE Veterinarians						
1.0 or Less	*	*		*	*	23
1.1 to 2.0	$10.00	$12.00		$11.91	$13.11	32
2.1 to 3.0	*	*		*	*	22
3.1 or More	$10.00	$12.50		$13.11	$16.50	41
Member Status						
Accredited Practice Member	$11.21	$13.58		$14.02	$17.00	48
Nonaccredited Member	*	*		*	*	20
Nonmember	$8.88	$11.00		$11.13	$13.09	46
Metropolitan Status						
Urban	*	*		*	*	7
Suburban	$11.00	$12.75		$13.90	$16.00	27
Second City	$9.75	$12.00		$12.30	$15.15	25
Town	$10.00	$12.00		$12.28	$15.00	50
Rural	*	*		*	*	10
Median Area Household Income						
$35,000 or Less	*	*		*	*	17
$35,000 to $49,999	$10.00	$12.00		$12.41	$14.50	48
$50,000 to $69,999	$11.25	$12.95		$13.98	$16.75	36
$70,000 or More	*	*		*	*	17

Note 1: An asterisk indicates that data was not reported due to an insufficient number of responses.
Note 2: 76% of respondents reported that they do not offer this service.

TABLE 5.19
BOARDING: SMALL MAMMAL

	25th Percentile	Median	Your Data	Average	75th Percentile	Number of Respondents
All Practices	**$10.00**	**$12.00**		**$12.32**	**$15.00**	**209**
Number of FTE Veterinarians						
1.0 or Less	$8.38	$10.05		$10.77	$13.00	46
1.1 to 2.0	$10.00	$12.00		$12.56	$14.00	59
2.1 to 3.0	$10.50	$13.10		$12.48	$15.00	44
3.1 or More	$10.00	$12.00		$13.04	$15.00	55
Member Status						
Accredited Practice Member	$11.05	$13.08		$13.44	$15.94	68
Nonaccredited Member	$10.64	$12.75		$13.58	$16.20	45
Nonmember	$8.50	$10.00		$10.73	$13.00	82
Metropolitan Status						
Urban	*	*		*	*	14
Suburban	$10.50	$13.00		$13.31	$16.00	56
Second City	$9.13	$11.34		$11.61	$14.23	36
Town	$10.00	$12.00		$12.25	$14.00	76
Rural	*	*		*	*	21
Median Area Household Income						
$35,000 or Less	*	*		*	*	23
$35,000 to $49,999	$9.88	$11.32		$11.71	$13.24	74
$50,000 to $69,999	$10.71	$12.75		$13.19	$15.75	73
$70,000 or More	$11.75	$13.20		$13.67	$16.00	33

Note 1: An asterisk indicates that data was not reported due to an insufficient number of responses.
Note 2: 60% of respondents reported that they do not offer this service.

TABLE 5.20
BOARDING: FERRET

	25th Percentile	Median	Your Data	Average	75th Percentile	Number of Respondents
All Practices	$10.00	$12.20		$12.84	$15.00	221
Number of FTE Veterinarians						
1.0 or Less	$9.50	$11.00		$11.39	$13.50	49
1.1 to 2.0	$10.18	$12.00		$12.94	$15.00	60
2.1 to 3.0	$11.48	$13.50		$13.26	$16.16	45
3.1 or More	$10.81	$13.08		$13.57	$15.94	60
Member Status						
Accredited Practice Member	$11.20	$13.15		$13.55	$16.00	71
Nonaccredited Member	$11.00	$13.25		$14.05	$16.26	50
Nonmember	$9.50	$11.06		$11.57	$13.88	84
Metropolitan Status						
Urban	*	*		*	*	12
Suburban	$12.00	$14.00		$14.26	$17.00	62
Second City	$9.75	$12.00		$12.11	$14.15	37
Town	$10.71	$12.00		$12.60	$14.23	80
Rural	*	*		*	*	24
Median Area Household Income						
$35,000 or Less	$8.00	$10.00		$10.12	$12.00	27
$35,000 to $49,999	$9.93	$11.40		$11.76	$13.08	77
$50,000 to $69,999	$11.94	$14.00		$14.05	$16.27	74
$70,000 or More	$12.25	$15.00		$14.68	$17.00	37

Note 1: An asterisk indicates that data was not reported due to an insufficient number of responses.
Note 2: 57% of respondents reported that they do not offer this service.

CHAPTER 6

DENTAL SERVICES

As consumer demand for dental services grows, it's important to consider ways to expand your dental offerings and create a marketing plan that encourages regular dental cleanings for pets. In this chapter, you'll learn different ways to educate clients about the importance of regular dental care and improve compliance with your recommendations.

THE IMPORTANCE OF DENTAL HEALTH

Brenda Richmond brings her six-year-old female cat, Whitney, to you for an annual comprehensive physical exam and vaccinations. She mentions that Whitney is eating less and has bad breath. When you perform an oral exam, you see yellowish brown teeth and receding gums. This advanced stage of periodontal disease could have been prevented with regular dental cleanings and brushing at home.

To help clients understand the importance of caring for their pets' teeth, all veterinary health-care team members must emphasize preventive care. Educational campaigns such as National Pet Dental Health Month each February have accelerated the demand for dental services. Share these facts about dental care to get clients to say yes to preventive dentistry:

- Studies at The Ohio State University and Cornell University found that 85% of dogs and cats aged six years or older have some form of periodontal disease.

- Periodontal disease is associated with chronic internal organ disease of the heart, kidneys, and liver, according to a Kansas State University study.

- Good oral health helps pets live longer, healthier lives—adding as much as five years to a pet's life.
- The American Veterinary Dental Society recommends a dental cleaning every six months for pets, just as human dentists recommend for people.

Dental cleanings and preventive care should be as much a part of your routine care as vaccinations. In fact, some dental specialists assert that if revenue from dental services is less than 10% of your gross revenue, you need to look very closely at how well you're promoting dental care.

EDUCATING CLIENTS ABOUT PREVENTIVE DENTISTRY

The first step in starting a preventive dental program is creating medical protocols. Have your doctors meet to decide how often you'll recommend dental cleanings, whether to require a preanesthetic blood test, what products and medications you'll carry, and which client-education tools you'll use. Write down your medical protocols so that every veterinarian and staff member understands your hospital's approach to preventive care. Update your dental protocols at least annually or when new products or medications are introduced. Your veterinarians should also seek ongoing continuing education in veterinary dentistry.

You can use a variety of teaching tools in your exam rooms to educate clients about preventive dentistry. Consider these educational resources:

Brochures

Help clients understand the potential problems associated with dental disease by giving them a handout to take home. AAHA offers a brochure titled "Your Pet's Dental Care" that describes dental disease, dental cleanings, and home care.

Dental Models

Using a model of a cat's or dog's mouth can help you teach a client how to properly brush her pet's teeth. You also can point out problem areas that may be difficult to see inside the patient's mouth.

Posters

Visual images help you graphically explain the damage of dental disease. Use pictures of various stages of dental disease so clients can identify their pets' current condition. A Hill's Pet Nutrition Inc. before-and-after poster shows one side of a dog's mouth after eating its Prescription Diet t/d and the other side with tartar-caked teeth. Virbac also offers a color poster that shows the various stages of dental disease in dogs and cats, as well as a brochure.

Digital Photography

If you own a digital camera, use a macro lens to take before-and-after pictures of a pet's teeth during an oral assessment treatment and prevention visit. Insert images in a report card to reinforce the client's decision and encourage future dental cleanings.

At All Pets Dental in Weston, Florida, Dr. Jan Bellows, Dipl. ABVP and AVDC, uses intraoral instant photography to help a client see inside her pet's mouth and explain dental pathology. He compares the pet's intraoral photo to a textbook representation of the disease. Intraoral instant cameras are available from many different suppliers—just use your favorite Internet search engine to find them and compare prices.

Disclosing Swabs

Disclosing swabs saturated with red dye can demonstrate residual plaque buildup. Most plaque and tartar accumulation in dogs and cats is usually visible and evidenced by bad breath, but it's far more dramatic when you show it in bright red. Swab a few teeth during the annual physical exam. For every 10 swabs, you'll likely get at least one client who will agree to a dental cleaning or home-care program who otherwise would not have. Technicians can also use disclosing swabs after a dental cleaning to check the thoroughness of their work.

CREATING BUNDLED SERVICES TO INCREASE COMPLIANCE

Every pet that visits your veterinary hospital needs some form of dental care—preventive or therapeutic. When you perform an oral exam during each physical, be sure to add dental information to the patient's medical record. An adequate dental record will provide subjective and objective information as well as an assessment and therapeutic plan. Use a dental diagram to note problem areas, your recommendations, and the treatment plan. You also can note whether the patient has mild gingivitis, moderate gingivitis, severe gingivitis, or periodontal disease.

Consider creating bundled packages that make routine dental cleanings affordable for clients. For example, provide an oral health package that includes a pre-anesthetic test, dental cleaning, and a pet toothbrush and toothpaste for home care. You can offer this preventive package for pets with mild gingivitis and create similar packages for more advanced dental disease.

Reinforce your recommendation and the oral health package with a client handout. At Paws & Claws Animal Hospital in Plano, Texas, Dr. Shawn P. Messonnier uses an educational handout after an oral exam to reinforce the fact that the pet needs a dental cleaning (see Figure 6.1).

FIGURE 6.1
EDUCATIONAL HANDOUT FOR DENTAL CLEANING

ORAL INFECTIONS IN DOGS AND CATS

Your pet has been diagnosed with an oral infection of his/her teeth and gums. Oral infections (also called periodontal infections or periodontitis) are among the most common diseases seen in dogs and cats. Thankfully, they also are among the LEAST EXPENSIVE to treat. These infections are painful and, if left untreated, will shorten your pet's life. Prompt treatment will relieve your pet's pain and suffering and help prevent kidney, heart, liver, lung, and gastrointestinal infections seen in pets with chronic oral infections.

Your veterinarian has prescribed a 10-day dose of antibiotics for your pet, which will help sterilize the mouth in preparation for the ultrasonic scaling that is needed to physically remove the infection from the teeth and gums. The antibiotics should be given for five days.

On day six, your pet will come to our hospital for a procedure called an ultrasonic scaling (please schedule prior to starting the antibiotics). Performed under a short-acting anesthetic, the scaling will be followed by an antibiotic rinse, polishing, and fluoride treatment to further remove the infection. You should continue the antibiotics after your pet goes home.

My pet _____ will take _____ twice daily for five days. On day six, _____ is scheduled for an ultrasonic scaling.

Please withhold food (but not water) after dinner on the night before the ultrasonic scaling. Please give your pet his/her antibiotic the morning and the evening of the procedure.

Source: Adapted from *Marketing Your Veterinary Practice*, Volume 2 (Mosby-Year Book Inc., 1997)

Case Study

How to Get Excellent Compliance for Dental Treatment

The key to a successful program is to prevent the development of disease rather than waiting for it to appear. Don't prejudge clients and assume that they won't accept your recommendation for annual dental cleanings for their pets.

At one practice in Colorado, 75% of clients bring their pets in for annual dental prophylactic treatments. The practice team enjoys such a high compliance rate because every health-care team member at the hospital passionately promotes the need for preventive care. To achieve this level of client compliance, use these strategies:

- **Speak with passion**. From the receptionist to the technician to the veterinarian, everyone must talk with clients about the importance of regular dental care.

- **Create a smile album**. Assemble a scrapbook with close-up photos of a pet's teeth before and after a dental cleaning. Also include photos of advanced procedures such as a root canal or extraction.

- **Educate with your newsletter**. Feature case studies and tips on home care such as an article on how to brush your pet's teeth.

- **Schedule the next oral exam at checkout.** After your receptionist collects payment for today's visit, she could say, "Dr. Smith asked to see Max again in six months for his oral exam. What day of the week in September works best for you?" This approach stresses the importance of the recheck.

- **Develop a recall system.** Create a system with categories for different levels of dental care such as minor, moderate, or major prophylactic care. These levels can be linked in your computer system by incrementally larger fees and decreased intervals between dental recall appointments.

Implementation Idea
Create a Dental Reminder Card

When creating reminder cards for dental cleanings, follow the guidelines of the American Veterinary Dental Society, which recommends a dental cleaning every six months for pets. Entering a six-month reminder in your computer will encourage clients to have oral exams twice a year so you can detect problems early. More frequent visits also give you the opportunity to reinforce the importance of brushing at home and feeding a dental diet.

Develop a reminder card with an educational message that stresses the importance of regular oral exams and dental cleanings. Avoid using veterinary jargon such as "dental prophy," which might confuse clients. Instead, use terms that clients understand from human dentistry such as "checkup" and "dental cleaning."

Also use the pet's name to personalize the message (see Figure 6.2). Boldface important information such as your hospital phone number and website. Include a P.S. message at the bottom of the reminder card because 80% of people will read it. Below the client's name and address you might add a short message about a dental special or extended appointment hours.

FIGURE 6.2
SAMPLE DENTAL REMINDER

Any Animal Hospital
123 Main Street
Your City, ST 00000
000/555-1212
www.youranimalhospital.com

Pet Name is now due for a dental checkup. During Pet Name's oral exam, your veterinarian will check for signs of dental disease:

- ✓ Bad breath
- ✓ Sensitivity around the mouth
- ✓ Loss of appetite or difficulty chewing food
- ✓ Yellow or brown teeth with tartar and plaque buildup
- ✓ Bleeding, inflamed, or receded gums
- ✓ Loose or missing teeth

Sincerely,
Your friends at Any Animal Hospital

P.S. Call us at 000/555-1212 today!

We are now open until 8 pm!

Did you know that regular dental care could add as much as five years to your pet's life?

Client i.d.
Client name
Address 1
Address 2
City, State postalcode

HELPFUL RESOURCES

The Academy of Veterinary Dentistry is an international organization of veterinarians with a special interest in the dental care of animals. Visit www.avdonline.org.

The American Society of Veterinary Dental Technicians offers a home-study course and qualifying exam. For information, call 800/613-3647 or visit www.asvdt.org.

The American Veterinary Dental College promotes the advancement of high standards in veterinary dentistry through the encouragement of all veterinary colleges to establish in-depth instruction and a high standard for training in veterinary dentistry. Call 215/898-5903 or visit www.avdc.org.

American Veterinary Dental Society membership is open to any veterinarian, dentist, dental hygienist, technician, or individual with an interest in veterinary dentistry. Call 800/332-AVDS or visit www.avds-online.org.

Canine Dental Record, AAHA. American Animal Hospital Association Press, 2002.

Dental Discharge Instructions Form, AAHA. American Animal Hospital Association Press, 2003.

Feline Dental Record, AAHA. American Animal Hospital Association Press, 2002.

The *Journal of Veterinary Dentistry* at www.jvdonline.org is the official publication of the American Veterinary Dental Society (AVDS), the Academy of Veterinary Dentistry (AVD), and the American Veterinary Dental College (AVDC).

"Your Pet's Dental Care" brochure, AAHA. American Animal Hospital Association Press, 2007.

CHAPTER 6

DATA TABLES

TABLE 6.1
DENTAL CASE: PREANESTHETIC EXAM

	25th Percentile	Median	Your Data	Average	75th Percentile	Number of Respondents
All Practices	$28.25	$37.26		$36.07	$44.00	147
Number of FTE Veterinarians						
1.0 or Less	$30.00	$36.38		$35.11	$41.50	48
1.1 to 2.0	$27.25	$37.63		$35.27	$44.00	44
2.1 to 3.0	*	*		*	*	24
3.1 or More	$24.63	$36.88		$34.67	$44.25	26
Member Status						
Accredited Practice Member	$26.50	$38.06		$36.67	$44.20	38
Nonaccredited Member	$26.93	$39.67		$35.02	$44.50	25
Nonmember	$26.50	$35.00		$33.16	$39.74	65
Metropolitan Status						
Urban	*	*		*	*	15
Suburban	$28.25	$38.00		$37.94	$45.00	39
Second City	$32.00	$40.25		$37.34	$44.95	40
Town	$25.00	$31.00		$33.27	$39.75	38
Rural	*	*		*	*	13
Median Area Household Income						
$35,000 or Less	*	*		*	*	16
$35,000 to $49,999	$28.25	$37.13		$34.77	$44.38	52
$50,000 to $69,999	$30.23	$38.00		$37.76	$42.75	57
$70,000 or More	*	*		*	*	17

Note 1: The patient is a six-year-old, 35-pound spayed female cocker spaniel that you examined and vaccinated one week ago. At that time, you found moderate dental calculus and slight gingivitis. The patient was admitted in the morning, anesthetized for routine dental scaling, subgingival curettage, and tooth polishing; no extractions were performed.
Note 2: An asterisk indicates that data was not reported due to an insufficient number of responses.
Note 3: 53% of the respondents reported that the fee for this service is included in the total dentistry fee and is not separately itemized.

TABLE 6.2
DENTAL CASE: CBC WITH DIFFERENTIAL

	25th Percentile	Median	Your Data	Average	75th Percentile	Number of Respondents
All Practices	**$32.56**	**$41.10**		**$47.46**	**$55.00**	**344**
Number of FTE Veterinarians						
1.0 or Less	$29.70	$36.90		$41.46	$48.63	101
1.1 to 2.0	$35.50	$45.00		$51.80	$68.00	103
2.1 to 3.0	$35.00	$42.00		$49.84	$56.77	69
3.1 or More	$31.00	$42.25		$45.64	$53.83	68
Member Status						
Accredited Practice Member	$35.00	$44.75		$49.56	$55.09	98
Nonaccredited Member	$30.00	$39.50		$49.61	$67.50	61
Nonmember	$30.00	$37.64		$43.46	$49.13	158
Metropolitan Status						
Urban	$34.38	$46.38		$49.08	$52.95	26
Suburban	$35.00	$45.00		$51.13	$57.93	96
Second City	$31.50	$40.50		$48.17	$58.42	72
Town	$33.25	$40.00		$45.51	$53.75	96
Rural	$25.50	$34.50		$40.60	$54.00	47
Median Area Household Income						
$35,000 or Less	$30.00	$37.28		$42.86	$44.25	43
$35,000 to $49,999	$30.00	$39.00		$43.52	$50.14	125
$50,000 to $69,999	$35.00	$42.50		$50.01	$58.88	113
$70,000 or More	$39.00	$49.00		$54.89	$66.80	55

Note 1: See the case description in the notes for Table 6.1.
Note 2: 10% of the respondents reported that the fee for this service is included in the total dentistry fee and is not separately itemized.

TABLE 6.3
DENTAL CASE: CHEMISTRY PANEL WITH EIGHT CHEMISTRIES

	25th Percentile	Median	Your Data	Average	75th Percentile	Number of Respondents
All Practices	**$45.00**	**$58.00**		**$61.86**	**$75.00**	**311**
Number of FTE Veterinarians						
1.0 or Less	$42.00	$54.00		$59.59	$72.00	95
1.1 to 2.0	$45.00	$59.50		$61.63	$75.30	90
2.1 to 3.0	$46.00	$58.75		$62.22	$78.15	62
3.1 or More	$54.25	$59.90		$64.88	$74.38	60
Member Status						
Accredited Practice Member	$48.00	$65.26		$65.92	$78.50	91
Nonaccredited Member	$48.25	$55.00		$62.47	$81.86	49
Nonmember	$42.00	$55.00		$58.17	$69.44	155
Metropolitan Status						
Urban	$49.13	$68.05		$71.52	$80.40	33
Suburban	$46.90	$59.00		$62.82	$75.75	85
Second City	$46.25	$60.93		$62.16	$73.53	60
Town	$44.75	$56.63		$61.09	$75.25	90
Rural	$40.00	$48.00		$52.36	$60.25	38
Median Area Household Income						
$35,000 or Less	$43.50	$50.75		$53.36	$65.25	38
$35,000 to $49,999	$42.75	$54.00		$60.99	$71.50	119
$50,000 to $69,999	$48.31	$63.75		$65.33	$85.00	104
$70,000 or More	$49.80	$60.00		$64.76	$76.56	45

Note 1: See the case description in the notes for Table 6.1.
Note 2: 22% of the respondents reported that the fee for this service is included in the total dentistry fee and is not separately itemized.

DENTAL SERVICES / 137

TABLE 6.4
DENTAL CASE: ANESTHESIA, 30 MINUTES

	25th Percentile	Median	Your Data	Average	75th Percentile	Number of Respondents
All Practices	**$50.04**	**$74.88**		**$76.47**	**$95.46**	**330**
Number of FTE Veterinarians						
1.0 or Less	$41.42	$70.45		$71.93	$95.00	102
1.1 to 2.0	$50.00	$70.00		$75.95	$92.00	95
2.1 to 3.0	$60.75	$80.58		$81.67	$100.70	70
3.1 or More	$58.13	$75.50		$79.28	$98.25	58
Member Status						
Accredited Practice Member	$60.00	$80.85		$82.31	$98.53	93
Nonaccredited Member	$55.35	$71.00		$76.70	$91.25	60
Nonmember	$45.00	$70.00		$72.20	$95.00	155
Metropolitan Status						
Urban	$65.00	$99.00		$91.02	$115.00	31
Suburban	$64.96	$81.75		$84.12	$101.19	88
Second City	$58.23	$75.00		$82.69	$104.25	69
Town	$45.85	$64.50		$67.79	$85.68	94
Rural	$39.50	$57.00		$58.29	$75.05	42
Median Area Household Income						
$35,000 or Less	$40.00	$56.00		$59.67	$75.99	37
$35,000 to $49,999	$45.00	$66.00		$70.69	$93.91	120
$50,000 to $69,999	$57.00	$80.00		$80.92	$99.90	109
$70,000 or More	$73.00	$90.00		$92.87	$109.61	57

Note 1: See the case description in the notes for Table 6.1.
Note 2: 21% of the respondents reported that the fee for this service is included in the total dentistry fee and is not separately itemized.

TABLE 6.5
DENTAL CASE: IV CATHETER AND PLACEMENT

	25th Percentile	Median	Your Data	Average	75th Percentile	Number of Respondents
All Practices	**$28.00**	**$38.53**		**$40.96**	**$49.00**	**322**
Number of FTE Veterinarians						
1.0 or Less	$25.00	$35.00		$38.02	$48.00	89
1.1 to 2.0	$27.18	$38.38		$41.48	$49.99	96
2.1 to 3.0	$30.00	$39.50		$42.02	$49.95	67
3.1 or More	$32.00	$40.00		$43.30	$49.00	65
Member Status						
Accredited Practice Member	$30.25	$39.38		$42.08	$48.13	94
Nonaccredited Member	$27.75	$36.00		$39.29	$46.80	63
Nonmember	$26.00	$35.50		$39.84	$49.96	138
Metropolitan Status						
Urban	$31.80	$47.65		$47.88	$60.93	34
Suburban	$30.24	$41.50		$43.12	$50.56	88
Second City	$30.00	$39.75		$45.57	$53.00	63
Town	$26.75	$35.00		$37.11	$45.00	91
Rural	$21.50	$29.75		$32.37	$44.00	41
Median Area Household Income						
$35,000 or Less	$25.00	$33.90		$34.78	$41.77	37
$35,000 to $49,999	$26.00	$37.00		$39.55	$48.88	116
$50,000 to $69,999	$30.00	$39.75		$42.29	$48.50	111
$70,000 or More	$30.69	$43.02		$46.37	$59.75	52

Note 1: See the case description in the notes for Table 6.1.
Note 2: 13% of the respondents reported that the fee for this service is included in the total dentistry fee and is not separately itemized.

TABLE 6.6
DENTAL CASE: IV FLUIDS

	25th Percentile	Median	Your Data	Average	75th Percentile	Number of Respondents
All Practices	$20.00	$30.00		$31.97	$40.00	248
Number of FTE Veterinarians						
1.0 or Less	$19.20	$25.00		$30.32	$35.70	71
1.1 to 2.0	$24.88	$31.00		$30.70	$38.00	73
2.1 to 3.0	$19.63	$30.00		$31.45	$40.70	57
3.1 or More	$24.80	$33.05		$37.34	$46.00	43
Member Status						
Accredited Practice Member	$20.00	$35.00		$34.92	$45.00	71
Nonaccredited Member	$23.50	$34.00		$34.36	$42.18	45
Nonmember	$20.00	$25.80		$28.65	$35.16	113
Metropolitan Status						
Urban	*	*		*	*	24
Suburban	$25.38	$35.41		$35.61	$45.00	66
Second City	$24.50	$33.08		$33.25	$42.59	50
Town	$16.00	$25.00		$28.57	$35.29	74
Rural	$15.00	$23.38		$23.98	$30.00	30
Median Area Household Income						
$35,000 or Less	$19.40	$25.48		$26.72	$31.31	32
$35,000 to $49,999	$18.70	$25.75		$29.44	$37.25	88
$50,000 to $69,999	$24.80	$35.00		$34.65	$43.00	83
$70,000 or More	$22.00	$34.75		$34.34	$46.00	39

Note 1: See the case description in the notes for Table 6.1.
Note 2: IV fluids assumed to be 1,000 ml Bag of Lactated Ringer's Solution.
Note 3: An asterisk indicates that data was not reported due to an insufficient number of responses.
Note 4: 31% of the respondents reported that the fee for this service is included in the total dentistry fee and is not separately itemized.

TABLE 6.7
DENTAL CASE: DENTAL SCALING AND POLISHING

	25th Percentile	Median	Your Data	Average	75th Percentile	Number of Respondents
All Practices	$55.00	$71.20		$86.40	$100.39	406
Number of FTE Veterinarians						
1.0 or Less	$55.00	$75.00		$86.15	$104.38	127
1.1 to 2.0	$53.79	$65.00		$84.67	$98.60	121
2.1 to 3.0	$55.00	$71.75		$82.60	$95.63	80
3.1 or More	$56.25	$76.00		$94.48	$139.19	73
Member Status						
Accredited Practice Member	$56.00	$73.00		$90.88	$112.50	115
Nonaccredited Member	$54.85	$65.48		$82.53	$94.75	69
Nonmember	$51.18	$70.00		$83.35	$100.00	193
Metropolitan Status						
Urban	$60.80	$79.50		$93.83	$123.85	38
Suburban	$55.00	$74.88		$93.36	$125.00	110
Second City	$53.79	$71.66		$83.69	$96.00	81
Town	$54.13	$66.50		$81.11	$87.24	114
Rural	$47.50	$57.00		$74.89	$81.40	55
Median Area Household Income						
$35,000 or Less	$46.88	$56.50		$64.86	$74.37	50
$35,000 to $49,999	$54.40	$68.00		$79.13	$93.13	150
$50,000 to $69,999	$55.00	$75.00		$92.94	$120.00	134
$70,000 or More	$60.88	$88.15		$103.42	$136.25	64

Note 1: See the case description in the notes for Table 6.1.
Note 2: 4% of the respondents reported that the fee for this service is included in the total dentistry fee and is not separately itemized.

TABLE 6.8
DENTAL CASE: SUBGINGIVAL CURETTAGE

	25th Percentile	Median	Your Data	Average	75th Percentile	Number of Respondents
All Practices	$20.00	$37.75		$54.00	$71.25	38

Note 1: See the case description in the notes for Table 6.1.
Note 2: Data was reported only for all practices due to an insufficient number of responses.
Note 3: 75% of the respondents reported that the fee for this service is included in the total dentistry fee and is not separately itemized.

TABLE 6.9
DENTAL CASE: FLUORIDE APPLICATION

	25th Percentile	Median	Your Data	Average	75th Percentile	Number of Respondents
All Practices	**$6.36**	**$10.75**		**$12.04**	**$15.00**	**88**
Number of FTE Veterinarians						
1.0 or Less	*	*		*	*	24
1.1 to 2.0	$5.25	$11.00		$12.44	$17.48	28
2.1 to 3.0	*	*		*	*	22
3.1 or More	*	*		*	*	14
Member Status						
Accredited Practice Member	$6.36	$10.13		$11.64	$14.65	36
Nonaccredited Member	*	*		*	*	15
Nonmember	$9.59	$11.72		$12.38	$15.75	30
Metropolitan Status						
Urban	*	*		*	*	8
Suburban	$7.50	$12.00		$12.40	$15.00	29
Second City	*	*		*	*	18
Town	*	*		*	*	23
Rural	*	*		*	*	9
Median Area Household Income						
$35,000 or Less	*	*		*	*	6
$35,000 to $49,999	$5.00	$10.00		$10.51	$14.70	31
$50,000 to $69,999	$7.83	$11.53		$12.69	$14.77	38
$70,000 or More	*	*		*	*	12

Note 1: See the case description in the notes for Table 6.1.
Note 2: An asterisk indicates that data was not reported due to an insufficient number of responses.
Note 3: 49% of the respondents reported that the fee for this service is included in the total dentistry fee and is not separately itemized.

TABLE 6.10
DENTAL CASE: ELECTRONIC MONITORING

	25th Percentile	Median	Your Data	Average	75th Percentile	Number of Respondents
All Practices	$13.95	$19.75		$20.99	$25.40	114
Number of FTE Veterinarians						
1.0 or Less	$14.00	$18.50		$18.78	$26.00	27
1.1 to 2.0	$12.38	$17.75		$19.61	$25.90	30
2.1 to 3.0	$15.25	$19.75		$22.89	$28.75	32
3.1 or More	*	*		*	*	23
Member Status						
Accredited Practice Member	$15.00	$20.00		$22.75	$28.25	42
Nonaccredited Member	$13.00	$18.50		$19.67	$25.50	25
Nonmember	$11.50	$17.50		$19.42	$25.00	42
Metropolitan Status						
Urban	*	*		*	*	14
Suburban	$16.50	$22.50		$24.93	$30.00	39
Second City	*	*		*	*	24
Town	$10.00	$15.00		$16.67	$24.00	32
Rural	*	*		*	*	5
Median Area Household Income						
$35,000 or Less	*	*		*	*	11
$35,000 to $49,999	$10.00	$17.75		$18.73	$25.00	42
$50,000 to $69,999	$15.00	$20.00		$21.55	$26.39	37
$70,000 or More	*	*		*	*	22

Note 1: See the case description in the notes for Table 6.1.
Note 2: An asterisk indicates that data was not reported due to an insufficient number of responses.
Note 3: 60% of the respondents reported that the fee for this service is included in the total dentistry fee and is not separately itemized.

TABLE 6.11
DENTAL CASE: POST-PROCEDURE PAIN MEDICATION

	25th Percentile	Median	Your Data	Average	75th Percentile	Number of Respondents
All Practices	$15.00	$20.16		$21.96	$27.50	318
Number of FTE Veterinarians						
1.0 or Less	$15.00	$20.00		$21.44	$27.63	89
1.1 to 2.0	$15.00	$20.60		$21.64	$26.00	95
2.1 to 3.0	$15.05	$20.00		$21.91	$27.65	72
3.1 or More	$16.63	$22.91		$23.36	$29.22	58
Member Status						
Accredited Practice Member	$17.93	$23.50		$24.28	$30.24	97
Nonaccredited Member	$15.50	$22.00		$22.50	$29.50	59
Nonmember	$15.00	$19.95		$20.13	$25.00	137
Metropolitan Status						
Urban	$19.00	$26.25		$25.91	$32.50	29
Suburban	$17.29	$24.38		$23.71	$29.75	91
Second City	$17.47	$24.31		$24.30	$30.15	70
Town	$14.65	$19.00		$19.03	$22.00	83
Rural	$12.00	$15.75		$17.60	$23.25	40
Median Area Household Income						
$35,000 or Less	$14.00	$18.00		$19.17	$23.38	37
$35,000 to $49,999	$14.58	$19.97		$20.54	$26.19	112
$50,000 to $69,999	$17.83	$21.00		$23.11	$28.13	110
$70,000 or More	$17.63	$25.00		$24.62	$30.33	52

Note 1: See the case description in the notes for Table 6.1.
Note 2: 14% of the respondents reported that the fee for this service is included in the total dentistry fee and is not separately itemized.

TABLE 6.12
DENTAL CASE: POST-PROCEDURE INJECTABLE ANTIBIOTICS

	25th Percentile	Median	Your Data	Average	75th Percentile	Number of Respondents
All Practices	**$15.00**	**$21.00**		**$21.01**	**$25.81**	**310**
Number of FTE Veterinarians						
1.0 or Less	$14.63	$20.25		$20.49	$26.77	92
1.1 to 2.0	$15.99	$21.33		$21.77	$26.00	94
2.1 to 3.0	$16.55	$21.00		$21.18	$25.50	61
3.1 or More	$15.00	$19.84		$20.43	$25.00	58
Member Status						
Accredited Practice Member	$15.99	$21.75		$21.43	$26.06	86
Nonaccredited Member	$16.00	$21.46		$21.65	$25.00	59
Nonmember	$14.00	$19.55		$20.03	$25.00	142
Metropolitan Status						
Urban	$21.25	$25.00		$26.34	$31.44	28
Suburban	$19.50	$24.01		$23.71	$28.61	88
Second City	$17.50	$22.50		$22.84	$27.63	61
Town	$12.49	$17.00		$17.21	$21.75	85
Rural	$10.38	$15.00		$16.22	$20.00	42
Median Area Household Income						
$35,000 or Less	$12.00	$17.00		$17.86	$24.56	40
$35,000 to $49,999	$14.37	$20.00		$19.77	$24.62	111
$50,000 to $69,999	$16.75	$21.20		$22.06	$26.13	101
$70,000 or More	$18.25	$24.51		$23.98	$29.00	52

Note 1: See the case description in the notes for Table 6.1.
Note 2: 18% of the respondents reported that the fee for this service is included in the total dentistry fee and is not separately itemized.

TABLE 6.13
DENTAL CASE: HOSPITALIZATION

	25th Percentile	Median	Your Data	Average	75th Percentile	Number of Respondents
All Practices	**$16.00**	**$22.85**		**$26.76**	**$32.48**	**181**
Number of FTE Veterinarians						
1.0 or Less	$14.77	$20.08		$24.21	$29.00	52
1.1 to 2.0	$17.13	$21.73		$25.45	$31.84	56
2.1 to 3.0	$15.75	$22.10		$24.89	$31.35	39
3.1 or More	$20.00	$29.50		$35.00	$48.45	31
Member Status						
Accredited Practice Member	$19.43	$26.00		$28.57	$34.00	49
Nonaccredited Member	$15.00	$25.00		$30.15	$44.00	31
Nonmember	$15.00	$20.00		$24.13	$29.13	86
Metropolitan Status						
Urban	*	*		*	*	18
Suburban	$18.00	$28.00		$30.78	$37.00	53
Second City	$18.00	$26.00		$27.02	$33.00	43
Town	$15.75	$18.20		$21.59	$26.67	46
Rural	*	*		*	*	16
Median Area Household Income						
$35,000 or Less	*	*		*	*	21
$35,000 to $49,999	$15.81	$20.40		$25.04	$29.88	60
$50,000 to $69,999	$16.00	$25.00		$28.71	$35.50	63
$70,000 or More	$20.25	$29.00		$31.62	$35.00	33

Note 1: See the case description in the notes for Table 6.1.
Note 2: An asterisk indicates that data was not reported due to an insufficient number of responses.
Note 3: 46% of the respondents reported that the fee for this service is included in the total dentistry fee and is not separately itemized.

TABLE 6.14
DENTAL CASE: ANTIBIOTICS, ONE-WEEK SUPPLY

	25th Percentile	Median	Your Data	Average	75th Percentile	Number of Respondents
All Practices	**$17.00**	**$22.30**		**$23.23**	**$29.74**	**409**
Number of FTE Veterinarians						
1.0 or Less	$15.00	$20.69		$22.52	$28.04	125
1.1 to 2.0	$17.00	$24.00		$23.89	$30.00	123
2.1 to 3.0	$17.72	$22.42		$22.96	$30.00	81
3.1 or More	$17.00	$21.17		$23.46	$29.81	76
Member Status						
Accredited Practice Member	$17.33	$23.92		$23.82	$30.00	118
Nonaccredited Member	$17.06	$22.60		$23.62	$29.78	72
Nonmember	$16.08	$21.13		$22.65	$28.00	190
Metropolitan Status						
Urban	$20.00	$25.00		$26.02	$30.99	38
Suburban	$20.00	$23.95		$25.22	$31.74	114
Second City	$16.75	$21.45		$23.34	$30.00	82
Town	$15.77	$21.00		$22.03	$28.00	113
Rural	$12.88	$16.75		$18.83	$25.00	54
Median Area Household Income						
$35,000 or Less	$15.00	$20.00		$21.08	$26.00	52
$35,000 to $49,999	$16.14	$21.00		$23.12	$30.00	142
$50,000 to $69,999	$18.00	$23.20		$23.68	$30.00	137
$70,000 or More	$20.00	$25.00		$24.75	$29.98	69

Note: See the case description in the notes for Table 6.1.

TABLE 6.15
DENTAL CASE: TOTAL

	25th Percentile	Median	Your Data	Average	75th Percentile	Number of Respondents
All Practices	$282.56	$356.10		$356.97	$428.90	436
Number of FTE Veterinarians						
1.0 or Less	$253.59	$338.50		$334.51	$409.84	133
1.1 to 2.0	$262.00	$340.12		$349.68	$416.00	131
2.1 to 3.0	$310.37	$381.88		$383.39	$441.54	84
3.1 or More	$302.51	$360.50		$373.40	$448.04	82
Member Status						
Accredited Practice Member	$307.00	$379.70		$385.70	$448.00	123
Nonaccredited Member	$301.00	$361.00		$365.49	$427.22	75
Nonmember	$247.30	$324.00		$329.47	$410.34	206
Metropolitan Status						
Urban	$304.69	$387.50		$397.73	$501.16	42
Suburban	$319.50	$388.00		$391.02	$461.76	119
Second City	$277.60	$381.95		$379.08	$475.80	87
Town	$278.10	$332.90		$325.49	$394.00	123
Rural	$213.20	$281.00		$279.05	$330.76	57
Median Area Household Income						
$35,000 or Less	$231.63	$297.11		$290.66	$358.75	52
$35,000 to $49,999	$262.00	$340.12		$343.01	$407.83	155
$50,000 to $69,999	$294.90	$364.00		$371.92	$441.19	149
$70,000 or More	$331.00	$407.00		$402.24	$467.46	71

Note 1: See the case description in the notes for Table 6.1.
Note 2: The total fee for the dentistry case is the sum of the fees for the preanesthetic exam, CBC with differential, chemistry panel with eight chemistries, anesthesia (30 minutes), IV catheter and placement, IV fluids, dental scaling and polishing, subgingival curettage, fluoride application, electronic monitoring, post-procedure pain medication, post-procedure injectable antibiotics, hospitalization, and one-week supply of antibiotics. Some respondents do not charge for some of the individual services provided in this case. In addition, some respondents reported that they do not offer some of the individual services. Therefore, the average total fee for the case may be significantly lower than the sum of the average fees for the individual services.

TABLE 6.16
EXTRACTION OF MODERATELY LOOSE PREMOLAR TOOTH

	25th Percentile	Median	Your Data	Average	75th Percentile	Number of Respondents
All Practices	$10.00	$18.00		$21.53	$26.50	399
Number of FTE Veterinarians						
1.0 or Less	$10.00	$15.68		$21.65	$29.18	120
1.1 to 2.0	$11.98	$15.48		$19.96	$25.00	124
2.1 to 3.0	$10.00	$20.00		$20.68	$25.00	77
3.1 or More	$10.00	$20.00		$24.05	$29.60	75
Member Status						
Accredited Practice Member	$13.00	$20.40		$25.42	$32.28	116
Nonaccredited Member	$10.18	$15.35		$20.44	$25.00	72
Nonmember	$10.00	$15.00		$19.20	$25.00	183
Metropolitan Status						
Urban	$10.49	$20.00		$23.77	$32.58	40
Suburban	$12.50	$20.00		$23.70	$29.70	110
Second City	$10.50	$20.00		$21.61	$25.00	85
Town	$10.00	$15.75		$20.24	$25.06	110
Rural	$10.00	$13.18		$17.62	$20.00	48
Median Area Household Income						
$35,000 or Less	$10.00	$12.00		$15.39	$23.00	47
$35,000 to $49,999	$10.00	$15.00		$18.52	$22.88	141
$50,000 to $69,999	$12.00	$20.00		$24.67	$30.00	133
$70,000 or More	$15.00	$21.50		$25.19	$33.06	69

Note 1: 4% of respondents reported that they would not assess an additional fee for this service if performed in conjunction with the case described in the notes for Table 6.1.
Note 2: 4% of the respondents reported that they do not offer this service.

TABLE 6.17
EXTRACTION OF FIRMLY IMPLANTED UPPER FOURTH PREMOLAR TOOTH

	25th Percentile	Median	Your Data	Average	75th Percentile	Number of Respondents
All Practices	$28.00	$49.50		$54.26	$75.00	415
Number of FTE Veterinarians						
1.0 or Less	$25.00	$40.40		$47.86	$62.10	126
1.1 to 2.0	$33.92	$53.00		$58.03	$75.00	131
2.1 to 3.0	$25.00	$45.65		$53.21	$65.00	79
3.1 or More	$30.00	$50.50		$57.40	$82.59	76
Member Status						
Accredited Practice Member	$35.50	$55.00		$61.74	$84.25	117
Nonaccredited Member	$27.75	$42.99		$50.61	$71.63	74
Nonmember	$25.00	$45.00		$51.74	$75.00	194
Metropolitan Status						
Urban	$30.00	$50.00		$57.13	$74.46	40
Suburban	$34.00	$54.00		$59.15	$75.30	113
Second City	$29.25	$50.00		$57.48	$75.00	85
Town	$25.00	$45.00		$51.97	$75.00	116
Rural	$22.50	$36.00		$43.25	$54.75	53
Median Area Household Income						
$35,000 or Less	$20.15	$31.90		$39.44	$50.00	50
$35,000 to $49,999	$25.00	$42.99		$47.51	$62.93	148
$50,000 to $69,999	$33.00	$53.00		$59.26	$75.00	139
$70,000 or More	$40.00	$60.00		$69.27	$89.35	69

Note: 3% of the respondents reported that they do not offer this service.

TABLE 6.18
ENDODONTIC TREATMENT OF UPPER FOURTH PREMOLAR TOOTH

	25th Percentile	Median	Your Data	Average	75th Percentile	Number of Respondents
All Practices	$25.00	$51.00	_____	$117.46	$178.75	74

Note 1: Data was reported only for all practices due to an insufficient number of responses.
Note 2: 5% of respondents reported that they would not assess an additional fee for this service if performed in conjunction with the case described in the notes for Table 6.1.
Note 3: 76% of the respondents reported that they do not offer this service.

TABLE 6.19
ENDODONTIC TREATMENT OF LOWER FIRST MOLAR

	25th Percentile	Median	Your Data	Average	75th Percentile	Number of Respondents
All Practices	$20.75	$45.00	_____	$116.06	$182.50	69

Note 1: Data was reported only for all practices due to an insufficient number of responses.
Note 2: 5% of respondents reported that they would not assess an additional fee for this service if performed in conjunction with the case described in the notes for Table 6.1.
Note 3: 78% of the respondents reported that they do not offer this service.

TABLE 6.20
ENDODONTIC TREATMENT OF UPPER CANINE TOOTH

	25th Percentile	Median	Your Data	Average	75th Percentile	Number of Respondents
All Practices	$22.41	$50.00	_____	$109.59	$190.00	71

Note 1: Data was reported only for all practices due to an insufficient number of responses.
Note 2: 6% of respondents reported that they would not assess an additional fee for this service if performed in conjunction with the case described in the notes for Table 6.1.
Note 3: 76% of the respondents reported that they do not offer this service.

CHAPTER 7

LABORATORY SERVICES

Good medicine as well as good business is driving the increase in diagnostic testing. In this chapter, you'll get advice on how you and your health-care team can educate clients about the importance of laboratory tests.

While scheduling dental cleanings for two six-year-old schnauzers, the receptionist asks the client, "Do you want to have that preop blood work done?" The client looks confused, so the receptionist continues, "Well, I guess your dogs probably don't need it because they had preop blood work as puppies when they were neutered."

Besides the obvious service slip-up, this scenario illustrates the need for staff training and the critical role everyone—from your front-office team to doctors—plays in educating clients about laboratory tests. Staff training and client education will make or break your clients' compliance level with laboratory testing, so you need to involve your team from the start.

THE LAB AS A PROFIT CENTER

More and more veterinarians are implementing mandatory preanesthetic tests and increasing the number of in-house and outside laboratory services they offer. Good medicine and good business are driving the trend. To benchmark your laboratory revenue, calculate your lab revenue as a percentage of the practice's total revenue. Then compare your number to industry standards. According to *Financial and Productivity Pulsepoints*, Fourth Edition (AAHA Press, 2006), the average lab revenue in small-animal practices for calendar year 2005 was 13.9%, and the 75th percentile was 17.3% (as compared to 13.4% and 15.9%, respectively, for calendar year 2003). If your lab revenue is below the average, consider:

- Are your lab fees at or below average?
- Are you recommending lab testing in every case where it would serve the patient well?
- How high are your compliance rates?

CREATING LAB TESTING PROTOCOLS

Before you can begin educating clients about the importance of lab testing, your team needs to develop specific medical protocols. For example, do you wish to create wellness screens for pediatric, adult, senior, and geriatric patients? You will need to set age guidelines and develop estimates for these screening packages.

Additional medical protocols you may want to address include laboratory testing for pets on long-term medications and pets on heartworm preventives. A final category to consider is sick patients. Although testing will be highly tailored based on each case, you can develop some baseline minimums for your protocols.

Every wellness screen should include various testing levels based on the pet's breed, history, and physical-exam findings. In Table 7.A, you'll find examples of wellness laboratory screens based on the patient's age.

EDUCATING CLIENTS ABOUT THE NEED FOR DIAGNOSTICS

Wellness screens for healthy pets help you ensure safer anesthetic procedures, detect diseases early, and increase client satisfaction with a more complete diagnosis. Even if a client spends $150 on laboratory tests and results are normal, she will likely feel the money was well spent as an investment in peace of mind—if the reasons for the tests were explained to her correctly. Add value to laboratory testing by clearly explaining results with handouts and vendor brochures.

Vendors offer in-clinic seminars, brochures, and posters to help you educate your staff and your clients. You can also show clients subtle changes in their pets' health over time when you compare a pet's preanesthetic test results for a spay procedure as a puppy to results when the pet has his first dental cleaning, and again when the pet reaches the early senior years.

Educating clients early about senior care is key, and laboratory testing is the foundation of an effective

TABLE 7.A

SAMPLE WELLNESS LABORATORY SCREENING PROTOCOLS

1 TO 5 YEARS OLD	6 TO 7 YEARS OLD	8 YEARS AND OLDER
CBC with differential	Comprehensive biochemical profile	Comprehensive biochemical profile
Albumin	CBC with differential	CBC with differential
BUN	Urinalysis	Urinalysis
Glucose	Thyroid test for large-breed dogs	Thyroid test
ALP		
Creatinine		
Total protein		
ALT		
Globulin		
Complete urinalysis		

senior-care program. Today, fully one-third of dogs and cats are considered seniors. When you perform complete histories, exams, and diagnostic tests, about 40% of senior pets will require additional tests and/or medication because you find underlying diseases, observes Dr. William D. Fortney, Director of Community Practice at Kansas State University's teaching hospital in Manhattan, Kansas. "Senior care focuses on client education, disease-prevention strategies, and early detection of medical and behavioral problems," he says. "The overall goal must be to improve the pet's quality of life, not just longevity."

GETTING CLIENTS TO SAY "YES" TO DIAGNOSTIC TESTS

Simplicity is fundamental when explaining laboratory tests to clients. You may want to run 13 chemical profiles, a urinalysis, and electrolytes, but explaining each one will leave the client's head swimming. Clients may look puzzled or feel overwhelmed if you rattle off an alphabet soup list of tests their pets need—BUN, CBC, UA, T4—and other medical jargon.

Limiting choices diminishes confusion. For example, one clinic in North Carolina offers three senior packages: basic, senior, and premium. The basic package includes a comprehensive exam, written evaluation, complete blood count, urinalysis, and glaucoma screen. The senior package offers the same services, plus three radiograph views, an EKG, and a blood-pressure screen. The premium package adds a dental cleaning.

Doctors recommend a specific package based on the pet's age and physical condition, but clients still have a choice. "Clients are very receptive to packages," says an associate at the practice. "When clients are educated about potential diseases that may creep up on their pets, they're receptive to screening for those."

When talking with clients, give them a choice between two "yes" options. Recommend what's best for the pet rather than assuming what the client can afford. Present two levels of laboratory testing based on the pet's age, health status, and physical exam findings.

For example, when presenting an estimate for a dental cleaning for a senior pet, your technician might say:

"Our hospital requires preanesthetic testing for every pet that undergoes anesthesia. Because Max is seven years old, he is considered a senior pet. We need to perform blood and urine tests to check the functions of his vital organs. These tests will help ensure his safety as well as uncover any potential health problems. You may choose either a Health Check Profile for $85 or a Health Check Plus Profile for $95 that will also check Max's thyroid function, which we discussed earlier. Which do you prefer today?" This scenario lets the client choose between two "yes" options, and the patient receives the care he needs. Give clients a written estimate and brochure that they can share with other family members (see Helpful Resources at the end of this chapter).

Case Study

How My Hospital Began Requiring Preanesthetic Testing

During a staff meeting at Sulphur Springs Veterinary Clinic in Manchester, Missouri, a part-time employee asked why preanesthetic testing was optional, because he knew the doctors were passionate about the need for diagnostics. In fact, the hospital enjoyed an 84% compliance rate for optional preanesthetic testing, thanks to its effective staff training and client education.

"Aren't we speaking with a forked tongue?" the staff member asked. A gut reaction told Dr. Jim Irwin that the staff member was right, but he wanted confirmation from the rest of his healthcare team. Dr. Irwin asked staff members to raise their hands if they thought preanesthetic testing should be mandatory for all anesthetic procedures. Most hands shot up immediately.

In 1996, the team at Sulphur Springs Veterinary Clinic implemented mandatory testing for all anesthetic procedures. Now Dr. Irwin gets 100% compliance on preanesthetic testing for his patients—

whether for a spay or neuter procedure, a dental cleaning, or a complex surgery. "If you make testing optional, it confuses the staff," Dr. Irwin says. "It's better to set a cut-and-dried policy."

Dr. Irwin and his team designed a program called Anesthesia ASAP: As Safe as Possible. Today's pet owners are better educated and have higher expectations for routine elective procedures such as dentistry and spay or neuter surgeries. The Anesthesia ASAP program educates clients, builds confidence, and demonstrates the hospital's high quality of care. Dr. Irwin bundles preanesthetic tests into procedures and offers three levels of testing based on the pet's age, breed, and health status.

To decide whether you want to move from optional to mandatory testing, track your current compliance rate and seek input from your staff. Medical quality should drive your decision, but you will also enjoy financial benefits. "Lab work is one of the most profitable revenue sources in any hospital," says Darin S. Nelson, Senior Vice President of Development for VCA Antech, Inc. in Irvine, California. "A test that costs you $30 in outside lab services can generate $150 in diagnostic revenue."

Follow a step-by-step approach, beginning with setting medical protocols, training your team, and developing client handouts, brochures, and consent forms. Once your health-care team is well prepared, implement your mandatory preanesthetic testing program. Be consistent with every client. Once clients understand the value and benefits of preanesthetic tests, they will accept your hospital policy of mandatory preanesthetic testing.

Implementation Idea

Ask Clients to Complete a Preanesthesia Check-in Questionnaire

In addition to consenting to a preanesthesia physical exam and blood and urine tests for her pet, every client whose pet will undergo anesthesia at Sulphur Springs Veterinary Clinic completes a check-in questionnaire. During the admission appointment on the morning of the procedure, a technician discusses the form with the pet owner and then draws blood and urine samples for preanesthetic tests.

This questionnaire helps gather information about the patient's history before a veterinarian performs a preanesthesia physical exam. "By the time I arrive at the clinic at 9 a.m., I have the preanesthetic test results and a completed history sheet," Dr. Jim Irwin says. "I do the preanesthetic exam, and the patient is ready for surgery."

Dr. Irwin and his team developed a preanesthesia questionnaire for healthy young animals that doctors complete and keep in the pet's medical record (see figure 7.1). Consider adapting this form for use in your hospital.

HELPFUL RESOURCES

First Choice Medical Protocols, Johnny D. Hoskins, DVM, PhD, Dipl. ACVIM and Ronald E. Whitford, DVM. Johnny D. Hoskins, 2001. Available through the American Animal Hospital Association.

"Lab Testing for Your Pet" brochure, AAHA. American Animal Hospital Association Press, 2007.

Legal Consent Forms for Veterinary Practices, Fourth Edition, James F. Wilson, DVM, JD. Priority Press, Ltd., 2006. Available through the American Animal Hospital Association.

Urinalysis Sticker for medical records, AAHA. American Animal Hospital Association Press.

FIGURE 7.1

SAMPLE PREANESTHESIA CLIENT QUESTIONNAIRE

PREANESTHESIA QUESTIONNAIRE FOR HEALTHY YOUNG ANIMALS

1. Has your pet been drinking and eating OK lately?
   _____

2. Has your pet had any recent weight change?
   _____

3. Have you noticed any behavior changes recently?
   _____

4. Has your pet previously been put under anesthesia (including at other clinics)?
   _____

   If yes, did your pet experience any problems?
   _____

5. Is your pet on any medication?
   _____

   If yes, was the medication given today?
   _____

6. Did you withhold food and water from your pet this morning?
   _____

7. Does your pet have any other problems that you are aware of?
   _____

8. Other comments
   _____
   _____

Initials of technician asking questions _____ Date _____

CHAPTER 7

DATA TABLES

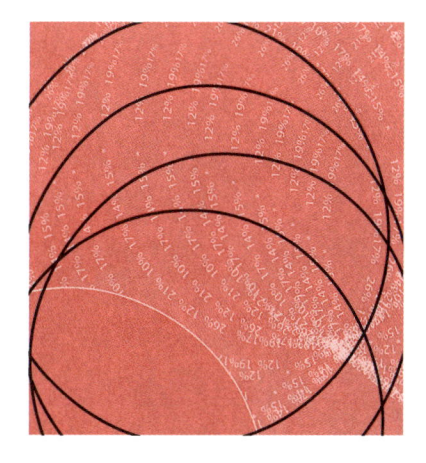

TABLE 7.1
TESTS PERFORMED IN-HOUSE OR BY OUTSIDE LAB

Test	In-house Lab	Outside Lab	Number of Respondents
ACTH Stimulation	9%	91%	504
Arterial/Venous Blood Gases	56%	44%	90
Avian Chlamydia	0%	100%	97
Avian Chromosomal Sexing	0%	100%	102
Avian ELISA Allergy Testing	0%	100%	18
Avian PBFDV	0%	100%	63
Avian Polyomavirus Test	0%	100%	79
Bacterial Culture and Sensitivity	7%	93%	516
Bladder Stone Analysis	0%	100%	506
Blood Parasite Test: Ehrlichia	30%	70%	383
Blood Parasite Test: Haemobartonella	12%	88%	303
Blood Parasite Test: Babesia	4%	96%	239
CBC with No Differential	66%	34%	217
CBC Automated	66%	34%	400
Health Check: CBC with 16–24 Chemistries and T4	16%	84%	449
Health Check: CBC with 8–12 Chemistries	44%	56%	398
CBC with Manual Differential	42%	58%	357
Chemistry Setup Fee	83%	17%	47
1 Chemistry	88%	12%	387
2 Chemistries	90%	10%	285
3 Chemistries	90%	10%	249
4 Chemistries	87%	13%	231
5–7 Chemistries	88%	12%	352
8–12 Chemistries	76%	24%	385
16–24 Chemistries	18%	82%	339
Cytology: Fine-Needle Aspirate	75%	25%	489
Cytology: Vaginal	94%	6%	454
Cytology: Ear Swab	97%	3%	503
Cytology: Skin Swab	95%	5%	455
Dex Suppression	10%	90%	441
Electrolytes	68%	32%	421
Fecal Dif-Quik Stain	90%	10%	288
Fecal Examination: Direct Smear	99%	1%	474
Fecal Examination: Flotation (Gravitational)	98%	2%	414
Fecal Examination: Flotation (Centrifugation, Zinc Sulfate)	64%	36%	306
Fecal Examination: Sedimentation (Baermann)	39%	61%	94
Fecal Examination: Giardia Wet Mount	91%	9%	296
Fecal Gram's Stain	76%	24%	191
Feline Leukemia (FeLV) Test	71%	29%	319
FeLV and FIV Test	93%	7%	522
FIV Test	31%	69%	235
Fructosamine Test	6%	94%	450
Fungal Culture	70%	30%	504
Giardia Antigen Test	59%	41%	392
Glucose Curve (6)	98%	2%	454
Glucose Single	97%	3%	492
Heartworm Test, Canine: Occult/Antigen	63%	37%	392
Heartworm Test, Canine: Occult/Antigen Plus Lyme and E. Canis	91%	9%	348
Heartworm Test, Feline: Occult/Antibody	17%	83%	275
Heartworm Test, Feline: Occult/Antigen	21%	79%	258
Heartworm Test, Feline: Occult/Antibody and Occult/Antigen	7%	93%	248
Histopathology: Multiple Tissues	2%	98%	452
Histopathology: Single Tissue	2%	98%	516
Lyme Testing	19%	81%	387
Pancreatic Evaluation	21%	79%	242
Trypsin-like Immunoreactivity (TLI)	1%	99%	381
Pancreatic Lipase Immunoreactivity (PLI)	2%	98%	246

TABLE 7.1 (CONTINUED)
TESTS PERFORMED IN-HOUSE OR BY OUTSIDE LAB

Test	In-house Lab	Outside Lab	Number of Respondents
Canine Pancreatic Lipase (CPL)	23%	77%	212
Specific Canine Pancreatic Lipase (Spec cal)	3%	97%	173
Parvovirus Test	77%	23%	496
PTH Assay	0%	100%	219
Reticulocyte Count	14%	86%	364
Serum Testing for Allergen-Specific IgE	3%	97%	256
T4	24%	76%	502
T4, T3, Free T4, and Free T4ED	1%	99%	393
TSH Level	0%	100%	311
Uric Acid: Avian	16%	84%	61
Uric Acid: Reptiles	19%	81%	53
Urinalysis: Complete	86%	14%	498
Urinalysis: Urine Specific Gravity	100%	0%	326
Urinalysis: Dipstick	100%	0%	369
Urinalysis: Sediment	95%	5%	241
Urinalysis: Microalbuminaria	47%	53%	189
Urinalysis: Urine Protein:Creatinine (UP:C) Ratio	20%	80%	403

TABLE 7.2
PERCENTAGE OF RESPONDENTS THAT DO NOT OFFER LAB TEST

Test	Percentage
ACTH Stimulation	8%
Arterial/Venous Blood Gases	87%
Avian Chlamydia	82%
Avian Chromosomal Sexing	82%
Avian ELISA Allergy Testing	98%
Avian PBFDV	89%
Avian Polyomavirus Test	86%
Bacterial Culture and Sensitivity	4%
Bladder Stone Analysis	4%
Blood Parasite Test: Ehrlichia	21%
Blood Parasite Test: Haemobartonella	35%
Blood Parasite Test: Babesia	51%
CBC Automated	15%
Health Check: CBC with 16–24 Chemistries and T4	10%
Health Check: CBC with 8–12 Chemistries	15%
CBC with Manual Differential	23%
CBC with No Differential	54%
Chemistry Setup Fee	84%
1 Chemistry	17%
2 Chemistries	33%
3 Chemistries	42%
4 Chemistries	47%
5–7 Chemistries	20%
8–12 Chemistries	15%
16–24 Chemistries	28%
Cytology: Fine-Needle Aspirate	2%
Cytology: Vaginal	7%
Cytology: Ear Swab	1%
Cytology: Skin Swab	5%
Dex Suppression	11%
Electrolytes	14%
Fecal Dif-Quik Stain	41%
Fecal Examination: Direct Smear	8%
Fecal Examination: Flotation (Gravitational)	19%
Fecal Examination: Flotation (Centrifugation, Zinc Sulfate)	38%
Fecal Examination: Sedimentation (Baermann)	81%
Fecal Examination: Giardia Wet Mount	35%
Fecal Gram's Stain	57%
Feline Leukemia (FeLV) Test	37%
FeLV and FIV Test	2%
FIV Test	50%
Fructosamine Test	12%
Fungal Culture	2%
Giardia Antigen Test	22%
Glucose Curve (6)	10%
Glucose Single	2%
Heartworm Test, Canine: Occult/Antigen	20%
Heartworm Test, Canine: Occult/Antigen Plus Lyme and E. Canis	29%
Heartworm Test, Feline: Occult/Antibody	46%
Heartworm Test, Feline: Occult/Antigen	47%
Heartworm Test, Feline: Occult/Antibody and Occult/Antigen	48%
Histopathology: Multiple Tissues	7%
Histopathology: Single Tissue	2%
Lyme Testing	18%
Pancreatic Evaluation	44%
Trypsin-like Immunoreactivity (TLI)	24%
Pancreatic Lipase Immunoreactivity (PLI)	52%
Canine Pancreatic Lipase (CPL)	60%

TABLE 7.2 (CONTINUED)
PERCENTAGE OF RESPONDENTS THAT DO NOT OFFER LAB TEST

Test	Percentage
Specific Canine Pancreatic Lipase (Spec cal)	66%
Parvovirus Test	6%
PTH Assay	54%
Reticulocyte Count	22%
Serum Testing for Allergen-Specific IgE	46%
T4	3%
T4, T3, Free T4, and Free T4ED	20%
TSH Level	34%
Uric Acid: Avian	90%
Uric Acid: Reptiles	92%
Urinalysis: Complete	1%
Urinalysis: Urine Specific Gravity	26%
Urinalysis: Dipstick	18%
Urinalysis: Sediment	40%
Urinalysis: Microalbuminaria	59%
Urinalysis: Urine Protein:Creatinine (UP:C) Ratio	18%

TABLE 7.3
LAB AND BLOOD COLLECTION FEE: CANINE

	25th Percentile	Median	Your Data	Average	75th Percentile	Number of Respondents
All Practices	$9.81	$15.23		$27.96	$26.99	124
Number of FTE Veterinarians						
1.0 or Less	$9.40	$16.00		$28.06	$36.00	35
1.1 to 2.0	$9.99	$11.90		$22.78	$27.50	29
2.1 to 3.0	$8.00	$12.05		$37.82	$76.38	26
3.1 or More	$10.86	$19.25		$26.76	$26.99	28
Member Status						
Accredited Practice Member	$9.50	$15.00		$23.75	$22.00	37
Nonaccredited Member	*	*		*	*	24
Nonmember	$9.86	$13.45		$26.05	$31.50	50
Metropolitan Status						
Urban	*	*		*	*	9
Suburban	$9.50	$18.00		$31.01	$31.23	41
Second City	$10.00	$14.23		$26.17	$34.50	26
Town	$8.73	$12.00		$26.11	$21.45	29
Rural	*	*		*	*	17
Median Area Household Income						
$35,000 or Less	*	*		*	*	13
$35,000 to $49,999	$9.45	$13.40		$28.40	$31.75	53
$50,000 to $69,999	$10.00	$14.23		$22.82	$24.25	36
$70,000 or More	*	*		*	*	19

Note 1: An asterisk indicates that data was not reported due to an insufficient number of responses.
Note 2: 73% of the respondents reported that this fee is included in the fee for the laboratory service.

TABLE 7.4
LAB AND BLOOD COLLECTION FEE: FELINE

	25th Percentile	Median	Your Data	Average	75th Percentile	Number of Respondents
All Practices	$9.99	$15.23	_____	$27.31	$25.00	124
Number of FTE Veterinarians						
1.0 or Less	$9.40	$16.00	_____	$27.97	$36.00	35
1.1 to 2.0	$10.00	$12.75	_____	$23.06	$25.00	29
2.1 to 3.0	$8.00	$11.50	_____	$34.49	$78.75	26
3.1 or More	$10.86	$19.25	_____	$26.76	$26.99	28
Member Status						
Accredited Practice Member	$9.50	$15.23	_____	$23.71	$22.06	38
Nonaccredited Member	*	*	_____	*	*	23
Nonmember	$9.98	$11.90	_____	$26.38	$30.00	51
Metropolitan Status						
Urban	*	*	_____	*	*	10
Suburban	$9.50	$18.00	_____	$28.68	$24.73	41
Second City	$10.00	$14.23	_____	$26.40	$29.13	26
Town	$9.59	$12.50	_____	$26.76	$21.73	28
Rural	*	*	_____	*	*	17
Median Area Household Income						
$35,000 or Less	*	*	_____	*	*	13
$35,000 to $49,999	$9.50	$13.40	_____	$27.79	$25.00	55
$50,000 to $69,999	$10.00	$15.00	_____	$23.25	$25.00	35
$70,000 or More	*	*	_____	*	*	18

Note 1: An asterisk indicates that data was not reported due to an insufficient number of responses.
Note 2: 74% of the respondents reported that this fee is included in the fee for the laboratory service.

TABLE 7.5
LAB AND BLOOD COLLECTION FEE FOR OTHER SPECIES

	25th Percentile	Median	Your Data	Average	75th Percentile	Number of Respondents
Birds	$9.86	$11.56	_____	$15.62	$19.25	30
Small Mammals	$9.49	$11.57	_____	$17.56	$19.13	42
Ferrets	$9.98	$15.00	_____	$26.28	$24.00	59

Note 1: Data was reported only for all practices due to an insufficient number of responses.
Note 2: 20% of the respondents reported that this fee is included in the fee for avian laboratory service.
Note 3: 74% of respondents reported that they do not offer this service for birds.
Note 4: 33% of the respondents reported that this fee is included in the fee for small mammal laboratory service.
Note 5: 59% of respondents reported that they do not offer this service for small mammals.
Note 6: 40% of the respondents reported that this fee is included in the fee for ferret laboratory service.
Note 7: 48% of respondents reported that they do not offer this service for ferrets.

LABORATORY TESTS PERFORMED IN-HOUSE

TABLE 7.6
IN-HOUSE LAB TEST FEES NOT REPORTED

Avian Chlamydia
Avian Chromosomal Sexing
Avian ELISA Allergy Testing
Avian PBFDV
Avian Polyomavirus
Bladder Stone Analysis
Blood Parasite Test: Babesia
Chemistry Setup Fee
Histopathology: Multiple Tissues
Histopathology: Single Tissue
Trypsin-like Immunoreactivity (TLI)
Pancreatic Lipase Immunoreactivity (PLI)
PTH Assay
Serum Testing for Allergen-Specific IgE
Specific Canine Pancreatic Lipase (Spec cal)
T4, T3, Free T4, and Free T4ED
TSH Level
Uric Acid: Avian
Uric Acid: Reptiles

Note: These in-house lab test fees were not reported due to an insufficient number of responses.

TABLE 7.7
FEES FOR SELECT IN-HOUSE LAB TESTS

	25th Percentile	Median	Your Data	Average	75th Percentile	Number of Respondents
ACTH Stimulation	$87.00	$109.50		$116.40	$149.00	45
Arterial/Venous Blood Gases	$30.00	$38.95		$41.93	$52.15	47
Bacterial Culture and Sensitivity	$40.00	$50.00		$51.34	$68.00	39
Blood Parasite Test: Haemobartonella	$20.25	$28.15		$27.32	$35.00	32
16–24 Chemistries	$57.25	$78.50		$78.80	$95.00	56
Dex Suppression	$76.50	$111.46		$100.09	$124.75	49
Fecal Examination: Sedimentation (Baermann)	$16.00	$19.00		$22.63	$28.00	29
FIV Test	$31.75	$38.00		$39.85	$45.00	61
Fructosamine Test	$24.33	$44.00		$42.37	$56.01	29
Heartworm Test, Feline: Occult/Antibody	$25.40	$31.50		$31.40	$36.50	47
Heartworm Test, Feline: Occult/Antigen	$25.00	$28.75		$28.46	$32.38	58
Lyme Testing	$32.00	$38.57		$44.07	$57.50	80
Pancreatic Evaluation	$31.08	$51.57		$57.24	$78.13	52
Canine Pancreatic Lipase (CPL)	$15.00	$20.50		$24.33	$25.96	50
Reticulocyte Count	$20.25	$28.00		$28.94	$35.60	53

Note: Data was reported only for all practices due to an insufficient number of responses.

TABLE 7.8
BLOOD PARASITE TEST: EHRLICHIA

	25th Percentile	Median	Your Data	Average	75th Percentile	Number of Respondents
All Practices	$31.50	$36.00		$40.07	$44.11	114
Number of FTE Veterinarians						
1.0 or Less	$30.04	$33.00		$37.46	$39.75	36
1.1 to 2.0	$32.00	$36.00		$39.08	$43.21	33
2.1 to 3.0	*	*		*	*	21
3.1 or More	*	*		*	*	22
Member Status						
Accredited Practice Member	$32.75	$41.45		$43.48	$44.57	30
Nonaccredited Member	*	*		*	*	20
Nonmember	$30.00	$33.25		$34.40	$36.75	52
Metropolitan Status						
Urban	*	*		*	*	6
Suburban	$33.50	$40.00		$43.40	$54.50	31
Second City	*	*		*	*	19
Town	$31.50	$35.50		$38.44	$42.00	44
Rural	*	*		*	*	13
Median Area Household Income						
$35,000 or Less	*	*		*	*	18
$35,000 to $49,999	$31.63	$36.00		$37.52	$41.45	41
$50,000 to $69,999	$32.00	$38.00		$42.55	$54.50	35
$70,000 or More	*	*		*	*	17

Note: An asterisk indicates that data was not reported due to an insufficient number of responses.

TABLE 7.9
CBC AUTOMATED

	25th Percentile	Median	Your Data	Average	75th Percentile	Number of Respondents
All Practices	$31.00	$35.75		$36.97	$42.60	275
Number of FTE Veterinarians						
1.0 or Less	$30.00	$35.00		$35.89	$42.60	67
1.1 to 2.0	$30.85	$35.00		$35.80	$40.70	85
2.1 to 3.0	$33.25	$38.00		$38.38	$43.72	54
3.1 or More	$31.48	$36.45		$37.90	$43.15	61
Member Status						
Accredited Practice Member	$32.00	$39.00		$39.43	$45.00	90
Nonaccredited Member	$31.40	$35.64		$37.52	$43.20	55
Nonmember	$30.00	$35.00		$35.02	$38.50	112
Metropolitan Status						
Urban	*	*		*	*	17
Suburban	$32.00	$38.00		$39.18	$45.10	59
Second City	$33.00	$36.88		$38.65	$46.43	65
Town	$30.00	$35.00		$35.59	$41.40	95
Rural	$26.25	$32.00		$33.80	$40.00	35
Median Area Household Income						
$35,000 or Less	$29.40	$33.50		$33.91	$37.50	43
$35,000 to $49,999	$30.10	$35.00		$36.15	$43.00	109
$50,000 to $69,999	$31.54	$38.00		$38.55	$43.50	79
$70,000 or More	$35.00	$38.00		$39.36	$44.95	36

Note: An asterisk indicates that data was not reported due to an insufficient number of responses.

TABLE 7.10
CBC WITH 8–12 CHEMISTRIES

	25th Percentile	Median	Your Data	Average	75th Percentile	Number of Respondents
All Practices	**$73.00**	**$87.00**		**$89.35**	**$104.00**	**188**
Number of FTE Veterinarians						
1.0 or Less	$69.50	$84.00		$87.09	$109.00	47
1.1 to 2.0	$73.00	$87.50		$90.61	$100.00	67
2.1 to 3.0	$69.50	$88.58		$88.07	$106.70	36
3.1 or More	$73.50	$85.00		$89.70	$104.00	35
Member Status						
Accredited Practice Member	$74.00	$90.45		$91.68	$107.38	52
Nonaccredited Member	$70.00	$89.80		$89.40	$108.00	39
Nonmember	$69.98	$82.13		$87.76	$100.05	82
Metropolitan Status						
Urban	*	*		*	*	10
Suburban	$75.50	$88.50		$91.07	$112.50	42
Second City	$73.15	$91.39		$92.22	$106.00	43
Town	$70.00	$80.55		$83.38	$94.44	64
Rural	$69.50	$84.00		$85.63	$100.00	27
Median Area Household Income						
$35,000 or Less	$75.00	$81.25		$88.79	$102.00	27
$35,000 to $49,999	$66.50	$80.00		$84.12	$99.38	73
$50,000 to $69,999	$76.00	$93.23		$94.17	$116.21	60
$70,000 or More	*	*		*	*	24

Note: An asterisk indicates that data was not reported due to an insufficient number of responses.

TABLE 7.11
CBC WITH 16–24 CHEMISTRIES AND T4

	25th Percentile	Median	Your Data	Average	75th Percentile	Number of Respondents
All Practices	**$88.19**	**$112.00**		**$115.52**	**$140.00**	**82**
Number of FTE Veterinarians						
1.0 or Less	*	*		*	*	19
1.1 to 2.0	$87.05	$117.50		$122.13	$151.50	30
2.1 to 3.0	*	*		*	*	14
3.1 or More	*	*		*	*	18
Member Status						
Accredited Practice Member	$100.67	$119.55		$122.85	$144.79	28
Nonaccredited Member	*	*		*	*	15
Nonmember	$83.00	$100.00		$112.12	$137.50	37
Metropolitan Status						
Urban	*	*		*	*	4
Suburban	*	*		*	*	16
Second City	*	*		*	*	19
Town	$87.70	$105.00		$113.63	$136.00	29
Rural	*	*		*	*	13
Median Area Household Income						
$35,000 or Less	*	*		*	*	15
$35,000 to $49,999	$78.50	$99.22		$109.54	$139.98	33
$50,000 to $69,999	*	*		*	*	24
$70,000 or More	*	*		*	*	9

Note: An asterisk indicates that data was not reported due to an insufficient number of responses.

TABLE 7.12
CBC WITH MANUAL DIFFERENTIAL

	25th Percentile	Median	Your Data	Average	75th Percentile	Number of Respondents
All Practices	$31.75	$37.00		$38.12	$42.80	161
Number of FTE Veterinarians						
1.0 or Less	$30.00	$35.00		$35.92	$42.00	37
1.1 to 2.0	$33.23	$37.00		$38.73	$42.00	36
2.1 to 3.0	$31.85	$36.88		$38.65	$47.05	40
3.1 or More	$31.63	$38.50		$39.12	$44.03	44
Member Status						
Accredited Practice Member	$31.90	$38.00		$38.41	$42.80	53
Nonaccredited Member	$31.16	$40.00		$40.65	$46.78	33
Nonmember	$31.53	$35.00		$36.54	$40.00	62
Metropolitan Status						
Urban	*	*		*	*	10
Suburban	$31.95	$36.88		$37.58	$42.38	42
Second City	$34.33	$38.00		$40.37	$47.69	37
Town	$31.49	$38.00		$39.23	$43.85	49
Rural	*	*		*	*	21
Median Area Household Income						
$35,000 or Less	*	*		*	*	18
$35,000 to $49,999	$30.50	$35.00		$36.79	$42.00	63
$50,000 to $69,999	$33.78	$38.10		$40.44	$48.03	53
$70,000 or More	*	*		*	*	24

Note: An asterisk indicates that data was not reported due to an insufficient number of responses.

TABLE 7.13
CBC WITH NO DIFFERENTIAL

	25th Percentile	Median	Your Data	Average	75th Percentile	Number of Respondents
All Practices	$27.65	$33.00		$33.90	$39.00	147
Number of FTE Veterinarians						
1.0 or Less	$23.58	$30.75		$31.56	$38.00	36
1.1 to 2.0	$27.88	$31.00		$32.36	$38.20	45
2.1 to 3.0	$30.24	$38.00		$37.42	$46.06	26
3.1 or More	$28.30	$35.00		$35.38	$42.80	35
Member Status						
Accredited Practice Member	$28.00	$35.88		$36.05	$43.75	38
Nonaccredited Member	$30.00	$34.50		$35.28	$42.30	25
Nonmember	$25.00	$31.25		$31.98	$38.00	69
Metropolitan Status						
Urban	*	*		*	*	9
Suburban	$29.75	$35.00		$35.38	$40.23	41
Second City	$29.38	$35.00		$35.91	$44.00	37
Town	$24.90	$30.98		$32.35	$40.00	47
Rural	*	*		*	*	11
Median Area Household Income						
$35,000 or Less	*	*		*	*	20
$35,000 to $49,999	$25.00	$31.25		$32.18	$38.00	61
$50,000 to $69,999	$26.66	$33.00		$33.66	$41.25	45
$70,000 or More	*	*		*	*	20

Note: An asterisk indicates that data was not reported due to an insufficient number of responses.

TABLE 7.14
1 CHEMISTRY

	25th Percentile	Median	Your Data	Average	75th Percentile	Number of Respondents
All Practices	**$15.00**	**$19.83**		**$20.25**	**$24.27**	**340**
Number of FTE Veterinarians						
1.0 or Less	$13.00	$16.50		$18.57	$22.00	79
1.1 to 2.0	$15.00	$19.73		$19.65	$23.13	102
2.1 to 3.0	$15.80	$22.00		$21.37	$25.00	71
3.1 or More	$17.88	$21.00		$21.65	$25.97	77
Member Status						
Accredited Practice Member	$17.50	$21.63		$22.45	$26.05	101
Nonaccredited Member	$15.00	$20.00		$19.91	$24.00	71
Nonmember	$14.00	$18.00		$18.85	$22.13	146
Metropolitan Status						
Urban	*	*		*	*	20
Suburban	$16.00	$21.69		$21.72	$25.44	90
Second City	$15.62	$20.00		$20.95	$25.15	77
Town	$14.00	$18.20		$18.84	$22.71	110
Rural	$12.75	$16.73		$18.71	$21.35	38
Median Area Household Income						
$35,000 or Less	$12.60	$15.80		$18.30	$20.00	39
$35,000 to $49,999	$15.00	$19.65		$19.73	$24.00	134
$50,000 to $69,999	$15.00	$21.00		$21.39	$25.89	105
$70,000 or More	$16.00	$20.00		$20.72	$25.00	55

Note: An asterisk indicates that data was not reported due to an insufficient number of responses.

TABLE 7.15
2 CHEMISTRIES

	25th Percentile	Median	Your Data	Average	75th Percentile	Number of Respondents
All Practices	**$25.50**	**$33.26**		**$33.59**	**$39.56**	**246**
Number of FTE Veterinarians						
1.0 or Less	$20.00	$30.00		$30.64	$36.19	63
1.1 to 2.0	$25.80	$32.00		$33.15	$38.34	66
2.1 to 3.0	$28.85	$36.00		$35.25	$44.00	52
3.1 or More	$28.00	$33.00		$35.18	$40.00	59
Member Status						
Accredited Practice Member	$29.68	$35.00		$36.14	$41.25	70
Nonaccredited Member	$28.63	$33.66		$34.15	$38.00	50
Nonmember	$23.00	$30.00		$31.78	$38.00	111
Metropolitan Status						
Urban	*	*		*	*	17
Suburban	$30.00	$34.98		$36.01	$45.00	70
Second City	$26.70	$32.00		$35.11	$40.50	58
Town	$21.30	$30.50		$30.27	$36.10	71
Rural	$24.50	$30.00		$31.00	$37.25	29
Median Area Household Income						
$35,000 or Less	$21.16	$30.00		$29.97	$36.00	30
$35,000 to $49,999	$25.00	$32.00		$32.55	$38.25	97
$50,000 to $69,999	$28.78	$35.00		$36.03	$43.45	76
$70,000 or More	$28.13	$33.06		$33.88	$39.94	40

Note: An asterisk indicates that data was not reported due to an insufficient number of responses.

TABLE 7.16
3 CHEMISTRIES

	25th Percentile	Median	Your Data	Average	75th Percentile	Number of Respondents
All Practices	**$34.50**	**$45.00**		**$44.12**	**$52.00**	**207**
Number of FTE Veterinarians						
1.0 or Less	$29.25	$42.49		$40.70	$49.50	48
1.1 to 2.0	$34.63	$44.13		$43.71	$51.50	60
2.1 to 3.0	$36.38	$45.95		$45.51	$51.29	44
3.1 or More	$37.50	$45.00		$46.44	$55.50	51
Member Status						
Accredited Practice Member	$38.95	$45.00		$45.90	$50.90	67
Nonaccredited Member	$37.80	$45.00		$44.03	$52.00	39
Nonmember	$30.00	$42.00		$42.51	$52.00	87
Metropolitan Status						
Urban	*	*		*	*	13
Suburban	$39.10	$45.00		$45.55	$52.39	61
Second City	$34.25	$45.00		$45.85	$54.00	52
Town	$29.75	$37.98		$40.41	$48.38	58
Rural	*	*		*	*	22
Median Area Household Income						
$35,000 or Less	*	*		*	*	22
$35,000 to $49,999	$32.24	$41.98		$41.75	$49.62	78
$50,000 to $69,999	$35.55	$45.00		$45.85	$54.00	71
$70,000 or More	$35.25	$45.09		$46.74	$51.25	33

Note: An asterisk indicates that data was not reported due to an insufficient number of responses.

TABLE 7.17
4 CHEMISTRIES

	25th Percentile	Median	Your Data	Average	75th Percentile	Number of Respondents
All Practices	**$40.00**	**$49.70**		**$51.98**	**$62.00**	**182**
Number of FTE Veterinarians						
1.0 or Less	$34.34	$42.49		$44.77	$54.72	42
1.1 to 2.0	$40.00	$49.70		$54.06	$66.00	56
2.1 to 3.0	$40.75	$50.00		$53.18	$63.93	41
3.1 or More	$43.00	$55.35		$55.14	$64.00	39
Member Status						
Accredited Practice Member	$41.99	$52.66		$54.38	$63.89	54
Nonaccredited Member	$42.09	$51.20		$53.18	$61.50	36
Nonmember	$36.00	$45.00		$49.64	$61.00	77
Metropolitan Status						
Urban	*	*		*	*	12
Suburban	$42.00	$53.00		$53.36	$60.60	53
Second City	$39.29	$52.20		$54.11	$70.00	46
Town	$36.00	$42.00		$46.89	$58.00	47
Rural	*	*		*	*	24
Median Area Household Income						
$35,000 or Less	*	*		*	*	19
$35,000 to $49,999	$38.39	$47.50		$49.32	$59.61	68
$50,000 to $69,999	$39.88	$51.50		$53.40	$64.51	62
$70,000 or More	$42.00	$52.00		$56.47	$66.50	31

Note: An asterisk indicates that data was not reported due to an insufficient number of responses.

TABLE 7.18
5-7 CHEMISTRIES

	25th Percentile	Median	Your Data	Average	75th Percentile	Number of Respondents
All Practices	**$39.99**	**$48.63**		**$50.33**	**$58.85**	**303**
Number of FTE Veterinarians						
1.0 or Less	$35.00	$42.79		$44.61	$55.00	73
1.1 to 2.0	$42.00	$47.50		$51.63	$58.92	95
2.1 to 3.0	$43.09	$50.00		$53.39	$63.19	62
3.1 or More	$40.00	$51.20		$51.01	$60.50	66
Member Status						
Accredited Practice Member	$41.75	$53.00		$53.86	$64.51	90
Nonaccredited Member	$42.00	$51.00		$51.12	$59.73	72
Nonmember	$35.25	$45.00		$47.43	$56.75	121
Metropolitan Status						
Urban	*	*		*	*	19
Suburban	$42.00	$50.00		$52.03	$62.01	89
Second City	$37.81	$49.00		$49.72	$58.10	59
Town	$37.64	$45.00		$48.41	$55.60	97
Rural	$36.50	$46.10		$48.34	$55.75	36
Median Area Household Income						
$35,000 or Less	$35.15	$45.50		$46.35	$52.19	36
$35,000 to $49,999	$38.38	$47.05		$48.75	$56.00	114
$50,000 to $69,999	$41.00	$51.40		$51.91	$59.00	99
$70,000 or More	$38.38	$48.60		$51.97	$62.01	48

Note: An asterisk indicates that data was not reported due to an insufficient number of responses.

TABLE 7.19
8-12 CHEMISTRIES

	25th Percentile	Median	Your Data	Average	75th Percentile	Number of Respondents
All Practices	**$55.88**	**$69.00**		**$71.07**	**$82.36**	**298**
Number of FTE Veterinarians						
1.0 or Less	$50.00	$66.50		$67.81	$78.22	70
1.1 to 2.0	$59.73	$68.00		$71.01	$82.00	95
2.1 to 3.0	$60.75	$75.00		$75.97	$86.42	70
3.1 or More	$52.75	$65.00		$69.13	$80.13	58
Member Status						
Accredited Practice Member	$62.63	$75.88		$79.45	$96.00	92
Nonaccredited Member	$57.26	$70.20		$70.37	$80.74	66
Nonmember	$53.00	$65.00		$66.32	$78.80	123
Metropolitan Status						
Urban	*	*		*	*	16
Suburban	$60.00	$75.00		$77.60	$96.00	79
Second City	$53.75	$64.85		$69.41	$82.40	59
Town	$56.90	$67.50		$67.91	$77.05	103
Rural	$46.00	$66.00		$64.92	$78.20	39
Median Area Household Income						
$35,000 or Less	$55.00	$60.00		$61.66	$71.65	36
$35,000 to $49,999	$55.00	$69.00		$69.42	$79.00	111
$50,000 to $69,999	$56.00	$71.80		$73.11	$85.00	100
$70,000 or More	$59.25	$71.00		$74.86	$87.77	46

Note: An asterisk indicates that data was not reported due to an insufficient number of responses.

TABLE 7.20
CYTOLOGY: FINE-NEEDLE ASPIRATE

	25th Percentile	Median	Your Data	Average	75th Percentile	Number of Respondents
All Practices	$20.00	$27.00		$29.81	$37.00	387
Number of FTE Veterinarians						
1.0 or Less	$18.50	$28.00		$29.79	$35.28	93
1.1 to 2.0	$20.00	$25.85		$28.05	$35.00	119
2.1 to 3.0	$21.30	$27.39		$31.65	$39.00	86
3.1 or More	$20.19	$27.98		$30.51	$38.17	80
Member Status						
Accredited Practice Member	$25.00	$29.00		$31.45	$39.00	110
Nonaccredited Member	$20.00	$26.18		$28.45	$36.00	79
Nonmember	$19.00	$25.00		$29.83	$35.40	167
Metropolitan Status						
Urban	*	*		*	*	23
Suburban	$22.50	$28.11		$30.06	$38.22	107
Second City	$22.00	$30.00		$32.12	$39.18	81
Town	$18.50	$25.00		$28.73	$35.00	120
Rural	$18.75	$24.25		$26.02	$32.23	50
Median Area Household Income						
$35,000 or Less	$16.00	$20.25		$23.93	$27.77	43
$35,000 to $49,999	$20.00	$26.33		$29.96	$36.00	142
$50,000 to $69,999	$22.00	$28.20		$30.71	$38.44	130
$70,000 or More	$22.50	$29.91		$31.96	$39.09	62

Note: An asterisk indicates that data was not reported due to an insufficient number of responses.

TABLE 7.21
CYTOLOGY: EAR SWAB

	25th Percentile	Median	Your Data	Average	75th Percentile	Number of Respondents
All Practices	$15.84	$20.00		$20.68	$25.00	484
Number of FTE Veterinarians						
1.0 or Less	$14.00	$19.50		$19.68	$24.01	122
1.1 to 2.0	$16.00	$19.80		$20.92	$25.00	145
2.1 to 3.0	$16.63	$20.00		$20.34	$24.00	108
3.1 or More	$16.50	$20.90		$21.95	$26.50	95
Member Status						
Accredited Practice Member	$18.00	$20.98		$21.95	$25.83	128
Nonaccredited Member	$17.00	$20.10		$21.36	$25.31	99
Nonmember	$15.00	$19.00		$19.66	$24.00	218
Metropolitan Status						
Urban	$16.00	$21.00		$22.58	$26.40	31
Suburban	$16.85	$21.60		$21.75	$26.38	144
Second City	$16.50	$20.90		$21.64	$25.00	100
Town	$15.00	$18.55		$19.60	$24.00	150
Rural	$12.25	$18.00		$17.79	$22.37	53
Median Area Household Income						
$35,000 or Less	$12.00	$16.25		$17.91	$24.00	58
$35,000 to $49,999	$15.00	$19.00		$20.43	$24.00	179
$50,000 to $69,999	$16.25	$20.04		$20.74	$24.96	157
$70,000 or More	$17.38	$22.00		$23.12	$27.37	78

TABLE 7.22
CYTOLOGY: SKIN SWAB

	25th Percentile	Median	Your Data	Average	75th Percentile	Number of Respondents
All Practices	**$17.96**	**$21.10**		**$22.76**	**$26.95**	**436**
Number of FTE Veterinarians						
1.0 or Less	$15.70	$20.00		$22.32	$27.30	108
1.1 to 2.0	$18.00	$21.30		$23.11	$27.75	133
2.1 to 3.0	$17.95	$20.90		$22.01	$25.31	95
3.1 or More	$18.00	$23.00		$23.06	$26.88	89
Member Status						
Accredited Practice Member	$19.54	$24.13		$24.72	$28.00	118
Nonaccredited Member	$18.00	$22.00		$23.55	$28.39	90
Nonmember	$16.00	$20.00		$21.24	$25.30	191
Metropolitan Status						
Urban	$18.00	$23.00		$23.44	$28.00	27
Suburban	$18.00	$23.00		$23.77	$28.00	128
Second City	$18.00	$22.00		$23.77	$28.00	91
Town	$17.70	$20.00		$21.74	$26.00	135
Rural	$15.00	$20.13		$20.68	$25.00	48
Median Area Household Income						
$35,000 or Less	$14.81	$19.28		$19.72	$25.00	54
$35,000 to $49,999	$18.15	$22.00		$23.48	$26.39	158
$50,000 to $69,999	$18.00	$21.43		$22.67	$27.88	144
$70,000 or More	$16.88	$22.62		$24.07	$28.50	69

TABLE 7.23
CYTOLOGY: VAGINAL

	25th Percentile	Median	Your Data	Average	75th Percentile	Number of Respondents
All Practices	**$20.00**	**$25.73**		**$27.86**	**$35.00**	**420**
Number of FTE Veterinarians						
1.0 or Less	$18.00	$24.11		$26.71	$34.93	100
1.1 to 2.0	$20.00	$25.00		$26.75	$33.30	125
2.1 to 3.0	$20.85	$28.00		$28.06	$35.10	97
3.1 or More	$22.69	$28.00		$30.14	$36.31	90
Member Status						
Accredited Practice Member	$24.11	$29.00		$31.22	$35.63	117
Nonaccredited Member	$18.65	$26.18		$27.22	$35.00	87
Nonmember	$18.50	$24.86		$26.04	$33.15	186
Metropolitan Status						
Urban	*	*		*	*	24
Suburban	$20.55	$28.20		$28.83	$34.78	118
Second City	$21.62	$30.00		$30.17	$37.57	89
Town	$18.50	$25.00		$26.05	$32.46	132
Rural	$18.00	$22.00		$23.91	$30.00	51
Median Area Household Income						
$35,000 or Less	$15.05	$22.00		$24.43	$30.45	49
$35,000 to $49,999	$19.00	$25.00		$26.33	$34.04	154
$50,000 to $69,999	$22.00	$29.00		$30.12	$35.88	137
$70,000 or More	$19.06	$29.00		$29.67	$35.42	69

Note: An asterisk indicates that data was not reported due to an insufficient number of responses.

TABLE 7.24
ELECTROLYTES

	25th Percentile	Median	Your Data	Average	75th Percentile	Number of Respondents
All Practices	**$20.80**	**$27.40**		**$30.90**	**$38.08**	**277**
Number of FTE Veterinarians						
1.0 or Less	$19.00	$28.00		$31.91	$39.98	53
1.1 to 2.0	$19.80	$25.00		$29.87	$38.08	85
2.1 to 3.0	$22.33	$31.85		$31.71	$38.88	64
3.1 or More	$22.00	$26.85		$29.99	$35.00	67
Member Status						
Accredited Practice Member	$22.55	$30.00		$32.35	$39.25	90
Nonaccredited Member	$20.50	$27.00		$30.32	$38.61	61
Nonmember	$19.50	$26.50		$29.93	$36.84	106
Metropolitan Status						
Urban	*	*		*	*	17
Suburban	$22.00	$28.00		$32.37	$40.00	74
Second City	$21.96	$29.75		$31.42	$36.00	66
Town	$19.25	$25.50		$29.87	$37.00	87
Rural	$20.00	$25.00		$27.18	$33.00	29
Median Area Household Income						
$35,000 or Less	$16.81	$25.50		$28.39	$35.25	26
$35,000 to $49,999	$20.00	$25.00		$28.90	$35.70	107
$50,000 to $69,999	$22.00	$28.50		$32.21	$39.95	95
$70,000 or More	$22.00	$35.00		$34.45	$41.00	45

Note 1: Assumed sodium, chloride, and potassium were tested.
Note 2: An asterisk indicates that data was not reported due to an insufficient number of responses.

TABLE 7.25
FECAL DIF-QUIK STAIN

	25th Percentile	Median	Your Data	Average	75th Percentile	Number of Respondents
All Practices	**$16.56**	**$20.05**		**$21.70**	**$25.26**	**244**
Number of FTE Veterinarians						
1.0 or Less	$15.00	$19.35		$21.30	$25.98	60
1.1 to 2.0	$16.78	$21.00		$21.78	$26.90	65
2.1 to 3.0	$17.64	$20.00		$21.44	$23.70	55
3.1 or More	$17.50	$19.50		$21.58	$25.00	58
Member Status						
Accredited Practice Member	$18.05	$22.00		$22.58	$25.56	72
Nonaccredited Member	$16.56	$20.55		$21.80	$26.35	48
Nonmember	$15.00	$20.00		$20.92	$25.63	105
Metropolitan Status						
Urban	$19.50	$21.70		$23.59	$24.63	13
Suburban	$18.00	$20.05		$22.10	$26.60	72
Second City	$17.95	$21.84		$22.97	$28.00	59
Town	$15.43	$19.08		$21.38	$24.75	64
Rural	$12.00	$18.00		$18.12	$22.80	31
Median Area Household Income						
$35,000 or Less	$12.88	$18.73		$19.30	$22.25	26
$35,000 to $49,999	$16.00	$20.00		$21.33	$24.00	87
$50,000 to $69,999	$17.16	$21.11		$22.15	$25.86	84
$70,000 or More	$18.00	$21.50		$23.62	$29.38	40

TABLE 7.26
FECAL EXAMINATION: DIRECT SMEAR

	25th Percentile	Median	Your Data	Average	75th Percentile	Number of Respondents
All Practices	**$12.65**	**$16.00**		**$16.62**	**$20.00**	**446**
Number of FTE Veterinarians						
1.0 or Less	$12.00	$15.80		$16.16	$19.43	112
1.1 to 2.0	$12.00	$16.00		$16.22	$19.50	141
2.1 to 3.0	$13.11	$17.43		$17.67	$21.25	94
3.1 or More	$14.00	$16.00		$16.51	$19.10	83
Member Status						
Accredited Practice Member	$14.94	$17.00		$18.19	$21.00	118
Nonaccredited Member	$12.85	$16.50		$17.32	$20.13	89
Nonmember	$11.00	$15.00		$15.25	$19.00	198
Metropolitan Status						
Urban	$14.00	$17.80		$17.04	$20.00	31
Suburban	$13.89	$17.85		$17.86	$20.83	124
Second City	$13.38	$16.63		$16.82	$19.83	92
Town	$12.10	$16.00		$15.97	$19.00	136
Rural	$10.00	$14.00		$14.63	$18.00	56
Median Area Household Income						
$35,000 or Less	$10.00	$14.00		$14.64	$17.90	61
$35,000 to $49,999	$13.00	$16.00		$16.03	$19.00	163
$50,000 to $69,999	$13.00	$16.90		$16.97	$20.10	143
$70,000 or More	$14.83	$18.88		$18.84	$22.00	66

TABLE 7.27
FECAL EXAMINATION: FLOTATION (GRAVITATIONAL)

	25th Percentile	Median	Your Data	Average	75th Percentile	Number of Respondents
All Practices	**$15.00**	**$17.76**		**$17.99**	**$20.70**	**394**
Number of FTE Veterinarians						
1.0 or Less	$14.18	$16.50		$17.10	$20.00	109
1.1 to 2.0	$15.00	$17.45		$17.71	$20.45	124
2.1 to 3.0	$15.35	$18.00		$18.61	$21.45	76
3.1 or More	$15.88	$18.00		$18.63	$20.80	73
Member Status						
Accredited Practice Member	$15.05	$18.00		$18.83	$21.85	90
Nonaccredited Member	$15.04	$18.00		$18.70	$22.49	80
Nonmember	$14.00	$17.00		$17.05	$20.00	185
Metropolitan Status						
Urban	$14.26	$19.00		$18.43	$20.25	26
Suburban	$16.00	$18.88		$19.59	$22.50	110
Second City	$15.00	$17.81		$18.53	$20.95	81
Town	$14.00	$16.50		$16.87	$19.25	127
Rural	$13.00	$15.90		$15.97	$18.00	43
Median Area Household Income						
$35,000 or Less	$12.00	$14.65		$15.15	$18.00	51
$35,000 to $49,999	$14.20	$16.00		$16.53	$19.00	143
$50,000 to $69,999	$16.00	$18.00		$19.49	$22.49	124
$70,000 or More	$17.00	$19.75		$20.34	$23.03	70

TABLE 7.28
FECAL EXAMINATION: FLOTATION (CENTRIFUGATION, ZINC SULFATE)

	25th Percentile	Median	Your Data	Average	75th Percentile	Number of Respondents
All Practices	$16.00	$18.60		$19.78	$23.35	190
Number of FTE Veterinarians						
1.0 or Less	$15.19	$18.49		$19.41	$21.63	50
1.1 to 2.0	$15.00	$18.00		$19.10	$22.00	55
2.1 to 3.0	$17.40	$19.00		$20.55	$23.80	41
3.1 or More	$16.35	$19.00		$20.45	$24.00	39
Member Status						
Accredited Practice Member	$16.26	$19.00		$20.36	$24.03	54
Nonaccredited Member	$15.85	$20.00		$20.60	$25.41	42
Nonmember	$16.00	$18.00		$19.05	$21.34	81
Metropolitan Status						
Urban	*	*		*	*	11
Suburban	$16.85	$22.25		$21.80	$25.23	52
Second City	$16.63	$19.80		$20.43	$23.00	41
Town	$16.50	$18.00		$18.71	$21.00	55
Rural	$12.50	$15.70		$16.40	$18.65	28
Median Area Household Income						
$35,000 or Less	$12.00	$15.75		$16.53	$19.93	26
$35,000 to $49,999	$15.94	$18.00		$18.48	$20.80	62
$50,000 to $69,999	$16.88	$20.50		$21.43	$25.00	62
$70,000 or More	$18.00	$21.50		$22.00	$25.56	32

Note: An asterisk indicates that data was not reported due to an insufficient number of responses.

TABLE 7.29
FECAL EXAMINATION: GIARDIA WET MOUNT

	25th Percentile	Median	Your Data	Average	75th Percentile	Number of Respondents
All Practices	$14.24	$17.81		$18.34	$21.30	252
Number of FTE Veterinarians						
1.0 or Less	$15.00	$18.00		$18.39	$20.90	73
1.1 to 2.0	$13.00	$16.00		$17.04	$20.00	76
2.1 to 3.0	$15.11	$17.92		$18.25	$22.54	48
3.1 or More	$15.38	$19.00		$19.50	$21.78	50
Member Status						
Accredited Practice Member	$15.00	$18.30		$18.65	$22.58	78
Nonaccredited Member	$13.88	$17.80		$17.93	$21.60	49
Nonmember	$13.00	$17.00		$17.47	$20.00	111
Metropolitan Status						
Urban	*	*		*	*	14
Suburban	$15.98	$19.47		$20.98	$23.68	65
Second City	$15.00	$17.55		$18.10	$20.75	56
Town	$12.93	$16.50		$16.98	$20.00	84
Rural	$11.38	$15.00		$15.64	$18.50	30
Median Area Household Income						
$35,000 or Less	$12.00	$15.00		$15.85	$19.00	35
$35,000 to $49,999	$13.83	$16.00		$16.77	$19.46	90
$50,000 to $69,999	$15.00	$18.00		$19.93	$23.60	79
$70,000 or More	$15.88	$20.00		$20.76	$23.19	42

Note: An asterisk indicates that data was not reported due to an insufficient number of responses.

TABLE 7.30
FECAL GRAM'S STAIN

	25th Percentile	Median	Your Data	Average	75th Percentile	Number of Respondents
All Practices	**$17.96**	**$22.96**		**$24.19**	**$29.00**	**140**
Number of FTE Veterinarians						
1.0 or Less	$17.18	$22.38		$23.06	$29.26	36
1.1 to 2.0	$18.50	$25.35		$26.40	$33.08	39
2.1 to 3.0	$17.77	$22.41		$24.72	$28.69	30
3.1 or More	$18.00	$22.65		$23.24	$27.95	31
Member Status						
Accredited Practice Member	$18.50	$24.30		$25.04	$29.78	45
Nonaccredited Member	$18.00	$23.23		$25.52	$28.52	30
Nonmember	$16.00	$21.92		$23.70	$29.26	56
Metropolitan Status						
Urban	*	*		*	*	9
Suburban	$17.64	$24.00		$25.15	$30.00	35
Second City	$17.95	$22.75		$23.70	$28.25	35
Town	$19.00	$22.82		$25.00	$29.13	41
Rural	*	*		*	*	17
Median Area Household Income						
$35,000 or Less	*	*		*	*	16
$35,000 to $49,999	$17.05	$23.20		$23.77	$28.25	43
$50,000 to $69,999	$17.91	$24.00		$25.13	$29.28	54
$70,000 or More	*	*		*	*	22

Note: An asterisk indicates that data was not reported due to an insufficient number of responses.

TABLE 7.31
FELINE LEUKEMIA (FeLV) TEST

	25th Percentile	Median	Your Data	Average	75th Percentile	Number of Respondents
All Practices	**$28.00**	**$32.00**		**$33.71**	**$38.50**	**207**
Number of FTE Veterinarians						
1.0 or Less	$27.00	$30.00		$32.13	$37.00	51
1.1 to 2.0	$27.13	$32.00		$32.94	$38.38	64
2.1 to 3.0	$25.86	$33.10		$33.78	$37.75	44
3.1 or More	$31.05	$35.00		$35.99	$40.50	41
Member Status						
Accredited Practice Member	$30.08	$34.32		$36.46	$40.38	56
Nonaccredited Member	$28.00	$34.50		$34.36	$40.00	42
Nonmember	$25.00	$30.00		$31.68	$35.19	92
Metropolitan Status						
Urban	*	*		*	*	17
Suburban	$30.00	$34.39		$36.39	$40.00	59
Second City	$27.75	$33.00		$33.48	$39.50	45
Town	$25.13	$30.00		$32.62	$37.73	56
Rural	$21.50	$31.50		$30.97	$35.83	26
Median Area Household Income						
$35,000 or Less	$23.60	$30.00		$29.73	$33.35	33
$35,000 to $49,999	$26.75	$32.00		$33.09	$39.13	78
$50,000 to $69,999	$28.26	$33.50		$34.89	$39.75	62
$70,000 or More	$29.38	$34.20		$36.82	$38.99	30

Note: An asterisk indicates that data was not reported due to an insufficient number of responses.

TABLE 7.32
FeLV AND FIV TEST

	25th Percentile	Median	Your Data	Average	75th Percentile	Number of Respondents
All Practices	**$38.00**	**$44.00**		**$44.85**	**$49.89**	**485**
Number of FTE Veterinarians						
1.0 or Less	$35.00	$43.33		$43.85	$50.75	118
1.1 to 2.0	$38.00	$42.60		$43.50	$47.80	153
2.1 to 3.0	$40.00	$44.64		$46.04	$50.00	106
3.1 or More	$39.00	$45.83		$46.30	$50.00	95
Member Status						
Accredited Practice Member	$39.99	$45.83		$46.56	$52.00	131
Nonaccredited Member	$39.50	$44.00		$45.15	$48.61	100
Nonmember	$35.00	$42.00		$43.34	$48.85	210
Metropolitan Status						
Urban	$40.00	$45.00		$45.37	$51.33	33
Suburban	$40.00	$45.47		$47.34	$55.00	138
Second City	$39.43	$45.00		$46.15	$50.75	105
Town	$38.00	$42.00		$43.17	$47.63	150
Rural	$35.00	$35.10		$39.39	$44.90	52
Median Area Household Income						
$35,000 or Less	$35.00	$40.00		$40.09	$44.76	57
$35,000 to $49,999	$37.50	$42.55		$43.68	$48.79	178
$50,000 to $69,999	$38.45	$45.47		$46.60	$53.21	162
$70,000 or More	$41.75	$46.35		$47.57	$51.63	74

TABLE 7.33
FUNGAL CULTURE

	25th Percentile	Median	Your Data	Average	75th Percentile	Number of Respondents
All Practices	**$23.73**	**$30.00**		**$31.93**	**$36.53**	**361**
Number of FTE Veterinarians						
1.0 or Less	$20.13	$29.00		$31.39	$37.63	85
1.1 to 2.0	$24.50	$30.00		$30.97	$35.31	115
2.1 to 3.0	$22.43	$28.78		$33.56	$39.26	80
3.1 or More	$24.50	$30.00		$32.51	$36.25	73
Member Status						
Accredited Practice Member	$26.25	$32.10		$35.20	$39.13	105
Nonaccredited Member	$25.00	$30.55		$34.72	$39.88	72
Nonmember	$20.29	$26.55		$28.83	$32.59	150
Metropolitan Status						
Urban	$25.00	$33.00		$34.60	$44.00	27
Suburban	$25.00	$30.91		$33.89	$38.75	104
Second City	$25.00	$30.00		$31.22	$35.25	75
Town	$22.14	$27.98		$30.93	$36.91	118
Rural	$19.00	$27.00		$29.25	$35.00	33
Median Area Household Income						
$35,000 or Less	$22.00	$25.38		$30.16	$34.00	46
$35,000 to $49,999	$24.00	$28.00		$29.55	$34.00	136
$50,000 to $69,999	$22.80	$31.50		$33.26	$39.00	123
$70,000 or More	$26.81	$33.86		$36.86	$45.38	52

TABLE 7.34
GIARDIA ANTIGEN TEST

	25th Percentile	Median	Your Data	Average	75th Percentile	Number of Respondents
All Practices	**$20.00**	**$25.00**		**$28.03**	**$34.88**	**232**
Number of FTE Veterinarians						
1.0 or Less	$18.43	$24.50		$26.94	$32.04	54
1.1 to 2.0	$19.98	$24.88		$27.72	$35.00	78
2.1 to 3.0	$19.83	$25.65		$26.66	$29.80	49
3.1 or More	$20.00	$27.00		$30.17	$37.73	46
Member Status						
Accredited Practice Member	$19.93	$23.50		$28.73	$36.50	65
Nonaccredited Member	$21.00	$27.00		$28.91	$33.82	48
Nonmember	$18.88	$24.47		$25.71	$30.40	102
Metropolitan Status						
Urban	*	*		*	*	19
Suburban	$19.80	$26.00		$28.00	$34.50	71
Second City	$19.90	$25.00		$27.98	$35.13	50
Town	$19.75	$25.00		$27.18	$32.25	77
Rural	*	*		*	*	13
Median Area Household Income						
$35,000 or Less	*	*		*	*	22
$35,000 to $49,999	$19.59	$23.33		$25.92	$29.88	88
$50,000 to $69,999	$20.00	$26.50		$28.56	$35.00	70
$70,000 or More	$20.00	$28.00		$31.39	$40.00	47

Note: An asterisk indicates that data was not reported due to an insufficient number of responses.

TABLE 7.35
GLUCOSE CURVE (6)

	25th Percentile	Median	Your Data	Average	75th Percentile	Number of Respondents
All Practices	**$62.55**	**$80.33**		**$85.03**	**$103.43**	**432**
Number of FTE Veterinarians						
1.0 or Less	$60.00	$75.00		$82.11	$104.50	99
1.1 to 2.0	$60.00	$80.00		$84.16	$101.00	135
2.1 to 3.0	$66.15	$82.70		$86.95	$108.31	94
3.1 or More	$65.00	$85.65		$87.68	$104.50	91
Member Status						
Accredited Practice Member	$69.05	$88.20		$92.24	$120.00	125
Nonaccredited Member	$68.00	$85.00		$84.99	$102.00	84
Nonmember	$58.60	$75.00		$79.63	$98.45	186
Metropolitan Status						
Urban	$84.50	$100.00		$101.40	$121.00	29
Suburban	$65.00	$84.00		$86.39	$104.75	113
Second City	$64.39	$85.19		$87.75	$106.50	94
Town	$59.10	$77.75		$81.53	$100.00	144
Rural	$56.25	$75.00		$76.84	$95.00	45
Median Area Household Income						
$35,000 or Less	$60.00	$72.00		$77.41	$100.00	51
$35,000 to $49,999	$60.00	$78.00		$83.91	$103.10	153
$50,000 to $69,999	$63.75	$85.00		$85.61	$106.50	145
$70,000 or More	$68.50	$90.00		$90.86	$112.32	71

TABLE 7.36
GLUCOSE SINGLE

	25th Percentile	Median	Your Data	Average	75th Percentile	Number of Respondents
All Practices	**$13.00**	**$17.00**		**$17.50**	**$21.15**	**474**
Number of FTE Veterinarians						
1.0 or Less	$12.00	$15.00		$16.39	$20.00	117
1.1 to 2.0	$14.00	$18.00		$18.08	$21.37	139
2.1 to 3.0	$13.50	$17.54		$18.05	$22.00	108
3.1 or More	$12.88	$18.00		$17.89	$21.80	94
Member Status						
Accredited Practice Member	$14.00	$19.00		$18.64	$22.70	127
Nonaccredited Member	$13.00	$16.58		$17.66	$22.00	102
Nonmember	$12.05	$15.46		$16.66	$20.00	200
Metropolitan Status						
Urban	$14.50	$20.00		$19.08	$22.00	33
Suburban	$14.35	$18.70		$18.76	$23.00	137
Second City	$14.05	$16.21		$17.48	$20.45	96
Town	$12.49	$16.00		$16.83	$20.44	149
Rural	$12.00	$15.00		$15.34	$18.50	51
Median Area Household Income						
$35,000 or Less	$11.40	$15.00		$15.07	$19.25	53
$35,000 to $49,999	$13.00	$17.30		$17.36	$21.37	171
$50,000 to $69,999	$14.00	$17.00		$17.74	$21.30	159
$70,000 or More	$14.88	$18.75		$18.88	$23.00	78

TABLE 7.37
HEARTWORM TEST, CANINE: OCCULT/ANTIGEN

	25th Percentile	Median	Your Data	Average	75th Percentile	Number of Respondents
All Practices	**$25.00**	**$29.00**		**$29.71**	**$34.50**	**255**
Number of FTE Veterinarians						
1.0 or Less	$23.00	$26.00		$27.14	$31.13	66
1.1 to 2.0	$25.13	$29.50		$30.79	$36.38	76
2.1 to 3.0	$26.30	$29.75		$29.95	$34.08	54
3.1 or More	$26.50	$31.00		$30.98	$34.25	54
Member Status						
Accredited Practice Member	$26.75	$30.00		$31.08	$35.00	65
Nonaccredited Member	$26.08	$31.00		$31.05	$35.00	50
Nonmember	$24.00	$27.50		$28.14	$32.00	119
Metropolitan Status						
Urban	*	*		*	*	18
Suburban	$26.63	$31.00		$31.03	$35.00	60
Second City	$25.88	$30.75		$30.98	$35.38	58
Town	$25.00	$28.46		$29.07	$32.00	85
Rural	$20.00	$27.00		$26.37	$31.75	27
Median Area Household Income						
$35,000 or Less	$22.63	$26.40		$28.16	$34.90	40
$35,000 to $49,999	$25.00	$29.00		$29.18	$32.85	99
$50,000 to $69,999	$26.00	$30.00		$30.25	$34.30	71
$70,000 or More	$28.00	$31.00		$31.65	$35.00	35

Note: An asterisk indicates that data was not reported due to an insufficient number of responses.

TABLE 7.38
HEARTWORM TEST, CANINE: OCCULT/ANTIGEN PLUS LYME AND E. CANIS

	25th Percentile	Median	Your Data	Average	75th Percentile	Number of Respondents
All Practices	**$31.45**	**$36.00**		**$38.75**	**$43.25**	**317**
Number of FTE Veterinarians						
1.0 or Less	$30.36	$35.00		$36.94	$39.75	72
1.1 to 2.0	$31.28	$36.53		$38.43	$45.00	102
2.1 to 3.0	$33.45	$36.50		$39.40	$44.00	70
3.1 or More	$30.30	$37.00		$39.55	$43.50	63
Member Status						
Accredited Practice Member	$31.93	$37.33		$39.38	$43.88	76
Nonaccredited Member	$31.63	$38.28		$40.67	$47.50	69
Nonmember	$30.00	$35.00		$37.20	$41.00	143
Metropolitan Status						
Urban	*	*		*	*	18
Suburban	$34.60	$39.00		$41.84	$47.50	83
Second City	$31.00	$35.00		$37.75	$44.42	55
Town	$31.00	$36.00		$38.34	$42.00	111
Rural	$29.88	$34.00		$33.96	$36.66	45
Median Area Household Income						
$35,000 or Less	$29.78	$32.13		$33.05	$36.75	36
$35,000 to $49,999	$30.83	$35.00		$36.38	$40.00	106
$50,000 to $69,999	$32.00	$36.50		$40.05	$44.71	109
$70,000 or More	$35.00	$42.00		$43.90	$49.63	58

Note: An asterisk indicates that data was not reported due to an insufficient number of responses.

TABLE 7.39
PARVOVIRUS TEST

	25th Percentile	Median	Your Data	Average	75th Percentile	Number of Respondents
All Practices	**$34.85**	**$40.00**		**$41.92**	**$48.00**	**381**
Number of FTE Veterinarians						
1.0 or Less	$32.00	$38.00		$40.22	$45.13	89
1.1 to 2.0	$35.00	$40.00		$40.30	$45.00	111
2.1 to 3.0	$34.49	$42.38		$44.29	$51.40	86
3.1 or More	$35.00	$40.00		$43.15	$48.89	83
Member Status						
Accredited Practice Member	$35.56	$41.00		$43.86	$51.00	100
Nonaccredited Member	$34.40	$40.15		$43.83	$50.73	77
Nonmember	$32.00	$37.99		$40.39	$46.00	175
Metropolitan Status						
Urban	$35.00	$40.00		$41.29	$47.50	29
Suburban	$36.00	$42.89		$45.01	$52.00	100
Second City	$35.00	$39.30		$42.81	$49.12	83
Town	$34.18	$40.00		$41.67	$46.80	120
Rural	$28.00	$35.00		$34.52	$40.26	42
Median Area Household Income						
$35,000 or Less	$30.00	$35.00		$37.06	$43.30	47
$35,000 to $49,999	$33.74	$40.00		$40.31	$45.16	150
$50,000 to $69,999	$35.00	$40.00		$43.74	$50.50	125
$70,000 or More	$37.20	$45.75		$47.77	$55.25	50

TABLE 7.40
T4

	25th Percentile	Median	Your Data	Average	75th Percentile	Number of Respondents
All Practices	**$34.37**	**$38.00**		**$40.74**	**$47.00**	**126**
Number of FTE Veterinarians						
1.0 or Less	$30.75	$39.03		$40.43	$47.13	36
1.1 to 2.0	$34.37	$38.50		$40.92	$47.50	42
2.1 to 3.0	*	*		*	*	22
3.1 or More	$31.00	$37.50		$41.01	$47.50	25
Member Status						
Accredited Practice Member	$35.00	$39.36		$42.65	$50.75	37
Nonaccredited Member	$33.61	$39.05		$40.46	$46.00	25
Nonmember	$33.00	$37.25		$39.61	$47.13	58
Metropolitan Status						
Urban	*	*		*	*	5
Suburban	*	*		*	*	22
Second City	$35.00	$38.68		$41.86	$48.25	34
Town	$33.48	$37.73		$38.90	$45.81	46
Rural	*	*		*	*	19
Median Area Household Income						
$35,000 or Less	$31.00	$35.72		$37.43	$44.02	25
$35,000 to $49,999	$33.24	$38.50		$40.12	$46.69	48
$50,000 to $69,999	$35.00	$37.73		$41.78	$47.88	42
$70,000 or More	*	*		*	*	11

Note: An asterisk indicates that data was not reported due to an insufficient number of responses.

TABLE 7.41
URINALYSIS: COMPLETE

	25th Percentile	Median	Your Data	Average	75th Percentile	Number of Respondents
All Practices	**$25.54**	**$30.00**		**$31.05**	**$35.25**	**447**
Number of FTE Veterinarians						
1.0 or Less	$24.44	$28.89		$29.82	$35.00	110
1.1 to 2.0	$25.63	$30.00		$31.14	$35.50	136
2.1 to 3.0	$26.00	$30.13		$31.15	$35.00	96
3.1 or More	$28.00	$30.56		$32.29	$36.41	89
Member Status						
Accredited Practice Member	$28.93	$32.67		$33.25	$36.41	120
Nonaccredited Member	$24.94	$30.00		$30.79	$36.36	93
Nonmember	$25.00	$28.64		$30.20	$34.96	197
Metropolitan Status						
Urban	$25.75	$33.31		$32.87	$36.34	26
Suburban	$27.00	$31.00		$32.27	$36.72	123
Second City	$26.71	$31.88		$32.80	$38.17	94
Town	$25.00	$28.50		$29.95	$35.00	147
Rural	$23.13	$27.80		$27.43	$32.25	50
Median Area Household Income						
$35,000 or Less	$21.00	$25.00		$26.47	$31.58	57
$35,000 to $49,999	$25.90	$30.00		$30.17	$34.50	165
$50,000 to $69,999	$26.98	$31.50		$32.43	$36.95	141
$70,000 or More	$28.00	$33.26		$34.61	$41.04	72

TABLE 7.42
URINALYSIS: SPECIFIC GRAVITY

	25th Percentile	Median	Your Data	Average	75th Percentile	Number of Respondents
All Practices	**$8.00**	**$11.00**		**$12.73**	**$16.00**	**275**
Number of FTE Veterinarians						
1.0 or Less	$8.00	$11.20		$13.44	$17.02	69
1.1 to 2.0	$8.09	$10.19		$12.11	$14.96	88
2.1 to 3.0	$7.50	$10.00		$11.68	$14.70	55
3.1 or More	$9.00	$12.50		$14.05	$18.13	54
Member Status						
Accredited Practice Member	$9.00	$11.81		$12.53	$15.00	77
Nonaccredited Member	$8.00	$10.00		$12.15	$13.75	59
Nonmember	$8.00	$10.00		$13.01	$16.28	118
Metropolitan Status						
Urban	*	*		*	*	22
Suburban	$7.60	$11.00		$12.27	$16.00	71
Second City	$7.50	$11.81		$13.94	$17.38	59
Town	$9.00	$10.00		$12.44	$15.00	87
Rural	$8.75	$11.00		$12.55	$13.63	33
Median Area Household Income						
$35,000 or Less	$6.65	$9.00		$10.42	$12.50	29
$35,000 to $49,999	$8.21	$10.50		$12.47	$15.00	93
$50,000 to $69,999	$8.00	$11.60		$13.43	$17.00	95
$70,000 or More	$8.00	$11.50		$13.14	$16.00	51

Note: An asterisk indicates that data was not reported due to an insufficient number of responses.

TABLE 7.43
URINALYSIS: DIP STICK

	25th Percentile	Median	Your Data	Average	75th Percentile	Number of Respondents
All Practices	**$13.00**	**$16.50**		**$17.70**	**$21.00**	**335**
Number of FTE Veterinarians						
1.0 or Less	$13.79	$15.88		$17.26	$20.21	90
1.1 to 2.0	$12.00	$16.11		$17.24	$21.00	104
2.1 to 3.0	$14.00	$18.00		$18.70	$22.47	68
3.1 or More	$13.23	$17.00		$18.27	$20.10	61
Member Status						
Accredited Practice Member	$12.19	$16.73		$17.80	$21.17	94
Nonaccredited Member	$12.00	$15.00		$17.77	$23.15	66
Nonmember	$13.50	$16.50		$17.55	$20.18	147
Metropolitan Status						
Urban	$12.00	$18.00		$18.44	$22.00	25
Suburban	$13.35	$16.50		$17.67	$21.00	95
Second City	$14.00	$18.00		$19.23	$24.00	71
Town	$13.00	$16.50		$17.45	$21.00	103
Rural	$11.63	$14.75		$15.27	$20.00	38
Median Area Household Income						
$35,000 or Less	$10.43	$15.00		$16.65	$21.00	41
$35,000 to $49,999	$13.50	$16.50		$17.59	$21.18	116
$50,000 to $69,999	$13.38	$17.19		$18.47	$21.54	116
$70,000 or More	$12.00	$16.50		$17.09	$20.40	55

TABLE 7.44
URINALYSIS: SEDIMENT

	25th Percentile	Median	Your Data	Average	75th Percentile	Number of Respondents
All Practices	**$13.00**	**$19.00**		**$20.40**	**$26.00**	**183**
Number of FTE Veterinarians						
1.0 or Less	$12.09	$15.50		$19.39	$25.00	54
1.1 to 2.0	$14.00	$20.00		$20.60	$24.94	64
2.1 to 3.0	$13.50	$18.00		$21.53	$28.50	29
3.1 or More	$14.55	$20.00		$22.24	$30.00	31
Member Status						
Accredited Practice Member	$15.63	$22.70		$24.31	$32.00	45
Nonaccredited Member	$12.50	$18.00		$19.23	$26.22	33
Nonmember	$12.00	$16.00		$18.42	$23.38	89
Metropolitan Status						
Urban	*	*		*	*	13
Suburban	$15.00	$21.00		$22.32	$29.00	47
Second City	$12.00	$17.25		$20.64	$27.69	43
Town	$14.14	$18.00		$20.20	$24.75	52
Rural	$12.00	$16.00		$17.53	$22.00	27
Median Area Household Income						
$35,000 or Less	*	*		*	*	23
$35,000 to $49,999	$12.63	$18.88		$19.15	$22.96	64
$50,000 to $69,999	$14.75	$20.00		$22.11	$28.25	62
$70,000 or More	$15.00	$20.50		$22.27	$29.76	30

Note: An asterisk indicates that data was not reported due to an insufficient number of responses.

TABLE 7.45
URINALYSIS: MICROALBUMINARIA

	25th Percentile	Median	Your Data	Average	75th Percentile	Number of Respondents
All Practices	**$20.16**	**$26.25**		**$25.66**	**$30.13**	**85**
Number of FTE Veterinarians						
1.0 or Less	*	*		*	*	24
1.1 to 2.0	$19.08	$25.00		$24.15	$28.85	28
2.1 to 3.0	*	*		*	*	12
3.1 or More	*	*		*	*	19
Member Status						
Accredited Practice Member	$24.69	$27.75		$27.74	$31.86	26
Nonaccredited Member	*	*		*	*	20
Nonmember	$18.40	$25.00		$24.63	$30.00	29
Metropolitan Status						
Urban	*	*		*	*	6
Suburban	*	*		*	*	20
Second City	*	*		*	*	20
Town	$19.50	$23.00		$24.47	$30.86	29
Rural	*	*		*	*	10
Median Area Household Income						
$35,000 or Less	*	*		*	*	14
$35,000 to $49,999	$21.16	$26.50		$26.62	$33.00	27
$50,000 to $69,999	$21.00	$27.75		$25.94	$31.75	34
$70,000 or More	*	*		*	*	8

Note: An asterisk indicates that data was not reported due to an insufficient number of responses.

TABLE 7.46
URINALYSIS: URINE PROTEIN:CREATININE (UP:C) RATIO

	25th Percentile	Median	Your Data	Average	75th Percentile	Number of Respondents
All Practices	$26.25	$38.52		$42.47	$56.00	84
Number of FTE Veterinarians						
1.0 or Less	*	*		*	*	23
1.1 to 2.0	*	*		*	*	22
2.1 to 3.0	*	*		*	*	17
3.1 or More	*	*		*	*	19
Member Status						
Accredited Practice Member	$28.50	$45.00		$47.77	$60.53	25
Nonaccredited Member	*	*		*	*	19
Nonmember	$28.50	$38.20		$40.48	$54.00	35
Metropolitan Status						
Urban	*	*		*	*	4
Suburban	*	*		*	*	17
Second City	*	*		*	*	20
Town	$30.75	$40.63		$41.66	$56.75	28
Rural	*	*		*	*	14
Median Area Household Income						
$35,000 or Less	*	*		*	*	9
$35,000 to $49,999	$23.50	$38.50		$39.21	$54.00	35
$50,000 to $69,999	$31.82	$40.00		$44.76	$60.00	29
$70,000 or More	*	*		*	*	10

Note: An asterisk indicates that data was not reported due to an insufficient number of responses.

LABORATORY TESTS PERFORMED BY OUTSIDE LAB

Note: The markup for all outside lab tests is calculated as follows: Markup = {[(Fee Charged to Client - Fee Paid to Lab) ÷ Fee Paid to Lab] x 100}.

TABLE 7.47
OUTSIDE LAB TEST FEES NOT REPORTED

Arterial/Venous Blood Gases
Avian ELISA Allergy Testing
2 Chemistries
3 Chemistries
4 Chemistries
Chemistry Setup Fee
Cytology: Ear Swab
Cytology: Skin Swab
Cytology: Vaginal
Fecal Dif-Quik Stain
Fecal Examination: Direct Smear
Fecal Examination: Flotation (Gravitational)
Fecal Examination: Giardia Wet Mount
Glucose Curve (6)
Glucose Single
Heartworm Test, Canine: Occult/Antigen Plus Lyme and E. Canis
Uric Acid: Reptiles
Urinalysis: Dipstick
Urinalysis: Sediment
Urinalysis: Specific Gravity

Note: These outside lab test fees were not reported due to an insufficient number of responses.

TABLE 7.48
FEES FOR SELECT OUTSIDE LAB TESTS

	25th Percentile	Median	Your Data	Average	75th Percentile	Number of Respondents
Avian PBFDV	$44.40	$68.00		$80.62	$115.50	49
Avian PBFDV (Fee Paid to Outside Lab)	$16.00	$50.50		$39.59	$55.50	50
Note: based on 45 respondents, the average markup for this lab test is 131%.						
Avian Polyomavirus Test	$43.50	$64.00		$73.64	$90.00	59
Avian Polyomavirus Test (Fee Paid to Outside Lab)	$16.00	$31.00		$33.73	$40.75	59
Note: based on 53 respondents, the average markup for this lab test is 131%.						
CBC with No Differential	$29.06	$34.56		$35.62	$43.78	44
CBC with No Differential (Fee Paid to Outside Lab)	$11.88	$16.00		$16.07	$17.88	53
Note: based on 45 respondents, the average markup for this lab test is 153%.						
1 Chemistry	$16.00	$26.44		$28.03	$35.25	27
1 Chemistry (Fee Paid to Outside Lab)	$7.25	$11.00		$13.79	$11.75	37
Note: based on 30 respondents, the average markup for this lab test is 163%.						
5–7 Chemistries	$41.50	$50.00		$50.08	$61.25	29
5–7 Chemistries (Fee Paid to Outside Lab)	$13.00	$18.40		$19.56	$23.25	31
Note: based on 29 respondents, the average markup for this lab test is 183%.						
Fecal Examination: Sedimentation (Baermann)	$42.40	$72.50		$68.51	$86.40	39
Fecal Examination: Sedimentation (Baermann) (Fee Paid to Outside Lab)	$30.50	$33.75		$31.86	$39.88	48
Note: based on 41 respondents, the average markup for this lab test is 123%.						
Fecal Gram's Stain	$35.13	$42.00		$41.71	$50.63	30
Fecal Gram's Stain (Fee Paid to Outside Lab)	$19.50	$21.00		$20.80	$22.00	37
Note: based on 32 respondents, the average markup for this lab test is 110%.						
Feline Leukemia (FeLV) Test	$27.00	$33.75		$35.92	$42.00	75
Feline Leukemia (FeLV) Test (Fee Paid to Outside Lab)	$12.25	$13.50		$15.39	$14.00	87
Note: based on 73 respondents, the average markup for this lab test is 144%.						
FeLV and FIV Test	$42.00	$46.25		$46.98	$53.90	31
FeLV and FIV Test (Fee Paid to Outside Lab)	$18.75	$22.00		$21.64	$23.25	35
Note: based on 34 respondents, the average markup for this lab test is 128%.						
Uric Acid: Avian	$23.50	$28.00		$34.91	$35.50	27
Uric Acid: Avian (Fee Paid to Outside Lab)	$10.49	$11.75		$15.48	$13.60	37
Note: based on 27 respondents, the average markup for this lab test is 147%.						
Urinalysis: Complete	$29.86	$33.20		$33.78	$39.35	62
Urinalysis: Complete (Fee Paid to Outside Lab)	$8.90	$14.25		$13.66	$14.50	62
Note: based on 57 respondents, the average markup for this lab test is 184%.						
Urinalysis: Microalbuminaria	$24.03	$27.50		$28.32	$32.38	62
Urinalysis: Microalbuminaria (Fee Paid to Outside Lab)	$11.95	$12.75		$14.55	$13.00	73
Note: based on 59 respondents, the average markup for this lab test is 128%.						

Note: Data was reported only for all practices due to an insufficient number of responses.

TABLE 7.49
ACTH STIMULATION

	25th Percentile	Median	Your Data	Average	75th Percentile	Number of Respondents
All Practices	**$92.00**	**$115.50**		**$128.31**	**$160.00**	**423**
Number of FTE Veterinarians						
1.0 or Less	$93.75	$115.50		$130.08	$165.00	98
1.1 to 2.0	$90.00	$111.88		$126.61	$150.00	128
2.1 to 3.0	$93.00	$123.08		$129.56	$162.00	91
3.1 or More	$93.06	$117.50		$130.22	$174.71	92
Member Status						
Accredited Practice Member	$92.19	$112.75		$123.54	$145.00	124
Nonaccredited Member	$91.75	$116.25		$132.92	$170.00	86
Nonmember	$90.00	$116.50		$128.14	$155.50	174
Metropolitan Status						
Urban	$94.13	$118.00		$127.67	$168.75	36
Suburban	$90.00	$110.00		$121.94	$152.00	127
Second City	$94.50	$114.50		$126.63	$149.63	84
Town	$95.88	$126.00		$136.82	$172.75	133
Rural	$89.00	$110.95		$127.60	$169.20	37
Median Area Household Income						
$35,000 or Less	$83.00	$110.00		$129.83	$182.80	33
$35,000 to $49,999	$90.00	$116.00		$125.63	$159.01	152
$50,000 to $69,999	$94.63	$115.73		$127.66	$153.50	148
$70,000 or More	$94.70	$119.88		$136.43	$172.84	80

TABLE 7.50
ACTH STIMULATION (FEE PAID TO OUTSIDE LAB)

	25th Percentile	Median	Your Data	Average	75th Percentile	Number of Respondents
All Practices	**$34.41**	**$45.00**		**$45.05**	**$45.00**	**368**
Number of FTE Veterinarians						
1.0 or Less	$36.00	$45.00		$45.60	$45.00	91
1.1 to 2.0	$35.00	$45.00		$46.56	$53.75	115
2.1 to 3.0	$28.00	$44.00		$42.62	$45.00	72
3.1 or More	$33.00	$45.00		$43.91	$50.75	78
Member Status						
Accredited Practice Member	$28.75	$45.00		$42.81	$45.00	109
Nonaccredited Member	$34.75	$45.00		$47.17	$51.25	78
Nonmember	$35.00	$45.00		$45.34	$47.00	158
Metropolitan Status						
Urban	$42.75	$45.00		$51.77	$60.25	27
Suburban	$28.13	$45.00		$42.96	$45.00	116
Second City	$35.00	$44.00		$45.64	$47.40	68
Town	$36.00	$45.00		$45.85	$45.69	117
Rural	$28.18	$45.00		$42.95	$46.25	34
Median Area Household Income						
$35,000 or Less	$39.00	$45.00		$48.53	$60.06	26
$35,000 to $49,999	$30.00	$45.00		$44.08	$45.75	132
$50,000 to $69,999	$39.50	$45.00		$45.85	$45.00	131
$70,000 or More	$33.25	$45.00		$44.72	$46.00	70

Note: Based on 352 respondents, the average markup for this lab test is 202%.

TABLE 7.51
AVIAN CHLAMYDIA

	25th Percentile	Median	Your Data	Average	75th Percentile	Number of Respondents
All Practices	$48.00	$70.00		$69.93	$91.00	81
Number of FTE Veterinarians						
1.0 or Less	*	*		*	*	10
1.1 to 2.0	*	*		*	*	19
2.1 to 3.0	*	*		*	*	21
3.1 or More	*	*		*	*	24
Member Status						
Accredited Practice Member	$48.00	$72.25		$70.39	$90.50	38
Nonaccredited Member	*	*		*	*	13
Nonmember	$49.00	$60.00		$66.29	$85.65	25
Metropolitan Status						
Urban	*	*		*	*	6
Suburban	*	*		*	*	17
Second City	$49.23	$70.00		$67.53	$80.68	26
Town	$45.75	$73.75		$69.90	$92.00	28
Rural	*	*		*	*	3
Median Area Household Income						
$35,000 or Less	*	*		*	*	7
$35,000 to $49,999	$51.50	$71.00		$69.58	$80.68	30
$50,000 to $69,999	$45.75	$66.25		$67.58	$91.50	28
$70,000 or More	*	*		*	*	12

Note: An asterisk indicates that data was not reported due to an insufficient number of responses.

TABLE 7.52
AVIAN CHLAMYDIA (FEE PAID TO OUTSIDE LAB)

	25th Percentile	Median	Your Data	Average	75th Percentile	Number of Respondents
All Practices	$20.00	$37.25		$34.63	$41.63	81
Number of FTE Veterinarians						
1.0 or Less	*	*		*	*	11
1.1 to 2.0	*	*		*	*	22
2.1 to 3.0	*	*		*	*	21
3.1 or More	*	*		*	*	21
Member Status						
Accredited Practice Member	$20.00	$37.00		$35.73	$41.25	35
Nonaccredited Member	*	*		*	*	16
Nonmember	$20.00	$35.00		$34.23	$47.00	27
Metropolitan Status						
Urban	*	*		*	*	5
Suburban	*	*		*	*	19
Second City	*	*		*	*	23
Town	$20.93	$37.75		$37.86	$47.00	29
Rural	*	*		*	*	4
Median Area Household Income						
$35,000 or Less	*	*		*	*	8
$35,000 to $49,999	$20.00	$37.50		$34.39	$40.00	29
$50,000 to $69,999	$16.00	$35.00		$32.01	$38.63	29
$70,000 or More	*	*		*	*	12

Note 1: Based on 74 respondents, the average markup for this lab test is 127%.
Note 2: An asterisk indicates that data was not reported due to an insufficient number of responses.

TABLE 7.53
AVIAN CHROMOSOMAL SEXING

	25th Percentile	Median	Your Data	Average	75th Percentile	Number of Respondents
All Practices	**$45.00**	**$55.95**		**$71.02**	**$81.50**	**85**
Number of FTE Veterinarians						
1.0 or Less	*	*		*	*	8
1.1 to 2.0	*	*		*	*	21
2.1 to 3.0	*	*		*	*	22
3.1 or More	$45.00	$64.45		$74.43	$89.00	27
Member Status						
Accredited Practice Member	$49.16	$61.40		$79.48	$117.00	43
Nonaccredited Member	*	*		*	*	14
Nonmember	*	*		*	*	24
Metropolitan Status						
Urban	*	*		*	*	7
Suburban	*	*		*	*	18
Second City	*	*		*	*	23
Town	$46.00	$55.00		$69.71	$83.00	31
Rural	*	*		*	*	4
Median Area Household Income						
$35,000 or Less	*	*		*	*	7
$35,000 to $49,999	$44.81	$55.90		$67.75	$86.75	28
$50,000 to $69,999	$45.00	$54.00		$72.72	$106.50	32
$70,000 or More	*	*		*	*	14

Note: An asterisk indicates that data was not reported due to an insufficient number of responses.

TABLE 7.54
AVIAN CHROMOSOMAL SEXING (FEE PAID TO OUTSIDE LAB)

	25th Percentile	Median	Your Data	Average	75th Percentile	Number of Respondents
All Practices	**$20.00**	**$20.00**		**$34.06**	**$61.33**	**81**
Number of FTE Veterinarians						
1.0 or Less	*	*		*	*	10
1.1 to 2.0	*	*		*	*	22
2.1 to 3.0	*	*		*	*	19
3.1 or More	*	*		*	*	24
Member Status						
Accredited Practice Member	$20.00	$28.00		$35.53	$62.65	35
Nonaccredited Member	*	*		*	*	18
Nonmember	$18.75	$20.00		$30.15	$27.50	25
Metropolitan Status						
Urban	*	*		*	*	7
Suburban	*	*		*	*	17
Second City	*	*		*	*	20
Town	$19.75	$22.37		$36.33	$63.00	29
Rural	*	*		*	*	6
Median Area Household Income						
$35,000 or Less	*	*		*	*	9
$35,000 to $49,999	$20.00	$22.37		$32.49	$48.88	25
$50,000 to $69,999	$19.13	$20.00		$30.58	$40.48	30
$70,000 or More	*	*		*	*	14

Note 1: Based on 74 respondents, the average markup for this lab test is 134%.
Note 2: An asterisk indicates that data was not reported due to an insufficient number of responses.

TABLE 7.55
BACTERIAL CULTURE AND SENSITIVITY

	25th Percentile	Median	Your Data	Average	75th Percentile	Number of Respondents
All Practices	**$63.99**	**$76.50**		**$77.95**	**$90.00**	**432**
Number of FTE Veterinarians						
1.0 or Less	$56.75	$71.75		$73.55	$87.63	118
1.1 to 2.0	$60.00	$80.19		$76.95	$89.63	126
2.1 to 3.0	$66.18	$75.75		$79.44	$90.00	88
3.1 or More	$67.08	$79.47		$81.11	$90.51	85
Member Status						
Accredited Practice Member	$64.50	$74.50		$78.35	$89.75	117
Nonaccredited Member	$65.00	$79.49		$78.61	$89.88	88
Nonmember	$62.60	$77.25		$77.26	$90.50	192
Metropolitan Status						
Urban	$62.68	$80.00		$80.63	$91.25	34
Suburban	$66.35	$81.50		$81.43	$90.18	129
Second City	$65.28	$79.00		$79.13	$88.96	85
Town	$61.00	$76.25		$77.48	$90.00	133
Rural	$50.50	$65.00		$66.27	$80.00	45
Median Area Household Income						
$35,000 or Less	$55.00	$68.08		$69.13	$82.75	46
$35,000 to $49,999	$60.50	$73.70		$75.22	$88.00	157
$50,000 to $69,999	$65.00	$80.63		$81.86	$91.50	140
$70,000 or More	$69.50	$81.50		$82.56	$90.35	79

TABLE 7.56
BACTERIAL CULTURE AND SENSITIVITY (FEE PAID TO OUTSIDE LAB)

	25th Percentile	Median	Your Data	Average	75th Percentile	Number of Respondents
All Practices	**$30.00**	**$35.75**		**$37.05**	**$43.50**	**405**
Number of FTE Veterinarians						
1.0 or Less	$25.85	$35.75		$34.94	$43.25	113
1.1 to 2.0	$32.50	$35.75		$36.68	$43.50	126
2.1 to 3.0	$27.55	$35.75		$37.59	$43.50	77
3.1 or More	$31.25	$36.85		$39.06	$45.00	75
Member Status						
Accredited Practice Member	$26.00	$35.75		$36.19	$43.50	107
Nonaccredited Member	$29.81	$38.00		$37.40	$43.50	90
Nonmember	$32.50	$36.00		$37.12	$43.50	182
Metropolitan Status						
Urban	$35.00	$42.75		$41.92	$55.00	31
Suburban	$30.19	$35.75		$36.83	$43.50	118
Second City	$30.00	$35.75		$36.63	$43.50	75
Town	$30.31	$35.88		$36.91	$43.50	128
Rural	$23.50	$35.75		$35.53	$43.50	45
Median Area Household Income						
$35,000 or Less	$26.13	$35.75		$34.83	$42.19	44
$35,000 to $49,999	$28.56	$35.75		$36.73	$43.50	148
$50,000 to $69,999	$30.23	$38.00		$38.01	$43.50	133
$70,000 or More	$34.00	$35.75		$37.17	$43.25	69

Note: Based on 376 respondents, the average markup for this lab test is 119%.

TABLE 7.57
BLADDER STONE ANALYSIS

	25th Percentile	Median	Your Data	Average	75th Percentile	Number of Respondents
All Practices	$46.35	$68.00		$74.95	$100.00	415
Number of FTE Veterinarians						
1.0 or Less	$45.00	$66.00		$72.76	$90.76	103
1.1 to 2.0	$45.00	$66.91		$75.75	$105.00	119
2.1 to 3.0	$44.43	$59.90		$70.27	$89.33	96
3.1 or More	$50.00	$75.00		$79.47	$106.25	85
Member Status						
Accredited Practice Member	$47.29	$67.46		$75.25	$95.00	116
Nonaccredited Member	$46.90	$67.00		$77.89	$110.75	89
Nonmember	$45.00	$65.00		$71.12	$92.75	172
Metropolitan Status						
Urban	$55.00	$84.14		$82.26	$119.00	31
Suburban	$48.55	$69.50		$76.63	$98.25	124
Second City	$48.00	$70.00		$80.68	$116.51	91
Town	$46.80	$65.85		$73.54	$98.00	127
Rural	$33.75	$48.33		$54.67	$70.94	38
Median Area Household Income						
$35,000 or Less	$36.85	$50.00		$63.37	$83.75	48
$35,000 to $49,999	$42.00	$68.00		$72.97	$98.00	147
$50,000 to $69,999	$48.11	$68.50		$77.14	$104.81	142
$70,000 or More	$52.00	$72.20		$81.74	$107.95	69

TABLE 7.58
BLADDER STONE ANALYSIS (FEE PAID TO OUTSIDE LAB)

	25th Percentile	Median	Your Data	Average	75th Percentile	Number of Respondents
All Practices	$23.50	$33.00		$39.68	$62.19	284
Number of FTE Veterinarians						
1.0 or Less	$25.00	$34.50		$40.59	$62.50	71
1.1 to 2.0	$20.38	$33.00		$39.56	$62.06	88
2.1 to 3.0	$21.25	$30.00		$36.23	$60.88	53
3.1 or More	$23.42	$38.50		$40.76	$60.00	66
Member Status						
Accredited Practice Member	$22.88	$33.00		$39.60	$62.25	73
Nonaccredited Member	$23.93	$40.00		$41.94	$62.06	64
Nonmember	$23.50	$32.50		$38.22	$60.00	123
Metropolitan Status						
Urban	*	*		*	*	20
Suburban	$23.50	$34.50		$40.03	$62.25	75
Second City	$22.38	$32.75		$40.19	$62.50	62
Town	$21.50	$32.50		$38.83	$62.00	95
Rural	$20.63	$30.00		$33.43	$42.00	28
Median Area Household Income						
$35,000 or Less	$19.50	$25.00		$35.35	$62.25	27
$35,000 to $49,999	$24.00	$33.00		$39.84	$60.00	107
$50,000 to $69,999	$23.34	$33.58		$39.28	$60.38	101
$70,000 or More	$23.50	$35.00		$41.20	$62.50	43

Note 1: Based on 256 respondents, the average markup for this lab test is 122%.
Note 2: An asterisk indicates that data was not reported due to an insufficient number of responses.

TABLE 7.59
BLOOD PARASITE TEST: BABESIA

	25th Percentile	Median	Your Data	Average	75th Percentile	Number of Respondents
All Practices	**$50.00**	**$102.28**		**$90.45**	**$118.00**	**163**
Number of FTE Veterinarians						
1.0 or Less	$44.30	$97.00		$85.73	$120.00	46
1.1 to 2.0	$50.00	$94.50		$87.62	$116.00	43
2.1 to 3.0	$52.00	$110.50		$96.09	$123.75	36
3.1 or More	$61.35	$90.00		$90.93	$116.75	33
Member Status						
Accredited Practice Member	$51.13	$113.00		$95.60	$125.00	42
Nonaccredited Member	$49.37	$75.00		$82.86	$118.00	31
Nonmember	$48.00	$107.00		$88.68	$115.00	75
Metropolitan Status						
Urban	*	*		*	*	11
Suburban	$65.88	$110.00		$98.57	$125.04	46
Second City	$46.00	$102.28		$88.46	$116.00	35
Town	$58.00	$96.50		$91.30	$118.00	55
Rural	*	*		*	*	11
Median Area Household Income						
$35,000 or Less	*	*		*	*	12
$35,000 to $49,999	$48.00	$86.50		$84.21	$113.00	51
$50,000 to $69,999	$59.44	$109.00		$98.00	$128.94	64
$70,000 or More	$64.88	$105.00		$93.30	$123.33	30

Note: An asterisk indicates that data was not reported due to an insufficient number of responses.

TABLE 7.60
BLOOD PARASITE TEST: BABESIA (FEE PAID TO OUTSIDE LAB)

	25th Percentile	Median	Your Data	Average	75th Percentile	Number of Respondents
All Practices	**$24.50**	**$56.50**		**$47.61**	**$56.50**	**185**
Number of FTE Veterinarians						
1.0 or Less	$19.75	$53.75		$43.78	$56.50	48
1.1 to 2.0	$36.00	$55.60		$46.85	$56.50	48
2.1 to 3.0	$30.00	$56.50		$49.27	$56.50	40
3.1 or More	$24.00	$53.95		$47.61	$56.50	43
Member Status						
Accredited Practice Member	$30.75	$56.50		$50.29	$58.21	48
Nonaccredited Member	$19.75	$56.25		$45.51	$56.50	38
Nonmember	$24.50	$54.00		$45.94	$56.50	85
Metropolitan Status						
Urban	*	*		*	*	14
Suburban	$40.80	$56.50		$48.29	$56.50	55
Second City	$22.25	$54.48		$45.69	$56.50	36
Town	$23.50	$56.50		$48.70	$56.50	61
Rural	*	*		*	*	14
Median Area Household Income						
$35,000 or Less	*	*		*	*	13
$35,000 to $49,999	$24.50	$56.50		$48.71	$56.50	60
$50,000 to $69,999	$23.00	$55.00		$46.50	$56.50	75
$70,000 or More	$31.50	$56.50		$47.13	$56.50	32

Note 1: Based on 162 respondents, the average markup for this lab test is 109%.
Note 2: An asterisk indicates that data was not reported due to an insufficient number of responses.

TABLE 7.61
BLOOD PARASITE TEST: EHRLICHIA

	25th Percentile	Median	Your Data	Average	75th Percentile	Number of Respondents
All Practices	**$68.00**	**$79.50**		**$83.38**	**$96.00**	**206**
Number of FTE Veterinarians						
1.0 or Less	$56.88	$76.00		$76.65	$90.11	49
1.1 to 2.0	$68.50	$78.00		$83.89	$93.75	59
2.1 to 3.0	$75.00	$85.75		$88.53	$99.75	43
3.1 or More	$66.75	$78.50		$84.35	$98.75	49
Member Status						
Accredited Practice Member	$72.60	$84.25		$86.80	$97.45	61
Nonaccredited Member	$68.50	$81.66		$87.97	$99.46	41
Nonmember	$60.00	$75.00		$80.08	$95.00	91
Metropolitan Status						
Urban	*	*		*	*	21
Suburban	$70.88	$86.15		$88.46	$97.80	56
Second City	$65.00	$75.35		$81.39	$93.63	49
Town	$68.00	$79.00		$79.82	$88.10	57
Rural	*	*		*	*	17
Median Area Household Income						
$35,000 or Less	*	*		*	*	18
$35,000 to $49,999	$65.00	$76.75		$81.31	$90.33	70
$50,000 to $69,999	$68.85	$85.68		$84.17	$97.92	72
$70,000 or More	$68.00	$87.20		$90.61	$107.20	39

Note: An asterisk indicates that data was not reported due to an insufficient number of responses.

TABLE 7.62
BLOOD PARASITE TEST: EHRLICHIA (FEE PAID TO OUTSIDE LAB)

	25th Percentile	Median	Your Data	Average	75th Percentile	Number of Respondents
All Practices	**$32.60**	**$39.50**		**$41.25**	**$46.26**	**228**
Number of FTE Veterinarians						
1.0 or Less	$32.60	$38.00		$39.69	$47.00	55
1.1 to 2.0	$33.60	$39.50		$40.22	$43.90	67
2.1 to 3.0	$32.90	$39.00		$40.39	$45.59	46
3.1 or More	$32.50	$39.50		$44.37	$52.76	53
Member Status						
Accredited Practice Member	$33.50	$39.50		$44.67	$51.75	67
Nonaccredited Member	$32.58	$36.75		$38.16	$39.50	50
Nonmember	$32.60	$39.25		$41.03	$49.88	98
Metropolitan Status						
Urban	*	*		*	*	21
Suburban	$34.09	$39.50		$42.79	$48.94	62
Second City	$31.25	$38.50		$41.08	$53.13	49
Town	$34.25	$39.25		$40.27	$39.50	68
Rural	*	*		*	*	22
Median Area Household Income						
$35,000 or Less	*	*		*	*	19
$35,000 to $49,999	$32.80	$39.50		$40.30	$39.75	81
$50,000 to $69,999	$32.67	$37.50		$41.03	$45.00	84
$70,000 or More	$32.60	$39.50		$43.40	$56.38	37

Note 1: Based on 199 respondents, the average markup for this lab test is 108%.
Note 2: An asterisk indicates that data was not reported due to an insufficient number of responses.

TABLE 7.63
BLOOD PARASITE TEST: HAEMOBARTONELLA

	25th Percentile	Median	Your Data	Average	75th Percentile	Number of Respondents
All Practices	**$27.00**	**$34.43**		**$48.21**	**$56.50**	**188**
Number of FTE Veterinarians						
1.0 or Less	$25.00	$30.57		$47.35	$58.50	50
1.1 to 2.0	$27.22	$33.10		$49.41	$57.00	47
2.1 to 3.0	$30.00	$37.50		$51.15	$65.00	43
3.1 or More	$27.00	$33.15		$43.01	$40.50	43
Member Status						
Accredited Practice Member	$28.50	$37.40		$49.04	$57.00	59
Nonaccredited Member	$28.50	$33.88		$44.11	$50.25	33
Nonmember	$25.00	$33.25		$51.27	$66.50	77
Metropolitan Status						
Urban	*	*		*	*	15
Suburban	$25.75	$33.10		$46.78	$57.50	49
Second City	$27.50	$35.00		$49.21	$57.00	39
Town	$27.50	$33.25		$44.80	$46.00	63
Rural	*	*		*	*	18
Median Area Household Income						
$35,000 or Less	*	*		*	*	20
$35,000 to $49,999	$28.25	$37.33		$51.18	$65.00	62
$50,000 to $69,999	$27.00	$33.10		$46.48	$47.80	67
$70,000 or More	$25.00	$35.00		$49.96	$65.25	32

Note: An asterisk indicates that data was not reported due to an insufficient number of responses.

TABLE 7.64
BLOOD PARASITE TEST: HAEMOBARTONELLA (FEE PAID TO OUTSIDE LAB)

	25th Percentile	Median	Your Data	Average	75th Percentile	Number of Respondents
All Practices	**$12.25**	**$14.25**		**$29.20**	**$42.74**	**222**
Number of FTE Veterinarians						
1.0 or Less	$12.38	$14.25		$31.21	$42.90	57
1.1 to 2.0	$11.80	$14.25		$26.90	$42.86	58
2.1 to 3.0	$12.50	$14.25		$30.45	$53.06	46
3.1 or More	$12.00	$13.50		$26.70	$27.00	54
Member Status						
Accredited Practice Member	$12.75	$14.25		$27.11	$36.50	65
Nonaccredited Member	$12.65	$14.25		$30.28	$53.25	45
Nonmember	$12.00	$14.50		$30.73	$49.25	95
Metropolitan Status						
Urban	*	*		*	*	18
Suburban	$12.33	$14.25		$29.15	$42.90	65
Second City	$12.31	$14.25		$29.88	$49.50	44
Town	$12.50	$14.25		$27.13	$30.13	69
Rural	*	*		*	*	22
Median Area Household Income						
$35,000 or Less	*	*		*	*	22
$35,000 to $49,999	$12.50	$14.25		$31.82	$49.69	70
$50,000 to $69,999	$12.33	$14.25		$28.97	$43.58	85
$70,000 or More	$12.19	$13.38		$30.58	$55.75	38

Note 1: Based on 186 respondents, the average markup for this lab test is 131%.
Note 2: An asterisk indicates that data was not reported due to an insufficient number of responses.

TABLE 7.65
CBC AUTOMATED

	25th Percentile	Median	Your Data	Average	75th Percentile	Number of Respondents
All Practices	$32.43	$38.00		$38.19	$44.38	118
Number of FTE Veterinarians						
1.0 or Less	$30.46	$36.00		$37.58	$42.88	34
1.1 to 2.0	$31.90	$37.50		$36.44	$41.00	30
2.1 to 3.0	$35.75	$40.88		$41.61	$47.25	30
3.1 or More	*	*		*	*	21
Member Status						
Accredited Practice Member	$34.38	$39.75		$41.12	$46.74	26
Nonaccredited Member	$38.00	$40.43		$40.54	$45.00	27
Nonmember	$30.00	$35.00		$35.73	$38.50	53
Metropolitan Status						
Urban	*	*		*	*	8
Suburban	$34.50	$38.13		$39.64	$44.53	54
Second City	*	*		*	*	14
Town	$30.75	$38.45		$38.78	$46.05	30
Rural	*	*		*	*	11
Median Area Household Income						
$35,000 or Less	*	*		*	*	9
$35,000 to $49,999	$28.10	$35.05		$34.75	$42.25	27
$50,000 to $69,999	$34.88	$38.50		$39.44	$45.25	46
$70,000 or More	$33.75	$38.25		$39.16	$44.19	33

Note: An asterisk indicates that data was not reported due to an insufficient number of responses.

TABLE 7.66
CBC AUTOMATED (FEE PAID TO OUTSIDE LAB)

	25th Percentile	Median	Your Data	Average	75th Percentile	Number of Respondents
All Practices	$11.38	$16.25		$15.86	$18.00	110
Number of FTE Veterinarians						
1.0 or Less	$14.75	$17.00		$16.55	$19.19	28
1.1 to 2.0	$11.50	$16.25		$15.90	$19.25	31
2.1 to 3.0	$9.30	$16.88		$16.05	$17.75	28
3.1 or More	*	*		*	*	19
Member Status						
Accredited Practice Member	$9.30	$15.38		$16.10	$18.88	28
Nonaccredited Member	$13.35	$17.00		$16.25	$19.00	28
Nonmember	$11.25	$16.13		$15.45	$17.81	46
Metropolitan Status						
Urban	*	*		*	*	9
Suburban	$9.30	$16.00		$14.86	$17.75	51
Second City	*	*		*	*	11
Town	$12.06	$17.75		$17.47	$19.25	30
Rural	*	*		*	*	9
Median Area Household Income						
$35,000 or Less	*	*		*	*	9
$35,000 to $49,999	*	*		*	*	21
$50,000 to $69,999	$12.01	$16.25		$15.36	$18.13	45
$70,000 or More	$9.30	$16.63		$15.21	$18.13	32

Note 1: Based on 103 respondents, the average markup for this lab test is 165%.
Note 2: An asterisk indicates that data was not reported due to an insufficient number of responses.

TABLE 7.67
CBC WITH 8–12 CHEMISTRIES

	25th Percentile	Median	Your Data	Average	75th Percentile	Number of Respondents
All Practices	$66.84	$78.50		$80.40	$91.22	193
Number of FTE Veterinarians						
1.0 or Less	$65.00	$78.20		$83.22	$94.58	46
1.1 to 2.0	$67.69	$77.50		$75.29	$85.91	52
2.1 to 3.0	$65.00	$81.70		$81.47	$97.54	40
3.1 or More	$69.75	$83.58		$83.20	$91.44	47
Member Status						
Accredited Practice Member	$67.38	$79.64		$80.65	$94.58	58
Nonaccredited Member	$66.38	$78.00		$83.81	$96.70	41
Nonmember	$67.07	$79.38		$79.54	$86.14	78
Metropolitan Status						
Urban	*	*		*	*	13
Suburban	$68.50	$81.25		$83.97	$97.53	61
Second City	$64.25	$75.05		$79.19	$92.50	39
Town	$68.81	$79.00		$80.41	$88.13	62
Rural	*	*		*	*	15
Median Area Household Income						
$35,000 or Less	*	*		*	*	15
$35,000 to $49,999	$65.00	$74.65		$76.24	$88.00	63
$50,000 to $69,999	$71.65	$82.00		$83.97	$94.40	75
$70,000 or More	$68.00	$80.77		$84.88	$94.10	35

Note: An asterisk indicates that data was not reported due to an insufficient number of responses.

TABLE 7.68
CBC WITH 8–12 CHEMISTRIES (FEE PAID TO OUTSIDE LAB)

	25th Percentile	Median	Your Data	Average	75th Percentile	Number of Respondents
All Practices	$23.75	$31.25		$31.15	$39.25	197
Number of FTE Veterinarians						
1.0 or Less	$25.75	$37.50		$33.76	$41.00	47
1.1 to 2.0	$23.33	$28.60		$29.86	$39.25	50
2.1 to 3.0	$22.25	$29.75		$30.36	$36.25	43
3.1 or More	$22.56	$31.37		$31.33	$39.25	48
Member Status						
Accredited Practice Member	$23.50	$31.13		$31.80	$40.75	64
Nonaccredited Member	$24.88	$31.77		$31.55	$39.31	42
Nonmember	$23.63	$30.00		$30.78	$39.00	76
Metropolitan Status						
Urban	*	*		*	*	14
Suburban	$20.95	$29.75		$30.19	$39.13	61
Second City	$23.63	$29.13		$30.95	$38.38	38
Town	$24.88	$32.75		$32.15	$39.69	62
Rural	*	*		*	*	17
Median Area Household Income						
$35,000 or Less	*	*		*	*	17
$35,000 to $49,999	$24.00	$28.60		$29.71	$38.00	61
$50,000 to $69,999	$22.75	$32.80		$31.69	$39.25	79
$70,000 or More	$24.33	$29.38		$30.64	$41.00	34

Note 1: Based on 181 respondents, the average markup for this lab test is 172%.
Note 2: An asterisk indicates that data was not reported due to an insufficient number of responses.

TABLE 7.69
CBC WITH 16–24 CHEMISTRIES AND T4

	25th Percentile	Median	Your Data	Average	75th Percentile	Number of Respondents
All Practices	**$88.50**	**$101.50**		**$106.36**	**$120.00**	**347**
Number of FTE Veterinarians						
1.0 or Less	$88.00	$97.12		$107.87	$125.00	85
1.1 to 2.0	$91.20	$101.70		$104.56	$120.00	101
2.1 to 3.0	$85.00	$105.40		$110.11	$130.00	75
3.1 or More	$86.50	$100.00		$103.23	$118.78	74
Member Status						
Accredited Practice Member	$91.25	$103.25		$108.61	$123.81	96
Nonaccredited Member	$87.50	$106.00		$108.95	$123.00	67
Nonmember	$87.25	$99.85		$104.60	$119.75	153
Metropolitan Status						
Urban	$100.28	$113.90		$113.92	$123.13	29
Suburban	$89.00	$110.00		$113.46	$132.26	113
Second City	$82.88	$98.13		$103.53	$116.74	69
Town	$87.45	$97.31		$101.39	$113.13	101
Rural	$86.75	$95.00		$96.14	$112.00	31
Median Area Household Income						
$35,000 or Less	$77.25	$97.13		$96.41	$117.03	28
$35,000 to $49,999	$88.00	$98.00		$104.59	$118.73	125
$50,000 to $69,999	$90.00	$101.85		$108.37	$122.89	121
$70,000 or More	$91.00	$102.50		$110.37	$128.25	65

TABLE 7.70
CBC WITH 16–24 CHEMISTRIES AND T4 (FEE PAID TO OUTSIDE LAB)

	25th Percentile	Median	Your Data	Average	75th Percentile	Number of Respondents
All Practices	**$32.35**	**$42.25**		**$41.03**	**$46.25**	**324**
Number of FTE Veterinarians						
1.0 or Less	$34.00	$43.25		$41.43	$46.25	81
1.1 to 2.0	$33.00	$43.25		$41.50	$46.25	93
2.1 to 3.0	$28.51	$41.25		$40.99	$46.25	67
3.1 or More	$31.35	$39.00		$39.69	$46.25	71
Member Status						
Accredited Practice Member	$28.50	$42.00		$40.23	$46.25	91
Nonaccredited Member	$33.98	$43.25		$41.13	$46.25	69
Nonmember	$32.06	$42.00		$40.77	$46.25	140
Metropolitan Status						
Urban	$40.25	$46.00		$44.66	$52.00	29
Suburban	$32.35	$43.25		$40.72	$46.25	97
Second City	$27.00	$37.00		$37.57	$46.56	62
Town	$34.81	$42.13		$42.25	$46.25	100
Rural	$34.00	$42.25		$42.31	$46.25	31
Median Area Household Income						
$35,000 or Less	$36.00	$43.63		$43.12	$51.25	26
$35,000 to $49,999	$28.50	$41.00		$40.35	$46.25	117
$50,000 to $69,999	$33.50	$42.25		$41.51	$46.25	117
$70,000 or More	$31.26	$41.13		$40.13	$46.25	58

Note: Based on 303 respondents, the average markup for this lab test is 170%.

TABLE 7.71
CBC WITH MANUAL DIFFERENTIAL

	25th Percentile	Median	Your Data	Average	75th Percentile	Number of Respondents
All Practices	$32.88	$38.00		$38.92	$45.03	166
Number of FTE Veterinarians						
1.0 or Less	$32.38	$37.47		$38.71	$43.75	40
1.1 to 2.0	$32.13	$37.49		$37.05	$41.17	64
2.1 to 3.0	$35.13	$43.00		$42.76	$49.56	32
3.1 or More	$29.47	$36.56		$37.10	$45.28	25
Member Status						
Accredited Practice Member	$35.12	$40.50		$41.33	$46.76	40
Nonaccredited Member	$33.50	$40.00		$40.03	$45.40	33
Nonmember	$31.00	$36.50		$36.81	$40.00	78
Metropolitan Status						
Urban	*	*		*	*	14
Suburban	$33.29	$39.00		$40.25	$46.07	49
Second City	$30.10	$38.00		$38.86	$46.93	29
Town	$35.00	$38.00		$38.86	$43.10	63
Rural	*	*		*	*	9
Median Area Household Income						
$35,000 or Less	*	*		*	*	16
$35,000 to $49,999	$31.36	$37.05		$37.90	$43.00	60
$50,000 to $69,999	$32.81	$38.00		$39.38	$46.96	58
$70,000 or More	$36.30	$41.00		$42.14	$46.26	31

Note 1: An asterisk indicates that data was not reported due to an insufficient number of responses.

TABLE 7.72
CBC WITH MANUAL DIFFERENTIAL (FEE PAID TO OUTSIDE LAB)

	25th Percentile	Median	Your Data	Average	75th Percentile	Number of Respondents
All Practices	$12.00	$16.75		$15.77	$18.25	174
Number of FTE Veterinarians						
1.0 or Less	$12.00	$16.50		$15.93	$18.19	44
1.1 to 2.0	$12.06	$16.75		$15.30	$17.75	64
2.1 to 3.0	$8.86	$15.63		$15.70	$18.50	34
3.1 or More	$14.08	$17.63		$16.54	$19.25	26
Member Status						
Accredited Practice Member	$9.30	$16.00		$14.51	$18.35	45
Nonaccredited Member	$15.00	$17.75		$17.22	$19.06	34
Nonmember	$10.94	$16.50		$15.36	$17.75	82
Metropolitan Status						
Urban	*	*		*	*	16
Suburban	$13.15	$16.75		$15.63	$18.25	53
Second City	$7.50	$16.00		$14.23	$16.75	28
Town	$12.25	$16.75		$15.45	$17.75	64
Rural	*	*		*	*	12
Median Area Household Income						
$35,000 or Less	*	*		*	*	16
$35,000 to $49,999	$9.30	$16.75		$16.38	$19.25	61
$50,000 to $69,999	$12.88	$16.25		$15.62	$17.94	64
$70,000 or More	$12.50	$17.75		$15.74	$18.25	31

Note 1: Based on 156 respondents, the average markup for this lab test is 167%.
Note 2: An asterisk indicates that data was not reported due to an insufficient number of responses.

TABLE 7.73
8–12 CHEMISTRIES

	25th Percentile	Median	Your Data	Average	75th Percentile	Number of Respondents
All Practices	**$50.00**	**$58.13**		**$61.38**	**$72.00**	**71**
Number of FTE Veterinarians						
1.0 or Less	*	*		*	*	21
1.1 to 2.0	*	*		*	*	18
2.1 to 3.0	*	*		*	*	14
3.1 or More	*	*		*	*	17
Member Status						
Accredited Practice Member	*	*		*	*	13
Nonaccredited Member	*	*		*	*	22
Nonmember	$41.25	$54.73		$57.22	$71.13	28
Metropolitan Status						
Urban	*	*		*	*	8
Suburban	$50.00	$58.13		$58.85	$71.50	25
Second City	*	*		*	*	16
Town	*	*		*	*	17
Rural	*	*		*	*	3
Median Area Household Income						
$35,000 or Less	*	*		*	*	4
$35,000 to $49,999	$50.78	$60.00		$63.29	$75.00	25
$50,000 to $69,999	$46.50	$55.00		$59.45	$70.00	30
$70,000 or More	*	*		*	*	11

Note: An asterisk indicates that data was not reported due to an insufficient number of responses.

TABLE 7.74
8–12 CHEMISTRIES (FEE PAID TO OUTSIDE LAB)

	25th Percentile	Median	Your Data	Average	75th Percentile	Number of Respondents
All Practices	**$19.23**	**$22.50**		**$24.16**	**$29.51**	**70**
Number of FTE Veterinarians						
1.0 or Less	*	*		*	*	19
1.1 to 2.0	*	*		*	*	18
2.1 to 3.0	*	*		*	*	13
3.1 or More	*	*		*	*	17
Member Status						
Accredited Practice Member	*	*		*	*	13
Nonaccredited Member	*	*		*	*	21
Nonmember	$15.85	$21.88		$24.10	$32.38	28
Metropolitan Status						
Urban	*	*		*	*	5
Suburban	$16.34	$22.25		$22.53	$28.25	25
Second City	*	*		*	*	16
Town	*	*		*	*	19
Rural	*	*		*	*	3
Median Area Household Income						
$35,000 or Less	*	*		*	*	5
$35,000 to $49,999	*	*		*	*	18
$50,000 to $69,999	$19.95	$22.25		$24.01	$26.50	33
$70,000 or More	*	*		*	*	13

Note 1: Based on 66 respondents, the average markup for this lab test is 179%.
Note 2: An asterisk indicates that data was not reported due to an insufficient number of responses.

TABLE 7.75
16–24 CHEMISTRIES

	25th Percentile	Median	Your Data	Average	75th Percentile	Number of Respondents
All Practices	$63.38	$75.58		$78.09	$92.00	242
Number of FTE Veterinarians						
1.0 or Less	$57.98	$75.00		$75.99	$90.25	66
1.1 to 2.0	$60.88	$75.75		$74.89	$89.75	72
2.1 to 3.0	$65.00	$75.00		$80.71	$97.75	52
3.1 or More	$66.75	$76.88		$80.29	$89.18	45
Member Status						
Accredited Practice Member	$66.75	$78.00		$82.49	$97.50	65
Nonaccredited Member	$69.30	$79.29		$82.64	$92.00	51
Nonmember	$57.76	$74.00		$73.13	$85.22	109
Metropolitan Status						
Urban	*	*		*	*	23
Suburban	$65.75	$78.13		$79.52	$90.00	82
Second City	$65.25	$75.00		$83.18	$106.63	53
Town	$63.50	$76.00		$77.16	$92.90	59
Rural	*	*		*	*	21
Median Area Household Income						
$35,000 or Less	$60.00	$75.00		$74.47	$92.25	25
$35,000 to $49,999	$60.00	$75.00		$76.64	$92.00	87
$50,000 to $69,999	$66.50	$78.00		$79.41	$90.00	81
$70,000 or More	$65.38	$77.94		$79.98	$93.38	46

Note: An asterisk indicates that data was not reported due to an insufficient number of responses.

TABLE 7.76
16–24 CHEMISTRIES (FEE PAID TO OUTSIDE LAB)

	25th Percentile	Median	Your Data	Average	75th Percentile	Number of Respondents
All Practices	$23.19	$33.00		$31.93	$37.00	240
Number of FTE Veterinarians						
1.0 or Less	$23.93	$33.00		$31.66	$37.00	64
1.1 to 2.0	$22.06	$34.90		$32.36	$37.00	70
2.1 to 3.0	$20.10	$33.00		$31.94	$37.00	53
3.1 or More	$24.64	$30.00		$30.94	$36.89	46
Member Status						
Accredited Practice Member	$21.25	$30.25		$30.64	$37.00	66
Nonaccredited Member	$28.56	$34.90		$34.24	$37.00	48
Nonmember	$20.30	$33.00		$30.94	$37.00	108
Metropolitan Status						
Urban	*	*		*	*	22
Suburban	$24.50	$32.00		$31.39	$37.00	79
Second City	$20.00	$33.00		$32.43	$37.00	53
Town	$25.00	$35.00		$32.73	$37.00	59
Rural	*	*		*	*	23
Median Area Household Income						
$35,000 or Less	$17.38	$34.63		$31.73	$37.00	26
$35,000 to $49,999	$20.25	$33.00		$31.44	$37.00	85
$50,000 to $69,999	$23.75	$33.00		$31.97	$37.00	82
$70,000 or More	$28.31	$33.00		$33.04	$37.00	44

Note 1: Based on 226 respondents, the average markup for this lab test is 173%.
Note 2: An asterisk indicates that data was not reported due to an insufficient number of responses.

TABLE 7.77
CYTOLOGY: FINE-NEEDLE ASPIRATE

	25th Percentile	Median	Your Data	Average	75th Percentile	Number of Respondents
All Practices	**$59.63**	**$75.00**		**$74.76**	**$92.73**	**102**
Number of FTE Veterinarians						
1.0 or Less	$57.00	$68.80		$72.81	$100.00	27
1.1 to 2.0	$48.88	$77.25		$74.39	$94.60	32
2.1 to 3.0	*	*		*	*	23
3.1 or More	*	*		*	*	16
Member Status						
Accredited Practice Member	*	*		*	*	20
Nonaccredited Member	*	*		*	*	24
Nonmember	$48.88	$70.00		$68.21	$85.69	48
Metropolitan Status						
Urban	*	*		*	*	11
Suburban	$62.75	$78.00		$79.31	$95.95	33
Second City	*	*		*	*	23
Town	$48.00	$75.00		$71.01	$95.30	31
Rural	*	*		*	*	3
Median Area Household Income						
$35,000 or Less	*	*		*	*	15
$35,000 to $49,999	$52.50	$75.00		$73.13	$92.70	37
$50,000 to $69,999	$65.38	$80.50		$80.54	$96.88	32
$70,000 or More	*	*		*	*	16

Note: An asterisk indicates that data was not reported due to an insufficient number of responses.

TABLE 7.78
CYTOLOGY: FINE-NEEDLE ASPIRATE (FEE PAID TO OUTSIDE LAB)

	25th Percentile	Median	Your Data	Average	75th Percentile	Number of Respondents
All Practices	**$30.05**	**$43.25**		**$40.39**	**$50.00**	**97**
Number of FTE Veterinarians						
1.0 or Less	$30.00	$42.00		$41.04	$50.00	27
1.1 to 2.0	$25.50	$37.00		$36.51	$50.00	29
2.1 to 3.0	*	*		*	*	20
3.1 or More	*	*		*	*	18
Member Status						
Accredited Practice Member	*	*		*	*	22
Nonaccredited Member	*	*		*	*	23
Nonmember	$30.08	$41.25		$39.65	$50.00	42
Metropolitan Status						
Urban	*	*		*	*	10
Suburban	$30.75	$43.63		$40.79	$50.00	32
Second City	*	*		*	*	21
Town	$30.10	$40.50		$39.38	$50.00	29
Rural	*	*		*	*	4
Median Area Household Income						
$35,000 or Less	*	*		*	*	11
$35,000 to $49,999	$27.19	$44.38		$40.38	$50.44	32
$50,000 to $69,999	$32.00	$40.00		$38.54	$44.75	35
$70,000 or More	*	*		*	*	17

Note 1: Based on 84 respondents, the average markup for this lab test is 102%.
Note 2: An asterisk indicates that data was not reported due to an insufficient number of responses.

TABLE 7.79
DEX SUPPRESSION

	25th Percentile	Median	Your Data	Average	75th Percentile	Number of Respondents
All Practices	**$92.00**	**$116.50**		**$115.58**	**$135.52**	**350**
Number of FTE Veterinarians						
1.0 or Less	$86.25	$118.75		$114.21	$137.72	80
1.1 to 2.0	$95.06	$117.64		$116.74	$137.88	108
2.1 to 3.0	$94.50	$115.00		$114.62	$135.00	75
3.1 or More	$90.92	$117.00		$115.89	$134.71	77
Member Status						
Accredited Practice Member	$92.75	$114.48		$115.64	$136.00	106
Nonaccredited Member	$97.75	$123.40		$123.24	$140.00	71
Nonmember	$90.00	$115.00		$111.50	$130.00	144
Metropolitan Status						
Urban	*	*		*	*	24
Suburban	$90.06	$116.75		$116.02	$134.66	112
Second City	$97.03	$115.50		$118.84	$142.38	68
Town	$95.25	$120.00		$116.31	$137.50	111
Rural	$76.50	$96.98		$105.09	$118.98	30
Median Area Household Income						
$35,000 or Less	$84.38	$112.50		$114.18	$145.25	32
$35,000 to $49,999	$90.00	$113.95		$112.29	$135.60	119
$50,000 to $69,999	$95.00	$120.00		$118.20	$135.00	123
$70,000 or More	$95.25	$117.00		$117.70	$135.50	69

Note: An asterisk indicates that data was not reported due to an insufficient number of responses.

TABLE 7.80
DEX SUPPRESSION (FEE PAID TO OUTSIDE LAB)

	25th Percentile	Median	Your Data	Average	75th Percentile	Number of Respondents
All Practices	**$38.96**	**$55.25**		**$49.97**	**$58.25**	**324**
Number of FTE Veterinarians						
1.0 or Less	$38.95	$55.25		$50.15	$58.25	75
1.1 to 2.0	$37.70	$55.25		$49.63	$58.25	103
2.1 to 3.0	$36.10	$50.00		$48.49	$58.25	64
3.1 or More	$40.50	$54.23		$50.85	$58.25	73
Member Status						
Accredited Practice Member	$38.88	$53.00		$49.32	$58.25	97
Nonaccredited Member	$36.10	$55.25		$50.31	$60.00	69
Nonmember	$39.38	$55.25		$50.18	$58.25	138
Metropolitan Status						
Urban	$45.00	$58.25		$54.37	$60.25	27
Suburban	$36.10	$53.25		$47.96	$58.25	101
Second City	$41.00	$55.25		$50.73	$58.69	58
Town	$42.31	$55.25		$51.33	$58.25	102
Rural	$34.90	$45.00		$47.05	$57.53	29
Median Area Household Income						
$35,000 or Less	$36.00	$50.00		$48.04	$58.69	26
$35,000 to $49,999	$40.10	$55.25		$50.19	$58.25	111
$50,000 to $69,999	$38.00	$55.23		$50.19	$58.25	120
$70,000 or More	$40.00	$55.25		$49.62	$58.25	59

Note: Based on 302 respondents, the average markup for this lab test is 131%.

TABLE 7.81
ELECTROLYTES

	25th Percentile	Median	Your Data	Average	75th Percentile	Number of Respondents
All Practices	**$32.00**	**$48.00**		**$44.69**	**$56.00**	**115**
Number of FTE Veterinarians						
1.0 or Less	$33.26	$43.89		$45.17	$60.50	33
1.1 to 2.0	$25.75	$50.00		$42.79	$56.13	37
2.1 to 3.0	$38.00	$48.50		$45.59	$54.56	27
3.1 or More	*	*		*	*	16
Member Status						
Accredited Practice Member	*	*		*	*	19
Nonaccredited Member	$38.00	$48.00		$45.80	$56.00	27
Nonmember	$31.25	$44.50		$43.12	$52.00	59
Metropolitan Status						
Urban	*	*		*	*	11
Suburban	$32.32	$51.00		$47.99	$58.29	38
Second City	*	*		*	*	23
Town	$31.63	$45.00		$43.20	$56.13	32
Rural	*	*		*	*	9
Median Area Household Income						
$35,000 or Less	*	*		*	*	10
$35,000 to $49,999	$31.13	$48.00		$43.04	$54.28	37
$50,000 to $69,999	$38.00	$48.50		$46.70	$63.00	43
$70,000 or More	*	*		*	*	22

Note 1: Assumed sodium, chloride, and potassium were tested.
Note 2: An asterisk indicates that data was not reported due to an insufficient number of responses.

TABLE 7.82
ELECTROLYTES (FEE PAID TO OUTSIDE LAB)

	25th Percentile	Median	Your Data	Average	75th Percentile	Number of Respondents
All Practices	**$14.63**	**$24.63**		**$20.00**	**$25.50**	**114**
Number of FTE Veterinarians						
1.0 or Less	$16.25	$24.63		$20.60	$25.50	32
1.1 to 2.0	$12.75	$24.25		$19.81	$25.50	37
2.1 to 3.0	*	*		*	*	24
3.1 or More	*	*		*	*	19
Member Status						
Accredited Practice Member	*	*		*	*	19
Nonaccredited Member	$11.00	$25.20		$19.96	$25.50	27
Nonmember	$14.63	$22.88		$19.68	$25.50	58
Metropolitan Status						
Urban	*	*		*	*	12
Suburban	$15.99	$25.50		$20.34	$25.50	38
Second City	*	*		*	*	22
Town	$14.63	$24.00		$19.89	$25.50	30
Rural	*	*		*	*	9
Median Area Household Income						
$35,000 or Less	*	*		*	*	12
$35,000 to $49,999	$16.45	$25.13		$21.08	$25.50	32
$50,000 to $69,999	$14.73	$24.25		$20.26	$25.50	45
$70,000 or More	*	*		*	*	21

Note 1: Based on 105 respondents, the average markup for this lab test is 136%.
Note 2: An asterisk indicates that data was not reported due to an insufficient number of responses.

TABLE 7.83
FECAL EXAMINATION: FLOTATION (CENTRIFUGATION, ZINC SULFATE)

	25th Percentile	Median	Your Data	Average	75th Percentile	Number of Respondents
All Practices	$19.00	$25.00		$25.58	$29.63	85
Number of FTE Veterinarians						
1.0 or Less	*	*		*	*	16
1.1 to 2.0	*	*		*	*	19
2.1 to 3.0	*	*		*	*	21
3.1 or More	$19.00	$25.00		$25.54	$29.50	27
Member Status						
Accredited Practice Member	$18.50	$25.00		$24.60	$29.50	37
Nonaccredited Member	*	*		*	*	18
Nonmember	*	*		*	*	22
Metropolitan Status						
Urban	*	*		*	*	10
Suburban	$20.00	$25.00		$25.86	$30.00	31
Second City	*	*		*	*	10
Town	$20.45	$26.88		$27.02	$30.00	29
Rural	*	*		*	*	4
Median Area Household Income						
$35,000 or Less	*	*		*	*	4
$35,000 to $49,999	$17.36	$24.50		$23.40	$28.20	38
$50,000 to $69,999	$20.00	$24.70		$26.04	$29.75	27
$70,000 or More	*	*		*	*	15

Note: An asterisk indicates that data was not reported due to an insufficient number of responses.

TABLE 7.84
FECAL EXAMINATION: FLOTATION (CENTRIFUGATION, ZINC SULFATE) (FEE PAID TO OUTSIDE LAB)

	25th Percentile	Median	Your Data	Average	75th Percentile	Number of Respondents
All Practices	$7.00	$9.42		$13.74	$15.00	98
Number of FTE Veterinarians						
1.0 or Less	*	*		*	*	23
1.1 to 2.0	*	*		*	*	20
2.1 to 3.0	*	*		*	*	19
3.1 or More	$7.00	$9.42		$11.65	$14.75	32
Member Status						
Accredited Practice Member	$6.62	$8.00		$11.44	$14.75	40
Nonaccredited Member	$7.00	$13.25		$15.39	$22.88	25
Nonmember	$6.99	$14.00		$16.19	$26.50	25
Metropolitan Status						
Urban	*	*		*	*	8
Suburban	$6.99	$8.00		$12.25	$14.75	33
Second City	*	*		*	*	17
Town	$7.00	$14.75		$15.71	$21.88	33
Rural	*	*		*	*	5
Median Area Household Income						
$35,000 or Less	*	*		*	*	5
$35,000 to $49,999	$6.97	$8.00		$12.44	$14.75	37
$50,000 to $69,999	$7.00	$13.25		$15.25	$23.38	37
$70,000 or More	*	*		*	*	18

Note 1: Based on 80 respondents, the average markup for this lab test is 177%.
Note 2: An asterisk indicates that data was not reported due to an insufficient number of responses.

TABLE 7.85
FIV TEST

	25th Percentile	Median	Your Data	Average	75th Percentile	Number of Respondents
All Practices	**$39.25**	**$45.50**		**$51.50**	**$55.00**	**133**
Number of FTE Veterinarians						
1.0 or Less	$39.00	$43.00		$46.18	$49.56	25
1.1 to 2.0	$37.63	$46.10		$53.82	$55.31	46
2.1 to 3.0	$36.13	$46.83		$48.51	$52.36	26
3.1 or More	$41.90	$50.00		$54.75	$57.50	31
Member Status						
Accredited Practice Member	$41.50	$50.00		$52.54	$56.75	39
Nonaccredited Member	$40.00	$45.75		$53.67	$50.38	32
Nonmember	$35.00	$42.50		$48.44	$52.95	52
Metropolitan Status						
Urban	*	*		*	*	11
Suburban	$40.25	$46.00		$54.33	$57.13	52
Second City	*	*		*	*	20
Town	$38.00	$42.00		$48.65	$52.80	39
Rural	*	*		*	*	9
Median Area Household Income						
$35,000 or Less	*	*		*	*	8
$35,000 to $49,999	$36.88	$42.00		$47.41	$51.61	40
$50,000 to $69,999	$38.00	$45.50		$52.69	$55.00	51
$70,000 or More	$41.62	$46.20		$51.26	$56.65	33

Note: An asterisk indicates that data was not reported due to an insufficient number of responses.

TABLE 7.86
FIV TEST (FEE PAID TO OUTSIDE LAB)

	25th Percentile	Median	Your Data	Average	75th Percentile	Number of Respondents
All Practices	**$18.25**	**$20.00**		**$24.69**	**$22.75**	**148**
Number of FTE Veterinarians						
1.0 or Less	$18.44	$20.00		$22.76	$22.75	32
1.1 to 2.0	$19.00	$20.00		$26.45	$24.13	49
2.1 to 3.0	$17.73	$20.00		$26.82	$26.83	29
3.1 or More	$16.18	$20.00		$22.66	$22.29	32
Member Status						
Accredited Practice Member	$17.99	$20.00		$25.61	$22.75	46
Nonaccredited Member	$19.50	$20.25		$27.34	$27.00	37
Nonmember	$16.65	$20.00		$22.45	$22.75	53
Metropolitan Status						
Urban	*	*		*	*	12
Suburban	$18.25	$20.00		$25.12	$22.75	59
Second City	*	*		*	*	22
Town	$18.25	$20.00		$23.33	$23.25	42
Rural	*	*		*	*	11
Median Area Household Income						
$35,000 or Less	*	*		*	*	10
$35,000 to $49,999	$18.25	$20.00		$24.65	$24.80	47
$50,000 to $69,999	$17.35	$20.00		$24.46	$22.75	53
$70,000 or More	$19.00	$20.00		$24.06	$22.75	37

Note 1: Based on 123 respondents, the average markup for this lab test is 125%.
Note 2: An asterisk indicates that data was not reported due to an insufficient number of responses.

TABLE 7.87
FRUCTOSAMINE TEST

	25th Percentile	Median	Your Data	Average	75th Percentile	Number of Respondents
All Practices	**$50.00**	**$55.00**		**$54.99**	**$64.25**	**376**
Number of FTE Veterinarians						
1.0 or Less	$47.13	$52.50		$54.81	$62.50	82
1.1 to 2.0	$50.00	$55.50		$54.78	$64.26	111
2.1 to 3.0	$48.00	$55.00		$54.26	$65.00	87
3.1 or More	$50.00	$55.00		$55.29	$60.50	83
Member Status						
Accredited Practice Member	$49.35	$55.10		$54.83	$64.50	117
Nonaccredited Member	$50.00	$57.25		$55.61	$64.95	82
Nonmember	$48.13	$53.00		$54.45	$62.25	149
Metropolitan Status						
Urban	$49.50	$50.50		$52.25	$56.77	26
Suburban	$47.63	$54.25		$53.16	$64.38	124
Second City	$50.00	$54.13		$56.35	$63.94	80
Town	$50.00	$56.50		$56.63	$65.00	112
Rural	$50.00	$55.00		$55.70	$63.00	31
Median Area Household Income						
$35,000 or Less	$47.88	$54.00		$55.64	$64.69	34
$35,000 to $49,999	$49.20	$53.55		$53.76	$60.70	129
$50,000 to $69,999	$48.38	$55.00		$54.99	$65.00	134
$70,000 or More	$51.25	$55.38		$56.81	$65.63	72

TABLE 7.88
FRUCTOSAMINE TEST (FEE PAID TO OUTSIDE LAB)

	25th Percentile	Median	Your Data	Average	75th Percentile	Number of Respondents
All Practices	**$23.81**	**$26.25**		**$24.12**	**$26.25**	**356**
Number of FTE Veterinarians						
1.0 or Less	$24.00	$26.25		$25.26	$26.25	79
1.1 to 2.0	$25.00	$26.25		$24.07	$26.25	110
2.1 to 3.0	$23.38	$26.25		$23.50	$26.25	77
3.1 or More	$21.50	$25.00		$23.48	$26.25	79
Member Status						
Accredited Practice Member	$21.00	$26.25		$23.19	$26.25	110
Nonaccredited Member	$23.88	$26.25		$24.14	$26.25	81
Nonmember	$24.00	$26.00		$24.59	$26.25	145
Metropolitan Status						
Urban	$21.13	$26.25		$23.82	$26.25	29
Suburban	$20.00	$25.00		$22.63	$26.25	115
Second City	$24.00	$26.25		$24.89	$26.25	71
Town	$25.00	$26.25		$25.15	$26.25	103
Rural	$25.00	$26.25		$24.78	$26.25	33
Median Area Household Income						
$35,000 or Less	$25.00	$26.25		$25.68	$26.25	32
$35,000 to $49,999	$24.00	$26.25		$24.48	$26.25	125
$50,000 to $69,999	$21.69	$25.80		$23.32	$26.25	126
$70,000 or More	$23.75	$26.25		$23.99	$26.25	67

Note: Based on 337 respondents, the average markup for this lab test is 128%.

TABLE 7.89
FUNGAL CULTURE

	25th Percentile	Median	Your Data	Average	75th Percentile	Number of Respondents
All Practices	$47.00	$74.30		$68.01	$85.65	121
Number of FTE Veterinarians						
1.0 or Less	$45.00	$70.00		$65.66	$86.35	35
1.1 to 2.0	$49.00	$74.78		$67.72	$86.30	38
2.1 to 3.0	*	*		*	*	21
3.1 or More	*	*		*	*	23
Member Status						
Accredited Practice Member	$41.05	$60.00		$62.00	$77.50	33
Nonaccredited Member	$68.06	$79.25		$75.76	$92.25	28
Nonmember	$55.00	$72.00		$68.66	$80.00	51
Metropolitan Status						
Urban	*	*		*	*	9
Suburban	$58.00	$78.00		$73.86	$87.83	41
Second City	*	*		*	*	23
Town	$53.00	$73.90		$70.00	$82.00	30
Rural	*	*		*	*	16
Median Area Household Income						
$35,000 or Less	*	*		*	*	13
$35,000 to $49,999	$42.40	$66.25		$62.95	$76.50	47
$50,000 to $69,999	$53.50	$78.50		$74.09	$93.19	33
$70,000 or More	*	*		*	*	23

Note: An asterisk indicates that data was not reported due to an insufficient number of responses.

TABLE 7.90
FUNGAL CULTURE (FEE PAID TO OUTSIDE LAB)

	25th Percentile	Median	Your Data	Average	75th Percentile	Number of Respondents
All Practices	$25.44	$38.25		$34.41	$39.25	128
Number of FTE Veterinarians						
1.0 or Less	$20.50	$37.15		$32.45	$39.25	39
1.1 to 2.0	$22.00	$38.25		$34.65	$39.25	38
2.1 to 3.0	*	*		*	*	22
3.1 or More	*	*		*	*	22
Member Status						
Accredited Practice Member	$22.00	$35.00		$30.60	$38.25	35
Nonaccredited Member	$37.18	$38.25		$36.55	$39.25	32
Nonmember	$29.25	$38.25		$35.70	$39.25	53
Metropolitan Status						
Urban	*	*		*	*	10
Suburban	$34.81	$38.25		$36.91	$39.25	40
Second City	$29.70	$38.25		$34.70	$39.25	27
Town	$25.00	$37.63		$33.72	$39.19	32
Rural	*	*		*	*	16
Median Area Household Income						
$35,000 or Less	*	*		*	*	12
$35,000 to $49,999	$22.75	$36.70		$33.39	$39.00	48
$50,000 to $69,999	$34.75	$38.25		$34.76	$39.25	39
$70,000 or More	*	*		*	*	24

Note 1: Based on 115 respondents, the average markup for this lab test is 111%.
Note 2: An asterisk indicates that data was not reported due to an insufficient number of responses.

TABLE 7.91
GIARDIA ANTIGEN TEST

	25th Percentile	Median	Your Data	Average	75th Percentile	Number of Respondents
All Practices	**$47.63**	**$57.50**		**$55.86**	**$69.10**	**130**
Number of FTE Veterinarians						
1.0 or Less	$35.75	$55.75		$51.83	$65.14	26
1.1 to 2.0	$55.00	$58.42		$57.74	$71.88	34
2.1 to 3.0	$48.42	$59.93		$57.62	$71.13	32
3.1 or More	$33.75	$56.00		$52.72	$65.00	31
Member Status						
Accredited Practice Member	$46.82	$55.75		$55.12	$66.98	42
Nonaccredited Member	$43.38	$61.05		$57.05	$71.90	27
Nonmember	$48.50	$57.50		$56.36	$69.38	47
Metropolitan Status						
Urban	*	*		*	*	9
Suburban	$49.58	$57.50		$58.84	$71.89	45
Second City	*	*		*	*	22
Town	$55.00	$58.50		$59.96	$71.28	37
Rural	*	*		*	*	14
Median Area Household Income						
$35,000 or Less	*	*		*	*	9
$35,000 to $49,999	$34.00	$56.00		$52.80	$63.50	43
$50,000 to $69,999	$53.94	$59.25		$57.75	$70.75	54
$70,000 or More	*	*		*	*	21

Note: An asterisk indicates that data was not reported due to an insufficient number of responses.

TABLE 7.92
GIARDIA ANTIGEN TEST (FEE PAID TO OUTSIDE LAB)

	25th Percentile	Median	Your Data	Average	75th Percentile	Number of Respondents
All Practices	**$25.00**	**$28.75**		**$25.48**	**$28.75**	**134**
Number of FTE Veterinarians						
1.0 or Less	$14.00	$27.75		$24.25	$28.75	31
1.1 to 2.0	$26.13	$28.75		$25.89	$28.75	33
2.1 to 3.0	$26.00	$28.75		$26.95	$29.08	33
3.1 or More	$21.90	$27.75		$24.47	$28.75	31
Member Status						
Accredited Practice Member	$22.00	$28.75		$25.10	$28.80	43
Nonaccredited Member	$26.30	$28.75		$25.51	$28.75	32
Nonmember	$25.25	$28.38		$25.91	$28.75	48
Metropolitan Status						
Urban	*	*		*	*	8
Suburban	$25.75	$28.75		$26.15	$29.08	49
Second City	*	*		*	*	23
Town	$27.50	$28.75		$27.45	$28.75	37
Rural	*	*		*	*	15
Median Area Household Income						
$35,000 or Less	*	*		*	*	8
$35,000 to $49,999	$24.25	$28.00		$25.10	$28.75	48
$50,000 to $69,999	$26.63	$28.75		$25.76	$29.08	53
$70,000 or More	*	*		*	*	23

Note 1: Based on 120 respondents, the average markup for this lab test is 121%.
Note 2: An asterisk indicates that data was not reported due to an insufficient number of responses.

TABLE 7.93
HEARTWORM TEST, CANINE: OCCULT/ANTIGEN

	25th Percentile	Median	Your Data	Average	75th Percentile	Number of Respondents
All Practices	**$24.58**	**$29.25**		**$28.79**	**$33.63**	**126**
Number of FTE Veterinarians						
1.0 or Less	$25.00	$29.53		$31.09	$36.00	31
1.1 to 2.0	$21.56	$29.18		$28.03	$32.00	40
2.1 to 3.0	*	*		*	*	23
3.1 or More	$20.75	$29.37		$28.95	$34.43	26
Member Status						
Accredited Practice Member	$27.25	$30.00		$30.47	$37.25	38
Nonaccredited Member	$23.30	$26.00		$28.17	$30.00	27
Nonmember	$20.20	$29.53		$27.92	$33.50	47
Metropolitan Status						
Urban	*	*		*	*	14
Suburban	$25.71	$30.00		$30.25	$35.32	46
Second City	*	*		*	*	22
Town	$21.25	$29.50		$29.38	$35.00	35
Rural	*	*		*	*	8
Median Area Household Income						
$35,000 or Less	*	*		*	*	7
$35,000 to $49,999	$18.19	$25.75		$26.47	$30.98	36
$50,000 to $69,999	$22.13	$29.25		$29.09	$35.08	50
$70,000 or More	$26.00	$30.00		$29.75	$34.00	31

Note: An asterisk indicates that data was not reported due to an insufficient number of responses.

TABLE 7.94
HEARTWORM TEST, CANINE: OCCULT/ANTIGEN (FEE PAID TO OUTSIDE LAB)

	25th Percentile	Median	Your Data	Average	75th Percentile	Number of Respondents
All Practices	**$4.00**	**$7.00**		**$6.78**	**$8.00**	**129**
Number of FTE Veterinarians						
1.0 or Less	$4.00	$6.75		$6.20	$7.50	31
1.1 to 2.0	$3.93	$7.00		$7.31	$8.13	42
2.1 to 3.0	*	*		*	*	23
3.1 or More	$4.40	$7.50		$6.71	$8.00	27
Member Status						
Accredited Practice Member	$3.25	$6.50		$6.35	$8.00	41
Nonaccredited Member	$4.25	$7.50		$6.97	$8.50	29
Nonmember	$4.29	$7.00		$6.76	$7.50	48
Metropolitan Status						
Urban	*	*		*	*	14
Suburban	$3.68	$5.38		$6.19	$7.50	50
Second City	*	*		*	*	19
Town	$6.00	$7.50		$7.19	$8.00	33
Rural	*	*		*	*	11
Median Area Household Income						
$35,000 or Less	*	*		*	*	7
$35,000 to $49,999	$4.00	$7.50		$7.47	$8.88	36
$50,000 to $69,999	$4.23	$7.00		$6.41	$8.00	53
$70,000 or More	$3.55	$6.75		$6.49	$7.50	32

Note 1: Based on 120 respondents, the average markup for this lab test is 411%.
Note 2: An asterisk indicates that data was not reported due to an insufficient number of responses.

TABLE 7.95
HEARTWORM TEST, FELINE: OCCULT/ANTIBODY

	25th Percentile	Median	Your Data	Average	75th Percentile	Number of Respondents
All Practices	**$37.30**	**$51.00**		**$49.77**	**$61.00**	**179**
Number of FTE Veterinarians						
1.0 or Less	$38.00	$50.00		$48.51	$59.50	44
1.1 to 2.0	$34.35	$51.00		$48.15	$60.16	58
2.1 to 3.0	$36.35	$48.00		$48.74	$60.26	37
3.1 or More	$39.13	$53.00		$52.08	$63.56	36
Member Status						
Accredited Practice Member	$42.33	$51.00		$51.65	$62.75	56
Nonaccredited Member	$39.00	$52.00		$51.22	$63.00	35
Nonmember	$35.00	$50.00		$47.16	$58.20	77
Metropolitan Status						
Urban	*	*		*	*	16
Suburban	$37.50	$50.00		$49.65	$63.00	53
Second City	$32.40	$46.33		$46.08	$56.63	32
Town	$40.00	$51.44		$51.92	$63.00	63
Rural	*	*		*	*	11
Median Area Household Income						
$35,000 or Less	*	*		*	*	17
$35,000 to $49,999	$36.85	$50.00		$48.29	$59.25	64
$50,000 to $69,999	$36.00	$51.00		$51.18	$63.00	63
$70,000 or More	$37.69	$51.50		$50.61	$63.19	30

Note: An asterisk indicates that data was not reported due to an insufficient number of responses.

TABLE 7.96
HEARTWORM TEST, FELINE: OCCULT/ANTIBODY (FEE PAID TO OUTSIDE LAB)

	25th Percentile	Median	Your Data	Average	75th Percentile	Number of Respondents
All Practices	**$15.00**	**$25.50**		**$22.63**	**$26.75**	**189**
Number of FTE Veterinarians						
1.0 or Less	$17.25	$25.00		$22.12	$25.50	44
1.1 to 2.0	$14.99	$25.50		$22.59	$26.50	63
2.1 to 3.0	$14.00	$22.25		$20.60	$25.50	39
3.1 or More	$17.06	$25.50		$24.14	$31.50	36
Member Status						
Accredited Practice Member	$15.00	$25.50		$23.01	$29.50	59
Nonaccredited Member	$15.00	$25.50		$23.09	$27.25	37
Nonmember	$15.19	$24.63		$22.30	$27.28	82
Metropolitan Status						
Urban	*	*		*	*	17
Suburban	$14.25	$25.50		$22.04	$25.50	55
Second City	$14.10	$24.25		$20.72	$25.50	31
Town	$16.75	$25.50		$24.29	$31.50	66
Rural	*	*		*	*	15
Median Area Household Income						
$35,000 or Less	*	*		*	*	19
$35,000 to $49,999	$18.50	$25.50		$23.55	$27.50	65
$50,000 to $69,999	$15.00	$25.50		$22.36	$29.53	68
$70,000 or More	$13.40	$25.50		$21.29	$25.50	32

Note 1: Based on 163 respondents, the average markup for this lab test is 135%.
Note 2: An asterisk indicates that data was not reported due to an insufficient number of responses.

TABLE 7.97
HEARTWORM TEST, FELINE: OCCULT/ANTIGEN

	25th Percentile	Median	Your Data	Average	75th Percentile	Number of Respondents
All Practices	$21.51	$29.00		$29.50	$35.00	157
Number of FTE Veterinarians						
1.0 or Less	$25.00	$28.50		$30.69	$37.50	40
1.1 to 2.0	$18.75	$26.00		$27.39	$32.50	55
2.1 to 3.0	$20.63	$31.00		$29.72	$37.63	28
3.1 or More	$19.63	$29.63		$31.64	$36.88	28
Member Status						
Accredited Practice Member	$18.94	$30.00		$30.98	$39.88	48
Nonaccredited Member	$20.00	$26.30		$28.96	$36.50	31
Nonmember	$24.00	$28.25		$29.22	$34.86	64
Metropolitan Status						
Urban	*	*		*	*	15
Suburban	$18.75	$30.00		$28.93	$34.80	51
Second City	*	*		*	*	22
Town	$21.42	$27.00		$29.53	$36.00	55
Rural	*	*		*	*	10
Median Area Household Income						
$35,000 or Less	*	*		*	*	14
$35,000 to $49,999	$24.29	$29.88		$30.55	$37.60	58
$50,000 to $69,999	$20.05	$28.00		$28.69	$34.19	52
$70,000 or More	$22.08	$30.00		$28.64	$32.25	29

Note: An asterisk indicates that data was not reported due to an insufficient number of responses.

TABLE 7.98
HEARTWORM TEST, FELINE: OCCULT/ANTIGEN (FEE PAID TO OUTSIDE LAB)

	25th Percentile	Median	Your Data	Average	75th Percentile	Number of Respondents
All Practices	$7.25	$7.50		$10.43	$13.00	169
Number of FTE Veterinarians						
1.0 or Less	$6.94	$7.50		$10.15	$13.00	38
1.1 to 2.0	$7.50	$7.50		$10.73	$13.00	61
2.1 to 3.0	$6.75	$7.50		$9.72	$13.00	33
3.1 or More	$7.50	$7.50		$10.92	$13.00	30
Member Status						
Accredited Practice Member	$7.00	$7.50		$10.65	$13.00	55
Nonaccredited Member	$7.50	$7.50		$9.67	$13.00	33
Nonmember	$7.00	$7.50		$10.80	$13.00	69
Metropolitan Status						
Urban	*	*		*	*	16
Suburban	$7.50	$7.50		$9.49	$13.00	52
Second City	$6.73	$7.50		$10.08	$13.00	25
Town	$7.50	$7.75		$11.44	$13.04	58
Rural	*	*		*	*	14
Median Area Household Income						
$35,000 or Less	*	*		*	*	15
$35,000 to $49,999	$7.00	$7.50		$10.13	$13.00	62
$50,000 to $69,999	$7.00	$7.50		$10.32	$13.00	57
$70,000 or More	$7.50	$7.50		$9.58	$13.00	31

Note 1: Based on 145 respondents, the average markup for this lab test is 219%.
Note 2: An asterisk indicates that data was not reported due to an insufficient number of responses.

TABLE 7.99
HEARTWORM TEST, FELINE: OCCULT/ANTIBODY AND OCCULT/ANTIGEN

	25th Percentile	Median	Your Data	Average	75th Percentile	Number of Respondents
All Practices	$53.00	$62.25		$64.46	$74.60	189
Number of FTE Veterinarians						
1.0 or Less	$58.50	$66.07		$70.11	$77.75	36
1.1 to 2.0	$50.98	$62.78		$62.22	$73.61	66
2.1 to 3.0	$49.75	$58.75		$59.35	$70.46	38
3.1 or More	$51.83	$61.88		$64.65	$74.31	38
Member Status						
Accredited Practice Member	$53.60	$62.00		$64.02	$73.25	61
Nonaccredited Member	$52.53	$68.00		$66.48	$77.00	36
Nonmember	$51.14	$60.00		$62.99	$73.25	78
Metropolitan Status						
Urban	*	*		*	*	17
Suburban	$55.15	$65.00		$66.34	$77.00	66
Second City	$47.95	$58.75		$61.13	$73.23	32
Town	$55.75	$62.88		$64.81	$76.25	62
Rural	*	*		*	*	9
Median Area Household Income						
$35,000 or Less	*	*		*	*	16
$35,000 to $49,999	$52.35	$60.60		$63.40	$73.93	64
$50,000 to $69,999	$52.00	$65.00		$65.08	$77.00	71
$70,000 or More	$56.27	$62.13		$67.72	$79.50	34

Note: An asterisk indicates that data was not reported due to an insufficient number of responses.

TABLE 7.100
HEARTWORM TEST, FELINE: OCCULT/ANTIBODY AND OCCULT/ANTIGEN (FEE PAID TO OUTSIDE LAB)

	25th Percentile	Median	Your Data	Average	75th Percentile	Number of Respondents
All Practices	$26.50	$29.25		$28.93	$36.38	200
Number of FTE Veterinarians						
1.0 or Less	$27.94	$29.25		$30.86	$38.50	42
1.1 to 2.0	$25.50	$29.25		$28.00	$31.75	73
2.1 to 3.0	$21.25	$29.25		$27.34	$29.50	38
3.1 or More	$23.19	$29.25		$29.07	$37.63	38
Member Status						
Accredited Practice Member	$25.00	$29.25		$28.96	$33.40	59
Nonaccredited Member	$26.38	$29.25		$28.80	$37.13	42
Nonmember	$25.56	$29.25		$28.63	$35.63	84
Metropolitan Status						
Urban	*	*		*	*	19
Suburban	$26.50	$29.25		$29.01	$38.50	69
Second City	$22.94	$29.25		$27.69	$30.50	29
Town	$26.50	$29.25		$30.09	$37.25	66
Rural	*	*		*	*	14
Median Area Household Income						
$35,000 or Less	*	*		*	*	19
$35,000 to $49,999	$27.75	$29.25		$29.51	$33.10	62
$50,000 to $69,999	$25.50	$29.25		$29.04	$38.50	81
$70,000 or More	$21.81	$29.25		$28.34	$36.13	34

Note 1: Based on 182 respondents, the average markup for this lab test is 124%.
Note 2: An asterisk indicates that data was not reported due to an insufficient number of responses.

TABLE 7.101
HISTOPATHOLOGY: MULTIPLE TISSUES

	25th Percentile	Median	Your Data	Average	75th Percentile	Number of Respondents
All Practices	$75.00	$101.88		$104.29	$135.00	382
Number of FTE Veterinarians						
1.0 or Less	$65.00	$97.00		$99.60	$136.63	92
1.1 to 2.0	$68.00	$101.00		$101.17	$135.00	121
2.1 to 3.0	$85.25	$104.00		$107.93	$141.69	82
3.1 or More	$79.00	$105.70		$104.48	$130.00	79
Member Status						
Accredited Practice Member	$71.88	$108.00		$105.80	$135.00	109
Nonaccredited Member	$67.00	$95.00		$100.00	$140.69	70
Nonmember	$75.00	$100.43		$103.35	$130.00	164
Metropolitan Status						
Urban	$90.50	$117.52		$114.69	$141.50	29
Suburban	$70.00	$109.00		$107.51	$147.66	111
Second City	$74.82	$100.00		$102.68	$138.13	76
Town	$75.50	$102.85		$104.34	$129.50	119
Rural	$65.00	$88.50		$92.94	$110.00	42
Median Area Household Income						
$35,000 or Less	$70.00	$108.00		$104.69	$138.00	39
$35,000 to $49,999	$71.88	$93.00		$93.47	$113.04	138
$50,000 to $69,999	$70.50	$109.74		$109.51	$146.13	134
$70,000 or More	$94.00	$110.50		$117.31	$150.00	63

TABLE 7.102
HISTOPATHOLOGY: MULTIPLE TISSUES (FEE PAID TO OUTSIDE LAB)

	25th Percentile	Median	Your Data	Average	75th Percentile	Number of Respondents
All Practices	$34.00	$52.00		$51.97	$70.00	363
Number of FTE Veterinarians						
1.0 or Less	$33.50	$50.00		$51.51	$70.25	83
1.1 to 2.0	$30.00	$52.00		$49.54	$69.00	119
2.1 to 3.0	$42.69	$51.50		$52.04	$70.56	72
3.1 or More	$36.00	$51.00		$52.66	$66.88	81
Member Status						
Accredited Practice Member	$36.00	$51.00		$51.41	$70.00	103
Nonaccredited Member	$27.61	$54.90		$51.16	$72.75	73
Nonmember	$36.00	$50.75		$52.72	$72.00	162
Metropolitan Status						
Urban	$46.75	$60.88		$58.28	$74.00	28
Suburban	$29.75	$51.50		$50.06	$67.35	102
Second City	$35.25	$52.00		$52.93	$66.63	68
Town	$38.25	$52.85		$52.94	$72.75	121
Rural	$30.00	$48.50		$49.78	$60.00	38
Median Area Household Income						
$35,000 or Less	$45.00	$53.50		$56.27	$73.06	42
$35,000 to $49,999	$34.00	$49.00		$48.71	$64.25	135
$50,000 to $69,999	$34.25	$55.00		$53.62	$72.75	121
$70,000 or More	$35.65	$55.95		$54.09	$72.75	57

Note: Based on 331 respondents, the average markup for this lab test is 109%.

TABLE 7.103
HISTOPATHOLOGY: SINGLE TISSUE

	25th Percentile	Median	Your Data	Average	75th Percentile	Number of Respondents
All Practices	$75.00	$90.00		$90.51	$107.85	447
Number of FTE Veterinarians						
1.0 or Less	$65.00	$85.00		$86.09	$107.00	111
1.1 to 2.0	$78.38	$92.13		$91.73	$108.00	140
2.1 to 3.0	$78.50	$90.77		$92.95	$106.13	98
3.1 or More	$76.99	$88.00		$90.13	$100.00	88
Member Status						
Accredited Practice Member	$80.60	$92.00		$93.91	$110.50	123
Nonaccredited Member	$75.10	$90.00		$91.99	$110.00	91
Nonmember	$73.38	$88.00		$87.88	$104.60	197
Metropolitan Status						
Urban	$82.25	$94.50		$95.64	$110.13	34
Suburban	$80.00	$93.25		$94.04	$109.50	128
Second City	$75.00	$85.75		$90.09	$107.89	94
Town	$75.00	$90.00		$88.97	$105.50	138
Rural	$65.00	$82.00		$82.98	$99.50	49
Median Area Household Income						
$35,000 or Less	$61.60	$77.00		$80.55	$97.60	49
$35,000 to $49,999	$75.00	$86.00		$87.11	$102.00	167
$50,000 to $69,999	$80.00	$93.00		$95.44	$111.58	149
$70,000 or More	$82.59	$93.50		$95.08	$112.57	73

TABLE 7.104
HISTOPATHOLOGY: SINGLE TISSUE (FEE PAID TO OUTSIDE LAB)

	25th Percentile	Median	Your Data	Average	75th Percentile	Number of Respondents
All Practices	$36.00	$45.25		$43.94	$54.00	423
Number of FTE Veterinarians						
1.0 or Less	$34.00	$43.75		$42.30	$51.75	110
1.1 to 2.0	$37.25	$46.75		$44.78	$54.00	133
2.1 to 3.0	$36.00	$45.48		$44.48	$53.81	84
3.1 or More	$35.75	$45.13		$43.24	$54.00	86
Member Status						
Accredited Practice Member	$36.75	$45.95		$44.41	$54.00	111
Nonaccredited Member	$37.13	$47.00		$45.14	$54.00	92
Nonmember	$35.00	$44.00		$43.10	$54.00	193
Metropolitan Status						
Urban	$42.00	$54.00		$50.30	$55.25	31
Suburban	$38.00	$45.95		$44.73	$54.00	122
Second City	$33.31	$45.25		$43.33	$54.00	85
Town	$35.00	$45.00		$43.30	$54.00	138
Rural	$35.00	$40.00		$40.91	$50.06	42
Median Area Household Income						
$35,000 or Less	$30.00	$43.50		$42.04	$54.00	51
$35,000 to $49,999	$34.00	$45.00		$42.63	$53.75	157
$50,000 to $69,999	$38.56	$46.00		$45.40	$54.00	140
$70,000 or More	$39.30	$46.46		$46.23	$54.00	68

Note: Based on 393 respondents, the average markup for this lab test is 112%.

TABLE 7.105
LYME TESTING

	25th Percentile	Median	Your Data	Average	75th Percentile	Number of Respondents
All Practices	$52.00	$68.00		$74.10	$89.25	253
Number of FTE Veterinarians						
1.0 or Less	$50.00	$63.75		$73.34	$87.75	57
1.1 to 2.0	$53.00	$64.76		$71.37	$84.00	79
2.1 to 3.0	$56.45	$72.00		$78.06	$93.45	53
3.1 or More	$49.83	$69.50		$73.98	$89.38	56
Member Status						
Accredited Practice Member	$52.00	$69.55		$74.10	$89.50	74
Nonaccredited Member	$52.95	$74.00		$77.80	$94.11	54
Nonmember	$50.00	$63.13		$70.88	$80.50	102
Metropolitan Status						
Urban	*	*		*	*	22
Suburban	$53.40	$70.00		$77.77	$92.35	73
Second City	$50.00	$64.00		$75.78	$91.74	49
Town	$53.00	$70.10		$74.33	$89.13	78
Rural	$45.00	$62.25		$66.30	$76.81	26
Median Area Household Income						
$35,000 or Less	$45.45	$55.13		$65.83	$80.00	27
$35,000 to $49,999	$49.33	$63.75		$72.38	$84.25	84
$50,000 to $69,999	$60.00	$71.10		$78.82	$91.17	86
$70,000 or More	$57.90	$69.85		$72.22	$89.50	47

Note: An asterisk indicates that data was not reported due to an insufficient number of responses.

TABLE 7.106
LYME TESTING (FEE PAID TO OUTSIDE LAB)

	25th Percentile	Median	Your Data	Average	75th Percentile	Number of Respondents
All Practices	$24.00	$31.50		$36.08	$44.75	264
Number of FTE Veterinarians						
1.0 or Less	$24.00	$29.20		$33.44	$41.44	64
1.1 to 2.0	$24.00	$31.38		$34.23	$44.00	78
2.1 to 3.0	$24.00	$28.40		$37.84	$44.76	53
3.1 or More	$24.00	$33.00		$39.24	$45.00	60
Member Status						
Accredited Practice Member	$24.00	$30.00		$34.91	$42.75	73
Nonaccredited Member	$24.00	$32.50		$37.91	$44.75	61
Nonmember	$24.00	$30.63		$35.74	$44.75	112
Metropolitan Status						
Urban	*	*		*	*	22
Suburban	$24.00	$31.75		$36.37	$44.75	82
Second City	$23.88	$30.00		$35.10	$44.75	50
Town	$24.00	$31.75		$36.47	$44.75	79
Rural	$24.00	$31.50		$34.34	$41.38	25
Median Area Household Income						
$35,000 or Less	$22.94	$24.00		$30.89	$40.69	26
$35,000 to $49,999	$24.00	$27.00		$37.01	$44.75	89
$50,000 to $69,999	$24.00	$31.75		$35.20	$44.56	92
$70,000 or More	$27.00	$34.38		$38.91	$44.75	48

Note 1: Based on 241 respondents, the average markup for this lab test is 119%.
Note 2: An asterisk indicates that data was not reported due to an insufficient number of responses.

LABORATORY SERVICES / 211

TABLE 7.107
PANCREATIC EVALUATION

	25th Percentile	Median	Your Data	Average	75th Percentile	Number of Respondents
All Practices	$60.00	$95.50		$98.12	$130.19	141
Number of FTE Veterinarians						
1.0 or Less	$74.85	$99.76		$103.40	$125.25	42
1.1 to 2.0	$48.88	$91.82		$93.74	$133.81	46
2.1 to 3.0	*	*		*	*	24
3.1 or More	*	*		*	*	24
Member Status						
Accredited Practice Member	$63.00	$122.00		$109.05	$143.88	37
Nonaccredited Member	$50.63	$90.50		$91.35	$133.08	37
Nonmember	$59.10	$95.25		$95.14	$121.18	58
Metropolitan Status						
Urban	*	*		*	*	13
Suburban	$51.81	$103.76		$101.07	$130.82	46
Second City	$62.48	$97.25		$98.94	$130.00	27
Town	$59.50	$93.50		$96.48	$139.29	38
Rural	*	*		*	*	15
Median Area Household Income						
$35,000 or Less	*	*		*	*	18
$35,000 to $49,999	$59.88	$94.50		$97.12	$131.22	50
$50,000 to $69,999	$57.80	$109.58		$100.75	$133.27	46
$70,000 or More	*	*		*	*	23

Note: An asterisk indicates that data was not reported due to an insufficient number of responses.

TABLE 7.108
PANCREATIC EVALUATION (FEE PAID TO OUTSIDE LAB)

	25th Percentile	Median	Your Data	Average	75th Percentile	Number of Respondents
All Practices	$26.78	$47.25		$45.84	$59.50	146
Number of FTE Veterinarians						
1.0 or Less	$35.75	$47.75		$49.92	$61.56	44
1.1 to 2.0	$18.19	$42.00		$42.68	$59.50	46
2.1 to 3.0	$18.31	$47.75		$44.43	$58.00	28
3.1 or More	*	*		*	*	23
Member Status						
Accredited Practice Member	$29.75	$47.13		$46.20	$61.00	32
Nonaccredited Member	$20.00	$46.13		$42.58	$58.00	36
Nonmember	$29.88	$47.50		$46.05	$59.50	69
Metropolitan Status						
Urban	*	*		*	*	12
Suburban	$21.13	$47.75		$46.12	$61.13	45
Second City	$32.38	$45.25		$44.34	$58.00	29
Town	$29.75	$47.00		$47.51	$60.75	41
Rural	*	*		*	*	16
Median Area Household Income						
$35,000 or Less	*	*		*	*	16
$35,000 to $49,999	$30.69	$47.75		$47.75	$62.00	52
$50,000 to $69,999	$19.00	$47.50		$44.42	$59.50	53
$70,000 or More	*	*		*	*	22

Note 1: Based on 133 respondents, the average markup for this lab test is 123%.
Note 2: An asterisk indicates that data was not reported due to an insufficient number of responses.

TABLE 7.109
TRYPSIN-LIKE IMMUNOREACTIVITY (TLI)

	25th Percentile	Median	Your Data	Average	75th Percentile	Number of Respondents
All Practices	$82.50	$100.00		$98.48	$114.00	315
Number of FTE Veterinarians						
1.0 or Less	$75.00	$100.00		$93.41	$110.00	75
1.1 to 2.0	$91.50	$100.00		$102.56	$115.00	89
2.1 to 3.0	$73.84	$100.00		$94.02	$114.38	65
3.1 or More	$87.00	$100.00		$101.05	$112.50	76
Member Status						
Accredited Practice Member	$84.00	$100.00		$99.94	$117.98	96
Nonaccredited Member	$81.84	$100.00		$100.44	$119.25	66
Nonmember	$79.83	$100.00		$95.10	$109.75	129
Metropolitan Status						
Urban	*	*		*	*	24
Suburban	$84.25	$100.00		$98.64	$113.88	104
Second City	$75.00	$100.00		$96.36	$113.25	66
Town	$83.35	$100.00		$101.24	$114.75	93
Rural	*	*		*	*	24
Median Area Household Income						
$35,000 or Less	$61.68	$95.00		$85.28	$101.25	25
$35,000 to $49,999	$79.83	$100.00		$95.39	$110.00	106
$50,000 to $69,999	$90.00	$100.00		$101.53	$117.15	115
$70,000 or More	$77.50	$105.60		$102.10	$121.75	65

Note: An asterisk indicates that data was not reported due to an insufficient number of responses.

TABLE 7.110
TRYPSIN-LIKE IMMUNOREACTIVITY (TLI) (FEE PAID TO OUTSIDE LAB)

	25th Percentile	Median	Your Data	Average	75th Percentile	Number of Respondents
All Practices	$44.00	$50.00		$47.26	$50.00	318
Number of FTE Veterinarians						
1.0 or Less	$42.88	$49.50		$45.76	$50.00	74
1.1 to 2.0	$43.94	$50.00		$49.00	$50.00	90
2.1 to 3.0	$31.85	$50.00		$43.91	$50.00	71
3.1 or More	$45.00	$50.00		$49.45	$50.00	74
Member Status						
Accredited Practice Member	$42.50	$50.00		$47.40	$50.00	97
Nonaccredited Member	$45.00	$50.00		$48.96	$50.00	72
Nonmember	$42.50	$50.00		$46.20	$50.00	129
Metropolitan Status						
Urban	*	*		*	*	24
Suburban	$44.00	$50.00		$46.84	$50.00	104
Second City	$41.75	$50.00		$47.38	$50.00	62
Town	$42.50	$50.00		$46.83	$50.00	95
Rural	$45.50	$50.00		$48.80	$50.00	28
Median Area Household Income						
$35,000 or Less	*	*		*	*	22
$35,000 to $49,999	$41.50	$50.00		$46.31	$50.00	106
$50,000 to $69,999	$44.25	$50.00		$47.81	$50.00	120
$70,000 or More	$44.25	$50.00		$49.11	$50.00	64

Note 1: Based on 283 respondents, the average markup for this lab test is 117%.
Note 2: An asterisk indicates that data was not reported due to an insufficient number of responses.

TABLE 7.111
PANCREATIC LIPASE IMMUNOREACTIVITY (PLI)

	25th Percentile	Median	Your Data	Average	75th Percentile	Number of Respondents
All Practices	$71.25	$90.73		$89.41	$104.50	167
Number of FTE Veterinarians						
1.0 or Less	$80.00	$94.50		$91.61	$105.51	38
1.1 to 2.0	$80.00	$95.50		$96.60	$109.40	39
2.1 to 3.0	$65.50	$83.94		$80.33	$95.50	36
3.1 or More	$66.38	$87.75		$88.98	$107.69	48
Member Status						
Accredited Practice Member	$66.40	$89.00		$89.02	$106.50	52
Nonaccredited Member	$69.00	$83.88		$86.03	$97.50	35
Nonmember	$82.25	$95.00		$90.73	$102.25	61
Metropolitan Status						
Urban	*	*		*	*	16
Suburban	$67.70	$90.75		$88.79	$104.13	56
Second City	$73.31	$90.50		$89.78	$109.50	36
Town	$74.06	$90.25		$90.56	$102.63	46
Rural	*	*		*	*	10
Median Area Household Income						
$35,000 or Less	*	*		*	*	11
$35,000 to $49,999	$65.00	$90.50		$86.54	$100.00	54
$50,000 to $69,999	$76.00	$92.00		$92.41	$107.50	71
$70,000 or More	$68.25	$92.25		$92.25	$100.00	28

Note: An asterisk indicates that data was not reported due to an insufficient number of responses.

TABLE 7.112
PANCREATIC LIPASE IMMUNOREACTIVITY (PLI) (FEE PAID TO OUTSIDE LAB)

	25th Percentile	Median	Your Data	Average	75th Percentile	Number of Respondents
All Practices	$32.00	$45.25		$42.86	$47.75	182
Number of FTE Veterinarians						
1.0 or Less	$45.19	$46.43		$44.21	$47.75	42
1.1 to 2.0	$35.00	$47.50		$44.33	$47.75	47
2.1 to 3.0	$25.00	$45.25		$38.74	$47.75	39
3.1 or More	$29.88	$45.25		$43.91	$47.75	49
Member Status						
Accredited Practice Member	$29.75	$45.25		$41.84	$47.75	54
Nonaccredited Member	$29.75	$45.25		$41.82	$47.75	41
Nonmember	$43.75	$46.43		$44.11	$47.75	70
Metropolitan Status						
Urban	*	*		*	*	17
Suburban	$31.25	$45.93		$43.35	$47.75	64
Second City	$29.75	$45.25		$40.86	$47.75	36
Town	$32.00	$45.25		$42.90	$47.75	51
Rural	*	*		*	*	10
Median Area Household Income						
$35,000 or Less	*	*		*	*	12
$35,000 to $49,999	$32.00	$47.00		$42.87	$47.75	59
$50,000 to $69,999	$31.50	$45.25		$42.76	$47.75	78
$70,000 or More	$32.50	$46.25		$43.03	$47.75	29

Note 1: Based on 154 respondents, the average markup for this lab test is 116%.
Note 2: An asterisk indicates that data was not reported due to an insufficient number of responses.

THE VETERINARY FEE REFERENCE / 214

TABLE 7.113
CANINE PANCREATIC LIPASE (CPL)

	25th Percentile	Median	Your Data	Average	75th Percentile	Number of Respondents
All Practices	**$26.00**	**$49.50**		**$59.18**	**$90.00**	**95**
Number of FTE Veterinarians						
1.0 or Less	$29.00	$60.00		$63.78	$95.00	27
1.1 to 2.0	*	*		*	*	23
2.1 to 3.0	*	*		*	*	21
3.1 or More	*	*		*	*	23
Member Status						
Accredited Practice Member	$29.50	$70.00		$67.87	$97.25	27
Nonaccredited Member	*	*		*	*	16
Nonmember	$23.38	$39.00		$49.17	$65.00	46
Metropolitan Status						
Urban	*	*		*	*	7
Suburban	$29.05	$45.10		$53.75	$72.00	33
Second City	*	*		*	*	23
Town	*	*		*	*	24
Rural	*	*		*	*	6
Median Area Household Income						
$35,000 or Less	*	*		*	*	9
$35,000 to $49,999	$25.00	$49.00		$54.15	$71.50	31
$50,000 to $69,999	$27.50	$66.00		$66.80	$95.00	36
$70,000 or More	*	*		*	*	18

Note: An asterisk indicates that data was not reported due to an insufficient number of responses.

TABLE 7.114
CANINE PANCREATIC LIPASE (CPL) (FEE PAID TO OUTSIDE LAB)

	25th Percentile	Median	Your Data	Average	75th Percentile	Number of Respondents
All Practices	**$11.75**	**$29.75**		**$28.99**	**$46.44**	**114**
Number of FTE Veterinarians						
1.0 or Less	$11.75	$22.25		$28.43	$45.81	32
1.1 to 2.0	$11.46	$24.50		$27.98	$46.50	33
2.1 to 3.0	*	*		*	*	22
3.1 or More	$22.00	$29.75		$33.05	$47.75	26
Member Status						
Accredited Practice Member	$13.31	$32.38		$33.71	$47.75	32
Nonaccredited Member	*	*		*	*	24
Nonmember	$11.75	$23.25		$25.15	$35.31	50
Metropolitan Status						
Urban	*	*		*	*	11
Suburban	$11.75	$23.25		$27.32	$45.75	36
Second City	$11.75	$36.25		$34.35	$47.81	26
Town	$11.75	$25.00		$27.64	$46.50	29
Rural	*	*		*	*	9
Median Area Household Income						
$35,000 or Less	*	*		*	*	11
$35,000 to $49,999	$11.75	$24.50		$28.41	$45.25	35
$50,000 to $69,999	$11.75	$29.75		$30.24	$47.75	47
$70,000 or More	*	*		*	*	17

Note 1: Based on 95 respondents, the average markup for this lab test is 128%.
Note 2: An asterisk indicates that data was not reported due to an insufficient number of responses.

TABLE 7.115
SPECIFIC CANINE PANCREATIC LIPASE (SPEC CPL)

	25th Percentile	Median	Your Data	Average	75th Percentile	Number of Respondents
All Practices	$57.68	$64.75		$73.07	$89.25	112
Number of FTE Veterinarians						
1.0 or Less	*	*		*	*	23
1.1 to 2.0	$59.50	$60.00		$75.43	$93.56	28
2.1 to 3.0	*	*		*	*	24
3.1 or More	$58.20	$67.00		$75.53	$97.00	31
Member Status						
Accredited Practice Member	$59.50	$66.94		$75.58	$94.88	44
Nonaccredited Member	$60.00	$69.50		$74.51	$80.03	27
Nonmember	$49.06	$59.75		$65.72	$76.93	36
Metropolitan Status						
Urban	*	*		*	*	7
Suburban	$58.20	$71.50		$74.81	$78.35	35
Second City	$50.00	$63.00		$67.30	$78.00	27
Town	$56.13	$61.50		$73.01	$79.62	32
Rural	*	*		*	*	9
Median Area Household Income						
$35,000 or Less	*	*		*	*	10
$35,000 to $49,999	$55.00	$60.00		$67.92	$76.48	34
$50,000 to $69,999	$58.85	$74.38		$77.95	$95.00	41
$70,000 or More	*	*		*	*	24

Note: An asterisk indicates that data was not reported due to an insufficient number of responses.

TABLE 7.116
SPECIFIC CANINE PANCREATIC LIPASE (SPEC CPL) (FEE PAID TO OUTSIDE LAB)

	25th Percentile	Median	Your Data	Average	75th Percentile	Number of Respondents
All Practices	$29.75	$29.75		$34.52	$35.75	118
Number of FTE Veterinarians						
1.0 or Less	*	*		*	*	20
1.1 to 2.0	$29.75	$29.75		$36.16	$40.00	35
2.1 to 3.0	$29.75	$29.75		$31.25	$29.85	25
3.1 or More	$29.75	$29.75		$36.31	$45.75	32
Member Status						
Accredited Practice Member	$29.75	$29.75		$35.48	$44.44	40
Nonaccredited Member	$29.75	$29.75		$33.77	$32.00	31
Nonmember	$29.75	$29.75		$33.75	$35.00	43
Metropolitan Status						
Urban	*	*		*	*	8
Suburban	$29.75	$29.75		$34.87	$35.00	39
Second City	*	*		*	*	23
Town	$29.75	$29.75		$34.90	$36.50	37
Rural	*	*		*	*	9
Median Area Household Income						
$35,000 or Less	*	*		*	*	10
$35,000 to $49,999	$29.75	$29.75		$34.86	$38.25	35
$50,000 to $69,999	$29.75	$29.75		$34.23	$35.00	47
$70,000 or More	*	*		*	*	24

Note 1: Based on 107 respondents, the average markup for this lab test is 116%.
Note 2: An asterisk indicates that data was not reported due to an insufficient number of responses.

TABLE 7.117
PARVOVIRUS TEST

	25th Percentile	Median	Your Data	Average	75th Percentile	Number of Respondents
All Practices	**$51.25**	**$61.14**		**$61.14**	**$68.57**	**90**
Number of FTE Veterinarians						
1.0 or Less	$57.63	$60.50		$62.62	$73.00	27
1.1 to 2.0	$47.05	$61.00		$58.83	$67.64	32
2.1 to 3.0	*	*		*	*	15
3.1 or More	*	*		*	*	14
Member Status						
Accredited Practice Member	$56.03	$63.10		$62.10	$70.38	29
Nonaccredited Member	*	*		*	*	17
Nonmember	$47.40	$56.44		$56.43	$63.00	32
Metropolitan Status						
Urban	*	*		*	*	4
Suburban	$56.38	$60.50		$61.87	$67.50	29
Second City	*	*		*	*	18
Town	$47.00	$63.10		$61.42	$72.00	31
Rural	*	*		*	*	7
Median Area Household Income						
$35,000 or Less	*	*		*	*	9
$35,000 to $49,999	*	*		*	*	22
$50,000 to $69,999	$55.81	$61.00		$62.43	$68.05	32
$70,000 or More	$48.94	$63.14		$62.22	$73.06	26

Note: An asterisk indicates that data was not reported due to an insufficient number of responses.

TABLE 7.118
PARVOVIRUS TEST (FEE PAID TO OUTSIDE LAB)

	25th Percentile	Median	Your Data	Average	75th Percentile	Number of Respondents
All Practices	**$25.00**	**$30.25**		**$29.38**	**$31.50**	**92**
Number of FTE Veterinarians						
1.0 or Less	$30.00	$30.88		$30.97	$31.50	30
1.1 to 2.0	$23.60	$30.25		$28.47	$31.50	31
2.1 to 3.0	*	*		*	*	12
3.1 or More	*	*		*	*	15
Member Status						
Accredited Practice Member	$30.25	$31.50		$32.62	$32.19	30
Nonaccredited Member	*	*		*	*	21
Nonmember	$23.00	$28.75		$27.18	$31.50	33
Metropolitan Status						
Urban	*	*		*	*	4
Suburban	$26.63	$30.25		$30.29	$31.50	33
Second City	*	*		*	*	13
Town	$27.00	$30.25		$29.04	$31.50	31
Rural	*	*		*	*	9
Median Area Household Income						
$35,000 or Less	*	*		*	*	10
$35,000 to $49,999	*	*		*	*	19
$50,000 to $69,999	$24.00	$30.25		$29.20	$31.50	35
$70,000 or More	$26.34	$30.25		$29.43	$31.50	26

Note 1: Based on 81 respondents, the average markup for this lab test is 120%.
Note 2: An asterisk indicates that data was not reported due to an insufficient number of responses.

TABLE 7.119
PTH ASSAY

	25th Percentile	Median	Your Data	Average	75th Percentile	Number of Respondents
All Practices	$96.00	$132.50		$122.78	$145.50	155
Number of FTE Veterinarians						
1.0 or Less	$90.00	$132.50		$121.99	$150.00	35
1.1 to 2.0	$96.50	$127.00		$121.64	$143.00	43
2.1 to 3.0	$102.38	$138.50		$134.14	$164.06	33
3.1 or More	$87.75	$126.00		$115.07	$144.35	40
Member Status						
Accredited Practice Member	$92.38	$132.50		$124.02	$145.25	52
Nonaccredited Member	$98.50	$132.50		$119.17	$144.63	29
Nonmember	$94.00	$130.00		$123.83	$150.00	63
Metropolitan Status						
Urban	*	*		*	*	16
Suburban	$86.20	$133.00		$119.13	$145.63	49
Second City	*	*		*	*	24
Town	$108.64	$130.50		$125.76	$144.05	54
Rural	*	*		*	*	11
Median Area Household Income						
$35,000 or Less	*	*		*	*	13
$35,000 to $49,999	$101.70	$130.00		$121.48	$142.00	45
$50,000 to $69,999	$91.75	$132.50		$122.22	$150.00	65
$70,000 or More	$104.78	$130.00		$126.42	$147.25	29

Note: An asterisk indicates that data was not reported due to an insufficient number of responses.

TABLE 7.120
PTH ASSAY (FEE PAID TO OUTSIDE LAB)

	25th Percentile	Median	Your Data	Average	75th Percentile	Number of Respondents
All Practices	$49.44	$66.25		$59.80	$66.50	170
Number of FTE Veterinarians						
1.0 or Less	$59.00	$66.25		$60.57	$66.63	42
1.1 to 2.0	$46.00	$66.50		$58.80	$68.38	48
2.1 to 3.0	$62.81	$66.50		$61.78	$68.56	36
3.1 or More	$52.00	$66.00		$58.22	$66.50	39
Member Status						
Accredited Practice Member	$47.31	$66.25		$59.53	$67.19	54
Nonaccredited Member	$46.00	$66.25		$58.19	$66.50	35
Nonmember	$61.00	$66.25		$61.32	$69.25	71
Metropolitan Status						
Urban	*	*		*	*	17
Suburban	$47.75	$66.25		$57.54	$66.50	54
Second City	$54.00	$66.25		$59.20	$69.25	27
Town	$62.75	$66.25		$62.90	$66.50	53
Rural	*	*		*	*	17
Median Area Household Income						
$35,000 or Less	*	*		*	*	16
$35,000 to $49,999	$50.00	$66.25		$60.48	$67.13	50
$50,000 to $69,999	$44.25	$66.25		$57.60	$66.50	71
$70,000 or More	$61.75	$66.38		$61.60	$66.50	30

Note 1: Based on 149 respondents, the average markup for this lab test is 112%.
Note 2: An asterisk indicates that data was not reported due to an insufficient number of responses.

TABLE 7.121
RETICULOCYTE COUNT

	25th Percentile	Median	Your Data	Average	75th Percentile	Number of Respondents
All Practices	**$25.00**	**$27.40**		**$29.50**	**$33.75**	**252**
Number of FTE Veterinarians						
1.0 or Less	$21.33	$26.50		$29.30	$35.23	65
1.1 to 2.0	$25.08	$27.25		$28.40	$31.98	76
2.1 to 3.0	$24.25	$27.35		$30.03	$33.69	52
3.1 or More	$25.00	$28.00		$29.90	$33.75	47
Member Status						
Accredited Practice Member	$25.00	$27.00		$30.68	$34.37	71
Nonaccredited Member	$26.00	$29.00		$30.20	$33.94	54
Nonmember	$24.00	$27.00		$28.25	$33.50	104
Metropolitan Status						
Urban	*	*		*	*	24
Suburban	$24.10	$27.50		$28.82	$34.06	77
Second City	$25.50	$27.30		$30.20	$33.75	47
Town	$25.00	$28.00		$29.15	$33.56	82
Rural	*	*		*	*	18
Median Area Household Income						
$35,000 or Less	*	*		*	*	22
$35,000 to $49,999	$23.80	$26.00		$27.83	$33.50	78
$50,000 to $69,999	$25.50	$29.00		$30.57	$35.00	95
$70,000 or More	$25.40	$29.00		$29.52	$34.50	53

Note: An asterisk indicates that data was not reported due to an insufficient number of responses.

TABLE 7.122
RETICULOCYTE COUNT (FEE PAID TO OUTSIDE LAB)

	25th Percentile	Median	Your Data	Average	75th Percentile	Number of Respondents
All Practices	**$12.00**	**$13.00**		**$12.21**	**$13.50**	**261**
Number of FTE Veterinarians						
1.0 or Less	$11.00	$12.95		$12.07	$13.50	64
1.1 to 2.0	$11.75	$13.00		$12.07	$13.50	79
2.1 to 3.0	$12.00	$13.00		$12.25	$13.50	56
3.1 or More	$12.25	$13.00		$12.38	$13.50	52
Member Status						
Accredited Practice Member	$11.81	$13.00		$12.17	$13.50	80
Nonaccredited Member	$12.00	$13.00		$12.30	$13.50	54
Nonmember	$12.25	$13.00		$12.22	$13.50	107
Metropolitan Status						
Urban	*	*		*	*	22
Suburban	$11.75	$13.00		$11.83	$13.50	80
Second City	$12.13	$13.00		$12.19	$13.50	49
Town	$12.25	$13.00		$12.48	$13.50	82
Rural	*	*		*	*	23
Median Area Household Income						
$35,000 or Less	*	*		*	*	21
$35,000 to $49,999	$11.94	$13.00		$12.29	$13.50	86
$50,000 to $69,999	$12.19	$13.00		$12.29	$13.50	102
$70,000 or More	$11.75	$13.00		$11.99	$13.50	47

Note 1: Based on 235 respondents, the average markup for this lab test is 140%.
Note 2: An asterisk indicates that data was not reported due to an insufficient number of responses.

TABLE 7.123
SERUM TESTING FOR ALLERGEN-SPECIFIC IgE

	25th Percentile	Median	Your Data	Average	75th Percentile	Number of Respondents
All Practices	$168.22	$195.00		$207.24	$238.75	193
Number of FTE Veterinarians						
1.0 or Less	$172.50	$220.00		$219.35	$265.38	44
1.1 to 2.0	$175.00	$196.03		$210.82	$238.75	56
2.1 to 3.0	$159.00	$194.75		$205.70	$235.00	49
3.1 or More	$155.75	$181.69		$193.40	$213.00	38
Member Status						
Accredited Practice Member	$173.70	$196.03		$207.94	$228.56	60
Nonaccredited Member	$169.75	$193.75		$215.80	$282.00	38
Nonmember	$162.00	$198.70		$204.73	$236.80	83
Metropolitan Status						
Urban	*	*		*	*	18
Suburban	$171.63	$208.00		$218.92	$245.85	49
Second City	$156.25	$195.68		$211.48	$259.19	40
Town	$162.28	$186.00		$198.49	$221.13	68
Rural	*	*		*	*	17
Median Area Household Income						
$35,000 or Less	*	*		*	*	19
$35,000 to $49,999	$170.00	$204.00		$217.84	$267.88	77
$50,000 to $69,999	$160.25	$195.00		$198.65	$226.38	64
$70,000 or More	$168.44	$195.00		$207.07	$235.00	31

Note: An asterisk indicates that data was not reported due to an insufficient number of responses.

TABLE 7.124
SERUM TESTING FOR ALLERGEN-SPECIFIC IgE (FEE PAID TO OUTSIDE LAB)

	25th Percentile	Median	Your Data	Average	75th Percentile	Number of Respondents
All Practices	$80.00	$96.25		$103.28	$125.00	170
Number of FTE Veterinarians						
1.0 or Less	$88.70	$98.75		$114.42	$145.00	48
1.1 to 2.0	$80.00	$94.88		$104.17	$133.75	44
2.1 to 3.0	$77.50	$92.50		$97.78	$107.00	43
3.1 or More	$79.50	$96.25		$96.73	$98.13	30
Member Status						
Accredited Practice Member	$77.50	$95.00		$93.55	$99.75	51
Nonaccredited Member	$75.90	$96.25		$102.63	$140.00	36
Nonmember	$85.00	$96.25		$111.36	$137.50	72
Metropolitan Status						
Urban	*	*		*	*	16
Suburban	$83.75	$95.63		$102.36	$128.75	42
Second City	$80.63	$96.25		$113.47	$140.38	36
Town	$80.00	$92.75		$100.42	$110.00	60
Rural	*	*		*	*	14
Median Area Household Income						
$35,000 or Less	*	*		*	*	23
$35,000 to $49,999	$87.50	$96.25		$108.38	$130.00	63
$50,000 to $69,999	$80.00	$90.00		$99.70	$122.50	60
$70,000 or More	*	*		*	*	21

Note 1: Based on 155 respondents, the average markup for this lab test is 113%.
Note 2: An asterisk indicates that data was not reported due to an insufficient number of responses.

TABLE 7.125
T4

	25th Percentile	Median	Your Data	Average	75th Percentile	Number of Respondents
All Practices	**$36.75**	**$42.00**		**$43.16**	**$49.00**	**345**
Number of FTE Veterinarians						
1.0 or Less	$31.40	$40.50		$40.89	$46.37	84
1.1 to 2.0	$39.00	$42.00		$43.48	$48.88	105
2.1 to 3.0	$37.63	$42.00		$42.53	$47.76	73
3.1 or More	$35.75	$43.00		$45.38	$53.09	70
Member Status						
Accredited Practice Member	$39.15	$44.00		$45.70	$50.94	97
Nonaccredited Member	$40.00	$43.00		$44.47	$48.94	72
Nonmember	$30.97	$40.00		$40.47	$47.10	142
Metropolitan Status						
Urban	$38.38	$41.58		$44.67	$50.21	28
Suburban	$37.13	$41.90		$42.33	$48.22	108
Second City	$39.00	$44.52		$44.95	$50.50	63
Town	$36.63	$42.75		$44.07	$50.50	108
Rural	$29.31	$38.25		$38.52	$44.32	32
Median Area Household Income						
$35,000 or Less	$29.31	$36.88		$36.61	$42.75	28
$35,000 to $49,999	$36.00	$42.00		$42.70	$48.73	121
$50,000 to $69,999	$37.00	$43.00		$44.63	$50.75	119
$70,000 or More	$38.84	$42.00		$43.99	$48.81	66

TABLE 7.126
T4 (FEE PAID TO OUTSIDE LAB)

	25th Percentile	Median	Your Data	Average	75th Percentile	Number of Respondents
All Practices	**$13.50**	**$19.00**		**$18.20**	**$20.08**	**320**
Number of FTE Veterinarians						
1.0 or Less	$15.00	$19.13		$17.85	$20.25	78
1.1 to 2.0	$14.50	$18.80		$18.89	$20.25	95
2.1 to 3.0	$12.60	$17.75		$17.09	$19.50	66
3.1 or More	$12.90	$18.75		$18.31	$20.25	70
Member Status						
Accredited Practice Member	$12.60	$18.50		$17.74	$19.50	92
Nonaccredited Member	$15.14	$19.25		$18.83	$20.25	72
Nonmember	$14.50	$18.50		$17.91	$20.00	132
Metropolitan Status						
Urban	$18.50	$19.50		$21.01	$21.00	25
Suburban	$12.60	$18.50		$16.76	$19.50	105
Second City	$12.83	$19.00		$18.80	$20.55	60
Town	$15.75	$19.00		$18.52	$20.05	93
Rural	$15.00	$19.50		$18.57	$20.13	29
Median Area Household Income						
$35,000 or Less	*	*		*	*	23
$35,000 to $49,999	$13.50	$19.25		$17.84	$20.10	111
$50,000 to $69,999	$13.18	$18.50		$18.59	$20.25	109
$70,000 or More	$13.00	$19.00		$17.59	$20.13	65

Note 1: Based on 299 respondents, the average markup for this lab test is 149%.
Note 2: An asterisk indicates that data was not reported due to an insufficient number of responses.

TABLE 7.127
T4, T3, FREE T4, AND FREE T4ED

	25th Percentile	Median	Your Data	Average	75th Percentile	Number of Respondents
All Practices	**$79.00**	**$94.95**		**$101.73**	**$121.00**	**315**
Number of FTE Veterinarians						
1.0 or Less	$75.00	$87.08		$94.73	$114.67	79
1.1 to 2.0	$76.00	$95.00		$100.83	$120.50	101
2.1 to 3.0	$84.00	$102.68		$111.47	$132.38	64
3.1 or More	$84.46	$94.50		$100.76	$117.56	64
Member Status						
Accredited Practice Member	$81.60	$95.00		$104.18	$125.00	95
Nonaccredited Member	$80.50	$100.00		$106.17	$128.22	65
Nonmember	$75.00	$90.00		$96.86	$116.54	133
Metropolitan Status						
Urban	$81.50	$105.43		$108.26	$126.23	26
Suburban	$80.75	$100.00		$108.57	$125.94	85
Second City	$76.00	$87.00		$96.71	$116.50	63
Town	$75.86	$90.64		$97.76	$120.00	95
Rural	$75.00	$87.50		$100.10	$120.00	43
Median Area Household Income						
$35,000 or Less	$67.05	$80.00		$84.98	$102.50	39
$35,000 to $49,999	$75.00	$90.64		$98.24	$118.20	117
$50,000 to $69,999	$83.23	$98.50		$109.06	$127.69	105
$70,000 or More	$84.23	$101.00		$108.38	$128.55	50

TABLE 7.128
T4, T3, FREE T4, AND FREE T4ED (FEE PAID TO OUTSIDE LAB)

	25th Percentile	Median	Your Data	Average	75th Percentile	Number of Respondents
All Practices	**$37.97**	**$45.00**		**$49.72**	**$59.00**	**317**
Number of FTE Veterinarians						
1.0 or Less	$35.00	$42.00		$47.22	$58.25	79
1.1 to 2.0	$36.38	$44.70		$48.46	$58.38	102
2.1 to 3.0	$39.00	$51.10		$52.31	$66.00	63
3.1 or More	$39.75	$48.92		$50.94	$58.25	66
Member Status						
Accredited Practice Member	$37.25	$44.00		$50.39	$63.25	87
Nonaccredited Member	$39.75	$49.00		$50.66	$58.25	79
Nonmember	$37.00	$43.10		$48.45	$58.25	135
Metropolitan Status						
Urban	$39.75	$53.00		$52.74	$58.25	27
Suburban	$39.00	$48.38		$51.02	$58.25	88
Second City	$35.50	$44.00		$48.88	$58.63	57
Town	$37.95	$44.50		$48.55	$59.75	104
Rural	$35.50	$43.00		$48.65	$66.50	37
Median Area Household Income						
$35,000 or Less	$35.00	$42.00		$45.59	$56.25	41
$35,000 to $49,999	$37.75	$45.00		$49.95	$62.63	117
$50,000 to $69,999	$39.00	$48.00		$50.58	$58.25	105
$70,000 or More	$38.00	$49.25		$51.13	$59.50	49

Note: Based on 287 respondents, the average markup for this lab test is 109%.

TABLE 7.129
TSH LEVEL

	25th Percentile	Median	Your Data	Average	75th Percentile	Number of Respondents
All Practices	**$53.59**	**$68.00**		**$65.23**	**$76.23**	**232**
Number of FTE Veterinarians						
1.0 or Less	$48.50	$64.50		$62.42	$75.50	55
1.1 to 2.0	$56.25	$68.00		$65.65	$75.75	65
2.1 to 3.0	$58.00	$68.00		$67.40	$77.19	52
3.1 or More	$54.25	$64.95		$65.32	$76.50	54
Member Status						
Accredited Practice Member	$52.79	$67.00		$65.48	$76.31	70
Nonaccredited Member	$51.75	$67.00		$65.41	$75.38	50
Nonmember	$54.75	$68.00		$64.88	$75.94	93
Metropolitan Status						
Urban	*	*		*	*	21
Suburban	$51.00	$68.00		$65.58	$76.34	73
Second City	$51.06	$63.50		$63.12	$74.15	48
Town	$55.00	$68.00		$65.61	$76.50	67
Rural	*	*		*	*	19
Median Area Household Income						
$35,000 or Less	*	*		*	*	24
$35,000 to $49,999	$55.00	$68.00		$64.76	$75.00	77
$50,000 to $69,999	$53.60	$68.00		$65.73	$76.50	84
$70,000 or More	$59.00	$66.50		$67.76	$77.85	43

Note: An asterisk indicates that data was not reported due to an insufficient number of responses.

TABLE 7.130
TSH LEVEL (FEE PAID TO OUTSIDE LAB)

	25th Percentile	Median	Your Data	Average	75th Percentile	Number of Respondents
All Practices	**$28.00**	**$32.25**		**$30.79**	**$34.00**	**253**
Number of FTE Veterinarians						
1.0 or Less	$25.00	$30.50		$29.10	$34.00	59
1.1 to 2.0	$29.38	$33.13		$31.64	$34.00	74
2.1 to 3.0	$27.63	$32.25		$30.05	$34.00	54
3.1 or More	$28.25	$32.25		$31.90	$34.00	59
Member Status						
Accredited Practice Member	$27.81	$31.00		$30.47	$34.00	74
Nonaccredited Member	$26.75	$31.00		$29.66	$34.00	60
Nonmember	$29.50	$32.25		$31.61	$34.00	102
Metropolitan Status						
Urban	*	*		*	*	22
Suburban	$28.00	$31.00		$30.42	$34.00	79
Second City	$23.88	$30.75		$28.45	$34.00	50
Town	$29.50	$34.00		$33.12	$34.00	74
Rural	*	*		*	*	22
Median Area Household Income						
$35,000 or Less	*	*		*	*	23
$35,000 to $49,999	$29.38	$33.13		$31.00	$34.00	86
$50,000 to $69,999	$28.00	$30.50		$31.02	$34.00	95
$70,000 or More	$29.50	$32.75		$31.61	$34.00	43

Note 1: Based on 221 respondents, the average markup for this lab test is 117%.
Note 2: An asterisk indicates that data was not reported due to an insufficient number of responses.

TABLE 7.131
URINALYSIS: URINE PROTEIN:CREATININE (UP:C) RATIO

	25th Percentile	Median	Your Data	Average	75th Percentile	Number of Respondents
All Practices	$49.00	$59.00		$57.56	$67.75	283
Number of FTE Veterinarians						
1.0 or Less	$45.00	$58.71		$55.43	$63.55	64
1.1 to 2.0	$44.88	$59.00		$55.25	$65.61	84
2.1 to 3.0	$50.88	$60.25		$59.99	$71.25	64
3.1 or More	$50.75	$59.00		$57.99	$68.44	62
Member Status						
Accredited Practice Member	$50.50	$59.00		$58.09	$69.00	87
Nonaccredited Member	$58.50	$62.00		$62.72	$73.13	59
Nonmember	$45.40	$58.00		$53.99	$62.50	113
Metropolitan Status						
Urban	$46.25	$58.50		$55.86	$63.73	25
Suburban	$49.75	$59.50		$57.87	$70.00	97
Second City	$52.00	$60.00		$60.06	$70.80	51
Town	$50.00	$59.68		$57.88	$66.37	87
Rural	*	*		*	*	20
Median Area Household Income						
$35,000 or Less	*	*		*	*	21
$35,000 to $49,999	$41.80	$58.50		$54.25	$63.00	91
$50,000 to $69,999	$50.00	$60.00		$58.99	$70.75	104
$70,000 or More	$51.63	$60.00		$59.35	$68.75	62

Note: An asterisk indicates that data was not reported due to an insufficient number of responses.

TABLE 7.132
URINALYSIS: URINE PROTEIN:CREATININE (UP:C) RATIO (FEE PAID TO OUTSIDE LAB)

	25th Percentile	Median	Your Data	Average	75th Percentile	Number of Respondents
All Practices	$24.00	$29.25		$26.10	$29.50	275
Number of FTE Veterinarians						
1.0 or Less	$22.75	$27.75		$25.40	$29.38	65
1.1 to 2.0	$24.00	$29.25		$25.92	$29.25	82
2.1 to 3.0	$21.82	$29.25		$25.74	$29.50	57
3.1 or More	$24.25	$29.25		$27.10	$29.50	61
Member Status						
Accredited Practice Member	$20.33	$27.75		$25.47	$29.50	81
Nonaccredited Member	$25.00	$29.25		$26.88	$29.50	62
Nonmember	$24.00	$28.00		$25.91	$29.25	111
Metropolitan Status						
Urban	*	*		*	*	22
Suburban	$21.88	$29.25		$25.81	$29.50	96
Second City	$25.16	$28.00		$26.51	$29.50	46
Town	$26.25	$29.25		$27.12	$29.50	89
Rural	*	*		*	*	19
Median Area Household Income						
$35,000 or Less	*	*		*	*	19
$35,000 to $49,999	$22.00	$27.75		$25.10	$29.25	87
$50,000 to $69,999	$24.04	$29.13		$26.34	$29.50	104
$70,000 or More	$23.66	$29.25		$26.44	$29.50	60

Note 1: Based on 260 respondents, the average markup for this lab test is 124%.
Note 2: An asterisk indicates that data was not reported due to an insufficient number of responses.

TABLE 7.133
INCREASES IN MARKUPS ON OUTSIDE LAB SERVICES FROM 2004 TO 2006

Type of Test	Average Markup in 2004	Average Markup in 2006	Statistically Significant Change	Minimum Number of Respondents
ACTH Stimulation	209%	257%	No	353
Avian Chlamydia	121%	390%	No	75
Avian Chromosomal Sexing	134%	137%	No	75
Avian PBFDV	135%	138%	No	46
Avian Polyomavirus Test	137%	140%	No	47
Bladder Stone Analysis	132%	134%	No	262
CBC with No Differential	177%	165%	No	46
CBC Automated	184%	173%	No	105
Health Check: CBC with 16–24 Chemistries and T4	176%	180%	No	310
Health Check: CBC with 8–12 Chemistries	180%	184%	No	186
CBC with Manual Differential	167%	176%	No	159
1 Chemistry	639%	181%	No	31
Cytology: Fine-Needle Aspirate	125%	108%	No	85
Dex Suppression	150%	139%	Yes	310
Fecal Gram's Stain	124%	110%	No	32
Feline Leukemia (FeLV) Test	188%	153%	Yes	75
FIV Test	144%	128%	No	126
Fructosamine Test	130%	133%	No	344
Fungal Culture	126%	111%	No	117
Histopathology: Multiple Tissues	132%	116%	Yes	336
Histopathology: Single Tissue	131%	115%	Yes	398
Trypsin-like Immunoreactivity (TLI)	121%	120%	No	288
PTH Assay	121%	115%	No	150
Serum Testing for Allergen-Specific IgE	104%	125%	No	156
T4	166%	161%	No	309
TSH Level	121%	120%	No	226
Urinalysis: Complete	179%	197%	No	58

Note 1: Yes = statistically significant difference at the .05 level.
Note 2: Each t-test tests the difference between the means of two groups. "Minimum Number of Respondents" is the size of the smaller of the two groups.
Note 3: Data was not reported if there were less than 25 observations in either group.

CHAPTER 8

DIAGNOSTIC AND IMAGING SERVICES

You can offer clients diagnostic services and imaging that help identify problems early and get an accurate diagnosis the first time rather than a wait-and-see approach. Here's how to make recommendations that lead to "yes."

Cindy McClain is back, and she's not happy. Her beloved cat, Ollie, has started urinating in the dining room again. You treated the 12-year-old Siamese male cat for a urinary tract infection last winter, and Cindy is feeding Ollie the therapeutic diet you recommended. But the problem appears to have returned. After examining Ollie, you decide to recommend a CBC, urinalysis, serum chemistry profile, and ultrasound to get an exact diagnosis of the complicated urinary problem. What would you say to Cindy to get her to say yes to these tests?

Ask yourself: "If this was my pet, would I do these tests and why?" advises Dr. Bill Liska, Dipl. ACVS of Gulf Coast Veterinary Specialists in Houston, Texas. "If you can answer this question with confidence and conviction, the rest is easy."

GETTING CLIENTS TO ACCEPT DIAGNOSTIC SERVICES AND IMAGING

Thanks to experiences in human medicine and media exposure, most clients know the same sophisticated diagnostics are available for their pets. Diagnostic services and imaging can help you give clients an accurate diagnosis immediately rather than an educated guess.

Taking a wait-and-see approach or trying various treatments until one finally works often disappoints clients and keeps expenses climbing. Instead, clearly explain the need for tests so clients will agree and you can reach a solution faster. Consider these benefits of testing now rather than later:

- You get a definitive diagnosis immediately.
- You have more treatment choices if a condition is identified early.
- Your client spends less money in the long run because you know how to treat the specific problem.
- Your client has more confidence in your medical expertise.
- Most important, the medical problem is resolved or the client has clear expectations of the prognosis.

To get clients to say yes, describe what the tests will tell you and explain the cost. When describing tests, you might use vendor brochures, models, before-and-after radiographs, customized handouts, posters, and other communication tools. "Say, 'This is what we need to do in order to make a diagnosis and do the best thing for your pet,'" Dr. Liska advises. "Teaching is important. Clients must have the perception and confidence that you know what you're doing. If they start to stare a little bit, you have to stop and explain the tests in easier-to-understand terms."

When talking with clients, use nonconfrontational body language. Dr. Liska keeps a short oak barstool in his exam room so he can sit and have an eye-to-eye conversation with the client. The lower height helps put clients at ease and sends a nonverbal cue that he's open to questions and discussion. Consider these tips on effective communication when you talk with clients about diagnostic services and imaging:

- Maintain positive eye contact and body language.
- Don't cross your arms.
- Stand up straight or lean in towards the client.
- Stand at an angle to the client rather than face-to-face across the exam table.
- Step out from behind the exam table or reception desk.
- Keep your best poker face. Don't smile or frown; just stay neutral.
- Stand still; don't shift back and forth.
- Look into your client's eyes.

Once you've recommended diagnostic tests, your client will want to know how much they will cost. Always have the client sign a written estimate before proceeding with tests or treatments. Consider a span of 20% with a low and high range. At Animal Clinic of North Clarksville in Clarksville, Tennessee, Dr. Ronald E. Whitford always quotes the high end of the range with what he calls a "fudge factor" built in. "I have yet to have a client upset that we charged her less than the original estimate," he says. "You exceed expectations because the amount is less than the estimate. The secret is to be proactive and tell clients what it will cost up front rather than having them get sticker shock at the front desk."

In addition to a written estimate, have clients sign a consent form. You should require a consent form for any anesthetic procedure, but also consider signed releases for drugs with known side effects, refused treatments, last-ditch-effort treatments, and experimental, radical, or nontraditional treatments. Consent forms explain risks, acknowledge that the client understands these risks, confirms that the client agrees to the treatment or procedure, and has discussed any questions or concerns with the doctor.

In *Legal Consent Forms for Veterinary Practices* (Priority Press, Ltd., 2006), you can access templates for more than 45 consents in six areas:

1. Consent forms for treatment and care, including hospital admission, dental care, surgery, anesthesia, emergency, and extraordinary care
2. Consent forms for variations in vaccination protocols

3. Consent forms for use of veterinary drugs, including adverse side effects, use in special species such as exotics, and extra-label use

4. Consent forms for difficult situations such as refusal of treatment, release of medical records, dispensing heartworm preventives without testing, euthanasia, and transfer of pet ownership

5. Consent forms for use of pets for educational purposes, including clinical trials at veterinary teaching hospitals, donation of pet remains, release of information for media or website publication, and other learning scenarios

6. Miscellaneous consent forms for petsitting, boarding, adoption, wildlife rehabilitation programs, employment testing consent, and other situations

Consent forms give clients confidence in your hospital and let them share responsibility for medical decisions. Your practice manager can get the program started and work with a veterinarian to identify which procedures need consent forms. Your manager also makes sure everyone uses consent forms consistently and understands their purpose to educate clients and limit the hospital's legal risks.

DEVELOPING EFFECTIVE REFERRAL RELATIONSHIPS WITH SPECIALISTS

If a patient needs diagnostic services or imaging that your hospital doesn't have, consider referring the case to a specialist at a referral center or university. "Clients *expect* the referring veterinarian not only to decide *when* a referral is indicated, but also *who* can provide the kind of care that is needed," Dr. Liska says. "The client's opinion of the veterinary specialist is a reflection on the referring veterinarian's ability or inability to find expertise and service." A worried client may ask, "Would it have made a difference if I had been referred sooner?"

Clients understand how referrals work in human medicine, and many pet owners expect the same level of care for their furry friends. Reasons for referrals include the need for a specialist's clinical or surgical expertise, advanced diagnostic capabilities, special equipment, and a staff to support intensive nursing care, Dr. Liska says. Referring cases to specialists can strengthen client loyalty and let you provide a link to the best care possible. To create win-win relationships, follow these steps:

1. **Refer cases early.** Don't wait to refer a case until the pet's health has deteriorated and the client is upset. Waiting too long can harm the doctor-client relationship, and the client may switch to another primary-care veterinarian.

2. **Make a personal introduction.** While the client is in your exam room, call the specialist, describe the case, and introduce the client. This dialogue sets the tone for the doctor-client relationship with the specialist.

3. **Be honest about finances.** Although specialty veterinary care seems a bargain when compared with human medicine, most clients spend discretionary income on pets' care. Tell clients about CareCredit (888/255-4426 or www.carecredit.com) as a financing option. Also ask the specialist for a ballpark estimate of the initial exam and diagnostic tests.

4. **Communicate often.** The specialist will share findings and recommendations with you within 24 to 48 hours. You should receive copies of lab results, radiographs, pathology reports, and discharge orders so you know precisely what was communicated to your client. Add these valuable documents to your patient's medical record. As the primary-care veterinarian, clients rely on you as the liaison to the specialist. Make sure you understand the specialist's recommendations and can answer additional questions the client may have.

Case Study

How I Educate Clients with Teaching Radiographs

When a referring veterinarian sends a client whose pet has hip dysplasia to

Gulf Coast Veterinary Specialists in Houston, Texas, Dr. Bill Liska, Dipl. ACVS, uses several tools during the initial consultation to educate the client. Most clients bring radiographs of their dogs' hips from referring veterinarians, and Dr. Liska has a double view box in each exam room to display two radiographs side by side. He places the dog's radiograph on one side and a teaching radiograph that shows various stages of hip dysplasia on the other. The 14-by-17-inch radiograph has nine images of hips that helps clients identify their dogs' stage of hip dysplasia and why surgery is needed.

After discussing radiographs, Dr. Liska gives clients a brochure with before-and-after radiographs of total hip replacement surgery. Written in a Q&A format, the brochure explains how the board-certified surgeon makes a diagnosis, options for treatment, necessary diagnostics, the surgical procedure, and the prognosis. He also has models of dogs' hips and the hip replacement in the exam room. "We also take radiographs after the procedure to show clients what we did," Dr. Liska says. If clients prefer, Dr. Liska takes a digital photo of the radiograph, grayscales it, and emails it to them. Clients often have the email of the dog's hips waiting for them when they return home.

Implementation Idea

Create an Educational Display that Features Diagnostic Services and Imaging

Most clients haven't seen what happens in the back of your veterinary hospital. In addition to offering tours, you can give clients a visual snapshot of your behind-the-scenes diagnostic services and imaging with framed photos in your reception area, exam hallways, and exam rooms.

At the Veterinary Referral Center of Colorado in Englewood, Colorado, clients see framed photos of pets getting ultrasound, radiographs, and other diagnostic procedures. The educational photos help put clients at ease and showcase the specialty hospital's sophisticated equipment. But why couldn't you do the same at your veterinary hospital?

Consider hanging pictures, creating a bulletin board, or assembling a scrapbook for your reception area of these diagnostic services and images:

- Technicians taking a radiograph
- Your lab technician running blood tests
- A technician or veterinarian using your microscope to view a slide
- A veterinarian performing a blood-pressure evaluation on a senior cat
- A dog getting an ECG with a veterinarian reviewing the results
- A pet receiving an intradermal skin test
- A veterinarian performing an advanced diagnostic procedure such as ultrasound or endoscopy

HELPFUL RESOURCES

Legal Consent Forms for Veterinary Practices, Fourth Edition, James F. Wilson, DVM, JD. Priority Press, Ltd., 2006. Available through the American Animal Hospital Association.

Radioquiz v.2.1 CD, Renee Leveille, MV, DACVR and Erin Norman Carmel, MV. University of Montreal, 2006. Available through the American Animal Hospital Association.

CHAPTER 8

DATA TABLES

DIAGNOSTIC SERVICES

TABLE 8.1
ROUTINE ECG: IN-HOUSE, SIX-LEAD

	25th Percentile	Median	Your Data	Average	75th Percentile	Number of Respondents
All Practices	**$45.00**	**$60.50**		**$63.07**	**$80.38**	**221**
Number of FTE Veterinarians						
1.0 or Less	$40.00	$55.00		$59.66	$83.89	47
1.1 to 2.0	$41.32	$57.00		$59.51	$74.50	57
2.1 to 3.0	$45.23	$64.13		$64.11	$79.95	48
3.1 or More	$49.65	$68.00		$67.96	$87.19	62
Member Status						
Accredited Practice Member	$48.47	$65.00		$66.22	$84.45	77
Nonaccredited Member	$40.00	$57.75		$57.78	$74.13	48
Nonmember	$40.00	$58.00		$60.91	$81.46	74
Metropolitan Status						
Urban	$50.60	$71.00		$71.04	$96.20	25
Suburban	$45.00	$62.00		$63.75	$75.00	55
Second City	$44.75	$63.25		$66.12	$84.09	49
Town	$40.00	$60.00		$59.34	$75.50	65
Rural	*	*		*	*	22
Median Area Household Income						
$35,000 or Less	$35.00	$55.00		$52.61	$65.00	27
$35,000 to $49,999	$41.99	$65.00		$64.51	$83.92	82
$50,000 to $69,999	$45.00	$63.25		$65.44	$83.05	73
$70,000 or More	$49.75	$64.00		$63.54	$77.29	33

Note 1: This case is an arrhythmia evaluation of a nine-year-old dachshund.
Note 2: The total fee charged for this service includes fees for the procedure, outside lab or diagnostic service charges, and interpretation. The total fee excludes the office call/exam fee.
Note 3: An asterisk indicates that data was not reported due to an insufficient number of responses.
Note 4: 53% of respondents reported that they do not offer this service.

TABLE 8.2
ROUTINE ECG: IN-HOUSE, LEAD II ONLY

	25th Percentile	Median	Your Data	Average	75th Percentile	Number of Respondents
All Practices	$25.00	$35.00		$39.72	$48.88	244
Number of FTE Veterinarians						
1.0 or Less	$24.99	$30.00		$38.29	$47.51	54
1.1 to 2.0	$25.00	$34.65		$37.37	$48.13	70
2.1 to 3.0	$29.75	$39.00		$41.08	$50.25	50
3.1 or More	$30.00	$35.00		$42.74	$55.00	59
Member Status						
Accredited Practice Member	$26.52	$35.38		$39.45	$50.00	78
Nonaccredited Member	$23.00	$33.00		$35.44	$45.25	49
Nonmember	$25.00	$34.50		$40.09	$45.00	100
Metropolitan Status						
Urban	*	*		*	*	18
Suburban	$30.00	$40.00		$40.98	$49.00	71
Second City	$30.00	$35.00		$43.52	$60.00	55
Town	$24.93	$30.00		$36.65	$45.00	74
Rural	*	*		*	*	23
Median Area Household Income						
$35,000 or Less	$25.00	$30.00		$41.37	$60.50	31
$35,000 to $49,999	$25.00	$33.00		$37.13	$45.00	73
$50,000 to $69,999	$25.61	$37.21		$42.02	$54.00	84
$70,000 or More	$28.00	$40.00		$40.20	$50.00	51

Note 1: This case is an arrhythmia evaluation of a nine-year-old dachshund.
Note 2: The total fee charged for this service includes fees for the procedure, outside lab or diagnostic service charges, and interpretation. The total fee excludes the office call/exam fee.
Note 3: An asterisk indicates that data was not reported due to an insufficient number of responses.
Note 4: 46% of respondents reported that they do not offer this service.

TABLE 8.3
ROUTINE ECG: OUTSIDE SERVICE

	25th Percentile	Median	Your Data	Average	75th Percentile	Number of Respondents
All Practices	$76.00	$89.00		$95.06	$105.00	223
Number of FTE Veterinarians						
1.0 or Less	$69.50	$86.25		$87.34	$102.50	34
1.1 to 2.0	$75.00	$85.00		$94.01	$101.19	72
2.1 to 3.0	$75.00	$90.00		$97.76	$108.50	53
3.1 or More	$80.00	$90.00		$96.80	$110.19	58
Member Status						
Accredited Practice Member	$81.00	$91.50		$98.34	$111.88	69
Nonaccredited Member	$79.00	$88.82		$92.47	$107.00	51
Nonmember	$75.00	$85.00		$94.78	$100.00	88
Metropolitan Status						
Urban	*	*		*	*	22
Suburban	$76.50	$84.25		$94.25	$112.25	52
Second City	$75.50	$93.76		$97.22	$114.69	48
Town	$74.00	$86.00		$91.77	$100.00	73
Rural	*	*		*	*	22
Median Area Household Income						
$35,000 or Less	*	*		*	*	18
$35,000 to $49,999	$75.75	$90.00		$96.82	$109.25	90
$50,000 to $69,999	$79.31	$86.75		$96.96	$108.50	74
$70,000 or More	$77.25	$90.00		$96.23	$113.50	34

Note 1: This case is an arrhythmia evaluation of a nine-year-old dachshund.
Note 2: The total fee charged for this service includes fees for the procedure, outside lab or diagnostic service charges, and interpretation. The total fee excludes the office call/exam fee.
Note 3: An asterisk indicates that data was not reported due to an insufficient number of responses.
Note 4: 48% of respondents reported that they do not offer this service.

TABLE 8.4
SCHIRMER TEAR TEST

	25th Percentile	Median	Your Data	Average	75th Percentile	Number of Respondents
All Practices	**$14.70**	**$17.89**		**$18.38**	**$21.26**	**529**
Number of FTE Veterinarians						
1.0 or Less	$12.41	$16.98		$17.56	$21.93	132
1.1 to 2.0	$15.00	$17.50		$18.31	$21.00	161
2.1 to 3.0	$15.00	$18.43		$18.81	$20.95	112
3.1 or More	$15.00	$18.00		$19.07	$23.31	106
Member Status						
Accredited Practice Member	$15.00	$18.00		$19.36	$23.84	146
Nonaccredited Member	$14.68	$18.00		$18.35	$21.57	109
Nonmember	$13.00	$17.00		$17.56	$20.27	228
Metropolitan Status						
Urban	$13.58	$16.48		$17.62	$23.00	42
Suburban	$15.00	$19.00		$19.50	$23.00	149
Second City	$15.00	$18.41		$19.45	$22.39	106
Town	$12.94	$16.39		$17.32	$20.25	166
Rural	$11.40	$16.00		$17.05	$19.65	57
Median Area Household Income						
$35,000 or Less	$12.00	$15.00		$16.36	$19.68	64
$35,000 to $49,999	$13.58	$17.00		$17.58	$20.48	193
$50,000 to $69,999	$15.12	$18.50		$19.23	$23.00	169
$70,000 or More	$15.00	$18.00		$19.88	$24.00	90

Note 1: The total fee charged for this service includes fees for the procedure, outside lab or diagnostic service charges, and interpretation. The total fee excludes the office call/exam fee.
Note 2: 3% of respondents reported that they do not offer this service.

TABLE 8.5
CORNEAL STAIN

	25th Percentile	Median	Your Data	Average	75th Percentile	Number of Respondents
All Practices	**$13.00**	**$17.50**		**$17.94**	**$21.80**	**535**
Number of FTE Veterinarians						
1.0 or Less	$12.00	$16.00		$17.45	$21.73	141
1.1 to 2.0	$13.00	$17.30		$17.64	$21.94	160
2.1 to 3.0	$15.00	$18.00		$18.45	$21.64	109
3.1 or More	$13.85	$17.00		$18.56	$22.00	107
Member Status						
Accredited Practice Member	$15.00	$18.00		$19.35	$23.04	144
Nonaccredited Member	$12.85	$18.00		$18.17	$22.43	110
Nonmember	$12.00	$16.00		$16.92	$20.00	234
Metropolitan Status						
Urban	$15.00	$19.12		$19.61	$24.85	39
Suburban	$15.00	$18.35		$19.46	$24.22	151
Second City	$15.00	$18.00		$19.11	$23.00	108
Town	$12.00	$15.08		$16.26	$19.27	170
Rural	$10.00	$15.00		$15.48	$19.10	57
Median Area Household Income						
$35,000 or Less	$10.00	$14.50		$14.70	$17.85	63
$35,000 to $49,999	$12.10	$16.95		$17.07	$20.00	196
$50,000 to $69,999	$14.94	$18.40		$19.20	$24.00	170
$70,000 or More	$15.00	$18.00		$19.56	$23.59	92

Note 1: The total fee charged for this service includes fees for the procedure, outside lab or diagnostic service charges, and interpretation. The total fee excludes the office call/exam fee.

Note 2: 2% of respondents reported that they do not offer this service.

TABLE 8.6
TONOMETRY

	25th Percentile	Median	Your Data	Average	75th Percentile	Number of Respondents
All Practices	**$20.00**	**$25.62**		**$26.02**	**$30.59**	**366**
Number of FTE Veterinarians						
1.0 or Less	$18.13	$25.00		$24.48	$29.97	76
1.1 to 2.0	$20.00	$25.20		$24.78	$29.15	101
2.1 to 3.0	$20.00	$26.00		$26.40	$32.00	87
3.1 or More	$22.00	$27.00		$27.98	$33.20	91
Member Status						
Accredited Practice Member	$22.58	$25.88		$27.20	$30.00	120
Nonaccredited Member	$20.00	$26.05		$25.91	$32.00	79
Nonmember	$19.00	$25.00		$24.80	$30.00	138
Metropolitan Status						
Urban	$20.63	$25.32		$26.61	$33.50	26
Suburban	$22.00	$27.04		$28.07	$32.48	105
Second City	$20.00	$26.00		$25.54	$29.65	81
Town	$20.00	$25.00		$24.57	$29.78	118
Rural	$18.60	$25.60		$25.08	$31.00	29
Median Area Household Income						
$35,000 or Less	$17.89	$25.00		$24.27	$31.75	36
$35,000 to $49,999	$20.00	$25.00		$24.42	$28.43	134
$50,000 to $69,999	$22.00	$27.40		$27.71	$33.00	122
$70,000 or More	$21.58	$26.00		$27.47	$33.73	65

Note 1: The total fee charged for this service includes fees for the procedure, outside lab or diagnostic service charges, and interpretation. The total fee excludes the office call/exam fee.
Note 2: 27% of respondents reported that they do not offer this service.

TABLE 8.7
BLOOD PRESSURE EVALUATION

	25th Percentile	Median	Your Data	Average	75th Percentile	Number of Respondents
All Practices	$20.00	$25.00		$26.78	$33.30	363
Number of FTE Veterinarians						
1.0 or Less	$19.13	$25.00		$26.74	$34.75	72
1.1 to 2.0	$19.93	$25.00		$25.77	$31.84	100
2.1 to 3.0	$20.00	$25.50		$27.82	$35.00	83
3.1 or More	$21.00	$25.80		$26.87	$33.20	93
Member Status						
Accredited Practice Member	$21.00	$27.00		$27.99	$34.00	123
Nonaccredited Member	$18.00	$25.00		$25.96	$30.00	79
Nonmember	$19.93	$25.00		$25.68	$31.81	136
Metropolitan Status						
Urban	$20.00	$25.00		$25.43	$30.00	26
Suburban	$22.03	$27.04		$29.46	$36.91	117
Second City	$20.00	$25.00		$26.60	$33.69	78
Town	$18.73	$25.00		$25.09	$30.78	109
Rural	$17.00	$24.00		$24.05	$34.00	29
Median Area Household Income						
$35,000 or Less	$16.80	$20.00		$22.73	$28.41	31
$35,000 to $49,999	$20.00	$25.00		$25.56	$30.00	124
$50,000 to $69,999	$20.00	$26.00		$27.42	$34.00	125
$70,000 or More	$21.50	$27.00		$29.45	$38.00	77

Note 1: The total fee charged for this service includes fees for the procedure, outside lab or diagnostic service charges, and interpretation. The total fee excludes the office call/exam fee.
Note 2: 28% of respondents reported that they do not offer this service.

TABLE 8.8
EAR SWAB EXAM/STAIN

	25th Percentile	Median	Your Data	Average	75th Percentile	Number of Respondents
All Practices	$15.70	$20.00		$20.87	$25.00	517
Number of FTE Veterinarians						
1.0 or Less	$15.00	$20.00		$20.23	$25.00	128
1.1 to 2.0	$15.00	$19.70		$20.37	$25.00	161
2.1 to 3.0	$17.00	$20.04		$20.58	$24.55	107
3.1 or More	$16.88	$21.00		$22.21	$26.50	106
Member Status						
Accredited Practice Member	$18.00	$21.50		$22.16	$25.63	141
Nonaccredited Member	$15.00	$20.00		$20.53	$25.36	106
Nonmember	$15.00	$19.68		$20.17	$25.00	230
Metropolitan Status						
Urban	$17.25	$21.00		$22.18	$26.00	40
Suburban	$17.00	$22.38		$22.54	$26.84	146
Second City	$16.00	$20.70		$21.05	$24.80	103
Town	$15.00	$18.78		$19.97	$24.09	164
Rural	$12.63	$17.00		$17.78	$21.81	56
Median Area Household Income						
$35,000 or Less	$12.00	$16.00		$17.72	$22.75	64
$35,000 to $49,999	$15.00	$19.00		$20.34	$24.00	187
$50,000 to $69,999	$16.93	$21.00		$21.49	$25.38	166
$70,000 or More	$17.55	$22.00		$22.96	$28.10	88

Note 1: The total fee charged for this service includes fees for the procedure, outside lab or diagnostic service charges, and interpretation. The total fee excludes the office call/exam fee.
Note 2: 2% of respondents reported that they do not offer this service.

TABLE 8.9
WOOD'S LAMP EXAMINATION

	25th Percentile	Median	Your Data	Average	75th Percentile	Number of Respondents
All Practices	**$8.53**	**$12.00**		**$12.25**	**$15.00**	**204**
Number of FTE Veterinarians						
1.0 or Less	$7.50	$10.52		$11.39	$15.00	56
1.1 to 2.0	$8.46	$10.98		$12.03	$15.00	68
2.1 to 3.0	$9.70	$13.75		$13.62	$18.00	39
3.1 or More	$10.00	$12.05		$12.46	$14.00	35
Member Status						
Accredited Practice Member	$10.00	$12.26		$12.79	$15.00	60
Nonaccredited Member	$5.20	$10.50		$11.18	$15.00	39
Nonmember	$9.64	$11.23		$12.11	$15.00	86
Metropolitan Status						
Urban	*	*		*	*	17
Suburban	$10.00	$13.51		$14.16	$17.00	47
Second City	$9.75	$12.64		$12.70	$15.00	38
Town	$8.30	$10.98		$11.62	$15.00	70
Rural	$5.40	$9.50		$9.37	$11.50	28
Median Area Household Income						
$35,000 or Less	$6.00	$10.00		$9.67	$12.38	30
$35,000 to $49,999	$8.48	$12.00		$11.95	$15.00	74
$50,000 to $69,999	$9.70	$12.00		$13.08	$16.22	67
$70,000 or More	$10.00	$12.54		$13.66	$16.78	28

Note 1: The total fee charged for this service includes fees for the procedure, outside lab or diagnostic service charges, and interpretation. The total fee excludes the office call/exam fee.
Note 2: An asterisk indicates that data was not reported due to an insufficient number of responses.
Note 3: 12% of respondents reported that this charge is included in the office call/exam fee.
Note 4: 33% of respondents reported that they do not offer this service.

THE VETERINARY FEE REFERENCE / 240

IMAGING SERVICES

TABLE 8.10
RADIOGRAPHIC SETUP FEE

	25th Percentile	Median	Your Data	Average	75th Percentile	Number of Respondents
All Practices	$35.00	$65.00		$66.81	$90.76	125
Number of FTE Veterinarians						
1.0 or Less	$37.50	$66.50		$71.21	$94.75	40
1.1 to 2.0	$35.00	$70.00		$69.64	$91.50	34
2.1 to 3.0	*	*		*	*	23
3.1 or More	*	*		*	*	23
Member Status						
Accredited Practice Member	$27.06	$55.60		$63.76	$85.00	36
Nonaccredited Member	*	*		*	*	18
Nonmember	$35.00	$65.00		$66.43	$87.75	53
Metropolitan Status						
Urban	*	*		*	*	6
Suburban	$29.50	$63.67		$68.68	$97.25	45
Second City	*	*		*	*	24
Town	$25.88	$65.00		$63.89	$89.01	33
Rural	*	*		*	*	14
Median Area Household Income						
$35,000 or Less	*	*		*	*	13
$35,000 to $49,999	$35.00	$58.00		$61.25	$85.00	51
$50,000 to $69,999	$22.75	$78.85		$71.89	$103.40	32
$70,000 or More	$40.00	$62.50		$66.99	$84.63	26

Note 1: This fee includes the use of the radiograph machine, initial setup of the machine, and recording in appropriate logs.
Note 2: An asterisk indicates that data was not reported due to an insufficient number of responses.
Note 3: 19% of the respondents reported that this charge is included in the radiograph fee.
Note 4: 50% of respondents reported that they do not offer this service.

TABLE 8.11
BASIS OF RADIOGRAPH CHARGES

	By Number of Films Taken	By Body Part	By Number of Films Taken and Body Part	Number of Respondents
All Practices	**90%**	**3%**	**8%**	**547**
Number of FTE Veterinarians				
1.0 or Less	88%	4%	11%	139
1.1 to 2.0	92%	3%	5%	169
2.1 to 3.0	92%	1%	7%	114
3.1 or More	88%	4%	10%	106
Member Status				
Accredited Practice Member	88%	3%	10%	144
Nonaccredited Member	90%	4%	7%	115
Nonmember	93%	1%	7%	238
Metropolitan Status				
Urban	88%	0%	13%	40
Suburban	88%	3%	10%	155
Second City	94%	3%	5%	110
Town	90%	3%	8%	172
Rural	92%	3%	5%	61
Median Area Household Income				
$35,000 or Less	87%	4%	9%	68
$35,000 to $49,999	93%	2%	6%	199
$50,000 to $69,999	91%	2%	7%	176
$70,000 or More	84%	4%	13%	91

Note: Some row totals do not equal 100% because some respondents selected more than one response.

Note: All radiograph fees in Tables 8.12-8.28 include the total fee charged to the client for a routine radiograph, including the fee for the film, procedure (taking and developing radiographs), and interpretation. The fee excludes the office call/exam fee.

TABLE 8.12
ROUTINE RADIOGRAPH: ONE VIEW, 8 X 10 CASSETTE, DOG OR CAT

	25th Percentile	Median	Your Data	Average	75th Percentile	Number of Respondents
All Practices	**$51.00**	**$62.50**		**$66.61**	**$78.80**	**451**
Number of FTE Veterinarians						
1.0 or Less	$45.00	$55.00		$61.14	$75.00	105
1.1 to 2.0	$50.00	$60.00		$65.48	$77.00	141
2.1 to 3.0	$55.86	$68.30		$70.19	$78.89	98
3.1 or More	$53.38	$66.50		$69.50	$88.38	93
Member Status						
Accredited Practice Member	$55.00	$70.90		$72.39	$88.00	121
Nonaccredited Member	$50.25	$60.75		$67.31	$79.58	100
Nonmember	$50.00	$60.00		$62.72	$74.75	189
Metropolitan Status						
Urban	$58.89	$66.08		$72.80	$90.00	32
Suburban	$55.00	$66.00		$71.40	$84.75	124
Second City	$51.75	$68.98		$69.15	$80.47	93
Town	$48.96	$59.68		$62.80	$75.00	144
Rural	$40.00	$55.13		$56.25	$67.88	50
Median Area Household Income						
$35,000 or Less	$45.00	$59.85		$56.48	$68.00	51
$35,000 to $49,999	$48.93	$60.00		$63.40	$75.00	169
$50,000 to $69,999	$54.00	$65.70		$69.47	$84.89	150
$70,000 or More	$56.00	$74.00		$74.51	$87.38	72

Note: 15% of respondents reported that they do not offer this service.

TABLE 8.13
ROUTINE RADIOGRAPH: ONE VIEW, 8 X 10 ADDITIONAL CASSETTE

	25th Percentile	Median	Your Data	Average	75th Percentile	Number of Respondents
All Practices	**$30.00**	**$40.00**		**$42.86**	**$50.00**	**450**
Number of FTE Veterinarians						
1.0 or Less	$27.50	$35.73		$40.79	$49.00	107
1.1 to 2.0	$30.00	$40.00		$43.26	$51.75	140
2.1 to 3.0	$30.24	$38.95		$42.26	$50.75	96
3.1 or More	$33.00	$41.25		$44.87	$50.00	92
Member Status						
Accredited Practice Member	$35.20	$44.85		$47.97	$55.00	127
Nonaccredited Member	$30.00	$43.00		$45.49	$55.10	97
Nonmember	$26.93	$35.00		$38.02	$47.78	188
Metropolitan Status						
Urban	$30.00	$40.00		$40.49	$48.25	31
Suburban	$34.25	$45.00		$45.99	$52.75	120
Second City	$30.00	$39.50		$41.69	$49.10	93
Town	$29.33	$40.35		$43.28	$52.73	148
Rural	$23.00	$35.00		$36.86	$48.90	51
Median Area Household Income						
$35,000 or Less	$20.00	$32.00		$34.35	$41.00	53
$35,000 to $49,999	$30.00	$38.84		$40.58	$48.60	170
$50,000 to $69,999	$32.96	$44.00		$45.32	$52.80	147
$70,000 or More	$34.25	$44.00		$48.57	$60.70	71

Note: 13% of respondents reported that they do not offer this service.

TABLE 8.14
ROUTINE RADIOGRAPH: ONE VIEW, 14 X 17 CASSETTE, DOG OR CAT

	25th Percentile	Median	Your Data	Average	75th Percentile	Number of Respondents
All Practices	$55.00	$65.25		$69.19	$81.13	477
Number of FTE Veterinarians						
1.0 or Less	$50.00	$62.00		$65.00	$79.25	117
1.1 to 2.0	$55.00	$65.00		$68.45	$79.63	150
2.1 to 3.0	$59.50	$70.00		$71.22	$80.00	105
3.1 or More	$56.50	$68.00		$72.54	$90.00	91
Member Status						
Accredited Practice Member	$59.13	$73.13		$74.25	$88.00	120
Nonaccredited Member	$58.10	$70.00		$70.63	$81.61	104
Nonmember	$52.00	$64.00		$65.21	$75.50	211
Metropolitan Status						
Urban	$55.00	$65.00		$69.09	$81.12	34
Suburban	$60.00	$71.50		$74.25	$87.00	128
Second City	$54.00	$70.00		$71.93	$90.00	95
Town	$52.00	$62.00		$66.31	$78.50	157
Rural	$45.75	$58.93		$59.77	$68.13	54
Median Area Household Income						
$35,000 or Less	$45.00	$60.00		$60.97	$75.00	59
$35,000 to $49,999	$52.00	$65.00		$66.07	$78.60	175
$50,000 to $69,999	$56.00	$68.75		$71.79	$86.25	155
$70,000 or More	$60.00	$74.75		$76.62	$88.00	77

Note: 9% of respondents reported that they do not offer this service.

TABLE 8.15
ROUTINE RADIOGRAPH: ONE VIEW, 14 X 17 ADDITIONAL CASSETTE

	25th Percentile	Median	Your Data	Average	75th Percentile	Number of Respondents
All Practices	$33.00	$42.83		$44.91	$53.88	480
Number of FTE Veterinarians						
1.0 or Less	$30.00	$40.98		$43.98	$55.00	117
1.1 to 2.0	$32.00	$45.00		$45.76	$55.00	150
2.1 to 3.0	$32.96	$40.00		$42.87	$52.80	103
3.1 or More	$35.00	$43.00		$46.92	$52.13	94
Member Status						
Accredited Practice Member	$35.31	$45.00		$48.35	$56.79	129
Nonaccredited Member	$32.40	$45.00		$47.93	$58.75	105
Nonmember	$31.00	$40.00		$40.80	$48.94	208
Metropolitan Status						
Urban	$32.25	$40.00		$42.77	$49.25	34
Suburban	$35.00	$45.00		$48.74	$58.00	127
Second City	$32.72	$42.15		$44.29	$54.25	98
Town	$32.48	$41.77		$44.54	$52.10	161
Rural	$24.00	$39.00		$39.42	$50.71	53
Median Area Household Income						
$35,000 or Less	$21.40	$35.00		$36.44	$45.00	59
$35,000 to $49,999	$32.50	$41.50		$43.07	$51.25	181
$50,000 to $69,999	$35.00	$45.00		$47.57	$57.06	152
$70,000 or More	$35.00	$45.60		$50.33	$60.18	78

Note: 8% of respondents reported that they do not offer this service.

TABLE 8.16
ROUTINE RADIOGRAPHS: TWO VIEWS, CHEST, 60-POUND DOG

	25th Percentile	Median	Your Data	Average	75th Percentile	Number of Respondents
All Practices	$87.91	$100.28		$103.52	$120.00	486
Number of FTE Veterinarians						
1.0 or Less	$82.63	$99.18		$101.13	$120.00	124
1.1 to 2.0	$90.00	$100.06		$103.62	$120.00	147
2.1 to 3.0	$89.00	$104.25		$104.06	$120.48	105
3.1 or More	$85.00	$99.83		$103.91	$118.76	96
Member Status						
Accredited Practice Member	$90.00	$105.18		$108.21	$124.60	131
Nonaccredited Member	$90.23	$105.00		$107.47	$124.95	105
Nonmember	$84.10	$98.00		$98.23	$115.00	212
Metropolitan Status						
Urban	$89.75	$107.00		$105.81	$121.32	37
Suburban	$95.28	$109.25		$112.70	$127.23	136
Second City	$88.14	$99.00		$104.36	$123.82	97
Town	$83.50	$98.00		$98.20	$115.00	155
Rural	$75.00	$94.13		$92.15	$106.98	54
Median Area Household Income						
$35,000 or Less	$65.75	$89.00		$87.47	$104.89	56
$35,000 to $49,999	$87.24	$99.83		$101.08	$118.63	182
$50,000 to $69,999	$90.00	$102.00		$106.37	$122.00	159
$70,000 or More	$98.00	$112.25		$115.23	$131.44	78

Note: 7% of respondents reported that they do not offer this service.

TABLE 8.17
ROUTINE RADIOGRAPHS: THREE VIEWS, CHEST, 60-POUND DOG

	25th Percentile	Median	Your Data	Average	75th Percentile	Number of Respondents
All Practices	$115.00	$137.00		$140.02	$167.08	477
Number of FTE Veterinarians						
1.0 or Less	$111.85	$131.97		$137.22	$165.00	122
1.1 to 2.0	$114.50	$140.55		$141.90	$168.00	143
2.1 to 3.0	$120.00	$137.80		$140.90	$165.00	104
3.1 or More	$112.50	$138.12		$140.02	$167.70	96
Member Status						
Accredited Practice Member	$115.71	$144.37		$145.84	$174.26	130
Nonaccredited Member	$117.50	$140.28		$144.96	$171.78	102
Nonmember	$112.00	$132.00		$133.88	$156.36	210
Metropolitan Status						
Urban	$125.63	$143.50		$145.24	$169.90	36
Suburban	$124.00	$145.00		$150.32	$179.00	131
Second City	$115.71	$131.65		$142.37	$169.40	98
Town	$108.25	$134.00		$134.42	$159.25	153
Rural	$95.00	$128.50		$122.39	$147.13	52
Median Area Household Income						
$35,000 or Less	$90.00	$116.51		$114.10	$139.13	56
$35,000 to $49,999	$117.58	$137.00		$138.27	$165.00	177
$50,000 to $69,999	$120.00	$143.07		$146.42	$170.13	158
$70,000 or More	$123.25	$148.00		$151.30	$178.50	75

Note: 8% of respondents reported that they do not offer this service.

TABLE 8.18
ROUTINE RADIOGRAPHS: TWO VIEWS, PELVIS, 60-POUND DOG

	25th Percentile	Median	Your Data	Average	75th Percentile	Number of Respondents
All Practices	**$86.79**	**$100.00**		**$103.30**	**$120.00**	**484**
Number of FTE Veterinarians						
1.0 or Less	$82.00	$99.00		$101.58	$120.78	123
1.1 to 2.0	$89.88	$100.03		$103.67	$119.25	146
2.1 to 3.0	$89.00	$105.00		$104.59	$121.48	105
3.1 or More	$85.00	$99.83		$103.31	$118.14	96
Member Status						
Accredited Practice Member	$90.00	$106.00		$108.87	$124.60	131
Nonaccredited Member	$90.00	$105.00		$107.52	$125.00	105
Nonmember	$84.00	$98.00		$97.97	$114.81	210
Metropolitan Status						
Urban	$89.75	$108.00		$107.38	$122.82	37
Suburban	$95.00	$107.00		$111.22	$127.30	135
Second City	$88.21	$98.26		$104.35	$123.73	98
Town	$83.75	$98.40		$98.43	$115.00	153
Rural	$75.00	$93.00		$92.07	$106.98	54
Median Area Household Income						
$35,000 or Less	$65.75	$89.00		$87.52	$104.89	56
$35,000 to $49,999	$87.83	$100.00		$101.46	$118.43	181
$50,000 to $69,999	$89.92	$102.03		$106.17	$122.99	158
$70,000 or More	$94.00	$111.00		$113.32	$131.44	78

Note: 7% of respondents reported that they do not offer this service.

TABLE 8.19
ROUTINE RADIOGRAPHS: TWO VIEWS, ABDOMEN, 60-POUND DOG

	25th Percentile	Median	Your Data	Average	75th Percentile	Number of Respondents
All Practices	**$87.08**	**$100.20**		**$103.50**	**$120.03**	**485**
Number of FTE Veterinarians						
1.0 or Less	$82.00	$100.00		$102.29	$122.00	123
1.1 to 2.0	$89.88	$100.03		$104.14	$120.09	146
2.1 to 3.0	$89.00	$104.63		$104.69	$122.16	106
3.1 or More	$85.00	$99.83		$102.57	$118.14	96
Member Status						
Accredited Practice Member	$90.00	$106.00		$108.69	$125.47	131
Nonaccredited Member	$90.00	$105.00		$107.69	$125.00	106
Nonmember	$84.00	$98.00		$98.34	$115.00	210
Metropolitan Status						
Urban	$89.75	$108.00		$106.52	$124.32	37
Suburban	$95.20	$108.00		$111.36	$126.00	135
Second City	$88.00	$99.00		$105.45	$125.00	99
Town	$82.75	$98.40		$98.35	$115.00	153
Rural	$75.00	$94.13		$92.29	$106.98	54
Median Area Household Income						
$35,000 or Less	$65.75	$89.00		$87.98	$106.50	56
$35,000 to $49,999	$86.00	$100.00		$101.65	$119.25	182
$50,000 to $69,999	$90.00	$102.03		$106.40	$123.11	158
$70,000 or More	$94.75	$111.00		$113.36	$131.25	78

Note: 7% of respondents reported that they do not offer this service.

TABLE 8.20
ROUTINE RADIOGRAPHS: TWO VIEWS, SPINE, DACHSHUND

	25th Percentile	Median	Your Data	Average	75th Percentile	Number of Respondents
All Practices	**$82.92**	**$98.00**		**$99.46**	**$117.52**	**486**
Number of FTE Veterinarians						
1.0 or Less	$80.00	$95.00		$97.78	$119.18	124
1.1 to 2.0	$83.75	$98.00		$98.97	$117.45	145
2.1 to 3.0	$85.00	$99.63		$101.13	$116.15	106
3.1 or More	$80.00	$96.75		$99.73	$116.34	97
Member Status						
Accredited Practice Member	$88.46	$102.25		$105.65	$122.43	132
Nonaccredited Member	$85.00	$99.88		$102.37	$120.00	106
Nonmember	$76.25	$93.00		$94.16	$112.63	210
Metropolitan Status						
Urban	$88.38	$104.78		$103.71	$119.75	36
Suburban	$90.00	$104.00		$106.92	$123.00	135
Second City	$86.13	$96.75		$102.42	$120.39	101
Town	$75.75	$91.50		$94.18	$112.50	153
Rural	$68.75	$91.00		$88.04	$104.06	54
Median Area Household Income						
$35,000 or Less	$60.19	$83.50		$81.93	$98.87	56
$35,000 to $49,999	$80.00	$97.75		$97.15	$115.18	181
$50,000 to $69,999	$86.26	$98.50		$103.30	$120.00	159
$70,000 or More	$90.00	$108.00		$110.27	$129.00	79

Note: 7% of respondents reported that they do not offer this service.

TABLE 8.21
ROUTINE RADIOGRAPHS: TWO VIEWS, ABDOMEN, CAT

	25th Percentile	Median	Your Data	Average	75th Percentile	Number of Respondents
All Practices	**$79.15**	**$95.00**		**$97.25**	**$116.00**	**491**
Number of FTE Veterinarians						
1.0 or Less	$75.00	$92.66		$96.37	$118.95	125
1.1 to 2.0	$78.50	$95.00		$95.78	$112.75	149
2.1 to 3.0	$84.00	$98.25		$99.67	$116.15	106
3.1 or More	$77.09	$95.00		$96.79	$115.29	96
Member Status						
Accredited Practice Member	$85.00	$100.00		$103.36	$120.00	135
Nonaccredited Member	$84.25	$98.25		$100.39	$120.00	104
Nonmember	$75.00	$90.00		$92.05	$111.13	213
Metropolitan Status						
Urban	$87.00	$104.23		$102.31	$119.00	39
Suburban	$85.00	$102.00		$105.56	$122.98	139
Second City	$83.75	$95.00		$99.92	$118.49	98
Town	$73.75	$90.16		$91.04	$108.50	154
Rural	$63.19	$86.50		$85.36	$102.34	54
Median Area Household Income						
$35,000 or Less	$60.00	$80.00		$79.63	$97.13	56
$35,000 to $49,999	$78.29	$95.00		$95.12	$114.31	184
$50,000 to $69,999	$85.00	$97.88		$100.86	$119.88	160
$70,000 or More	$85.00	$104.75		$108.31	$128.12	80

Note: 6% of respondents reported that they do not offer this service.

TABLE 8.22
ROUTINE RADIOGRAPHS: TWO VIEWS, CHEST, CAT

	25th Percentile	Median	Your Data	Average	75th Percentile	Number of Respondents
All Practices	$78.00	$95.00		$97.14	$116.00	491
Number of FTE Veterinarians						
1.0 or Less	$75.00	$92.00		$95.70	$118.50	125
1.1 to 2.0	$79.25	$95.00		$95.89	$112.75	149
2.1 to 3.0	$84.00	$98.49		$99.31	$116.00	107
3.1 or More	$77.09	$95.00		$97.26	$115.29	96
Member Status						
Accredited Practice Member	$86.31	$100.00		$103.92	$120.13	134
Nonaccredited Member	$83.50	$98.00		$99.93	$120.00	105
Nonmember	$75.00	$90.00		$91.99	$111.13	213
Metropolitan Status						
Urban	$87.00	$104.23		$102.49	$120.00	39
Suburban	$85.00	$101.00		$105.03	$122.98	139
Second City	$81.75	$95.00		$99.19	$118.49	98
Town	$74.68	$90.35		$91.50	$108.50	154
Rural	$63.19	$87.00		$85.61	$104.06	54
Median Area Household Income						
$35,000 or Less	$60.00	$80.00		$79.29	$97.50	55
$35,000 to $49,999	$78.00	$95.00		$95.02	$112.13	185
$50,000 to $69,999	$84.00	$96.88		$100.65	$119.88	160
$70,000 or More	$85.00	$104.75		$108.31	$128.12	80

Note: 6% of respondents reported that they do not offer this service.

TABLE 8.23
ROUTINE RADIOGRAPHS: TWO VIEWS, FOREARM, CAT

	25th Percentile	Median	Your Data	Average	75th Percentile	Number of Respondents
All Practices	$75.00	$93.81		$94.83	$115.00	492
Number of FTE Veterinarians						
1.0 or Less	$69.50	$90.00		$93.50	$116.33	126
1.1 to 2.0	$75.25	$95.00		$93.53	$110.25	149
2.1 to 3.0	$80.56	$96.25		$97.25	$116.00	106
3.1 or More	$72.99	$93.98		$94.66	$115.29	96
Member Status						
Accredited Practice Member	$84.00	$98.00		$101.23	$118.35	135
Nonaccredited Member	$75.88	$96.25		$97.12	$120.00	104
Nonmember	$71.98	$89.05		$90.05	$110.00	214
Metropolitan Status						
Urban	$86.00	$98.00		$100.03	$119.00	39
Suburban	$82.69	$100.00		$102.38	$120.00	139
Second City	$79.41	$93.92		$96.92	$118.10	98
Town	$71.90	$90.00		$89.80	$106.00	155
Rural	$60.56	$85.00		$82.64	$101.00	54
Median Area Household Income						
$35,000 or Less	$59.89	$77.50		$78.13	$97.13	56
$35,000 to $49,999	$73.05	$91.83		$91.25	$109.63	184
$50,000 to $69,999	$82.78	$95.40		$99.80	$120.00	161
$70,000 or More	$84.00	$101.25		$105.51	$128.12	80

Note: 6% of respondents reported that they do not offer this service.

TABLE 8.24
ROUTINE RADIOGRAPH: ONE VIEW, 8 X 10 CASSETTE, BIRD/REPTILE/SMALL MAMMAL

	25th Percentile	Median	Your Data	Average	75th Percentile	Number of Respondents
All Practices	**$52.00**	**$63.00**		**$66.66**	**$78.94**	**298**
Number of FTE Veterinarians						
1.0 or Less	$47.26	$55.00		$63.01	$80.00	53
1.1 to 2.0	$50.00	$60.00		$62.91	$75.00	103
2.1 to 3.0	$56.05	$69.65		$69.66	$79.30	64
3.1 or More	$55.00	$67.90		$70.86	$88.88	73
Member Status						
Accredited Practice Member	$55.00	$70.00		$70.72	$85.00	87
Nonaccredited Member	$52.50	$62.00		$67.21	$79.80	73
Nonmember	$50.00	$58.13		$64.03	$75.00	120
Metropolitan Status						
Urban	*	*		*	*	17
Suburban	$54.25	$65.00		$67.06	$78.56	84
Second City	$52.40	$68.60		$71.22	$86.25	54
Town	$50.00	$60.00		$65.92	$80.46	108
Rural	$47.00	$57.45		$61.23	$73.13	32
Median Area Household Income						
$35,000 or Less	$46.00	$60.00		$55.92	$66.50	33
$35,000 to $49,999	$51.00	$60.00		$63.95	$76.98	113
$50,000 to $69,999	$55.00	$70.00		$71.32	$85.00	99
$70,000 or More	$52.59	$67.32		$71.10	$88.94	48

Note 1: An asterisk indicates that data was not reported due to an insufficient number of responses.
Note 2: 40% of respondents reported that they do not offer this service.

TABLE 8.25
ROUTINE RADIOGRAPH: ONE VIEW, 14 X 17 CASSETTE, BIRD/REPTILE/SMALL MAMMAL

	25th Percentile	Median	Your Data	Average	75th Percentile	Number of Respondents
All Practices	**$55.00**	**$66.25**		**$69.72**	**$80.00**	**302**
Number of FTE Veterinarians						
1.0 or Less	$50.00	$60.00		$66.44	$81.50	53
1.1 to 2.0	$55.00	$65.00		$66.84	$77.50	105
2.1 to 3.0	$59.25	$72.00		$72.53	$82.65	68
3.1 or More	$56.00	$68.00		$72.81	$90.00	71
Member Status						
Accredited Practice Member	$56.43	$71.00		$72.00	$84.89	86
Nonaccredited Member	$59.00	$72.00		$71.25	$83.53	71
Nonmember	$54.81	$64.73		$67.43	$78.59	125
Metropolitan Status						
Urban	*	*		*	*	17
Suburban	$56.15	$70.00		$71.21	$80.00	86
Second City	$55.75	$72.00		$75.27	$90.23	54
Town	$52.88	$64.00		$68.00	$83.64	108
Rural	$50.45	$60.00		$61.17	$67.25	33
Median Area Household Income						
$35,000 or Less	$46.50	$60.00		$58.76	$70.50	30
$35,000 to $49,999	$54.81	$65.00		$66.74	$78.25	109
$50,000 to $69,999	$56.50	$70.00		$72.54	$85.31	106
$70,000 or More	$60.00	$74.75		$77.17	$91.00	51

Note 1: An asterisk indicates that data was not reported due to an insufficient number of responses.
Note 2: 40% of respondents reported that they do not offer this service.

TABLE 8.26
ROUTINE RADIOGRAPHS: TWO VIEWS, 8 X 10 CASSETTE, BIRD/REPTILE/SMALL MAMMAL

	25th Percentile	Median	Your Data	Average	75th Percentile	Number of Respondents
All Practices	$75.00	$92.29		$94.65	$114.26	306
Number of FTE Veterinarians						
1.0 or Less	$70.00	$90.00		$92.66	$114.75	59
1.1 to 2.0	$75.25	$90.00		$92.38	$108.65	101
2.1 to 3.0	$79.34	$98.70		$97.82	$116.00	69
3.1 or More	$73.00	$93.95		$94.64	$115.39	71
Member Status						
Accredited Practice Member	$79.58	$95.00		$98.16	$116.68	89
Nonaccredited Member	$77.25	$96.25		$96.92	$116.77	72
Nonmember	$71.90	$88.00		$90.18	$110.00	127
Metropolitan Status						
Urban	*	*		*	*	18
Suburban	$83.02	$95.25		$99.02	$112.38	88
Second City	$78.26	$93.24		$96.79	$115.25	58
Town	$72.50	$91.00		$91.83	$113.00	107
Rural	$60.50	$82.50		$82.50	$105.55	32
Median Area Household Income						
$35,000 or Less	$59.89	$72.50		$74.78	$93.13	32
$35,000 to $49,999	$75.50	$93.95		$93.71	$110.50	115
$50,000 to $69,999	$80.67	$93.74		$97.72	$118.59	100
$70,000 or More	$81.42	$104.75		$104.31	$121.88	52

Note 1: An asterisk indicates that data was not reported due to an insufficient number of responses.
Note 2: 39% of respondents reported that they do not offer this service.

TABLE 8.27
ROUTINE RADIOGRAPHS: TWO VIEWS, 14 X 17 CASSETTE, BIRD/REPTILE/SMALL MAMMAL

	25th Percentile	Median	Your Data	Average	75th Percentile	Number of Respondents
All Practices	$80.00	$97.75		$98.22	$115.20	317
Number of FTE Veterinarians						
1.0 or Less	$75.00	$92.00		$93.78	$114.25	59
1.1 to 2.0	$80.00	$96.00		$98.16	$115.00	107
2.1 to 3.0	$85.00	$101.10		$101.47	$116.30	73
3.1 or More	$79.36	$95.75		$97.17	$115.29	72
Member Status						
Accredited Practice Member	$80.50	$98.33		$100.29	$119.00	92
Nonaccredited Member	$85.00	$100.00		$102.99	$116.54	73
Nonmember	$75.00	$91.65		$93.58	$110.68	132
Metropolitan Status						
Urban	*	*		*	*	19
Suburban	$85.25	$101.05		$103.09	$117.63	92
Second City	$79.33	$93.78		$99.06	$119.18	57
Town	$75.00	$96.00		$95.42	$114.75	111
Rural	$67.25	$90.50		$88.45	$106.98	34
Median Area Household Income						
$35,000 or Less	$60.00	$75.00		$77.77	$93.13	32
$35,000 to $49,999	$82.00	$98.50		$98.94	$115.00	115
$50,000 to $69,999	$83.13	$96.25		$99.38	$118.59	108
$70,000 or More	$90.75	$106.00		$108.01	$121.46	54

Note 1: An asterisk indicates that data was not reported due to an insufficient number of responses.
Note 2: 37% of respondents reported that they do not offer this service.

TABLE 8.28
ROUTINE RADIOGRAPHY: DENTAL

	25th Percentile	Median	Your Data	Average	75th Percentile	Number of Respondents
All Practices	$38.59	$63.90		$68.48	$96.38	237
Number of FTE Veterinarians						
1.0 or Less	$38.33	$55.00		$62.80	$83.00	47
1.1 to 2.0	$37.00	$65.00		$73.61	$101.00	63
2.1 to 3.0	$40.00	$75.00		$72.39	$98.00	55
3.1 or More	$37.50	$55.00		$63.10	$93.25	65
Member Status						
Accredited Practice Member	$33.93	$52.79		$61.92	$85.75	90
Nonaccredited Member	$35.00	$60.00		$65.92	$98.00	51
Nonmember	$46.32	$80.00		$79.12	$110.00	77
Metropolitan Status						
Urban	*	*		*	*	23
Suburban	$34.25	$55.00		$64.03	$95.75	76
Second City	$36.18	$59.09		$67.38	$90.23	46
Town	$39.96	$65.00		$68.24	$98.13	66
Rural	*	*		*	*	20
Median Area Household Income						
$35,000 or Less	*	*		*	*	20
$35,000 to $49,999	$37.25	$53.12		$63.66	$89.23	84
$50,000 to $69,999	$36.75	$71.20		$70.18	$100.00	82
$70,000 or More	$39.00	$72.52		$72.47	$101.00	43

Note 1: Fee includes the following views: rostral and caudal mandible and rostral and caudal maxilla.
Note 2: An asterisk indicates that data was not reported due to an insufficient number of responses.
Note 3: 52% of respondents reported that they do not offer this service.

TABLE 8.29
MYELOGRAMS: TWO VIEWS, CERVICAL SPINE

	25th Percentile	Median	Your Data	Average	75th Percentile	Number of Respondents
All Practices	$84.38	$102.55		$109.44	$124.25	60

Note 1: Data was reported only for all practices due to an insufficient number of responses.
Note 2: 87% of respondents reported that they do not offer this service.

TABLE 8.30
ULTRASOUND UNIT IN-HOUSE

	Yes	No	Number of Respondents
All Practices	**45%**	**55%**	**564**
Number of FTE Veterinarians			
1.0 or Less	21%	79%	154
1.1 to 2.0	40%	60%	167
2.1 to 3.0	53%	47%	117
3.1 or More	77%	23%	108
Member Status			
Accredited Practice Member	60%	40%	147
Nonaccredited Member	51%	49%	116
Nonmember	34%	66%	250
Metropolitan Status			
Urban	44%	56%	41
Suburban	52%	48%	161
Second City	47%	53%	114
Town	42%	58%	173
Rural	35%	65%	65
Median Area Household Income			
$35,000 or Less	25%	75%	67
$35,000 to $49,999	44%	56%	209
$50,000 to $69,999	52%	48%	180
$70,000 or More	49%	51%	94

TABLE 8.31
ULTRASOUND: CHEST AND ABDOMEN

	25th Percentile	Median	Your Data	Average	75th Percentile	Number of Respondents
All Practices	**$166.25**	**$275.00**		**$271.33**	**$357.50**	**197**
Number of FTE Veterinarians						
1.0 or Less	$178.63	$280.00		$286.17	$355.00	27
1.1 to 2.0	$189.50	$275.00		$273.32	$350.00	55
2.1 to 3.0	$148.75	$237.22		$257.20	$371.25	46
3.1 or More	$163.75	$275.00		$262.63	$321.25	62
Member Status						
Accredited Practice Member	$179.10	$284.75		$285.56	$373.75	72
Nonaccredited Member	$140.00	$260.00		$273.29	$353.25	46
Nonmember	$125.00	$275.00		$251.52	$350.00	59
Metropolitan Status						
Urban	*	*		*	*	15
Suburban	$200.00	$285.00		$299.76	$415.00	59
Second City	$183.38	$293.30		$287.15	$380.00	48
Town	$139.75	$261.25		$250.95	$333.61	53
Rural	*	*		*	*	21
Median Area Household Income						
$35,000 or Less	*	*		*	*	15
$35,000 to $49,999	$166.48	$275.00		$267.61	$357.50	76
$50,000 to $69,999	$191.88	$265.50		$272.44	$355.00	70
$70,000 or More	$184.75	$320.00		$305.39	$430.00	33

Note 1: The fee includes procedure and interpretation but excludes the office call/exam fee.
Note 2: An asterisk indicates that data was not reported due to an insufficient number of responses.
Note 3: 15% of respondents reported that they do not offer this service.

TABLE 8.32
ULTRASOUND: CHEST ONLY

	25th Percentile	Median	Your Data	Average	75th Percentile	Number of Respondents
All Practices	**$130.38**	**$195.00**		**$201.30**	**$259.13**	**216**
Number of FTE Veterinarians						
1.0 or Less	$155.44	$200.00		$224.06	$280.00	31
1.1 to 2.0	$122.50	$215.00		$198.71	$261.89	58
2.1 to 3.0	$143.50	$194.37		$204.95	$271.35	52
3.1 or More	$125.00	$175.00		$185.04	$234.30	67
Member Status						
Accredited Practice Member	$142.25	$194.00		$206.88	$250.00	76
Nonaccredited Member	$120.00	$200.00		$206.59	$275.00	51
Nonmember	$106.63	$192.87		$188.26	$250.00	68
Metropolitan Status						
Urban	*	*		*	*	15
Suburban	$150.00	$223.06		$221.52	$295.00	71
Second City	$143.53	$196.90		$204.32	$246.25	50
Town	$139.88	$192.50		$196.31	$250.00	58
Rural	*	*		*	*	21
Median Area Household Income						
$35,000 or Less	*	*		*	*	14
$35,000 to $49,999	$125.00	$189.50		$193.80	$250.00	81
$50,000 to $69,999	$145.00	$204.50		$212.61	$258.00	79
$70,000 or More	$150.00	$220.00		$215.96	$295.00	39

Note 1: The fee includes procedure and interpretation but excludes the office call/exam fee.
Note 2: An asterisk indicates that data was not reported due to an insufficient number of responses.
Note 3: 11% of respondents reported that they do not offer this service.

TABLE 8.33
ULTRASOUND: ABDOMEN ONLY

	25th Percentile	Median	Your Data	Average	75th Percentile	Number of Respondents
All Practices	$127.00	$181.65		$192.13	$250.00	244
Number of FTE Veterinarians						
1.0 or Less	$150.00	$190.00		$211.75	$271.59	32
1.1 to 2.0	$117.50	$207.50		$190.89	$256.00	64
2.1 to 3.0	$130.75	$185.00		$189.00	$250.00	61
3.1 or More	$123.75	$165.75		$180.97	$225.50	78
Member Status						
Accredited Practice Member	$135.80	$191.20		$196.61	$240.00	87
Nonaccredited Member	$120.50	$179.75		$191.84	$262.35	56
Nonmember	$100.06	$178.20		$183.12	$250.00	79
Metropolitan Status						
Urban	*	*		*	*	16
Suburban	$146.25	$221.53		$215.73	$275.00	80
Second City	$132.00	$183.30		$195.70	$244.35	53
Town	$132.25	$164.33		$181.22	$242.50	70
Rural	*	*		*	*	23
Median Area Household Income						
$35,000 or Less	*	*		*	*	15
$35,000 to $49,999	$100.03	$165.00		$179.56	$246.85	89
$50,000 to $69,999	$143.17	$200.00		$205.52	$250.00	89
$70,000 or More	$155.00	$221.00		$214.85	$277.50	45

Note 1: The fee includes procedure and interpretation but excludes the office call/exam fee.
Note 2: An asterisk indicates that data was not reported due to an insufficient number of responses.
Note 3: 2% of respondents reported that they do not offer this service.

TABLE 8.34
ULTRASOUND: GUIDED BIOPSY COLLECTION, LIVER

	25th Percentile	Median	Your Data	Average	75th Percentile	Number of Respondents
All Practices	$78.41	$121.00		$150.32	$198.31	158
Number of FTE Veterinarians						
1.0 or Less	*	*		*	*	19
1.1 to 2.0	$65.00	$117.50		$133.23	$177.50	42
2.1 to 3.0	$77.50	$120.00		$148.19	$177.50	41
3.1 or More	$77.83	$110.00		$159.38	$198.88	49
Member Status						
Accredited Practice Member	$71.50	$110.00		$142.53	$175.00	61
Nonaccredited Member	$75.00	$100.47		$131.10	$175.00	38
Nonmember	$85.07	$125.00		$159.67	$200.00	43
Metropolitan Status						
Urban	*	*		*	*	13
Suburban	$91.88	$130.63		$162.46	$211.25	48
Second City	$77.24	$103.00		$159.68	$217.50	36
Town	$75.00	$120.00		$132.64	$175.00	47
Rural	*	*		*	*	13
Median Area Household Income						
$35,000 or Less	*	*		*	*	10
$35,000 to $49,999	$78.41	$121.00		$147.63	$198.31	62
$50,000 to $69,999	$80.00	$129.75		$159.56	$211.25	56
$70,000 or More	$80.00	$120.00		$143.83	$175.00	27

Note 1: The fee includes procedure and interpretation but excludes the office call/exam fee.
Note 2: An asterisk indicates that data was not reported due to an insufficient number of responses.
Note 3: 28% of respondents reported that they do not offer this service.

DIAGNOSTIC IMAGING CASES

TABLE 8.35
DIAGNOSTIC IMAGING CASE ONE: TWO FILMS AND YOUR INTERPRETATION

	25th Percentile	Median	Your Data	Average	75th Percentile	Number of Respondents
All Practices	$87.00	$105.00		$108.95	$127.47	510
Number of FTE Veterinarians						
1.0 or Less	$82.68	$99.75		$106.26	$127.93	129
1.1 to 2.0	$85.00	$105.00		$107.67	$125.85	156
2.1 to 3.0	$88.07	$106.00		$108.18	$124.90	108
3.1 or More	$90.00	$115.00		$113.44	$132.23	101
Member Status						
Accredited Practice Member	$90.25	$110.00		$114.62	$131.91	136
Nonaccredited Member	$86.25	$107.50		$107.91	$124.98	104
Nonmember	$81.00	$100.00		$104.54	$123.53	231
Metropolitan Status						
Urban	$88.38	$113.00		$113.54	$134.65	36
Suburban	$98.00	$113.00		$117.85	$135.00	143
Second City	$90.00	$112.68		$112.95	$129.45	106
Town	$80.00	$99.75		$102.65	$120.05	163
Rural	$72.64	$89.52		$94.42	$107.33	56
Median Area Household Income						
$35,000 or Less	$74.50	$94.75		$93.36	$114.10	62
$35,000 to $49,999	$85.00	$103.05		$106.56	$126.00	187
$50,000 to $69,999	$90.00	$112.00		$113.69	$130.00	164
$70,000 or More	$97.25	$112.75		$118.24	$140.00	88

Note 1: The patient is a two-year-old, 25-pound, male, mixed-breed dog with a lame right forelimb. Fee includes two views (of the radius and ulna) and interpretation and excludes fees for office call/exam, sedation, anesthesia, and hospitalization.
Note 2: 4% of respondents reported that they do not offer this service.

TABLE 8.36
DIAGNOSTIC IMAGING CASE ONE: TWO FILMS AND SPECIALIST INTERPRETATION FEE

	25th Percentile	Median	Your Data	Average	75th Percentile	Number of Respondents
All Practices	$129.50	$160.00		$160.79	$194.00	237
Number of FTE Veterinarians						
1.0 or Less	$130.00	$156.50		$155.47	$180.00	43
1.1 to 2.0	$127.88	$157.25		$157.07	$177.63	70
2.1 to 3.0	$131.75	$166.17		$165.97	$200.24	62
3.1 or More	$121.50	$153.13		$159.30	$199.04	56
Member Status						
Accredited Practice Member	$131.10	$170.00		$175.73	$217.80	71
Nonaccredited Member	$129.50	$163.57		$158.33	$180.38	50
Nonmember	$125.00	$154.80		$149.39	$176.00	91
Metropolitan Status						
Urban	*	*		*	*	23
Suburban	$145.14	$168.25		$169.19	$197.83	70
Second City	$136.50	$168.80		$169.79	$203.55	47
Town	$120.75	$142.20		$153.05	$189.90	74
Rural	*	*		*	*	21
Median Area Household Income						
$35,000 or Less	*	*		*	*	20
$35,000 to $49,999	$125.75	$155.00		$156.20	$186.05	98
$50,000 to $69,999	$143.40	$169.50		$172.23	$207.50	75
$70,000 or More	$123.28	$163.00		$159.38	$207.00	39

Note 1: The patient is a two-year-old, 25-pound, male, mixed-breed dog with a lame right forelimb. Fee includes two views (of the radius and ulna) and specialist's interpretation and excludes fees for office call/exam, sedation, anesthesia, and hospitalization.
Note 2: An asterisk indicates that data was not reported due to an insufficient number of responses.

TABLE 8.37
DIAGNOSTIC IMAGING CASE TWO: CYSTOGRAM PROCEDURE

	25th Percentile	Median	Your Data	Average	75th Percentile	Number of Respondents
All Practices	$50.00	$75.00		$94.54	$125.40	183
Number of FTE Veterinarians						
1.0 or Less	$50.00	$68.50		$86.63	$120.00	40
1.1 to 2.0	$36.38	$67.50		$81.26	$120.05	50
2.1 to 3.0	$48.99	$75.00		$97.19	$146.32	44
3.1 or More	$53.88	$88.00		$110.25	$143.38	42
Member Status						
Accredited Practice Member	$42.07	$83.25		$97.44	$136.57	46
Nonaccredited Member	$50.00	$75.00		$95.21	$132.50	31
Nonmember	$50.00	$75.50		$91.99	$125.00	92
Metropolitan Status						
Urban	*	*		*	*	17
Suburban	$50.13	$80.00		$99.82	$143.25	52
Second City	$50.00	$79.50		$96.84	$129.50	36
Town	$34.76	$60.00		$78.44	$100.00	57
Rural	*	*		*	*	18
Median Area Household Income						
$35,000 or Less	*	*		*	*	20
$35,000 to $49,999	$40.25	$73.00		$87.58	$120.15	68
$50,000 to $69,999	$50.00	$80.00		$93.11	$133.89	61
$70,000 or More	$53.75	$87.50		$119.89	$191.75	30

Note 1: An asterisk indicates that data was not reported due to an insufficient number of responses.
Note 2: 46% of respondents reported that they do not offer this service.

TABLE 8.38
DIAGNOSTIC IMAGING CASE TWO: CONTRAST MATERIALS

	25th Percentile	Median	Your Data	Average	75th Percentile	Number of Respondents
All Practices	**$20.00**	**$27.00**		**$31.69**	**$38.00**	**163**
Number of FTE Veterinarians						
1.0 or Less	$20.00	$25.00		$34.39	$35.93	36
1.1 to 2.0	$16.88	$23.68		$29.33	$35.75	50
2.1 to 3.0	$20.00	$25.60		$30.87	$35.50	33
3.1 or More	$20.00	$32.50		$32.13	$45.50	39
Member Status						
Accredited Practice Member	$20.00	$27.82		$32.48	$38.20	43
Nonaccredited Member	$19.50	$25.00		$28.14	$35.75	34
Nonmember	$20.00	$27.78		$33.36	$40.00	74
Metropolitan Status						
Urban	*	*		*	*	13
Suburban	$20.00	$27.00		$30.92	$38.20	47
Second City	$19.50	$26.45		$28.58	$35.00	34
Town	$16.63	$22.10		$27.84	$34.88	48
Rural	*	*		*	*	19
Median Area Household Income						
$35,000 or Less	*	*		*	*	15
$35,000 to $49,999	$16.60	$25.00		$33.92	$39.38	61
$50,000 to $69,999	$21.25	$28.19		$32.42	$40.00	52
$70,000 or More	$19.25	$25.00		$29.39	$35.00	32

Note 1: An asterisk indicates that data was not reported due to an insufficient number of responses.
Note 2: 14% of the respondents reported that this charge is included in the total charge for radiography.
Note 3: 42% of respondents reported that they do not offer this service.

TABLE 8.39
DIAGNOSTIC IMAGING CASE TWO: BLADDER CATHETERIZATION

	25th Percentile	Median	Your Data	Average	75th Percentile	Number of Respondents
All Practices	$20.00	$30.00		$32.90	$41.89	234
Number of FTE Veterinarians						
1.0 or Less	$20.00	$30.00		$34.06	$44.15	57
1.1 to 2.0	$18.13	$27.63		$31.71	$43.00	68
2.1 to 3.0	$18.82	$27.00		$32.29	$36.50	53
3.1 or More	$24.25	$31.09		$33.30	$39.75	48
Member Status						
Accredited Practice Member	$20.50	$30.00		$33.35	$39.97	61
Nonaccredited Member	$20.00	$29.00		$32.79	$42.43	49
Nonmember	$19.50	$28.00		$32.42	$42.55	107
Metropolitan Status						
Urban	*	*		*	*	20
Suburban	$20.00	$30.00		$33.94	$45.00	66
Second City	$21.15	$30.00		$38.67	$45.62	49
Town	$18.00	$26.50		$28.39	$35.00	71
Rural	$19.38	$28.50		$34.26	$42.64	26
Median Area Household Income						
$35,000 or Less	*	*		*	*	23
$35,000 to $49,999	$18.00	$27.75		$32.12	$39.01	88
$50,000 to $69,999	$20.00	$30.00		$33.43	$43.55	76
$70,000 or More	$25.25	$32.25		$36.55	$44.70	44

Note 1: An asterisk indicates that data was not reported due to an insufficient number of responses.
Note 2: 9% of the respondents reported that this charge is included in the total charge for radiography.
Note 3: 31% of respondents reported that they do not offer this service.

TABLE 8.40
DIAGNOSTIC IMAGING CASE TWO: DOUBLE CONTRAST CYSTOGRAM PROCEDURE

	25th Percentile	Median	Your Data	Average	75th Percentile	Number of Respondents
All Practices	**$45.00**	**$75.00**		**$102.64**	**$150.00**	**167**
Number of FTE Veterinarians						
1.0 or Less	$39.25	$55.00		$92.97	$147.24	33
1.1 to 2.0	$33.95	$55.00		$84.86	$125.75	45
2.1 to 3.0	$50.00	$82.50		$111.39	$159.27	42
3.1 or More	$50.00	$85.25		$114.14	$176.25	40
Member Status						
Accredited Practice Member	$50.00	$98.50		$118.63	$183.64	48
Nonaccredited Member	$50.00	$90.00		$104.53	$154.35	32
Nonmember	$35.00	$66.85		$91.89	$132.49	75
Metropolitan Status						
Urban	*	*		*	*	15
Suburban	$45.25	$73.55		$100.12	$150.00	44
Second City	$50.00	$100.00		$115.08	$169.75	37
Town	$33.00	$50.00		$81.80	$117.00	55
Rural	*	*		*	*	14
Median Area Household Income						
$35,000 or Less	*	*		*	*	24
$35,000 to $49,999	$40.00	$55.00		$83.44	$119.50	59
$50,000 to $69,999	$50.00	$95.00		$119.68	$172.28	49
$70,000 or More	$48.00	$90.00		$121.70	$199.25	33

Note 1: An asterisk indicates that data was not reported due to an insufficient number of responses.
Note 2: 7% of the respondents reported that this charge is included in the total charge for radiography.
Note 3: 46% of respondents reported that they do not offer this service.

TABLE 8.41
DIAGNOSTIC IMAGING CASE TWO: ALL FILMS AND YOUR INTERPRETATION

	25th Percentile	Median	Your Data	Average	75th Percentile	Number of Respondents
All Practices	$120.50	$155.86	_____	$166.48	$202.30	241
Number of FTE Veterinarians						
1.0 or Less	$105.00	$142.30	_____	$157.28	$183.23	60
1.1 to 2.0	$119.00	$159.12	_____	$162.02	$202.60	67
2.1 to 3.0	$121.48	$155.86	_____	$166.55	$203.44	53
3.1 or More	$127.39	$158.70	_____	$178.35	$208.00	50
Member Status						
Accredited Practice Member	$128.00	$165.00	_____	$170.46	$200.00	67
Nonaccredited Member	$119.00	$158.62	_____	$171.42	$223.00	51
Nonmember	$114.21	$150.00	_____	$162.40	$200.00	106
Metropolitan Status						
Urban	*	*	_____	*	*	16
Suburban	$133.00	$165.00	_____	$182.48	$215.15	73
Second City	$119.10	$158.19	_____	$169.25	$206.00	48
Town	$105.13	$153.00	_____	$155.31	$199.25	73
Rural	$91.25	$130.00	_____	$133.08	$165.00	28
Median Area Household Income						
$35,000 or Less	$81.00	$129.26	_____	$126.79	$173.25	29
$35,000 to $49,999	$117.19	$150.00	_____	$156.86	$189.25	84
$50,000 to $69,999	$131.00	$165.00	_____	$182.57	$223.84	79
$70,000 or More	$132.00	$185.00	_____	$186.71	$246.00	43

Note 1: An asterisk indicates that data was not reported due to an insufficient number of responses.
Note 2: 5% of the respondents reported that this charge is included in the total charge for radiography.
Note 3: 31% of respondents reported that they do not offer this service.

TABLE 8.42
DIAGNOSTIC IMAGING CASE TWO: TOTAL

	25th Percentile	Median	Your Data	Average	75th Percentile	Number of Respondents
All Practices	**$181.61**	**$255.85**		**$283.85**	**$359.91**	**308**
Number of FTE Veterinarians						
1.0 or Less	$150.00	$220.00		$258.78	$295.13	74
1.1 to 2.0	$169.63	$235.95		$266.19	$300.92	84
2.1 to 3.0	$187.75	$268.65		$284.89	$382.88	72
3.1 or More	$222.00	$285.90		$312.92	$400.00	67
Member Status						
Accredited Practice Member	$203.28	$269.25		$294.45	$391.26	85
Nonaccredited Member	$187.25	$249.05		$275.05	$336.00	64
Nonmember	$168.70	$246.50		$279.05	$371.25	138
Metropolitan Status						
Urban	*	*		*	*	23
Suburban	$210.36	$283.31		$289.24	$376.75	92
Second City	$180.90	$255.70		$297.38	$405.75	63
Town	$155.00	$222.00		$249.05	$301.15	95
Rural	$167.88	$239.50		$259.07	$327.13	32
Median Area Household Income						
$35,000 or Less	$127.00	$207.50		$242.36	$343.59	36
$35,000 to $49,999	$164.33	$242.51		$253.95	$300.00	114
$50,000 to $69,999	$200.78	$285.62		$305.00	$399.00	99
$70,000 or More	$221.12	$293.50		$343.21	$403.50	53

Note 1: The total fee for diagnostic imaging case two is the sum of the fees for the cystogram procedure, contrast materials, bladder catheterization, double contrast cystogram procedure, films, and interpretation. The total fee does not include fees for office call/examination, sedation, anesthesia, or hospitalization. Some respondents do not charge for some of the individual services included in this package, though the services are provided as part of the service. In addition, some respondents reported that they do not offer some of the individual services. Therefore, the average total fee for the case may be significantly lower than the sum of the average fees for the individual services.

Note 2: An asterisk indicates that data was not reported due to an insufficient number of responses.

CHAPTER 9

PRESCRIPTION MEDICATIONS

Medicines and supplies constitute the second-largest expense in veterinary hospitals—and one of the more profitable centers of your practice. Use these tips to show clients the value of your pharmacy.

EDUCATING CLIENTS PAYS OFF

You can easily justify a higher fee for medical services based on how you differentiate your practice from colleagues', but it's nearly impossible to justify higher prices for medicines and supplies. Internet pharmacies and pet-supply companies are aggressively soliciting consumers who like bargains, but do your clients realize that those "bargains" often come with hidden fees?

It might be tempting to lower your prices to stay competitive, but leading practice-management consultants and savvy practice owners realize the best defense is client education. The doctors at Veterinary Medical Clinic in Tampa, Florida, find ways to work with clients who are looking for a good deal without lowering prices. Owner Dr. Eddie Garcia and his staff charge a dispensing fee for prescriptions filled, offer free delivery and discounts for clients who purchase a year's worth of product up front, and take advantage of vendors' rebate offers for clients. Most of all, they stress the importance of manufacturer guarantees for veterinarian-dispensed medications. One tactic they don't use, however, is lowering their prices.

Dr. Garcia also stresses the importance of staying one step ahead of the clients who are filling prescriptions via the Internet. Any request that comes into his clinic is handled by the staff. They pull the medical record, call the client, and explain their protocols. Dr. Garcia offers these additional suggestions:

- Offer free delivery of the product
- Offer to send email reminders to give the product to the pet(s)
- Offer one year of the product at a 10% discount plus any vendor coupons or promotions
- Explain the benefit of product guarantees offered by the manufacturer that may be available only if the product is dispensed by the veterinary clinic

"Don't try to compete on price," advises Denise Tumblin, CPA, president and owner of Wutchiett Tumblin and Associates in Columbus, Ohio. "Educate your clients about the added value they receive when they purchase from you, particularly product quality control, reputable vendors, and addressing adverse reactions." Tumblin notes that most of the practices in the Wutchiett Tumblin and Associates Well-Managed Companion Animal Practices database apply a 100% markup on heartworm and flea preventive products and either exclude the dispensing fee or charge one that is lower than the dispensing fee for counted and packaged medications.

Markups for other medications, she says, range from 135% to 175%, plus a dispensing fee. Dispensing fees average around $8. "We don't recommend a discount rate or elimination of the dispensing fee for clients whose pets need multiple prescriptions," she adds, "but if a client buys in bulk, the dispensing fee is commonly excluded."

BUYING STRATEGIES

Most veterinarians agree that the potential lost sales aren't worth the risk of losing a valued client. Buying groups, such as AAHA MARKETLink, give smaller practices an edge in purchasing pharmaceuticals and supplies so they can remain competitive. For example, AAHA MARKETLink stocks about 23,000 products from more than 350 vendors.

Judi Kodner Casey, the marketing coordinator for AAHA MARKETLink, says the buying group saves practice owners even more by freeing up staff time. She points to studies that indicate that placing an order takes four to five hours from start to finish, including the time it takes to call the vendor, open boxes, stock the product, and pay the bill. Shopping around for deals takes even more time, and the cost adds up. With a buying group, one phone call takes care of it all.

This is where establishing a relationship with vendors makes good business sense. Many vendors willingly will delay billing for you when you buy in bulk, and you can pocket the savings as net profit.

Implementation Idea

Plan Ahead for Profit and Cash Flow

Heartworm and flea and tick preventives are dispensed year round in many practices. Let's say you average $20,000 in sales for heartworm and flea preventives each month. If you purchase a year's worth of products for $120,000 on January 1 and your vendor delays billing until September 1, you can enjoy significant profit. As you can see in Table 9.A, if you save $13,500 per month for nine months (assumes 0% interest rate), you'll make a profit and have plenty of cash to pay the bill when it's due.

With delayed billing, it's vital to be disciplined about saving the money you'll need in the future to pay the bill. Each month, set aside a certain amount of money in a separate account so that you have the money when the balance is due. It makes even more sense if you put the money in an interest-bearing money-market mutual fund.

TABLE 9.A

HEARTWORM AND FLEA PREVENTIVE COMBINED SALES

Month	Sales	Savings Account Balance	Payment Due	Cumulative Profit
January	$20,000	$13,500		$6,500
February	$20,000	$27,000		$13,000
March	$20,000	$40,500		$19,500
April	$20,000	$54,000		$26,000
May	$20,000	$67,500		$32,500
June	$20,000	$81,000		$39,000
July	$20,000	$94,500		$45,500
August	$20,000	$108,000		$52,000
September	$20,000	$121,500	-$120,000	$58,500
October	$20,000	$1,500		$78,500
November	$20,000	$1,500		$98,500
December	$20,000	$1,500		$118,500

VALUE YOUR EXPERTISE

It's true that with convenience comes complacency. It's easy for clients to administer many medicines to their pets, and they may inadvertently downplay the fact that they are introducing foreign chemicals into their pets' systems. By educating clients about possible adverse reactions, interaction risks with multiple prescriptions, and the manufacturer incentives that come with veterinarian-dispensed medications, you further establish yourself as the medical expert.

It is a relationship that is especially crucial with an aging pet population and their growing medical needs. You can stay competitive and still make a profit. It's all about knowledge: The more clients know about your medical expertise, the more value they will place in your recommendations.

Case Study

How AAHA MARKETLink Helps Provide Better Medical Care

At Fur and Feathers Animal Clinic in Detroit, Michigan, owner Dr. Justin Anderson knows professional services are his practice's most valuable asset. He says being a member of AAHA's MARKETLink buying group helps. Dr. Anderson has reduced his medical and supply inventory by $14,000 a month. Hours he and his staff once spent shopping for vendor deals are now spent with patients and clients. Best of all, space between his practice's two exam rooms that the pharmacy once took up now houses a mini-laboratory where technicians can perform fecal analysis, urinalysis, and heartworm tests. "I can go to one provider and know I'm getting the best price," he says. "I order my supplies and medications in two weekly shipments. If I needed it, I could get daily shipments."

One aspect of AAHA MARKETLink that makes it stand out from other buying groups, Dr. Anderson notes, is that order representatives are registered veterinary technicians. "The people who handle inventory for me are registered technicians, and they all speak the same language," he says. "They recommend products or suggest new ways to use them." It's service that he can take to the bank.

Note: The name of the practice and the doctor who was interviewed for this case study have been changed.

HELPFUL RESOURCES

"Managing Your Inventory Investment" seminar, offered by AAHA MARKETLink. Visit www.aahamarketlink.com for more information.

Practice Made Perfect, Marsha L. Heinke, DVM, EA, CPA, CVPM and John B. McCarthy, DVM, MBA. American Animal Hospital Association Press, 2001.

CHAPTER 9

DATA TABLES

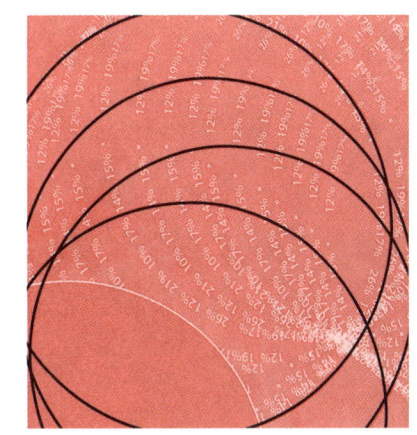

TABLE 9.1
PRESCRIPTION FEE FOR MEDICATION DISPENSED FROM YOUR HOSPITAL

	25th Percentile	Median	Your Data	Average	75th Percentile	Number of Respondents
All Practices	**$5.00**	**$8.00**		**$7.95**	**$10.00**	**476**
Number of FTE Veterinarians						
1.0 or Less	$5.00	$7.15		$7.53	$9.50	123
1.1 to 2.0	$6.00	$8.00		$8.64	$10.00	139
2.1 to 3.0	$5.00	$7.50		$7.97	$10.00	101
3.1 or More	$5.00	$7.83		$7.53	$9.28	98
Member Status						
Accredited Practice Member	$6.50	$8.00		$8.66	$10.00	139
Nonaccredited Member	$5.00	$8.00		$8.32	$10.00	96
Nonmember	$5.00	$7.00		$7.51	$10.00	200
Metropolitan Status						
Urban	$5.00	$7.50		$7.40	$9.25	31
Suburban	$6.00	$8.00		$8.84	$11.09	133
Second City	$5.00	$8.00		$8.01	$9.00	97
Town	$5.00	$7.50		$7.51	$9.96	150
Rural	$4.00	$7.00		$7.26	$9.50	57
Median Area Household Income						
$35,000 or Less	$5.00	$7.00		$7.06	$9.50	57
$35,000 to $49,999	$5.00	$7.22		$7.49	$9.00	181
$50,000 to $69,999	$5.50	$8.00		$8.50	$10.42	148
$70,000 or More	$6.00	$8.00		$8.57	$10.00	77

Note: 9% of respondents reported that they do not offer this service.

TABLE 9.2
PRESCRIPTION FEE ASSESSED IF CLIENT HAS PRESCRIPTION FILLED ELSEWHERE

	Yes	No	Number of Respondents
All Practices	**18%**	**82%**	**561**
Number of FTE Veterinarians			
1.0 or Less	17%	83%	151
1.1 to 2.0	20%	80%	169
2.1 to 3.0	21%	79%	117
3.1 or More	12%	88%	109
Member Status			
Accredited Practice Member	16%	84%	147
Nonaccredited Member	14%	86%	116
Nonmember	21%	79%	248
Metropolitan Status			
Urban	17%	83%	41
Suburban	10%	90%	157
Second City	16%	84%	114
Town	21%	79%	175
Rural	25%	75%	64
Median Area Household Income			
$35,000 or Less	26%	74%	69
$35,000 to $49,999	20%	80%	205
$50,000 to $69,999	13%	87%	181
$70,000 or More	9%	91%	92

TABLE 9.3
PRESCRIPTION FEE IF CLIENT HAS PRESCRIPTION FILLED ELSEWHERE

	25th Percentile	Median	Your Data	Average	75th Percentile	Number of Respondents
All Practices	$5.00	$8.00		$8.91	$10.00	96

Note: Data was reported only for all practices due to an insufficient number of responses.

TABLE 9.4
AVERAGE MARKUP: PRESCRIPTION MEDICATIONS

	25th Percentile	Median	Your Data	Average	75th Percentile	Number of Respondents
All Practices	100%	100%		127%	150%	540
Number of FTE Veterinarians						
1.0 or Less	100%	100%		124%	150%	144
1.1 to 2.0	100%	103%		129%	150%	164
2.1 to 3.0	100%	100%		125%	150%	112
3.1 or More	100%	120%		131%	150%	104
Member Status						
Accredited Practice Member	100%	130%		135%	150%	142
Nonaccredited Member	100%	100%		130%	150%	114
Nonmember	100%	100%		125%	150%	240
Metropolitan Status						
Urban	100%	100%		130%	160%	35
Suburban	100%	100%		127%	150%	152
Second City	100%	120%		138%	150%	111
Town	100%	100%		127%	150%	169
Rural	100%	100%		115%	130%	63
Median Area Household Income						
$35,000 or Less	100%	100%		123%	150%	66
$35,000 to $49,999	100%	100%		126%	150%	195
$50,000 to $69,999	100%	110%		130%	150%	175
$70,000 or More	100%	123%		133%	150%	90

Note: Markup is defined as [(Selling Price - Cost of Product)/Cost of Product] x 100.

TABLE 9.5
AVERAGE MARKUP: HEARTWORM PREVENTIVES

	25th Percentile	Median	Your Data	Average	75th Percentile	Number of Respondents
All Practices	**80%**	**100%**		**94%**	**100%**	**506**
Number of FTE Veterinarians						
1.0 or Less	80%	100%		94%	100%	135
1.1 to 2.0	75%	100%		92%	100%	152
2.1 to 3.0	71%	100%		92%	100%	104
3.1 or More	100%	100%		103%	105%	102
Member Status						
Accredited Practice Member	89%	100%		99%	100%	128
Nonaccredited Member	73%	100%		91%	100%	105
Nonmember	80%	100%		96%	100%	231
Metropolitan Status						
Urban	70%	100%		87%	100%	35
Suburban	76%	100%		99%	100%	138
Second City	80%	100%		98%	100%	104
Town	80%	100%		92%	100%	159
Rural	80%	100%		90%	100%	61
Median Area Household Income						
$35,000 or Less	75%	100%		86%	100%	61
$35,000 to $49,999	89%	100%		96%	100%	188
$50,000 to $69,999	75%	100%		95%	100%	165
$70,000 or More	75%	100%		96%	100%	80

Note: Markup is defined as [(Selling Price - Cost of Product)/Cost of Product] x 100.

TABLE 9.6
AVERAGE MARKUP: FOOD

	25th Percentile	Median	Your Data	Average	75th Percentile	Number of Respondents
All Practices	**30%**	**40%**		**42%**	**45%**	**483**
Number of FTE Veterinarians						
1.0 or Less	25%	38%		41%	45%	128
1.1 to 2.0	30%	40%		42%	50%	151
2.1 to 3.0	30%	41%		40%	45%	94
3.1 or More	30%	45%		46%	50%	94
Member Status						
Accredited Practice Member	38%	45%		45%	50%	126
Nonaccredited Member	30%	40%		43%	50%	98
Nonmember	26%	37%		41%	45%	220
Metropolitan Status						
Urban	35%	45%		46%	50%	35
Suburban	30%	44%		41%	50%	138
Second City	32%	42%		46%	50%	98
Town	30%	40%		42%	45%	148
Rural	25%	35%		36%	45%	56
Median Area Household Income						
$35,000 or Less	27%	39%		37%	45%	58
$35,000 to $49,999	30%	40%		43%	45%	177
$50,000 to $69,999	30%	40%		44%	50%	156
$70,000 or More	30%	43%		41%	49%	81

Note: Markup is defined as [(Selling Price - Cost of Product)/Cost of Product] x 100.

TABLE 9.7
AVERAGE MARKUP: FLEA AND TICK PRODUCTS

	25th Percentile	Median	Your Data	Average	75th Percentile	Number of Respondents
All Practices	**66%**	**100%**		**87%**	**100%**	**507**
Number of FTE Veterinarians						
1.0 or Less	56%	100%		83%	100%	134
1.1 to 2.0	67%	100%		87%	100%	152
2.1 to 3.0	60%	90%		83%	100%	106
3.1 or More	78%	100%		94%	100%	101
Member Status						
Accredited Practice Member	80%	100%		92%	100%	131
Nonaccredited Member	64%	100%		86%	100%	105
Nonmember	60%	100%		87%	100%	229
Metropolitan Status						
Urban	54%	100%		79%	100%	35
Suburban	68%	100%		86%	100%	139
Second City	75%	100%		94%	100%	105
Town	70%	100%		87%	100%	159
Rural	50%	90%		80%	100%	60
Median Area Household Income						
$35,000 or Less	59%	95%		82%	100%	62
$35,000 to $49,999	70%	100%		89%	100%	187
$50,000 to $69,999	62%	100%		86%	100%	166
$70,000 or More	70%	100%		87%	100%	80

Note: Markup is defined as [(Selling Price - Cost of Product)/Cost of Product] x 100.

TABLE 9.8
AVERAGE MARKUP: OVER-THE-COUNTER PRODUCTS

	25th Percentile	Median	Your Data	Average	75th Percentile	Number of Respondents
All Practices	**80%**	**100%**		**102%**	**109%**	**497**
Number of FTE Veterinarians						
1.0 or Less	64%	100%		97%	100%	128
1.1 to 2.0	75%	100%		101%	120%	158
2.1 to 3.0	71%	100%		100%	100%	100
3.1 or More	100%	100%		104%	121%	98
Member Status						
Accredited Practice Member	100%	100%		109%	125%	132
Nonaccredited Member	88%	100%		106%	125%	107
Nonmember	75%	100%		99%	100%	218
Metropolitan Status						
Urban	55%	100%		91%	100%	30
Suburban	76%	100%		107%	129%	140
Second City	84%	100%		107%	110%	100
Town	90%	100%		101%	110%	162
Rural	50%	100%		89%	100%	55
Median Area Household Income						
$35,000 or Less	58%	100%		95%	100%	62
$35,000 to $49,999	90%	100%		102%	110%	180
$50,000 to $69,999	80%	100%		103%	125%	163
$70,000 or More	71%	100%		103%	120%	80

Note: Markup is defined as [(Selling Price - Cost of Product)/Cost of Product] x 100.

CHAPTER 10

FLUID THERAPY SERVICES

In this chapter, you'll learn about the benefits of fluid therapy, when and how you might incorporate more fluid therapy into your practice to increase the quality of care, and how it can increase your bottom line.

THE IMPORTANCE OF FLUID THERAPY

Fluid therapy plays a vital role in the management of the critical-care or emergency patient. But some practitioners don't give fluid therapy the attention it needs in the management of anesthetized pets, whether for a short dental cleaning or a lengthy hip replacement.

"A paradigm shift is needed," states Dr. Elisa Mazzafarro, PhD, DACVECC, of Wheat Ridge Animal Hospital in Wheat Ridge, Colorado. "Any type of anesthesia administration requires that a vascular access by intravenous catheter be placed in the animal." Dr. Mazzafarro says that practices currently incorporate fluid therapy into the surgical protocol for less than half of all surgeries. "Most clinics are focusing fluid therapy on the more critically ill and dehydrated patients."

The benefits of fluid therapy are numerous and as a result can be used for a variety of situations other than the most obvious (cases such as dehydration). *The Veterinary ICU Book* (Teton NewMedia, 2002) lists 37 different scenarios where fluid therapy could be implemented, including:

- Abdominal trauma
- Acute renal failure
- Acute respiratory distress syndrome
- Acute tumor lysis syndrome
- Aspiration pneumonitis
- Black-widow spider bite
- Brain trauma
- Brown-recluse spider bite
- Cardiovascular dysfunction
- Chemotherapy-induced GI toxicity
- Cholecalciferol rodenticide poisoning
- Diabetic ketoacidosis
- Diaphragmatic hernia
- Disseminated intravascular coagulation
- GI disorders with radiation exposure
- Heatstroke
- Hemorrhagic diarrhea
- Hemostasis disorders
- Hyperosmolar coma
- Hypoadrenocorticism
- Insect stings
- Multiorgan failure
- Oliguric renal failure
- Optimizing fluid volume and blood pressure
- Pancreatitis
- Pneumonia
- Portosystemic shunt
- Projectile thoracic injury
- Pulmonary contusion
- Pulmonary edema
- Scorpion stings
- Sepsis
- Severe liver disease
- Snakebites
- Strychnine poisoning
- Thermal injury
- Uremic gastrointestinal disorders

The most common reasons for including fluid therapy in patient care in any scenario are numerous and vary by case. Some of the most common reasons are:

- Improves tissue perfusion
- Replaces fluids lost as a result of dehydration
- Meets the needs of patients not consuming enough fluids on their own
- Replaces fluids lost as a result of diarrhea or vomiting
- Increases vascular volume and tissue perfusion during anesthesia administration
- Increases vascular volume due to blood loss

INCORPORATING FLUID THERAPY INTO YOUR PROTOCOLS

Fluid therapy is important to incorporate into any scenario where an emergency can occur during a routine procedure, as well as all critical-care and emergency scenarios. "Successful anesthesia happens at the microscopic level to keep organs healthy, maintain adequate blood pressure, and ensure perfusion, especially of the kidney," states Dr. Mazzafarro. "If this does not happen and the kidney suffers damage during surgery, that damage can cause cumulative insults over time, leading to kidney failure." To avoid this, she follows this protocol in her practice:

- Any pet that is administered any type of anesthesia requires vascular access by IV catheter.
- Keep the infusion pump on standby mode to help keep the catheter from clotting (also called "keep vein open" or "keep line open").
- Closely monitor blood pressure during surgeries for any changes.
- Provide IV fluids during surgery.

Case Study

Safety During Surgery

An IV catheter that is ready and in place during a procedure is like an insurance policy for the life of the pet. Consider the following scenario:

A healthy, six-month-old Siberian husky puppy comes into the practice for a traditional ovariohysterectomy. The practice just hired an associate who graduated from veterinary school a month earlier. The new graduate volunteers to perform the surgery. During surgery, he drops the ovarian pedicle, and the dog hemorrhages.

This causes the blood pressure to drop, but the associate followed the practice's standard of care to place an IV catheter in any patient receiving anesthesia. This allows him to react quickly, and he is able to administer fluid therapy immediately. This raises the blood pressure and maintains vital organ perfusion. Had the IV catheter not been in place, the puppy might have lost organ function and died.

Dr. Bernie Hansen, Associate Professor of Critical Care at North Carolina State University College of Veterinary Medicine, concurs. "The industry does not have a standard protocol for fluid therapy because the fluid protocol is completely situational," said Dr. Hansen. "The most important recommendation I can make is to insert an IV catheter for any surgery—elective or emergency—because it will help you to react immediately to an unforeseen emergency."

Certain procedures, like spays and neuters, have almost been made into commodity services because clients shop around for the best price. Some veterinary hospitals try to win new customers by providing the lowest price, but it may be at the expense of quality of care—by eliminating fluid therapy, for example.

Implementation Idea

Set a Standard of Care for Fluid Therapy

To set high standards for yourself, your staff, and your practice, implement a standard of care for fluid therapy:

1. Require an IV catheter to be placed in any animal that receives anesthesia.

2. Administer fluid therapy to every surgical patient.

3. Educate your staff, including receptionists, technicians, and associates, about the fees for IV catheter, IV fluids, and IV drip so that they are sure to include them on surgical estimates and invoices.

Train your staff to explain to the client what is included in a procedure, that your price is higher because you include fluid therapy in your surgical protocol, and how your practice's philosophy is to provide the best medicine for the health and safety of the pet. If the entire staff is in agreement, the information presented to the client will be much more powerful and credible.

ESTABLISHING FLUID THERAPY FEES

Dr. Mazzafarro suggests itemizing the fluid therapy fees on the invoice so that the client can see exactly how much and what was used during the procedures. At Wheat Ridge Animal Hospital, they break it down into three categories: IV catheter, IV fluid (noting amount used), and IV drip set. Although it is okay to categorize them together in one anesthetic fluid category, breaking them out keeps the client informed about the specifics of the procedure.

Implementing a standard of care for fluid therapy in all procedures requiring anesthesia in your practice would increase revenue and improve the standard of care for and well-being of your patients. For example, an invoice for a dental cleaning that now requires fluid therapy under your new standard of care could include $67.60, which includes $35.73 for the IV indwelling catheter and $31.87 for a liter of IV fluids.

You may also itemize the fee for placement of the IV indwelling catheter (average fee of $36.78), though many fee survey respondents reported that they include the fee for that service in the fee for the fluids.

If your practice performed just one dental cleaning a week, that would add $3,515 in revenue over the course of one year for dental cleanings alone. And if you calculate that your clinic did not include fluid therapy on 50% of the 500 cases requiring anesthesia over the course of one year (using $67.60 as the average amount for fluid therapy per case), then your clinic would add an additional $16,900 in annual revenue. But most importantly, your standard of care will improve, and clients will continue to be satisfied with your service knowing you are putting the best interests of the patients first.

HELPFUL RESOURCES

Fluid Therapy for Nurses and Technicians, by Paula Jane Houston-Moore, VN. Elsevier, 2004.

The Veterinary ICU Book, by Wayne E. Wingfield and Marc R. Raffe. Teton NewMedia, 2002.

Veterinary Emergency and Critical Care Manual, Second Edition, by Larry P. Tilley, DVM, DACVIM. Lifelearn, 2006. Available through the American Animal Hospital Association.

CHAPTER 10

DATA TABLES

TABLE 10.1
BUTTERFLY CATHETER

	25th Percentile	Median	Your Data	Average	75th Percentile	Number of Respondents
All Practices	$15.00	$25.00		$26.63	$36.19	216
Number of FTE Veterinarians						
1.0 or Less	$15.00	$25.00		$27.73	$38.95	68
1.1 to 2.0	$14.25	$24.20		$25.32	$32.63	65
2.1 to 3.0	$15.63	$28.73		$27.76	$39.75	34
3.1 or More	$15.00	$22.10		$25.72	$34.75	47
Member Status						
Accredited Practice Member	$12.63	$27.68		$27.64	$38.09	64
Nonaccredited Member	$7.50	$18.00		$20.70	$31.00	31
Nonmember	$15.75	$25.00		$26.34	$35.25	106
Metropolitan Status						
Urban	*	*		*	*	19
Suburban	$17.50	$29.00		$30.30	$42.00	63
Second City	$15.50	$24.50		$24.39	$35.25	41
Town	$12.19	$20.60		$23.71	$31.75	60
Rural	$10.00	$22.68		$23.14	$33.75	28
Median Area Household Income						
$35,000 or Less	$10.00	$24.20		$22.93	$31.50	25
$35,000 to $49,999	$10.00	$20.45		$23.41	$35.00	77
$50,000 to $69,999	$20.00	$29.70		$30.75	$40.00	74
$70,000 or More	$15.21	$27.60		$28.06	$36.39	36

Note 1: An asterisk indicates that data was not reported due to an insufficient number of responses.
Note 2: 14% of the respondents reported that this service is included in the fee for the fluid.
Note 3: 34% of respondents reported that they do not offer this service.

TABLE 10.2
IV INDWELLING CATHETER

	25th Percentile	Median	Your Data	Average	75th Percentile	Number of Respondents
All Practices	**$23.00**	**$35.00**		**$35.73**	**$45.25**	**343**
Number of FTE Veterinarians						
1.0 or Less	$19.60	$31.49		$33.36	$44.50	110
1.1 to 2.0	$23.15	$35.00		$34.77	$45.44	102
2.1 to 3.0	$26.94	$37.66		$37.02	$45.10	60
3.1 or More	$26.31	$39.18		$39.47	$47.55	66
Member Status						
Accredited Practice Member	$26.01	$37.28		$38.29	$47.20	100
Nonaccredited Member	$22.00	$35.00		$33.71	$42.50	55
Nonmember	$20.00	$32.00		$34.64	$47.00	167
Metropolitan Status						
Urban	$29.45	$39.95		$43.21	$59.09	34
Suburban	$28.00	$37.00		$38.85	$47.40	101
Second City	$21.25	$34.00		$33.59	$44.21	64
Town	$20.63	$34.25		$34.75	$45.00	92
Rural	$14.25	$24.20		$27.92	$37.79	46
Median Area Household Income						
$35,000 or Less	$18.56	$26.75		$29.88	$38.64	37
$35,000 to $49,999	$18.60	$30.00		$31.80	$42.00	125
$50,000 to $69,999	$28.00	$39.50		$38.72	$47.40	115
$70,000 or More	$30.92	$39.35		$42.43	$55.68	60

Note 1: 18% of the respondents reported that this service is included in the fee for the fluid.
Note 2: 4% of respondents reported that they do not offer this service.

TABLE 10.3
JUGULAR CATHETER

	25th Percentile	Median	Your Data	Average	75th Percentile	Number of Respondents
All Practices	**$28.00**	**$45.00**		**$46.48**	**$60.00**	**158**
Number of FTE Veterinarians						
1.0 or Less	$25.00	$39.00		$42.99	$58.00	51
1.1 to 2.0	$25.75	$45.05		$46.09	$57.50	46
2.1 to 3.0	$39.58	$45.87		$49.44	$62.60	26
3.1 or More	$33.00	$45.50		$49.31	$65.25	33
Member Status						
Accredited Practice Member	$34.00	$45.32		$50.91	$65.13	50
Nonaccredited Member	*	*		*	*	22
Nonmember	$27.50	$42.00		$45.02	$58.45	80
Metropolitan Status						
Urban	*	*		*	*	16
Suburban	$28.75	$45.00		$46.08	$64.25	42
Second City	$27.50	$46.00		$46.07	$60.00	27
Town	$27.00	$42.00		$43.12	$52.50	45
Rural	*	*		*	*	23
Median Area Household Income						
$35,000 or Less	*	*		*	*	18
$35,000 to $49,999	$22.12	$39.00		$40.54	$53.00	53
$50,000 to $69,999	$37.01	$49.65		$51.53	$62.63	57
$70,000 or More	$36.91	$46.80		$53.47	$70.30	26

Note 1: An asterisk indicates that data was not reported due to an insufficient number of responses.
Note 2: 11% of the respondents reported that this service is included in the fee for the fluid.
Note 3: 51% of respondents reported that they do not offer this service.

TABLE 10.4
CATHETER PLACEMENT: BUTTERFLY CATHETER

	25th Percentile	Median	Your Data	Average	75th Percentile	Number of Respondents
All Practices	$18.88	$26.49		$29.73	$38.73	132
Number of FTE Veterinarians						
1.0 or Less	$15.00	$24.00		$24.72	$32.00	41
1.1 to 2.0	$17.75	$26.00		$28.72	$38.04	41
2.1 to 3.0	$25.55	$36.93		$36.23	$44.63	32
3.1 or More	*	*		*	*	17
Member Status						
Accredited Practice Member	$20.23	$31.00		$32.21	$39.80	41
Nonaccredited Member	$15.00	$25.00		$26.98	$34.50	27
Nonmember	$18.88	$25.00		$27.86	$35.00	60
Metropolitan Status						
Urban	*	*		*	*	11
Suburban	$20.00	$30.00		$31.36	$39.66	33
Second City	$21.60	$30.00		$32.49	$40.00	31
Town	$17.03	$25.00		$26.98	$35.00	37
Rural	*	*		*	*	15
Median Area Household Income						
$35,000 or Less	*	*		*	*	16
$35,000 to $49,999	$15.00	$25.00		$27.23	$39.25	48
$50,000 to $69,999	$20.75	$29.00		$32.95	$40.93	42
$70,000 or More	*	*		*	*	21

Note 1: An asterisk indicates that data was not reported due to an insufficient number of responses.
Note 2: 29% of the respondents reported that this service is included in the fee for the fluid.
Note 3: 37% of respondents reported that they do not offer this service.

TABLE 10.5
CATHETER PLACEMENT: IV INDWELLING CATHETER

	25th Percentile	Median	Your Data	Average	75th Percentile	Number of Respondents
All Practices	$25.00	$35.00		$36.78	$46.03	215
Number of FTE Veterinarians						
1.0 or Less	$20.00	$30.00		$32.95	$40.00	71
1.1 to 2.0	$24.10	$36.00		$36.49	$46.65	61
2.1 to 3.0	$27.05	$35.15		$36.79	$44.63	48
3.1 or More	$31.50	$42.00		$45.06	$50.83	33
Member Status						
Accredited Practice Member	$27.00	$36.65		$38.82	$48.91	62
Nonaccredited Member	$25.00	$34.50		$36.42	$45.90	41
Nonmember	$23.11	$34.00		$35.66	$45.87	97
Metropolitan Status						
Urban	$28.00	$39.50		$42.65	$53.93	25
Suburban	$27.75	$36.90		$38.78	$48.00	52
Second City	$26.00	$36.00		$38.33	$50.00	53
Town	$23.50	$30.00		$32.14	$41.50	53
Rural	$22.00	$33.50		$31.75	$40.00	25
Median Area Household Income						
$35,000 or Less	$20.94	$26.38		$30.08	$37.25	26
$35,000 to $49,999	$24.05	$32.75		$34.96	$44.07	80
$50,000 to $69,999	$27.10	$37.91		$40.14	$49.61	65
$70,000 or More	$26.25	$37.72		$39.45	$45.87	37

Note 1: 37% of the respondents reported that this service is included in the fee for the fluid.
Note 2: 9% of respondents reported that they do not offer this service.

TABLE 10.6
CATHETER PLACEMENT: JUGULAR CATHETER

	25th Percentile	Median	Your Data	Average	75th Percentile	Number of Respondents
All Practices	$30.00	$45.55		$49.19	$65.00	101
Number of FTE Veterinarians						
1.0 or Less	$20.75	$36.75		$38.22	$48.80	26
1.1 to 2.0	$30.00	$46.50		$50.54	$64.00	31
2.1 to 3.0	$34.90	$60.00		$54.28	$68.61	29
3.1 or More	*	*		*	*	15
Member Status						
Accredited Practice Member	$37.51	$52.00		$53.17	$68.50	33
Nonaccredited Member	*	*		*	*	18
Nonmember	$24.63	$40.00		$45.40	$62.80	48
Metropolitan Status						
Urban	*	*		*	*	10
Suburban	$31.13	$48.50		$49.34	$65.88	28
Second City	*	*		*	*	22
Town	$24.75	$37.91		$42.32	$53.49	25
Rural	*	*		*	*	9
Median Area Household Income						
$35,000 or Less	*	*		*	*	13
$35,000 to $49,999	$34.25	$42.00		$45.49	$62.88	33
$50,000 to $69,999	$37.43	$51.75		$54.49	$66.50	34
$70,000 or More	*	*		*	*	16

Note 1: An asterisk indicates that data was not reported due to an insufficient number of responses.
Note 2: 23% of the respondents reported that this service is included in the fee for the fluid.
Note 3: 51% of respondents reported that they do not offer this service.

TABLE 10.7
INITIAL BAG OF FLUIDS: 1,000 ML LACTATED RINGER'S

	25th Percentile	Median	Your Data	Average	75th Percentile	Number of Respondents
All Practices	**$20.00**	**$30.00**		**$31.87**	**$41.07**	**401**
Number of FTE Veterinarians						
1.0 or Less	$20.00	$28.00		$30.69	$40.00	121
1.1 to 2.0	$20.88	$30.61		$32.38	$39.99	124
2.1 to 3.0	$20.00	$30.00		$30.64	$40.21	77
3.1 or More	$20.75	$31.41		$34.21	$46.00	73
Member Status						
Accredited Practice Member	$18.43	$30.00		$31.11	$42.81	110
Nonaccredited Member	$22.00	$30.00		$32.62	$40.71	73
Nonmember	$20.00	$29.29		$31.73	$41.13	191
Metropolitan Status						
Urban	$25.00	$35.00		$38.18	$50.50	39
Suburban	$25.00	$34.68		$34.24	$43.48	114
Second City	$20.13	$30.00		$32.38	$42.86	84
Town	$16.50	$25.00		$28.64	$38.17	108
Rural	$15.00	$22.30		$25.96	$35.56	47
Median Area Household Income						
$35,000 or Less	$20.00	$29.00		$29.39	$35.05	47
$35,000 to $49,999	$19.96	$28.00		$30.48	$39.89	144
$50,000 to $69,999	$22.08	$32.00		$32.71	$41.10	132
$70,000 or More	$18.61	$34.90		$33.24	$45.00	68

TABLE 10.8
FLUID INFUSION PUMP SETUP

	25th Percentile	Median	Your Data	Average	75th Percentile	Number of Respondents
All Practices	**$13.65**	**$19.95**		**$20.60**	**$25.00**	**99**
Number of FTE Veterinarians						
1.0 or Less	$12.00	$19.00		$20.28	$25.00	35
1.1 to 2.0	$12.50	$16.55		$18.52	$23.75	29
2.1 to 3.0	*	*		*	*	18
3.1 or More	*	*		*	*	16
Member Status						
Accredited Practice Member	$14.89	$20.85		$22.06	$26.50	32
Nonaccredited Member	*	*		*	*	11
Nonmember	$12.00	$17.50		$19.36	$25.00	45
Metropolitan Status						
Urban	*	*		*	*	12
Suburban	$12.41	$20.75		$21.22	$25.00	36
Second City	*	*		*	*	19
Town	*	*		*	*	24
Rural	*	*		*	*	6
Median Area Household Income						
$35,000 or Less	*	*		*	*	4
$35,000 to $49,999	$12.83	$18.50		$19.27	$23.35	37
$50,000 to $69,999	$12.24	$17.00		$19.24	$25.14	32
$70,000 or More	*	*		*	*	24

Note 1: An asterisk indicates that data was not reported due to an insufficient number of responses.
Note 2: 35% of the respondents reported that this service is included in the fee for the fluid.
Note 3: 40% of respondents reported that they do not offer this service.

TABLE 10.9
FLUID INFUSION PUMP USE (EIGHT HOURS)

	25th Percentile	Median	Your Data	Average	75th Percentile	Number of Respondents
All Practices	**$10.48**	**$16.53**		**$19.40**	**$23.94**	**152**
Number of FTE Veterinarians						
1.0 or Less	$12.00	$16.00		$22.88	$26.00	39
1.1 to 2.0	$11.00	$16.55		$18.41	$24.00	47
2.1 to 3.0	$10.08	$20.00		$17.78	$21.93	28
3.1 or More	$9.75	$13.80		$17.36	$20.00	36
Member Status						
Accredited Practice Member	$12.00	$18.03		$20.78	$25.75	48
Nonaccredited Member	*	*		*	*	23
Nonmember	$10.00	$15.64		$18.78	$22.63	70
Metropolitan Status						
Urban	*	*		*	*	20
Suburban	$11.50	$17.50		$21.99	$24.50	53
Second City	$10.00	$16.50		$18.21	$23.75	35
Town	$10.00	$12.75		$15.14	$19.63	28
Rural	*	*		*	*	12
Median Area Household Income						
$35,000 or Less	*	*		*	*	10
$35,000 to $49,999	$10.00	$16.55		$19.59	$23.00	51
$50,000 to $69,999	$10.00	$14.85		$17.70	$20.00	53
$70,000 or More	$10.80	$20.00		$20.98	$28.11	36

Note 1: An asterisk indicates that data was not reported due to an insufficient number of responses.
Note 2: 31% of the respondents reported that this service is included in the fee for the fluid.
Note 3: 32% of respondents reported that they do not offer this service.

TABLE 10.10
LACTATED RINGER'S SOLUTION (1,000 ML)

	25th Percentile	Median	Your Data	Average	75th Percentile	Number of Respondents
All Practices	**$20.00**	**$29.85**		**$32.47**	**$40.00**	**411**
Number of FTE Veterinarians						
1.0 or Less	$20.00	$29.15		$32.90	$41.44	126
1.1 to 2.0	$20.00	$30.00		$31.59	$38.99	124
2.1 to 3.0	$20.00	$28.00		$31.59	$37.50	79
3.1 or More	$18.71	$28.90		$33.96	$44.50	78
Member Status						
Accredited Practice Member	$19.15	$30.00		$32.56	$43.25	109
Nonaccredited Member	$20.77	$29.50		$33.97	$40.86	72
Nonmember	$20.00	$28.00		$31.11	$39.78	202
Metropolitan Status						
Urban	$27.50	$35.00		$40.78	$53.00	37
Suburban	$24.58	$33.53		$34.92	$45.00	118
Second City	$20.00	$30.00		$33.01	$40.00	85
Town	$17.75	$25.00		$29.44	$36.88	114
Rural	$15.00	$21.15		$25.36	$31.05	50
Median Area Household Income						
$35,000 or Less	$20.00	$30.00		$31.11	$37.53	47
$35,000 to $49,999	$18.59	$26.21		$30.50	$39.70	154
$50,000 to $69,999	$21.63	$30.00		$33.60	$40.78	137
$70,000 or More	$22.00	$34.63		$34.56	$40.63	65

Note 1: 5% of the respondents reported that they do not offer this service.

TABLE 10.11
RINGER'S SOLUTION (1,000 ML)

	25th Percentile	Median	Your Data	Average	75th Percentile	Number of Respondents
All Practices	$20.00	$30.00		$32.53	$43.21	194
Number of FTE Veterinarians						
1.0 or Less	$20.00	$29.48		$33.71	$41.35	70
1.1 to 2.0	$21.88	$30.00		$33.08	$45.00	57
2.1 to 3.0	$20.00	$25.65		$27.95	$34.91	36
3.1 or More	$18.13	$30.00		$33.97	$47.75	28
Member Status						
Accredited Practice Member	$20.00	$31.50		$33.38	$45.00	44
Nonaccredited Member	$20.00	$28.00		$33.63	$44.50	35
Nonmember	$20.00	$28.00		$30.80	$40.00	103
Metropolitan Status						
Urban	*	*		*	*	15
Suburban	$21.69	$31.00		$32.80	$44.63	66
Second City	$21.63	$32.38		$36.06	$45.75	42
Town	$18.88	$25.00		$28.71	$35.25	46
Rural	*	*		*	*	21
Median Area Household Income						
$35,000 or Less	*	*		*	*	16
$35,000 to $49,999	$20.00	$27.21		$33.09	$45.00	74
$50,000 to $69,999	$20.00	$28.94		$30.47	$40.00	72
$70,000 or More	$21.00	$34.75		$33.57	$44.00	27

Note 1: An asterisk indicates that data was not reported due to an insufficient number of responses
Note 2: 54% of the respondents reported that they do not provide this service

TABLE 10.12
NORMOSOL-R (1,000 ML)

	25th Percentile	Median	Your Data	Average	75th Percentile	Number of Respondents
All Practices	$20.52	$30.00		$33.12	$42.00	243
Number of FTE Veterinarians						
1.0 or Less	$22.88	$30.00		$35.71	$45.00	70
1.1 to 2.0	$24.31	$30.00		$32.20	$39.70	74
2.1 to 3.0	$20.00	$27.50		$29.34	$35.00	43
3.1 or More	$19.25	$29.85		$33.93	$45.00	53
Member Status						
Accredited Practice Member	$19.50	$28.50		$31.48	$43.93	65
Nonaccredited Member	$24.89	$33.90		$36.70	$44.50	43
Nonmember	$20.00	$29.90		$32.43	$40.00	122
Metropolitan Status						
Urban	*	*		*	*	21
Suburban	$24.58	$31.11		$33.96	$44.63	78
Second City	$25.00	$31.00		$36.09	$45.00	47
Town	$19.50	$25.00		$28.74	$35.00	67
Rural	*	*		*	*	24
Median Area Household Income						
$35,000 or Less	*	*		*	*	22
$35,000 to $49,999	$20.00	$30.00		$33.52	$44.05	84
$50,000 to $69,999	$23.00	$29.00		$33.03	$41.13	91
$70,000 or More	$18.88	$29.52		$30.95	$41.00	42

Note 1: An asterisk indicates that data was not reported due to an insufficient number of responses.
Note 2: 43% of the respondents reported that they do not provide this service.

THE VETERINARY FEE REFERENCE / 288

TABLE 10.13
0.9% NaCl (1,000 ML)

	25th Percentile	Median	Your Data	Average	75th Percentile	Number of Respondents
All Practices	**$20.00**	**$29.85**		**$32.43**	**$41.19**	**397**
Number of FTE Veterinarians						
1.0 or Less	$20.00	$29.95		$33.33	$43.43	113
1.1 to 2.0	$19.75	$30.00		$31.28	$39.80	117
2.1 to 3.0	$20.23	$28.00		$31.56	$38.53	81
3.1 or More	$18.71	$28.90		$33.65	$44.50	82
Member Status						
Accredited Practice Member	$19.11	$29.40		$32.43	$43.59	114
Nonaccredited Member	$21.26	$29.50		$33.34	$40.11	70
Nonmember	$20.00	$28.00		$31.30	$40.00	188
Metropolitan Status						
Urban	$27.00	$35.00		$41.38	$56.28	35
Suburban	$23.00	$34.00		$34.61	$45.00	117
Second City	$20.13	$30.34		$33.39	$42.25	80
Town	$16.69	$25.00		$29.29	$36.13	106
Rural	$15.00	$22.65		$25.47	$34.80	52
Median Area Household Income						
$35,000 or Less	$21.50	$30.00		$31.08	$36.29	45
$35,000 to $49,999	$18.00	$26.71		$30.53	$42.25	144
$50,000 to $69,999	$22.30	$29.95		$33.03	$40.78	137
$70,000 or More	$20.00	$34.95		$35.19	$44.00	63

Note 1: 7% of the respondents reported that they do not provide this fluid.

TABLE 10.14
D5W (5% DEXTROSE) (1,000 ML)

	25th Percentile	Median	Your Data	Average	75th Percentile	Number of Respondents
All Practices	**$20.00**	**$30.00**		**$33.98**	**$42.00**	**328**
Number of FTE Veterinarians						
1.0 or Less	$20.52	$30.00		$33.65	$41.25	91
1.1 to 2.0	$20.00	$30.00		$32.04	$39.45	100
2.1 to 3.0	$20.00	$30.00		$32.65	$39.66	66
3.1 or More	$23.05	$32.53		$38.67	$48.00	68
Member Status						
Accredited Practice Member	$20.00	$32.00		$34.54	$46.09	93
Nonaccredited Member	$25.00	$33.48		$36.62	$44.88	56
Nonmember	$20.00	$29.85		$32.60	$40.00	159
Metropolitan Status						
Urban	$25.00	$35.00		$41.40	$59.70	29
Suburban	$25.85	$35.00		$36.56	$46.14	92
Second City	$21.63	$31.13		$35.60	$43.50	70
Town	$20.00	$27.50		$32.04	$39.09	90
Rural	$15.00	$19.15		$23.27	$29.65	41
Median Area Household Income						
$35,000 or Less	$20.00	$30.00		$33.70	$38.41	35
$35,000 to $49,999	$19.04	$27.75		$31.08	$39.90	116
$50,000 to $69,999	$22.16	$30.00		$34.90	$44.01	118
$70,000 or More	$24.75	$35.00		$37.93	$45.05	54

Note 1: 21% of the respondents reported that they do not provide this service.

TABLE 10.15
DEXTRAN (1,000 ML)

	25th Percentile	Median	Your Data	Average	75th Percentile	Number of Respondents
All Practices	$25.00	$39.00		$44.84	$57.63	69

Note 1: Data was reported only for all practices due to an insufficient number of responses.
Note 2: 83% of the respondents reported that they do not provide this service.

TABLE 10.16
HETASTARCH (1,000 ML)

	25th Percentile	Median	Your Data	Average	75th Percentile	Number of Respondents
All Practices	$45.00	$60.00		$71.37	$89.44	198
Number of FTE Veterinarians						
1.0 or Less	$40.00	$54.40		$66.73	$89.00	46
1.1 to 2.0	$34.50	$60.00		$67.04	$80.62	53
2.1 to 3.0	$46.11	$55.05		$70.66	$92.00	49
3.1 or More	$53.61	$70.55		$79.43	$97.13	48
Member Status						
Accredited Practice Member	$49.00	$60.00		$75.85	$88.29	63
Nonaccredited Member	$42.88	$61.00		$67.85	$77.85	34
Nonmember	$40.00	$58.00		$69.01	$90.00	87
Metropolitan Status						
Urban	*	*		*	*	20
Suburban	$45.00	$60.00		$71.02	$89.81	68
Second City	$36.74	$55.28		$69.07	$97.78	30
Town	$51.00	$62.50		$75.29	$92.50	47
Rural	$30.00	$50.00		$63.57	$75.45	28
Median Area Household Income						
$35,000 or Less	*	*		*	*	13
$35,000 to $49,999	$38.71	$58.95		$71.39	$96.65	68
$50,000 to $69,999	$48.50	$60.50		$73.30	$88.12	73
$70,000 or More	$45.79	$57.36		$71.33	$91.50	38

Note 1: An asterisk indicates that data was not reported due to an insufficient number of responses
Note 2: 52% of the respondents reported that they do not provide this service

CHAPTER 11

HOSPITALIZATION SERVICES

Are you charging clients for all of the services that you provide hospitalized patients? Find out how you can track charges accurately and provide extra comforts for hospitalized pets and their owners.

USING ESTIMATES TO INCREASE COMPLIANCE

When a distraught client brought his aging golden retriever to a veterinarian to diagnose the dog's chronic gastritis, the doctor examined the pet, discussed potential causes, and recommended tests that would require one day of hospitalization. The veterinarian explained the need for extensive diagnostic tests to determine the specific cause, including a complete blood count, serum chemistry profile and electrolytes, abdominal radiographs, and possibly an ultrasound. The client replied, "Whatever you recommend, doctor. Money is not a concern." The veterinarian proceeded with the best diagnostic and treatment options, which totaled $1,100.

When the client picked up his dog the next morning, he was pleased with the findings and treatment plan but shocked by the price tag. He asked to speak with the hospital manager. When the hospital manager learned that the doctor forgot to give the client a written estimate, she offered a sincere apology and reduced the fee by 20%. Although the doctor learned a valuable lesson about using written estimates, the hospital lost $220.

OFFERING OPTIMAL CARE FOR HOSPITALIZED PATIENTS

Improve your quality of care by setting medical protocols that emphasize quality medicine. To get the latest information on veterinary protocols, you may want to consult the *AAHA Standards of Accreditation* and reference books, including *First Choice Medical Protocols* by Johnny D. Hoskins, DVM, Ph.D., Dipl. ACVIM and Ronald E. Whitford, DVM (AVLS, 2001).

Pain management has low cost and high client satisfaction. In fact, Dr. Ronald E. Whitford says most veterinarians don't do justice to pain management because of a subconscious thought that the client cannot afford three to six pain injections for their pets. "The difference between OK care and the best care is typically tens of dollars, not hundreds or thousands," says Dr. Whitford, who owns R-W Consulting and three hospitals in Clarksville, Tennessee. "My recommendation is to charge a flat rate for pain management for 24 hours, not by the number of injections. Clients don't care how it's itemized—they just want to know the total."

In addition to pain medication, you can provide comforts that enhance healing. For example, hospitalized patients can rest in cages with heated floors, and staff members can line cages with an ulcer dressing used in human medicine. The bedding is soft and absorbent with a thin rubber bottom that wicks away moisture. At his three hospitals, Dr. Whitford uses slotted grates to elevate cage floors so pets stay cleaner and dryer.

Although your practice may be closed at night, a doctor or technician can check on the patient every two hours to make sure IVs are still running and the patient is comfortable. Clients appreciate this attentive level of care for their hospitalized pets.

CHARGING APPROPRIATELY FOR HOSPITALIZATION

Veterinarians follow two schools of thought on hospitalization charges: 1) Charge for every item and service individually, or 2) Create hospitalization packages based on the level of care. The fault of the first method is that charges are often missed in the rush to discharge patients. Instead, Dr. Whitford suggests developing hospitalization packages based on Level 1, 2, and 3 care.

For example, Level 1 care is for minimal daily care such as a patient admitted for lab tests or a cyst removal. Level 2 requires more patient monitoring and case management, such as needed after a spay procedure. Level 3 is for intensive- and critical-care patients, such as a dog with parvovirus or a pet that was hit by a car. Each level includes a per-day hospitalization fee as well as the doctor and technician's time to administer treatments, monitor vital signs, and provide care. Dr. Whitford recommends setting the daily hospitalization package fee based on how many times during the day a doctor will need to work with the patient. The minimum fee must be at least as much as an outpatient recheck exam, he says.

To avoid missing charges, have a technician enter charges into the computer daily so you capture each service, injection, and supply you used for the patient. The daily hospitalization package fee also should be entered to give you a chronological order that tells the client what happened each day. The client will look at the total dollars but also be able to see how the pet was cared for.

Entering hospitalization charges daily will help you avoid overlooking items such as a sedative given before radiographs, bag of IV fluids, or a pain injection. Train your staff to make sure they know it's their responsibility to enter a medication in the computer if they dispense it. If it's not on the invoice, the client assumes it wasn't given.

You want to consider creating an inventory system that helps prevent missed charges on supplies and medications for hospitalized patients. Each item—from a syringe to the can of food that a hospitalized patient eats—gets an inventory sticker before it goes on the shelf or in a drawer. When a staff member or doctor uses the item, he can remove the inventory sticker and place it on the travel sheet. When you close out the bill, you then double-check for all of those charges.

Dr. Whitford also recommends having a technician or staff member who is familiar with the case reconcile the tracking sheet and medical record when the patient is discharged. "I've seen a lot of missed charges because the person collecting the money has no idea about medical services and therefore cannot appropriately interpret the medical record."

Case Study

Practicing Preventive Pain Management

"Pain prevention is a critical element of veterinary healthcare delivery," says Dr. Thomas E. Catanzaro, MHA, Dipl. FACHE, of Catanzaro & Associates and Veterinary Practice Consultants in Golden, Colorado. "Pain management is what happens if we don't prevent it."

When you walk through a hospital ward, wouldn't it be nice not to hear any barking or whining? This could be the case at your practice if every staff member and doctor takes a preventive approach with a pain-scoring system.

"Pain scoring helps track the patient's progress," says Dr. Thomas E. Catanzaro, MHA, Dipl. FACHE, of Catanzaro & Associates and Veterinary Practice Consultants in Golden, Colorado. "For example, if a patient has a pain score of 5 at 8 a.m., and it's down to a score of 2 by 10 a.m., you'd be pleased with the amount of pain medication you're giving. Pain scoring lets staff members talk about a patient's progress quickly and track it easily."

Because a painful state is catabolic, it leads to reduced immunity and mobility and increased infection and morbidity. Pain is a stressful event that interferes with patient healing and recovery, according to *Essentials of Small Animal Anesthesia and Analgesia* (Lippincott, Williams and Wilkins, 1999). Optimal timing of analgesia is before surgery, and before the patient is exposed to noxious stimuli.

Preemptive analgesia reduces the likelihood of developing peripheral and central nervous system hypersensitization, according to Pfizer Animal Health literature. If analgesics are administered after the animal awakens in pain, the pain is more difficult to treat and the patient requires significantly more medication than it would have if it received a preemptive dose.

Consider including a pain medication injection in the bundle of services for each surgery, just as anesthesia is included. Then offer an upgrade to a Fentanyl patch, Dr. Catanzaro advises. When presenting the healthcare plan with estimated expenses, your technician might say: "Pain control is essential with this procedure, so we included a 12- to 24-hour pain control injection. But for an additional $29.50, we can use a patch that extends the pain control for three to five days. Which do you prefer today?" This lets the client choose from two "yes" options.

Implementation Idea

Audit Your Medical Records for Missed Charges

Pull a medical record for a patient that was hospitalized within the last 10 days. Use the checklist in Figure 11.1 to see if the applicable items were recorded in the medical record and charged to the client.

FIGURE 11.1

CHECKLIST FOR MEDICAL-RECORD AUDIT

Service or Product (Check only applicable items)	Recorded in Medical Record	Charged to Client	Amount Lost if Not Charged
CBC, chemistry profile, urinalysis, fecal, and other lab tests			
Dispensed medications including analgesics			
Elizabethan collar			
Injections (look for documentation of first injection as well as additional injections)			
Inpatient food			
IV setup line, IV catheter, IV pump, and IV fluids			
Number of days the patient was hospitalized			
Nursing care			
Oxygen			
Physical exam (unless included in daily hospitalization fees)			
Radiographs			
Recheck			
Supplies (gauze, topical antibacterials, etc.)			
Ultrasound			
Total			

HELPFUL RESOURCES

First Choice Medical Protocols, Johnny D. Hoskins, DVM, PhD, Dipl. ACVIM and Ronald E. Whitford, DVM. Johnny D. Hoskins, 2001. Available through the American Animal Hospital Association.

AAHA Standards of Accreditation, American Animal Hospital Association. Available on www.aahanet.org.

CHAPTER 11

DATA TABLES

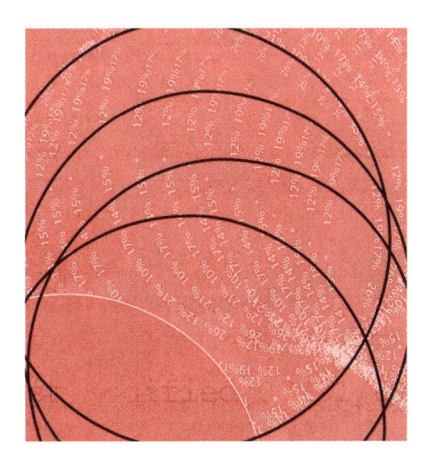

Note: Hospitalization fees are reported for a routine medical case, not including fees for examination, veterinarian supervision, or treatment. Included in the fees for hospitalization with IV are fees for IV fluids and IV catheter and placement.

TABLE 11.1
HOSPITALIZATION WITH IV: NO OVERNIGHT STAY, 10-POUND CAT

	25th Percentile	Median	Your Data	Average	75th Percentile	Number of Respondents
All Practices	$40.43	$70.79		$76.00	$105.00	406
Number of FTE Veterinarians						
1.0 or Less	$35.71	$69.25		$71.21	$102.00	118
1.1 to 2.0	$41.40	$63.75		$75.59	$105.00	123
2.1 to 3.0	$40.00	$82.00		$78.90	$111.45	82
3.1 or More	$43.00	$73.00		$81.25	$112.00	79
Member Status						
Accredited Practice Member	$48.35	$79.33		$82.82	$117.19	116
Nonaccredited Member	$41.69	$71.00		$81.39	$114.80	71
Nonmember	$35.95	$62.00		$70.29	$99.00	191
Metropolitan Status						
Urban	$51.57	$75.00		$87.76	$119.15	37
Suburban	$43.00	$82.00		$83.84	$116.17	119
Second City	$37.82	$76.50		$78.24	$106.11	82
Town	$40.38	$63.50		$69.58	$95.00	110
Rural	$31.23	$51.00		$59.55	$82.00	50
Median Area Household Income						
$35,000 or Less	$34.25	$60.00		$62.76	$85.88	46
$35,000 to $49,999	$39.63	$73.00		$78.40	$109.25	142
$50,000 to $69,999	$41.69	$68.50		$73.75	$105.00	139
$70,000 or More	$53.50	$88.50		$87.33	$114.00	69

Note: 4% of respondents reported that they do not offer this service.

TABLE 11.2
HOSPITALIZATION WITHOUT IV: NO OVERNIGHT STAY, 10-POUND CAT

	25th Percentile	Median	Your Data	Average	75th Percentile	Number of Respondents
All Practices	$18.00	$25.35		$29.29	$36.94	412
Number of FTE Veterinarians						
1.0 or Less	$19.00	$25.15		$28.51	$36.69	124
1.1 to 2.0	$18.05	$25.40		$30.15	$38.25	125
2.1 to 3.0	$15.00	$25.00		$27.40	$34.50	79
3.1 or More	$18.00	$27.03		$31.10	$42.81	78
Member Status						
Accredited Practice Member	$19.53	$26.00		$29.10	$36.13	116
Nonaccredited Member	$21.00	$27.30		$32.56	$41.75	73
Nonmember	$15.00	$22.25		$27.84	$35.48	197
Metropolitan Status						
Urban	$21.58	$26.33		$31.64	$39.25	38
Suburban	$19.82	$29.88		$33.36	$45.00	122
Second City	$19.50	$27.28		$31.22	$40.25	86
Town	$16.38	$23.13		$26.00	$32.03	110
Rural	$13.23	$19.55		$21.71	$25.72	48
Median Area Household Income						
$35,000 or Less	$14.91	$20.00		$22.75	$32.00	47
$35,000 to $49,999	$17.96	$25.00		$29.23	$38.88	144
$50,000 to $69,999	$18.00	$27.00		$29.83	$36.42	137
$70,000 or More	$22.36	$29.38		$33.14	$42.39	74

Note: 3% of respondents reported that they do not offer this service.

TABLE 11.3
HOSPITALIZATION WITH IV: NO OVERNIGHT STAY, 25-POUND DOG

	25th Percentile	Median	Your Data	Average	75th Percentile	Number of Respondents
All Practices	$41.40	$72.50		$79.56	$111.00	399
Number of FTE Veterinarians						
1.0 or Less	$40.00	$72.12		$73.86	$105.53	113
1.1 to 2.0	$40.50	$65.85		$78.89	$113.61	120
2.1 to 3.0	$40.15	$79.00		$80.33	$113.08	81
3.1 or More	$43.50	$74.00		$86.29	$112.75	80
Member Status						
Accredited Practice Member	$50.13	$81.80		$89.62	$122.10	119
Nonaccredited Member	$41.69	$75.50		$82.19	$114.80	67
Nonmember	$37.38	$62.63		$72.80	$100.09	186
Metropolitan Status						
Urban	$53.25	$75.25		$94.28	$124.05	36
Suburban	$42.81	$83.80		$88.18	$120.50	116
Second City	$39.88	$80.00		$83.78	$113.56	82
Town	$41.90	$63.75		$70.95	$96.10	109
Rural	$35.00	$51.75		$61.16	$82.75	48
Median Area Household Income						
$35,000 or Less	$37.75	$60.50		$65.83	$92.33	44
$35,000 to $49,999	$40.00	$74.50		$81.50	$114.95	140
$50,000 to $69,999	$40.20	$71.00		$76.74	$110.00	135
$70,000 or More	$55.25	$90.90		$92.95	$118.58	70

Note: 6% of respondents reported that they do not offer this service.

TABLE 11.4
HOSPITALIZATION WITHOUT IV: NO OVERNIGHT STAY, 25-POUND DOG

	25th Percentile	Median	Your Data	Average	75th Percentile	Number of Respondents
All Practices	**$19.38**	**$26.05**		**$30.25**	**$36.76**	**402**
Number of FTE Veterinarians						
1.0 or Less	$20.00	$26.00		$29.80	$37.27	117
1.1 to 2.0	$19.00	$26.10		$31.38	$37.50	123
2.1 to 3.0	$17.36	$26.00		$27.68	$34.88	80
3.1 or More	$20.00	$27.68		$31.74	$42.06	76
Member Status						
Accredited Practice Member	$20.00	$28.00		$30.54	$36.69	116
Nonaccredited Member	$22.89	$28.75		$33.14	$39.63	68
Nonmember	$17.00	$25.00		$28.73	$35.59	190
Metropolitan Status						
Urban	$24.25	$29.18		$33.65	$43.50	36
Suburban	$20.00	$30.00		$34.15	$45.00	117
Second City	$20.00	$29.00		$31.76	$39.50	85
Town	$17.00	$25.00		$27.41	$34.00	111
Rural	$15.75	$20.00		$22.29	$24.95	45
Median Area Household Income						
$35,000 or Less	$17.00	$20.00		$24.32	$32.50	47
$35,000 to $49,999	$18.15	$25.15		$30.01	$38.32	140
$50,000 to $69,999	$19.00	$28.00		$30.62	$36.80	135
$70,000 or More	$23.70	$30.00		$34.67	$45.00	70

Note: 4% of respondents reported that they do not offer this service.

TABLE 11.5
HOSPITALIZATION WITH IV: NO OVERNIGHT STAY, 60-POUND DOG

	25th Percentile	Median	Your Data	Average	75th Percentile	Number of Respondents
All Practices	**$42.00**	**$75.00**		**$80.85**	**$112.47**	**394**
Number of FTE Veterinarians						
1.0 or Less	$40.00	$75.00		$77.22	$110.23	113
1.1 to 2.0	$41.10	$71.10		$79.51	$112.58	118
2.1 to 3.0	$41.85	$84.23		$82.68	$114.61	80
3.1 or More	$45.00	$77.70		$86.66	$113.00	79
Member Status						
Accredited Practice Member	$51.50	$80.35		$88.84	$122.16	115
Nonaccredited Member	$42.30	$79.00		$84.33	$115.83	66
Nonmember	$40.00	$69.75		$74.77	$104.51	186
Metropolitan Status						
Urban	$54.00	$75.00		$91.58	$130.00	35
Suburban	$42.94	$86.30		$88.22	$121.80	114
Second City	$41.25	$83.00		$84.73	$117.33	81
Town	$42.00	$66.25		$74.04	$100.53	108
Rural	$35.00	$53.50		$64.13	$86.50	48
Median Area Household Income						
$35,000 or Less	$40.00	$60.00		$67.70	$93.77	43
$35,000 to $49,999	$41.00	$75.50		$83.94	$116.09	141
$50,000 to $69,999	$42.63	$75.75		$78.21	$110.23	133
$70,000 or More	$54.50	$93.00		$90.55	$120.50	67

Note: 6% of respondents reported that they do not offer this service.

TABLE 11.6
HOSPITALIZATION WITHOUT IV: NO OVERNIGHT STAY, 60-POUND DOG

	25th Percentile	Median	Your Data	Average	75th Percentile	Number of Respondents
All Practices	$20.00	$28.00		$31.85	$38.50	401
Number of FTE Veterinarians						
1.0 or Less	$20.48	$28.00		$31.63	$38.75	116
1.1 to 2.0	$19.85	$29.50		$32.75	$40.00	123
2.1 to 3.0	$18.00	$27.05		$29.16	$35.86	80
3.1 or More	$20.25	$27.75		$33.22	$42.38	76
Member Status						
Accredited Practice Member	$21.93	$30.00		$32.56	$40.00	116
Nonaccredited Member	$22.85	$30.30		$34.11	$41.27	67
Nonmember	$18.00	$26.00		$30.28	$36.30	190
Metropolitan Status						
Urban	$24.00	$30.00		$35.39	$45.00	35
Suburban	$21.27	$31.00		$36.19	$46.00	117
Second City	$21.45	$30.00		$33.19	$40.00	85
Town	$18.00	$25.00		$28.51	$35.00	111
Rural	$16.60	$21.35		$24.35	$29.00	45
Median Area Household Income						
$35,000 or Less	$17.88	$22.25		$25.52	$32.81	46
$35,000 to $49,999	$20.00	$26.00		$31.38	$38.38	140
$50,000 to $69,999	$19.50	$28.50		$31.91	$39.00	135
$70,000 or More	$24.75	$32.53		$37.08	$45.50	70

Note: 4% of respondents reported that they do not offer this service.

TABLE 11.7
HOSPITALIZATION WITH IV: NO OVERNIGHT STAY, 100-POUND DOG

	25th Percentile	Median	Your Data	Average	75th Percentile	Number of Respondents
All Practices	$44.50	$76.00		$83.63	$114.80	391
Number of FTE Veterinarians						
1.0 or Less	$45.00	$75.88		$80.39	$114.50	112
1.1 to 2.0	$42.00	$72.00		$83.03	$113.00	119
2.1 to 3.0	$42.45	$79.00		$83.50	$117.63	77
3.1 or More	$45.00	$79.65		$89.79	$115.00	79
Member Status						
Accredited Practice Member	$51.64	$80.18		$89.77	$122.12	114
Nonaccredited Member	$43.63	$79.00		$85.93	$120.38	64
Nonmember	$41.70	$73.75		$78.87	$107.39	186
Metropolitan Status						
Urban	$57.00	$77.58		$97.45	$132.00	34
Suburban	$43.00	$88.00		$91.36	$129.55	114
Second City	$45.25	$80.00		$88.66	$124.56	80
Town	$42.00	$68.75		$74.92	$101.15	107
Rural	$35.00	$62.50		$66.76	$86.50	48
Median Area Household Income						
$35,000 or Less	$41.40	$60.00		$70.44	$95.00	43
$35,000 to $49,999	$42.38	$76.63		$86.88	$120.25	138
$50,000 to $69,999	$43.38	$76.50		$80.00	$112.33	132
$70,000 or More	$58.50	$96.50		$94.72	$121.70	68

Note: 7% of respondents reported that they do not offer this service.

TABLE 11.8
HOSPITALIZATION WITHOUT IV: NO OVERNIGHT STAY, 100-POUND DOG

	25th Percentile	Median	Your Data	Average	75th Percentile	Number of Respondents
All Practices	$20.00	$30.00		$34.59	$41.96	402
Number of FTE Veterinarians						
1.0 or Less	$22.00	$29.60		$33.08	$41.96	114
1.1 to 2.0	$19.92	$29.50		$33.77	$42.00	123
2.1 to 3.0	$18.00	$28.50		$34.08	$38.50	81
3.1 or More	$22.00	$31.21		$38.36	$45.00	78
Member Status						
Accredited Practice Member	$22.03	$30.60		$35.79	$43.00	117
Nonaccredited Member	$22.96	$33.69		$35.24	$43.85	66
Nonmember	$19.00	$27.93		$33.82	$40.00	192
Metropolitan Status						
Urban	$25.40	$36.00		$42.21	$48.45	35
Suburban	$22.34	$33.19		$38.96	$47.05	118
Second City	$22.03	$32.95		$36.26	$44.00	85
Town	$18.75	$26.00		$30.44	$36.00	111
Rural	$18.05	$22.00		$25.62	$31.00	45
Median Area Household Income						
$35,000 or Less	$18.00	$24.45		$26.10	$32.81	46
$35,000 to $49,999	$20.26	$29.75		$36.03	$44.60	140
$50,000 to $69,999	$20.00	$29.00		$34.19	$40.38	136
$70,000 or More	$25.00	$32.98		$38.51	$48.38	70

Note: 5% of respondents reported that they do not offer this service.

TABLE 11.9
HOSPITALIZATION WITH IV: NO OVERNIGHT STAY, BIRD

	25th Percentile	Median	Your Data	Average	75th Percentile	Number of Respondents
All Practices	$29.78	$66.94		$66.37	$97.13	68

Note 1: Data was reported only for all practices due to an insufficient number of responses.
Note 2: 83% of respondents reported that they do not offer this service.

TABLE 11.10
HOSPITALIZATION WITHOUT IV: NO OVERNIGHT STAY, BIRD

	25th Percentile	Median	Your Data	Average	75th Percentile	Number of Respondents
All Practices	$16.70	$26.00		$29.91	$36.80	107
Number of FTE Veterinarians						
1.0 or Less	*	*		*	*	21
1.1 to 2.0	$17.81	$29.59		$35.01	$49.63	30
2.1 to 3.0	$14.61	$21.04		$27.16	$36.44	30
3.1 or More	$15.50	$28.50		$29.13	$36.50	25
Member Status						
Accredited Practice Member	$17.25	$28.50		$29.37	$40.00	35
Nonaccredited Member	*	*		*	*	14
Nonmember	$14.61	$22.00		$27.81	$35.96	50
Metropolitan Status						
Urban	*	*		*	*	10
Suburban	$18.06	$31.00		$34.91	$46.00	27
Second City	*	*		*	*	20
Town	$15.00	$25.00		$27.21	$36.20	35
Rural	*	*		*	*	13
Median Area Household Income						
$35,000 or Less	*	*		*	*	12
$35,000 to $49,999	$15.75	$26.00		$32.92	$42.75	40
$50,000 to $69,999	$18.00	$29.80		$30.57	$36.79	44
$70,000 or More	*	*		*	*	7

Note 1: An asterisk indicates that data was not reported due to an insufficient number of responses.
Note 2: 71% of respondents reported that they do not offer this service.

TABLE 11.11
HOSPITALIZATION WITH IV: NO OVERNIGHT STAY, REPTILE

	25th Percentile	Median	Your Data	Average	75th Percentile	Number of Respondents
All Practices	$27.83	$57.50		$63.64	$88.40	76

Note 1: Data was reported only for all practices due to an insufficient number of responses.
Note 2: An asterisk indicates that data was not reported due to an insufficient number of responses.
Note 3: 80% of respondents reported that they do not offer this service.

TABLE 11.12
HOSPITALIZATION WITHOUT IV: NO OVERNIGHT STAY, REPTILE

	25th Percentile	Median	Your Data	Average	75th Percentile	Number of Respondents
All Practices	$17.00	$25.17		$29.52	$37.38	114
Number of FTE Veterinarians						
1.0 or Less	*	*		*	*	22
1.1 to 2.0	$17.11	$28.84		$32.48	$44.23	30
2.1 to 3.0	$17.00	$22.63		$28.59	$38.51	30
3.1 or More	$15.00	$24.75		$27.41	$34.00	31
Member Status						
Accredited Practice Member	$18.39	$28.35		$29.08	$37.26	34
Nonaccredited Member	*	*		*	*	17
Nonmember	$14.06	$22.00		$27.28	$35.95	55
Metropolitan Status						
Urban	*	*		*	*	12
Suburban	$18.55	$30.50		$34.30	$45.49	28
Second City	*	*		*	*	19
Town	$15.00	$24.30		$27.06	$34.50	39
Rural	*	*		*	*	14
Median Area Household Income						
$35,000 or Less	*	*		*	*	12
$35,000 to $49,999	$15.00	$24.75		$30.86	$43.00	47
$50,000 to $69,999	$18.88	$28.75		$30.52	$36.46	44
$70,000 or More	*	*		*	*	8

Note 1: An asterisk indicates that data was not reported due to an insufficient number of responses.
Note 2: 69% of respondents reported that they do not offer this service.

TABLE 11.13
HOSPITALIZATION WITH IV: NO OVERNIGHT STAY, SMALL MAMMAL

	25th Percentile	Median	Your Data	Average	75th Percentile	Number of Respondents
All Practices	$30.00	$58.00		$66.23	$88.30	121
Number of FTE Veterinarians						
1.0 or Less	$32.00	$58.00		$60.08	$80.38	25
1.1 to 2.0	$30.00	$63.25		$65.58	$90.00	31
2.1 to 3.0	$21.25	$60.50		$64.46	$102.48	30
3.1 or More	$35.00	$59.00		$73.31	$87.24	34
Member Status						
Accredited Practice Member	$33.25	$53.25		$68.23	$96.65	37
Nonaccredited Member	*	*		*	*	19
Nonmember	$31.95	$66.00		$66.63	$84.00	53
Metropolitan Status						
Urban	*	*		*	*	15
Suburban	$29.85	$71.50		$76.17	$108.72	29
Second City	$28.75	$58.50		$64.59	$92.59	26
Town	$31.50	$54.38		$58.94	$82.65	34
Rural	*	*		*	*	15
Median Area Household Income						
$35,000 or Less	*	*		*	*	11
$35,000 to $49,999	$27.45	$57.50		$72.55	$103.84	44
$50,000 to $69,999	$33.50	$58.00		$63.12	$81.88	49
$70,000 or More	*	*		*	*	15

Note 1: An asterisk indicates that data was not reported due to an insufficient number of responses.
Note 2: 69% of respondents reported that they do not offer this service.

TABLE 11.14
HOSPITALIZATION WITHOUT IV: NO OVERNIGHT STAY, SMALL MAMMAL

	25th Percentile	Median	Your Data	Average	75th Percentile	Number of Respondents
All Practices	$16.34	$23.18		$28.19	$35.96	162
Number of FTE Veterinarians						
1.0 or Less	$15.00	$22.25		$25.77	$32.00	35
1.1 to 2.0	$19.39	$27.75		$32.49	$44.23	42
2.1 to 3.0	$15.00	$20.04		$26.83	$35.25	46
3.1 or More	$15.00	$23.88		$27.52	$34.25	38
Member Status						
Accredited Practice Member	$19.65	$26.00		$28.54	$35.92	49
Nonaccredited Member	$20.00	$27.00		$33.73	$47.00	27
Nonmember	$12.40	$20.00		$24.99	$32.00	75
Metropolitan Status						
Urban	*	*		*	*	16
Suburban	$18.05	$25.38		$30.71	$40.75	42
Second City	$17.00	$25.33		$31.13	$45.00	35
Town	$15.06	$23.63		$26.28	$32.75	48
Rural	*	*		*	*	19
Median Area Household Income						
$35,000 or Less	*	*		*	*	18
$35,000 to $49,999	$14.75	$22.82		$28.93	$38.00	61
$50,000 to $69,999	$18.50	$27.00		$28.46	$35.95	59
$70,000 or More	*	*		*	*	21

Note 1: An asterisk indicates that data was not reported due to an insufficient number of responses.
Note 2: 57% of respondents reported that they do not offer this service.

TABLE 11.15
HOSPITALIZATION WITH IV: NO OVERNIGHT STAY, FERRET

	25th Percentile	Median	Your Data	Average	75th Percentile	Number of Respondents
All Practices	$30.00	$58.00		$65.39	$93.48	141
Number of FTE Veterinarians						
1.0 or Less	$32.13	$45.50		$64.56	$100.61	28
1.1 to 2.0	$30.00	$61.75		$66.43	$94.66	40
2.1 to 3.0	$24.65	$75.25		$66.42	$110.64	34
3.1 or More	$35.00	$59.00		$64.31	$82.65	38
Member Status						
Accredited Practice Member	$38.13	$62.95		$71.22	$98.86	44
Nonaccredited Member	*	*		*	*	20
Nonmember	$28.00	$63.00		$64.81	$96.43	67
Metropolitan Status						
Urban	*	*		*	*	13
Suburban	$29.78	$63.00		$69.60	$102.86	32
Second City	$31.95	$70.58		$71.96	$104.03	29
Town	$33.25	$60.50		$62.21	$85.08	45
Rural	*	*		*	*	19
Median Area Household Income						
$35,000 or Less	*	*		*	*	16
$35,000 to $49,999	$28.00	$55.13		$66.80	$98.68	52
$50,000 to $69,999	$35.00	$66.79		$68.74	$95.45	55
$70,000 or More	*	*		*	*	15

Note 1: An asterisk indicates that data was not reported due to an insufficient number of responses.
Note 2: 64% of respondents reported that they do not offer this service.

TABLE 11.16
HOSPITALIZATION WITHOUT IV: NO OVERNIGHT STAY, FERRET

	25th Percentile	Median	Your Data	Average	75th Percentile	Number of Respondents
All Practices	$16.85	$23.25		$28.58	$35.98	169
Number of FTE Veterinarians						
1.0 or Less	$17.31	$22.54		$27.29	$33.80	38
1.1 to 2.0	$17.35	$27.00		$31.74	$43.49	45
2.1 to 3.0	$14.98	$21.04		$26.41	$35.00	42
3.1 or More	$17.00	$24.75		$28.72	$35.00	43
Member Status						
Accredited Practice Member	$19.65	$26.00		$28.71	$35.92	53
Nonaccredited Member	*	*		*	*	24
Nonmember	$12.98	$21.04		$25.99	$33.08	83
Metropolitan Status						
Urban	*	*		*	*	13
Suburban	$18.06	$26.00		$31.84	$42.00	43
Second City	$17.00	$26.00		$30.95	$41.25	38
Town	$15.19	$23.78		$26.72	$33.38	50
Rural	*	*		*	*	22
Median Area Household Income						
$35,000 or Less	*	*		*	*	19
$35,000 to $49,999	$15.00	$22.84		$28.43	$36.00	62
$50,000 to $69,999	$18.50	$27.15		$30.04	$36.20	63
$70,000 or More	*	*		*	*	21

Note 1: An asterisk indicates that data was not reported due to an insufficient number of responses.
Note 2: 55% of respondents reported that they do not offer this service.

TABLE 11.17
HOSPITALIZATION WITH IV: OVERNIGHT STAY, 10-POUND CAT

	25th Percentile	Median	Your Data	Average	75th Percentile	Number of Respondents
All Practices	$45.00	$80.00		$87.98	$122.00	399
Number of FTE Veterinarians						
1.0 or Less	$40.30	$75.00		$78.82	$113.23	117
1.1 to 2.0	$46.70	$76.00		$89.61	$129.80	120
2.1 to 3.0	$49.25	$98.00		$95.06	$131.00	81
3.1 or More	$45.89	$83.00		$92.55	$135.87	77
Member Status						
Accredited Practice Member	$56.00	$101.07		$100.00	$135.60	107
Nonaccredited Member	$47.00	$83.00		$91.84	$137.00	69
Nonmember	$42.00	$70.60		$81.03	$112.18	196
Metropolitan Status						
Urban	$47.00	$88.00		$102.93	$159.85	37
Suburban	$51.25	$90.50		$98.41	$139.67	112
Second City	$46.60	$95.00		$91.04	$128.65	83
Town	$44.89	$74.25		$78.84	$107.13	109
Rural	$32.50	$64.00		$70.53	$100.52	49
Median Area Household Income						
$35,000 or Less	$36.25	$67.00		$76.60	$104.19	44
$35,000 to $49,999	$44.33	$78.71		$88.84	$120.19	142
$50,000 to $69,999	$45.00	$77.35		$84.35	$119.30	136
$70,000 or More	$65.70	$105.00		$102.45	$138.32	67

Note: 6% of respondents reported that they do not offer this service.

TABLE 11.18
HOSPITALIZATION WITHOUT IV: OVERNIGHT STAY, 10-POUND CAT

	25th Percentile	Median	Your Data	Average	75th Percentile	Number of Respondents
All Practices	**$25.00**	**$34.50**		**$38.53**	**$48.00**	**404**
Number of FTE Veterinarians						
1.0 or Less	$22.62	$32.01		$36.53	$47.06	122
1.1 to 2.0	$25.00	$35.46		$38.19	$49.77	123
2.1 to 3.0	$26.00	$34.00		$39.28	$47.23	81
3.1 or More	$26.35	$37.95		$41.90	$57.13	74
Member Status						
Accredited Practice Member	$27.79	$40.98		$42.15	$52.11	110
Nonaccredited Member	$26.00	$36.00		$41.94	$54.00	71
Nonmember	$20.00	$30.00		$35.40	$47.17	196
Metropolitan Status						
Urban	$29.00	$39.00		$41.57	$51.07	33
Suburban	$27.50	$41.69		$43.11	$57.50	115
Second City	$27.00	$38.75		$40.78	$50.00	86
Town	$22.38	$30.00		$35.65	$45.38	114
Rural	$18.00	$24.50		$28.46	$35.23	49
Median Area Household Income						
$35,000 or Less	$17.70	$28.50		$32.28	$43.00	46
$35,000 to $49,999	$24.00	$30.00		$36.63	$45.99	140
$50,000 to $69,999	$25.00	$35.00		$39.46	$48.00	141
$70,000 or More	$32.85	$45.00		$45.28	$57.50	67

Note: 4% of respondents reported that they do not offer this service.

TABLE 11.19
HOSPITALIZATION WITH IV: OVERNIGHT STAY, 25-POUND DOG

	25th Percentile	Median	Your Data	Average	75th Percentile	Number of Respondents
All Practices	**$47.00**	**$80.58**		**$91.02**	**$127.56**	**392**
Number of FTE Veterinarians						
1.0 or Less	$41.65	$74.63		$80.87	$116.70	114
1.1 to 2.0	$46.80	$80.00		$92.58	$129.60	117
2.1 to 3.0	$50.00	$92.50		$96.43	$134.85	79
3.1 or More	$51.10	$82.50		$98.70	$139.73	78
Member Status						
Accredited Practice Member	$60.00	$101.07		$106.25	$137.70	107
Nonaccredited Member	$46.25	$84.00		$91.87	$136.80	66
Nonmember	$42.00	$69.81		$83.49	$116.49	192
Metropolitan Status						
Urban	$48.50	$86.12		$104.28	$152.65	36
Suburban	$53.06	$91.00		$100.16	$141.00	105
Second City	$49.61	$87.50		$96.28	$134.85	83
Town	$45.00	$74.25		$83.30	$113.00	111
Rural	$35.00	$68.95		$72.54	$104.26	48
Median Area Household Income						
$35,000 or Less	$37.50	$68.00		$77.09	$103.38	45
$35,000 to $49,999	$45.00	$78.80		$91.31	$125.79	141
$50,000 to $69,999	$46.00	$80.65		$89.47	$120.55	133
$70,000 or More	$68.00	$113.00		$105.33	$138.47	63

Note: 7% of respondents reported that they do not offer this service.

TABLE 11.20
HOSPITALIZATION WITHOUT IV: OVERNIGHT STAY, 25-POUND DOG

	25th Percentile	Median	Your Data	Average	75th Percentile	Number of Respondents
All Practices	$25.89	$36.78	_____	$40.58	$49.70	396
Number of FTE Veterinarians						
1.0 or Less	$24.00	$35.00	_____	$38.18	$48.00	117
1.1 to 2.0	$27.13	$37.00	_____	$40.86	$50.10	120
2.1 to 3.0	$27.13	$36.00	_____	$40.36	$48.15	80
3.1 or More	$27.00	$39.00	_____	$44.45	$57.50	75
Member Status						
Accredited Practice Member	$30.13	$42.50	_____	$44.01	$53.88	108
Nonaccredited Member	$27.30	$37.95	_____	$43.48	$57.75	69
Nonmember	$22.63	$32.00	_____	$37.94	$48.88	192
Metropolitan Status						
Urban	$32.00	$40.00	_____	$46.28	$52.90	35
Suburban	$29.73	$42.50	_____	$45.79	$61.13	109
Second City	$29.00	$40.00	_____	$41.62	$50.00	83
Town	$25.00	$32.75	_____	$37.69	$48.00	114
Rural	$20.00	$25.50	_____	$29.75	$35.35	48
Median Area Household Income						
$35,000 or Less	$19.69	$31.25	_____	$33.85	$43.00	46
$35,000 to $49,999	$25.00	$32.01	_____	$38.35	$46.75	140
$50,000 to $69,999	$25.00	$37.00	_____	$41.46	$52.00	135
$70,000 or More	$33.49	$48.00	_____	$48.98	$60.00	65

Note: 6% of respondents reported that they do not offer this service.

TABLE 11.21
HOSPITALIZATION WITH IV: OVERNIGHT STAY, 60-POUND DOG

	25th Percentile	Median	Your Data	Average	75th Percentile	Number of Respondents
All Practices	$48.00	$84.75	_____	$94.58	$132.00	391
Number of FTE Veterinarians						
1.0 or Less	$44.00	$75.75	_____	$84.85	$121.22	113
1.1 to 2.0	$46.80	$82.00	_____	$95.35	$131.00	117
2.1 to 3.0	$55.00	$103.75	_____	$100.61	$138.00	79
3.1 or More	$49.90	$86.00	_____	$102.00	$139.73	78
Member Status						
Accredited Practice Member	$65.94	$102.63	_____	$108.55	$144.06	108
Nonaccredited Member	$51.00	$86.00	_____	$95.79	$138.95	65
Nonmember	$43.25	$74.50	_____	$87.45	$119.95	192
Metropolitan Status						
Urban	$55.00	$77.48	_____	$110.72	$158.18	34
Suburban	$57.50	$102.13	_____	$105.78	$146.53	106
Second City	$49.61	$88.73	_____	$97.87	$138.47	83
Town	$45.00	$77.00	_____	$85.59	$114.45	111
Rural	$35.00	$70.45	_____	$75.34	$114.81	48
Median Area Household Income						
$35,000 or Less	$37.75	$67.25	_____	$76.74	$101.31	44
$35,000 to $49,999	$47.48	$79.40	_____	$95.02	$135.82	140
$50,000 to $69,999	$48.50	$86.00	_____	$93.10	$124.95	133
$70,000 or More	$69.75	$118.00	_____	$110.45	$142.00	64

Note: 7% of respondents reported that they do not offer this service.

TABLE 11.22
HOSPITALIZATION WITHOUT IV: OVERNIGHT STAY, 60-POUND DOG

	25th Percentile	Median	Your Data	Average	75th Percentile	Number of Respondents
All Practices	**$28.00**	**$39.00**		**$44.03**	**$55.00**	**399**
Number of FTE Veterinarians						
1.0 or Less	$25.13	$35.00		$40.71	$55.00	116
1.1 to 2.0	$30.00	$40.00		$45.25	$55.00	123
2.1 to 3.0	$28.50	$39.50		$42.34	$52.24	80
3.1 or More	$27.25	$40.64		$48.98	$63.38	76
Member Status						
Accredited Practice Member	$34.00	$44.90		$48.55	$57.00	111
Nonaccredited Member	$30.00	$39.50		$46.21	$58.88	68
Nonmember	$25.00	$35.00		$41.23	$52.36	193
Metropolitan Status						
Urban	$36.39	$43.10		$51.15	$56.59	34
Suburban	$31.51	$45.98		$50.79	$62.97	112
Second City	$30.00	$42.50		$45.46	$55.23	85
Town	$25.00	$35.00		$39.28	$49.25	113
Rural	$21.89	$28.20		$32.12	$35.79	48
Median Area Household Income						
$35,000 or Less	$20.34	$30.25		$34.78	$46.13	44
$35,000 to $49,999	$26.66	$35.00		$42.60	$50.00	143
$50,000 to $69,999	$29.46	$40.94		$44.50	$55.88	136
$70,000 or More	$36.31	$50.89		$53.21	$61.25	66

Note: 6% of respondents reported that they do not offer this service.

TABLE 11.23
HOSPITALIZATION WITH IV: OVERNIGHT STAY, 100-POUND DOG

	25th Percentile	Median	Your Data	Average	75th Percentile	Number of Respondents
All Practices	**$49.90**	**$87.38**		**$96.73**	**$135.15**	**382**
Number of FTE Veterinarians						
1.0 or Less	$45.00	$76.88		$88.48	$126.33	112
1.1 to 2.0	$47.65	$84.59		$96.73	$129.60	113
2.1 to 3.0	$54.25	$103.63		$104.27	$144.14	78
3.1 or More	$54.00	$90.60		$101.83	$138.47	75
Member Status						
Accredited Practice Member	$68.31	$102.78		$110.58	$145.95	106
Nonaccredited Member	$49.63	$84.00		$95.05	$140.88	62
Nonmember	$45.00	$78.00		$89.81	$121.75	187
Metropolitan Status						
Urban	$61.63	$86.03		$110.99	$157.25	33
Suburban	$58.50	$103.00		$108.11	$151.05	105
Second City	$49.71	$98.04		$100.49	$143.76	80
Town	$45.58	$78.88		$86.12	$114.75	108
Rural	$40.00	$71.00		$81.46	$121.00	47
Median Area Household Income						
$35,000 or Less	$40.35	$68.95		$82.06	$114.13	44
$35,000 to $49,999	$48.00	$81.00		$97.65	$140.06	136
$50,000 to $69,999	$50.00	$87.13		$93.41	$124.48	130
$70,000 or More	$73.50	$119.13		$113.26	$145.70	62

Note: 8% of respondents reported that they do not offer this service.

TABLE 11.24
HOSPITALIZATION WITHOUT IV: OVERNIGHT STAY, 100-POUND DOG

	25th Percentile	Median	Your Data	Average	75th Percentile	Number of Respondents
All Practices	$29.69	$40.34		$44.81	$56.47	394
Number of FTE Veterinarians						
1.0 or Less	$26.00	$37.00		$42.11	$56.46	115
1.1 to 2.0	$30.00	$42.00		$46.25	$57.63	122
2.1 to 3.0	$30.00	$41.50		$43.94	$52.25	79
3.1 or More	$29.50	$42.00		$47.56	$60.38	74
Member Status						
Accredited Practice Member	$36.00	$45.00		$49.47	$58.25	110
Nonaccredited Member	$30.15	$41.15		$47.86	$63.25	66
Nonmember	$25.00	$35.46		$41.59	$52.72	191
Metropolitan Status						
Urban	$37.18	$45.00		$48.43	$58.00	31
Suburban	$32.30	$46.00		$51.49	$64.00	111
Second City	$33.08	$44.00		$47.41	$56.72	84
Town	$26.00	$36.00		$40.14	$50.19	113
Rural	$21.89	$30.00		$33.74	$38.68	48
Median Area Household Income						
$35,000 or Less	$21.55	$32.00		$35.89	$48.00	45
$35,000 to $49,999	$28.00	$37.57		$43.66	$50.80	140
$50,000 to $69,999	$29.81	$41.98		$45.83	$57.88	136
$70,000 or More	$39.50	$52.00		$52.26	$60.87	63

Note: 6% of respondents reported that they do not offer this service.

TABLE 11.25
HOSPITALIZATION WITH IV AND TUBE FEEDING: OVERNIGHT STAY, BIRD

	25th Percentile	Median	Your Data	Average	75th Percentile	Number of Respondents
All Practices	$32.00	$69.00		$79.73	$98.50	71

Note 1: Data was reported only for all practices due to an insufficient number of responses.
Note 2: 81% of respondents reported that they do not offer this service.

TABLE 11.26
HOSPITALIZATION WITHOUT IV AND TUBE FEEDING: OVERNIGHT STAY, BIRD

	25th Percentile	Median	Your Data	Average	75th Percentile	Number of Respondents
All Practices	**$22.62**	**$32.50**		**$39.77**	**$46.25**	**90**
Number of FTE Veterinarians						
1.0 or Less	*	*		*	*	21
1.1 to 2.0	$23.75	$38.25		$48.09	$60.09	26
2.1 to 3.0	*	*		*	*	22
3.1 or More	*	*		*	*	20
Member Status						
Accredited Practice Member	$19.50	$35.65		$40.78	$51.80	26
Nonaccredited Member	*	*		*	*	12
Nonmember	$22.00	$32.00		$37.80	$43.00	43
Metropolitan Status						
Urban	*	*		*	*	8
Suburban	*	*		*	*	20
Second City	*	*		*	*	19
Town	$19.00	$32.00		$35.69	$36.78	29
Rural	*	*		*	*	12
Median Area Household Income						
$35,000 or Less	*	*		*	*	11
$35,000 to $49,999	$18.50	$31.00		$43.93	$47.50	36
$50,000 to $69,999	$25.00	$33.04		$37.39	$46.25	34
$70,000 or More	*	*		*	*	7

Note 1: An asterisk indicates that data was not reported due to an insufficient number of responses.
Note 2: 75% of respondents reported that they do not offer this service.

TABLE 11.27
HOSPITALIZATION WITH IV: OVERNIGHT STAY, REPTILE

	25th Percentile	Median	Your Data	Average	75th Percentile	Number of Respondents
All Practices	**$28.00**	**$61.63**		**$72.07**	**$104.65**	**82**
Number of FTE Veterinarians						
1.0 or Less	*	*		*	*	18
1.1 to 2.0	*	*		*	*	23
2.1 to 3.0	*	*		*	*	18
3.1 or More	*	*		*	*	23
Member Status						
Accredited Practice Member	*	*		*	*	22
Nonaccredited Member	*	*		*	*	14
Nonmember	$26.28	$47.29		$61.67	$93.63	38
Metropolitan Status						
Urban	*	*		*	*	10
Suburban	*	*		*	*	23
Second City	*	*		*	*	13
Town	$31.55	$70.54		$68.38	$92.88	26
Rural	*	*		*	*	9
Median Area Household Income						
$35,000 or Less	*	*		*	*	8
$35,000 to $49,999	$23.74	$43.00		$64.99	$115.55	29
$50,000 to $69,999	$42.74	$83.00		$81.02	$109.33	34
$70,000 or More	*	*		*	*	9

Note 1: An asterisk indicates that data was not reported due to an insufficient number of responses.
Note 2: 78% of respondents reported that they do not offer this service.

TABLE 11.28
HOSPITALIZATION WITHOUT IV: OVERNIGHT STAY, REPTILE

	25th Percentile	Median	Your Data	Average	75th Percentile	Number of Respondents
All Practices	$24.25	$33.00		$40.52	$49.13	92
Number of FTE Veterinarians						
1.0 or Less	*	*		*	*	20
1.1 to 2.0	$24.75	$37.75		$44.69	$58.88	26
2.1 to 3.0	*	*		*	*	21
3.1 or More	$25.10	$37.00		$44.08	$57.75	25
Member Status						
Accredited Practice Member	$25.20	$37.00		$43.09	$56.80	27
Nonaccredited Member	*	*		*	*	13
Nonmember	$22.13	$30.00		$37.66	$43.73	44
Metropolitan Status						
Urban	*	*		*	*	11
Suburban	*	*		*	*	23
Second City	*	*		*	*	15
Town	$20.00	$30.50		$35.55	$36.80	32
Rural	*	*		*	*	10
Median Area Household Income						
$35,000 or Less	*	*		*	*	8
$35,000 to $49,999	$20.00	$29.70		$43.25	$51.00	35
$50,000 to $69,999	$25.20	$34.00		$38.67	$52.00	39
$70,000 or More	*	*		*	*	8

Note 1: An asterisk indicates that data was not reported due to an insufficient number of responses.
Note 2: 75% of respondents reported that they do not offer this service.

TABLE 11.29
HOSPITALIZATION WITH IV: OVERNIGHT STAY, SMALL MAMMAL

	25th Percentile	Median	Your Data	Average	75th Percentile	Number of Respondents
All Practices	$28.50	$64.30		$72.88	$107.62	116
Number of FTE Veterinarians						
1.0 or Less	$26.28	$49.04		$66.91	$92.13	26
1.1 to 2.0	$35.55	$66.84		$74.75	$102.35	32
2.1 to 3.0	$25.55	$62.80		$72.82	$129.66	28
3.1 or More	$27.88	$71.55		$76.13	$110.00	30
Member Status						
Accredited Practice Member	$34.50	$73.20		$86.93	$132.00	31
Nonaccredited Member	*	*		*	*	21
Nonmember	$25.85	$48.08		$63.36	$89.00	53
Metropolitan Status						
Urban	*	*		*	*	15
Suburban	$27.80	$87.00		$84.76	$121.85	30
Second City	*	*		*	*	22
Town	$31.55	$69.94		$69.22	$92.88	34
Rural	*	*		*	*	14
Median Area Household Income						
$35,000 or Less	*	*		*	*	12
$35,000 to $49,999	$25.00	$47.25		$62.17	$87.50	42
$50,000 to $69,999	$34.50	$80.00		$82.94	$110.00	43
$70,000 or More	*	*		*	*	17

Note 1: An asterisk indicates that data was not reported due to an insufficient number of responses.
Note 2: 70% of respondents reported that they do not offer this service.

TABLE 11.30
HOSPITALIZATION WITHOUT IV: OVERNIGHT STAY, SMALL MAMMAL

	25th Percentile	Median	Your Data	Average	75th Percentile	Number of Respondents
All Practices	**$22.00**	**$32.00**		**$38.04**	**$49.00**	**137**
Number of FTE Veterinarians						
1.0 or Less	$18.25	$29.25		$32.39	$41.61	32
1.1 to 2.0	$24.13	$31.81		$38.42	$52.88	36
2.1 to 3.0	$25.00	$32.00		$39.00	$50.00	35
3.1 or More	$25.00	$34.60		$41.95	$54.88	34
Member Status						
Accredited Practice Member	$27.18	$39.60		$44.70	$60.14	41
Nonaccredited Member	*	*		*	*	22
Nonmember	$20.00	$29.00		$32.95	$42.08	63
Metropolitan Status						
Urban	*	*		*	*	16
Suburban	$20.50	$36.00		$41.19	$58.95	36
Second City	$22.82	$40.21		$42.13	$56.70	27
Town	$21.25	$30.73		$35.27	$39.25	40
Rural	*	*		*	*	17
Median Area Household Income						
$35,000 or Less	*	*		*	*	15
$35,000 to $49,999	$18.50	$29.35		$36.60	$47.50	48
$50,000 to $69,999	$27.04	$34.25		$38.91	$50.75	52
$70,000 or More	*	*		*	*	20

Note 1: An asterisk indicates that data was not reported due to an insufficient number of responses.
Note 2: 63% of respondents reported that they do not offer this service.

TABLE 11.31
HOSPITALIZATION WITH IV: OVERNIGHT STAY, FERRET

	25th Percentile	Median	Your Data	Average	75th Percentile	Number of Respondents
All Practices	**$29.00**	**$60.00**		**$73.03**	**$110.00**	**129**
Number of FTE Veterinarians						
1.0 or Less	$27.63	$44.75		$66.15	$107.88	28
1.1 to 2.0	$34.50	$63.25		$73.93	$109.90	35
2.1 to 3.0	$26.66	$51.00		$70.47	$119.29	30
3.1 or More	$30.05	$74.10		$79.64	$110.00	36
Member Status						
Accredited Practice Member	$36.25	$73.20		$83.68	$128.12	37
Nonaccredited Member	*	*		*	*	23
Nonmember	$26.70	$48.08		$64.49	$100.00	59
Metropolitan Status						
Urban	*	*		*	*	12
Suburban	$27.11	$76.00		$81.08	$116.82	33
Second City	$32.95	$92.00		$84.56	$127.58	27
Town	$30.20	$60.00		$65.22	$91.00	39
Rural	*	*		*	*	15
Median Area Household Income						
$35,000 or Less	*	*		*	*	14
$35,000 to $49,999	$26.90	$54.13		$66.84	$109.53	48
$50,000 to $69,999	$34.50	$86.00		$83.60	$112.44	51
$70,000 or More	*	*		*	*	13

Note 1: An asterisk indicates that data was not reported due to an insufficient number of responses.
Note 2: 66% of respondents reported that they do not offer this service.

TABLE 11.32
HOSPITALIZATION WITHOUT IV: OVERNIGHT STAY, FERRET

	25th Percentile	Median	Your Data	Average	75th Percentile	Number of Respondents
All Practices	$22.38	$32.00		$37.19	$47.00	142
Number of FTE Veterinarians						
1.0 or Less	$19.25	$32.50		$34.60	$42.06	32
1.1 to 2.0	$23.25	$30.00		$37.44	$47.75	37
2.1 to 3.0	$23.75	$32.00		$36.68	$46.75	37
3.1 or More	$25.00	$34.60		$39.74	$51.50	36
Member Status						
Accredited Practice Member	$25.10	$37.00		$40.81	$55.40	45
Nonaccredited Member	*	*		*	*	21
Nonmember	$20.00	$29.50		$32.61	$42.08	67
Metropolitan Status						
Urban	*	*		*	*	12
Suburban	$24.00	$32.00		$40.06	$49.50	37
Second City	$26.59	$41.50		$42.14	$54.70	30
Town	$20.00	$30.00		$34.05	$37.00	43
Rural	*	*		*	*	18
Median Area Household Income						
$35,000 or Less	*	*		*	*	16
$35,000 to $49,999	$20.00	$28.45		$35.69	$45.95	51
$50,000 to $69,999	$25.65	$34.75		$39.58	$53.50	56
$70,000 or More	*	*		*	*	16

Note 1: An asterisk indicates that data was not reported due to an insufficient number of responses.
Note 2: 61% of respondents reported that they do not offer this service.

TABLE 11.33
STAFF PHYSICALLY PRESENT IN HOSPITAL 24 HOURS A DAY

	Yes	No	Number of Respondents
All Practices	8%	92%	463
Number of FTE Veterinarians			
1.0 or Less	4%	96%	141
1.1 to 2.0	4%	96%	140
2.1 to 3.0	8%	92%	87
3.1 or More	21%	79%	89
Member Status			
Accredited Practice Member	12%	88%	127
Nonaccredited Member	5%	95%	80
Nonmember	8%	92%	223
Metropolitan Status			
Urban	14%	86%	44
Suburban	10%	90%	131
Second City	8%	92%	95
Town	6%	94%	126
Rural	5%	95%	58
Median Area Household Income			
$35,000 or Less	8%	92%	53
$35,000 to $49,999	8%	92%	166
$50,000 to $69,999	7%	93%	156
$70,000 or More	9%	91%	78

Note: Of those who responded "no" to this question, 8% said that their practices were solely outpatient clinics.

CHAPTER 12

ANESTHESIA SERVICES

Convey the necessity of preanesthetic testing and recovery care by using parallels to human medicine, and watch your clients' perception of value and compliance grow. In this chapter, you'll learn how to educate clients about anesthesia and calm their nerves.

When a family brought their miniature schnauzer to a veterinary hospital for a routine dental cleaning, the veterinarian determined that several of the elderly animal's teeth needed to be pulled. The only problem: No one could reach family members to consult with them. The veterinarian decided to pull the teeth rather than prolong the anesthetic procedure or risk putting the pet under anesthesia a second time.

Although the doctor knew her decision was medically sound, the client did not. The angry client believed she should have had the opportunity to consult with the doctor first. If the client had been properly educated beforehand, the bond between the doctor and the client would not have been strained.

"Effective communication during any procedure requiring anesthesia is the principal element to success," says Dr. Cecelia Soares, MS, a veterinarian, licensed marriage and family therapist, and director of The OTHER End of the Leash™ in Walnut Creek, California. "Think of your clients' pets as their children, because they do," Dr. Soares advises. In fact, you may find that comparisons to human medicine make it easier for your clients to understand the value of everything from preanesthetic blood tests to blood-pressure monitoring to pain management to the recovery process. "I use human medicine as a bolster because clients are more familiar with that," Dr. Soares says. "Most people understand it but just don't apply it to the veterinary side."

EDUCATING CLIENTS ABOUT ANESTHESIA

When preparing a client for a pet's surgery or anesthetic procedure, it's important to cover all the bases. You should expect and be prepared for questions along the way. More important, as a veterinarian you must decide ahead of time what you require as part of your standard of care. Speak with authority and confidence, Dr. Soares recommends. If you believe in the services you offer and can effectively communicate their value, clients won't challenge you.

Stress the Importance of Preanesthetic Blood Work

Performing preanesthetic blood work not only alerts you to potential health problems, but it also helps ensure a safer surgical or anesthetic procedure. Communicating the value and importance of diagnostic tests to clients, however, can sometimes prove difficult. Dr. Soares frequently tells clients her "crystal ball is broken."

"Nobody is clairvoyant in this business," she says. "Tell clients there are no guarantees with anesthesia even though it is relatively low risk, and one of the elements that adds to the low-risk factor is to have preanesthetic testing. It's just another tool in the tool box to make sure we can get as much information as possible."

When clients question whether they should spend money on preanesthetic tests, start by telling them that a veterinary hospital is much like a human hospital. You are medical professionals who perform medical procedures and surgeries, and just as physicians or surgeons recommend—if not require—presurgical blood screens for people, you recommend the same for pets.

Educate Clients About Blood-Pressure Monitoring

As with preanesthetic blood tests, blood-pressure monitoring is crucial during the surgical or anesthetic procedure. It allows you to continually monitor the animal's vital signs and head off potential problems before it's too late. Blood-pressure monitoring is well worth the additional expense simply because it gives you more information about the pet's condition during surgery. And the more information you have, the better care your client's pet will receive. Remind your clients that they would never go into surgery without blood-pressure monitoring, and neither should their pets. "The species is different, but the concerns are the same," Dr. Soares says.

Use Consent Forms

Consent forms help you proactively promote IV fluids and pain management. Consent forms serve many purposes, especially in today's changing legal climate. Not only do they protect you and your hospital from potential lawsuits, they also outline the risks of the procedure and give clients another opportunity to ask questions and then agree to those risks. Dr. Soares recommends including statements about pain management and the use of IV fluids in consent forms.

"There is no such thing as a practice without pain management," she says. "The pet gets pain medication, period. I don't see a reason for it to be optional. It's a well-known fact that animals recover much faster if you provide pain management."

CALMING NERVOUS CLIENTS

The day of surgery is here, and your client seems uneasy about her pet's procedure. Remember to simplify your language and slow down your speech. Clients may be anxious and have trouble processing information as a result. Most important, develop communication techniques that calm clients. The more information clients receive, the more at ease they will feel. Try following these steps:

1. Schedule surgery check-in and discharge appointments. This gives the client a timeline for the hospitalization period, provides another opportunity for the client to ask questions, and lets the veterinarian or technician discuss any last-minute details or take-home instructions.

2. Designate one staff member as a client's main contact and provide a business card when checking in. Depending on the severity of the case, a technician may be the client's primary contact, while a receptionist might be the secondary contact. In serious cases, the veterinarian should be the primary contact.

3. Gather several contact numbers for the client. Be sure to get home, work, mobile, and pager numbers if available. The more numbers you have, the easier it will be to get in touch with a client.

4. Schedule a postsurgery call. Have a technician or veterinarian call the client after the surgery is complete and the animal is going into recovery. Also confirm the time of the discharge appointment.

5. If dealing with a serious case, call clients during the exploratory phase to let them know your findings. If complications develop, call the client immediately, clearly explain options, and offer to let the client say goodbye to the pet if it's a life-threatening situation.

6. Implement an open-hospital policy. Always give your clients as much access to their pets as possible. If your hospital is set up accordingly, offer clients the option to observe part of their pets' surgery, and let them be present during recovery. Even though clients may choose not to observe their pets in surgery or be with them in recovery, the fact that the option is there will calm them.

Case Study

Waking Up in a Comforting Environment

Atom Gardiner spends his days in a softly lit room intensely monitoring patients, speaking calmly to them, and offering each a reassuring touch—what he describes as "loving them back to consciousness." Gardiner knows no other job except the one he finds in the recovery room at the Veterinary Referral Center of Colorado in Englewood, Colorado. To Gardiner, a veterinary technician, postanesthetic care is just as important as preanesthetic tests.

With 14 years of experience in emergency and critical care, Gardiner became frustrated with the stories he heard of how animals were neglected after surgery. And as a technician in a busy specialty hospital he found himself unable to devote the attention needed to each animal because of other demands. "I wanted to sit with them for two hours until they were calm," Gardiner says. "I wanted to have a position where caring for patients' postoperative needs was my only job."

As a result, Gardiner became owner and operator of Veterinary Anesthesia Resources (VAR) at the Veterinary Referral Center of Colorado nearly four years ago. His only job now is to care for animals coming out of surgery. As each patient enters the recovery room, he measures the pet's blood glucose, packed cell volume, total protein, and arterial blood pressure levels. He administers analgesics before the animal can wake and experience discomfort.

A heated room, coupled with recirculating warm water blankets and warmed parenteral fluids administered via constant-rate infusion pumps, help ensure patients' body temperatures and fluid levels return to normal. In addition, animals aren't extubated until the chewing/swallowing reflex begins to reduce risk of aspiration. To accommodate more serious cases, slowly recovering patients are laterally rotated every 30 to 60 minutes to prevent hypostatic lung congestion. Gardiner monitors pulse rate, body temperature, respiratory rate, mucous membrane color, and capillary refill time and records those values every 30 minutes.

"You can avert many problems and additional medical expenses by constant monitoring," Gardiner says. "Most of these extra steps may be considered 'over the top,' but when people are willing to spend thousands to fix an animal's leg, they usually don't have a problem spending a little more to ensure their pets' safe and comfortable recovery."

To explain the importance of postanesthetic care, Gardiner developed a client-education letter (see Figure 12.1). Before admitting their pets, clients visit the recovery room and observe the care animals receive. Gardiner and his staff are available to answer questions.

FIGURE 12.1

SAMPLE POSTANESTHESIA-CARE CLIENT-EDUCATION LETTER

Date

Dear Client:

Imagine yourself undergoing an anesthetic procedure. You are admitted to preop, an anesthesia mask is placed over your face, and before you can count back from 10, you are waking up in the hospital's recovery room. You're cold, disoriented, and struggling to comprehend what has just happened to you. Now imagine how your pet must feel!

Postanesthesia support is instrumental in securing the safety and comfort of your pet following anesthesia. The recovery technician is there to calm, reassure, and intensely monitor your pet throughout the recovery process. Tender, loving care can go a long way to help smooth your pet's recovery from anesthesia. To ensure a safer recovery, your pet is under constant observation until fully recuperated from the effects of anesthesia.

Immediately following the anesthetic procedure, intense monitoring is provided in an isolated recovery suite. The recovery technician's only responsibility is the care of your pet. Upon presentation to recovery, specific tests are performed and therapeutic modifications are implemented to ensure your pet's safe recovery.

Environmental/supplemental heat therapy is provided with recirculating warm-water blankets. Warmed IV fluids are administered via constant-rate infusion pumps to compensate for metabolic needs and losses during surgery. Medications to relieve pain are administered while your pet is still sedated, prior to your pet's recognition of discomfort. Your pet's vital signs are measured and documented every 30 minutes until normalized and stabilized. This allows even the most subtle physiologic changes to be identified as they occur. Treatment can then be initiated before these changes become problematic. In the rare event of complications, board-certified veterinary specialists are on hand, and the resources of the entire hospital are available.

Our goal is to prevent postanesthesia complications and to make your pet as comfortable as possible. Dedicated medical management, while providing a reassuring word and a gentle hand, will enable us to do just that.

Sincerely,

The staff of Veterinary Anesthesia Resources at the Veterinary Referral Center of Colorado

Implementation Idea

How to Create a "Comfort Recovery" Package for Your Patients

Although Veterinary Anesthesia Resources at the Veterinary Referral Center of Colorado provides an impressive level of service, it may not be realistic for you to dedicate an entire room to recovery, let alone extra staff members. However, there are ways to promote the value of postanesthetic care without the resources available to a larger hospital.

For example, offer clients a special recovery package option. Here are a few items your recovery package could include:

- Your lab technician running blood tests
- A staff member's presence during the first hour of recovery
- Additional monitoring of the animal's vital signs to ensure a safer and quicker recovery
- Heated blankets and orthopedic beds
- Warm IV fluids
- A care package to take home, including instructions on how to make the pet more comfortable and a healthy snack or special diet

HELPFUL RESOURCES

Anesthesia Assessment and Plan Form, AAHA. American Animal Hospital Association Press, 2003.

Anesthesia Record, AAHA. American Animal Hospital Association Press, 2003.

Minor Surgical/Anesthetic Procedure Sticker for medical records, AAHA. American Animal Hospital Association Press, 2005.

Veterinary Anesthesia Update, Second Edition, Nancy Brock, DVM, Dipl. ACVA. Nancy Brock, 2007.

CHAPTER 12

DATA TABLES

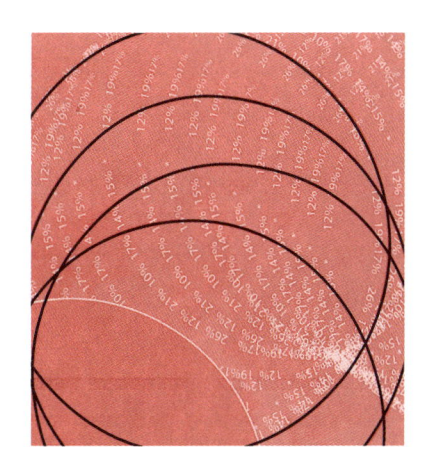

ANESTHESIA SERVICES / 323

TABLE 12.1
IV SEDATIVE

	25th Percentile	Median	Your Data	Average	75th Percentile	Number of Respondents
All Practices	$33.53	$45.00		$47.93	$59.73	421
Number of FTE Veterinarians						
1.0 or Less	$30.00	$42.00		$44.91	$55.20	126
1.1 to 2.0	$34.13	$47.44		$50.35	$61.74	128
2.1 to 3.0	$34.25	$44.75		$48.18	$62.13	80
3.1 or More	$35.50	$46.00		$47.78	$54.22	81
Member Status						
Accredited Practice Member	$35.00	$46.63		$49.44	$60.98	116
Nonaccredited Member	$33.50	$45.00		$47.05	$55.47	73
Nonmember	$31.59	$44.50		$46.96	$57.80	203
Metropolitan Status						
Urban	$44.50	$61.05		$62.70	$84.05	37
Suburban	$37.88	$49.00		$50.06	$61.40	114
Second City	$35.00	$48.25		$50.65	$60.63	90
Town	$30.00	$42.00		$44.08	$54.88	117
Rural	$26.00	$35.15		$36.84	$45.00	55
Median Area Household Income						
$35,000 or Less	$28.75	$40.60		$42.10	$50.25	50
$35,000 to $49,999	$32.19	$44.00		$45.87	$55.60	156
$50,000 to $69,999	$35.00	$45.00		$48.23	$57.80	143
$70,000 or More	$41.93	$55.00		$57.12	$73.94	62

Note 1: Fee is for IV sedative for a 30-minute radiology procedure for a 30-pound dog with no intubation or inhalant.
Note 2: 6% of respondents reported that they do not offer this service.

TABLE 12.2
IM SEDATIVE

	25th Percentile	Median	Your Data	Average	75th Percentile	Number of Respondents
All Practices	$28.70	$39.00		$42.13	$52.20	417
Number of FTE Veterinarians						
1.0 or Less	$25.00	$35.00		$39.93	$50.00	127
1.1 to 2.0	$28.88	$40.65		$43.00	$56.12	128
2.1 to 3.0	$31.93	$40.00		$42.74	$50.00	79
3.1 or More	$30.00	$39.50		$43.38	$52.00	78
Member Status						
Accredited Practice Member	$32.23	$39.50		$43.21	$49.56	110
Nonaccredited Member	$28.65	$40.49		$43.25	$52.25	71
Nonmember	$25.00	$36.44		$40.62	$53.31	208
Metropolitan Status						
Urban	$35.75	$50.00		$54.02	$68.90	40
Suburban	$33.00	$42.00		$44.32	$55.72	115
Second City	$30.00	$40.38		$44.85	$58.38	86
Town	$25.00	$33.05		$38.31	$46.61	112
Rural	$23.25	$32.42		$32.32	$41.28	56
Median Area Household Income						
$35,000 or Less	$22.50	$31.00		$33.03	$42.40	50
$35,000 to $49,999	$25.38	$34.43		$40.30	$49.63	154
$50,000 to $69,999	$31.74	$41.03		$43.53	$55.00	138
$70,000 or More	$36.63	$48.00		$50.04	$64.12	65

Note 1: Fee is for IM sedative for an abscess treatment for a fractious cat.
Note 2: 7% of respondents reported that they do not offer this service.

TABLE 12.3
CLIENT CHARGED FOR PREANESTHETIC EXAM

	Yes	No	Number of Respondents
All Practices	**15%**	**85%**	**458**
Number of FTE Veterinarians			
1.0 or Less	14%	86%	139
1.1 to 2.0	14%	86%	139
2.1 to 3.0	18%	82%	85
3.1 or More	17%	83%	89
Member Status			
Accredited Practice Member	13%	87%	127
Nonaccredited Member	14%	86%	80
Nonmember	16%	84%	218
Metropolitan Status			
Urban	14%	86%	44
Suburban	16%	84%	129
Second City	15%	85%	94
Town	13%	87%	126
Rural	18%	82%	56
Median Area Household Income			
$35,000 or Less	17%	83%	53
$35,000 to $49,999	16%	84%	164
$50,000 to $69,999	16%	84%	154
$70,000 or More	10%	90%	77

Note: Respondents answered the question: Do you charge for a preanesthetic exam today assuming that you examined a patient a week ago, charged an exam fee, and recommended that the patient be dropped off today for a minor surgical procedure?

TABLE 12.4
PREANESTHETIC EXAM

	25th Percentile	Median	Your Data	Average	75th Percentile	Number of Respondents
All Practices	**$18.20**	**$30.00**	_____	**$30.14**	**$39.00**	**67**

Note: Data was reported only for all practices due to an insufficient number of responses.

TABLE 12.5
PREANESTHETIC SEDATION

	25th Percentile	Median	Your Data	Average	75th Percentile	Number of Respondents
All Practices	**$18.00**	**$26.00**		**$30.59**	**$39.00**	**221**
Number of FTE Veterinarians						
1.0 or Less	$16.45	$25.88		$29.34	$40.00	70
1.1 to 2.0	$16.13	$26.00		$29.28	$32.38	64
2.1 to 3.0	$18.50	$27.50		$32.29	$40.63	42
3.1 or More	$20.40	$25.00		$32.09	$40.98	41
Member Status						
Accredited Practice Member	$17.19	$26.00		$31.10	$40.49	66
Nonaccredited Member	$20.00	$28.13		$35.84	$45.00	38
Nonmember	$15.00	$25.60		$28.59	$35.00	103
Metropolitan Status						
Urban	*	*		*	*	22
Suburban	$21.83	$28.98		$33.89	$40.47	74
Second City	$22.00	$26.00		$32.46	$37.00	39
Town	$16.50	$23.33		$24.69	$30.70	55
Rural	$9.63	$15.13		$24.12	$29.38	28
Median Area Household Income						
$35,000 or Less	*	*		*	*	24
$35,000 to $49,999	$17.63	$24.87		$27.64	$32.38	76
$50,000 to $69,999	$21.14	$28.73		$35.63	$44.53	80
$70,000 or More	$13.50	$26.20		$27.53	$35.28	36

Note 1: Fee is for a 40-pound dog.
Note 2: An asterisk indicates that data was not reported due to an insufficient number of responses.
Note 3: 40% of the respondents reported that the charge is included in the total fee for administering anesthesia.
Note 4: 8% of respondents reported that they do not offer this service.

TABLE 12.6
IV INDUCTION

	25th Percentile	Median	Your Data	Average	75th Percentile	Number of Respondents
All Practices	$25.00	$37.97		$44.20	$55.73	205
Number of FTE Veterinarians						
1.0 or Less	$25.00	$38.25		$44.31	$58.64	63
1.1 to 2.0	$25.00	$36.00		$46.76	$59.10	63
2.1 to 3.0	$24.59	$32.50		$40.65	$48.75	37
3.1 or More	$26.20	$40.00		$43.14	$50.00	40
Member Status						
Accredited Practice Member	$28.00	$43.30		$45.78	$57.38	58
Nonaccredited Member	$22.20	$31.00		$36.51	$41.00	37
Nonmember	$25.00	$38.75		$44.49	$57.25	100
Metropolitan Status						
Urban	$25.88	$39.00		$45.08	$56.29	25
Suburban	$28.19	$39.00		$44.16	$54.50	56
Second City	$26.50	$42.00		$49.56	$61.13	45
Town	$23.50	$29.00		$38.79	$45.00	49
Rural	$20.41	$32.95		$39.32	$41.25	26
Median Area Household Income						
$35,000 or Less	*	*		*	*	18
$35,000 to $49,999	$24.73	$33.00		$41.70	$49.63	74
$50,000 to $69,999	$28.00	$39.00		$43.76	$50.00	71
$70,000 or More	$26.50	$45.00		$51.63	$64.07	37

Note 1: Fee is for a 40-pound dog.
Note 2: An asterisk indicates that data was not reported due to an insufficient number of responses.
Note 3: 46% of the respondents reported that the charge is included in the total fee for administering anesthesia.
Note 4: 6% of respondents reported that they do not offer this service.

TABLE 12.7
INTUBATION

	25th Percentile	Median	Your Data	Average	75th Percentile	Number of Respondents
All Practices	$15.15	$30.00		$44.46	$57.50	29

Note 1: Fee is for a 40-pound dog.
Note 2: Data was reported only for all practices due to an insufficient number of responses.
Note 3: 81% of the respondents reported that the charge is included in the total fee for administering anesthesia.
Note 4: 12% of respondents reported that they do not offer this service.

TABLE 12.8
INHALANT: 30 MINUTES ISOFLURANE

	25th Percentile	Median	Your Data	Average	75th Percentile	Number of Respondents
All Practices	$46.00	$70.00		$72.40	$92.70	387
Number of FTE Veterinarians						
1.0 or Less	$39.00	$60.00		$64.70	$86.00	115
1.1 to 2.0	$50.00	$68.63		$72.65	$92.99	112
2.1 to 3.0	$53.69	$80.00		$79.39	$97.75	76
3.1 or More	$49.69	$75.50		$74.97	$92.89	80
Member Status						
Accredited Practice Member	$55.25	$76.00		$79.09	$99.39	104
Nonaccredited Member	$48.00	$71.50		$69.67	$87.27	68
Nonmember	$44.00	$64.00		$69.00	$92.00	189
Metropolitan Status						
Urban	$61.00	$96.50		$94.12	$120.50	41
Suburban	$46.00	$76.00		$74.35	$95.00	107
Second City	$53.00	$75.00		$77.99	$95.30	83
Town	$41.12	$60.00		$62.52	$80.28	101
Rural	$35.00	$50.00		$56.56	$77.40	46
Median Area Household Income						
$35,000 or Less	$40.00	$60.00		$61.38	$82.00	42
$35,000 to $49,999	$41.30	$61.50		$67.05	$86.23	148
$50,000 to $69,999	$51.88	$77.00		$76.69	$99.16	126
$70,000 or More	$62.00	$82.75		$83.19	$99.00	61

Note 1: Fee is for a 40-pound dog.
Note 2: 11% of the respondents reported that the charge is included in the total fee for administering anesthesia.
Note 3: 3% of respondents reported that they do not offer this service.

TABLE 12.9
INHALANT: 60 MINUTES ISOFLURANE

	25th Percentile	Median	Your Data	Average	75th Percentile	Number of Respondents
All Practices	**$61.80**	**$92.75**		**$102.00**	**$130.50**	**398**
Number of FTE Veterinarians						
1.0 or Less	$54.81	$81.00		$91.57	$122.87	122
1.1 to 2.0	$73.98	$95.00		$103.98	$128.75	115
2.1 to 3.0	$62.00	$101.00		$107.03	$135.00	79
3.1 or More	$74.75	$105.00		$108.41	$134.20	78
Member Status						
Accredited Practice Member	$78.00	$100.00		$111.26	$136.00	105
Nonaccredited Member	$66.13	$97.50		$104.45	$132.13	72
Nonmember	$60.00	$90.00		$96.35	$128.00	193
Metropolitan Status						
Urban	$99.00	$127.50		$132.66	$181.38	40
Suburban	$65.75	$99.50		$108.11	$140.00	109
Second City	$65.00	$93.05		$104.98	$135.33	88
Town	$60.00	$85.00		$90.44	$120.00	103
Rural	$44.45	$75.00		$80.02	$108.50	49
Median Area Household Income						
$35,000 or Less	$48.69	$76.50		$90.16	$121.00	46
$35,000 to $49,999	$60.00	$87.50		$93.63	$121.00	151
$50,000 to $69,999	$70.75	$100.87		$108.64	$140.56	128
$70,000 or More	$80.00	$106.00		$116.59	$135.00	63

Note 1: Fee is for a 40-pound dog.
Note 2: 8% of the respondents reported that the charge is included in the total fee for administering anesthesia.
Note 3: 3% of respondents reported that they do not offer this service.

TABLE 12.10
INHALANT: ADDITIONAL 60 MINUTES ISOFLURANE

	25th Percentile	Median	Your Data	Average	75th Percentile	Number of Respondents
All Practices	$50.00	$80.00		$89.60	$120.00	368
Number of FTE Veterinarians						
1.0 or Less	$40.00	$61.60		$76.02	$92.43	106
1.1 to 2.0	$52.20	$78.37		$85.33	$110.00	111
2.1 to 3.0	$59.93	$92.50		$99.33	$135.60	76
3.1 or More	$60.00	$92.25		$103.00	$128.00	72
Member Status						
Accredited Practice Member	$60.00	$100.00		$106.62	$137.25	100
Nonaccredited Member	$48.00	$78.00		$85.84	$104.00	71
Nonmember	$50.00	$65.00		$81.35	$108.00	171
Metropolitan Status						
Urban	$62.00	$92.40		$95.05	$120.00	39
Suburban	$59.50	$89.50		$100.74	$139.29	98
Second City	$51.25	$84.00		$91.90	$123.50	85
Town	$45.30	$77.88		$83.41	$111.55	94
Rural	$45.75	$60.00		$71.64	$90.00	44
Median Area Household Income						
$35,000 or Less	$38.60	$65.00		$79.82	$100.00	41
$35,000 to $49,999	$50.00	$74.75		$84.37	$110.58	138
$50,000 to $69,999	$57.00	$82.00		$98.44	$126.00	121
$70,000 or More	$60.00	$90.00		$94.16	$133.10	59

Note 1: Fee is for a 40-pound dog.
Note 2: 9% of the respondents reported that the charge is included in the total fee for administering anesthesia.
Note 3: 5% of respondents reported that they do not offer this service.

TABLE 12.11
ANESTHETIC MONITORING: ELECTRONIC

	25th Percentile	Median	Your Data	Average	75th Percentile	Number of Respondents
All Practices	$15.00	$21.61		$26.19	$31.75	173
Number of FTE Veterinarians						
1.0 or Less	$15.00	$22.00		$25.35	$29.50	45
1.1 to 2.0	$12.40	$19.75		$22.29	$29.00	45
2.1 to 3.0	$13.31	$22.31		$26.52	$30.00	44
3.1 or More	$17.75	$23.50		$30.68	$43.08	36
Member Status						
Accredited Practice Member	$18.00	$24.00		$26.27	$32.63	61
Nonaccredited Member	$15.38	$21.75		$26.56	$33.75	33
Nonmember	$12.50	$20.00		$24.81	$29.00	67
Metropolitan Status						
Urban	*	*		*	*	19
Suburban	$19.02	$24.10		$29.26	$36.13	60
Second City	$12.56	$20.00		$26.77	$29.50	40
Town	$9.75	$17.00		$20.29	$25.40	38
Rural	*	*		*	*	15
Median Area Household Income						
$35,000 or Less	*	*		*	*	19
$35,000 to $49,999	$12.08	$22.75		$25.36	$34.00	60
$50,000 to $69,999	$16.13	$22.25		$27.80	$30.67	58
$70,000 or More	$17.38	$23.60		$28.44	$38.38	32

Note 1: An asterisk indicates that data was not reported due to an insufficient number of responses.
Note 2: 50% of the respondents reported that the charge is included in the total fee for administering anesthesia.
Note 3: 9% of respondents reported that they do not offer this service.

TABLE 12.12
ANESTHETIC MONITORING: MANUAL

	25th Percentile	Median	Your Data	Average	75th Percentile	Number of Respondents
All Practices	$15.00	$23.75		$26.76	$32.00	48

Note 1: Data was reported only for all practices due to an insufficient number of responses.
Note 2: 68% of the respondents reported that the charge is included in the total fee for administering anesthesia.
Note 3: 19% of respondents reported that they do not offer this service.

ANESTHESIA SERVICES / 331

TABLE 12.13
MASK INHALATION ANESTHESIA: SMALL MAMMAL

	25th Percentile	Median	Your Data	Average	75th Percentile	Number of Respondents
All Practices	$32.00	$50.00		$56.76	$75.00	275
Number of FTE Veterinarians						
1.0 or Less	$30.00	$47.50		$51.00	$67.70	67
1.1 to 2.0	$30.13	$50.00		$59.06	$77.48	88
2.1 to 3.0	$34.81	$50.00		$59.13	$76.50	57
3.1 or More	$32.63	$51.75		$56.45	$75.50	60
Member Status						
Accredited Practice Member	$39.00	$57.75		$63.52	$82.50	83
Nonaccredited Member	$32.28	$58.23		$59.53	$82.50	40
Nonmember	$30.00	$45.00		$51.18	$68.06	137
Metropolitan Status						
Urban	$39.50	$57.35		$65.07	$89.08	30
Suburban	$35.00	$55.50		$59.38	$76.75	68
Second City	$45.00	$68.75		$66.17	$83.00	61
Town	$30.00	$45.50		$49.91	$64.25	80
Rural	$30.00	$35.00		$41.57	$50.00	31
Median Area Household Income						
$35,000 or Less	$30.00	$42.25		$45.10	$54.85	32
$35,000 to $49,999	$30.88	$45.50		$54.23	$75.00	98
$50,000 to $69,999	$30.00	$55.00		$59.70	$77.75	94
$70,000 or More	$39.13	$60.20		$63.73	$82.59	45

Note: 32% of respondents reported that they do not offer this service.

TABLE 12.14
MASK INHALATION ANESTHESIA: BIRD

	25th Percentile	Median	Your Data	Average	75th Percentile	Number of Respondents
All Practices	$30.00	$51.00		$56.53	$77.50	125
Number of FTE Veterinarians						
1.0 or Less	$29.44	$51.00		$55.33	$73.78	29
1.1 to 2.0	$28.94	$50.00		$53.48	$74.41	41
2.1 to 3.0	$33.26	$52.00		$62.05	$86.30	27
3.1 or More	$37.00	$51.13		$56.17	$73.66	26
Member Status						
Accredited Practice Member	$45.60	$65.00		$65.29	$88.25	37
Nonaccredited Member	*	*		*	*	15
Nonmember	$29.37	$48.00		$51.17	$71.00	63
Metropolitan Status						
Urban	*	*		*	*	12
Suburban	$37.00	$62.50		$61.87	$84.50	28
Second City	$38.35	$69.50		$65.58	$85.25	26
Town	$30.00	$50.00		$53.27	$73.66	42
Rural	*	*		*	*	16
Median Area Household Income						
$35,000 or Less	*	*		*	*	13
$35,000 to $49,999	$30.00	$50.00		$54.04	$78.44	45
$50,000 to $69,999	$30.00	$56.25		$59.24	$84.50	48
$70,000 or More	*	*		*	*	16

Note 1: An asterisk indicates that data was not reported due to an insufficient number of responses.
Note 2: 69% of respondents reported that they do not offer this service.

TABLE 12.15
PAIN MANAGEMENT INCLUDED IN ELECTIVE PROCEDURE FEES

	Yes, for No Additional Fee	Yes for an Additional Fee	Not Routinely Provided	Number of Respondents
All Practices	**27%**	**72%**	**4%**	**458**
Number of FTE Veterinarians				
1.0 or Less	25%	71%	5%	138
1.1 to 2.0	26%	75%	4%	140
2.1 to 3.0	30%	66%	6%	86
3.1 or More	26%	76%	1%	88
Member Status				
Accredited Practice Member	31%	72%	0%	127
Nonaccredited Member	21%	83%	3%	80
Nonmember	25%	71%	6%	218
Metropolitan Status				
Urban	25%	73%	2%	44
Suburban	24%	78%	2%	129
Second City	25%	74%	5%	95
Town	25%	71%	5%	125
Rural	41%	57%	7%	56
Median Area Household Income				
$35,000 or Less	30%	64%	6%	53
$35,000 to $49,999	25%	72%	6%	166
$50,000 to $69,999	25%	75%	3%	154
$70,000 or More	29%	72%	1%	75

Note: Some row totals do not equal 100% because some respondents selected more than one answer.

TABLE 12.16
PAIN MANAGEMENT INCLUDED IN NONELECTIVE PROCEDURE FEES

	Yes, for No Additional Fee	Yes for an Additional Fee	Not Routinely Provided	Number of Respondents
All Practices	**12%**	**87%**	**3%**	**456**
Number of FTE Veterinarians				
1.0 or Less	12%	82%	7%	138
1.1 to 2.0	13%	88%	2%	139
2.1 to 3.0	14%	86%	1%	85
3.1 or More	8%	94%	1%	89
Member Status				
Accredited Practice Member	12%	90%	0%	125
Nonaccredited Member	11%	93%	1%	80
Nonmember	12%	84%	5%	219
Metropolitan Status				
Urban	9%	86%	5%	43
Suburban	11%	90%	1%	126
Second City	10%	89%	4%	94
Town	14%	84%	3%	126
Rural	16%	83%	3%	58
Median Area Household Income				
$35,000 or Less	10%	90%	0%	51
$35,000 to $49,999	11%	85%	6%	166
$50,000 to $69,999	12%	89%	1%	156
$70,000 or More	16%	85%	3%	73

Note: Some row totals do not equal 100% because some respondents selected more than one answer.

TABLE 12.17
FENTANYL PATCH USE

	Yes	No	Number of Respondents
All Practices	45%	55%	454
Number of FTE Veterinarians			
1.0 or Less	24%	76%	136
1.1 to 2.0	44%	56%	140
2.1 to 3.0	64%	36%	85
3.1 or More	61%	39%	87
Member Status			
Accredited Practice Member	58%	42%	125
Nonaccredited Member	38%	62%	78
Nonmember	40%	60%	218
Metropolitan Status			
Urban	48%	52%	44
Suburban	52%	48%	128
Second City	47%	53%	95
Town	38%	62%	121
Rural	37%	63%	57
Median Area Household Income			
$35,000 or Less	32%	68%	50
$35,000 to $49,999	42%	58%	164
$50,000 to $69,999	49%	51%	152
$70,000 or More	50%	50%	78

TABLE 12.18
FENTANYL PATCH

	25th Percentile	Median	Your Data	Average	75th Percentile	Number of Respondents
All Practices	$40.00	$54.75		$56.80	$70.00	151
Number of FTE Veterinarians						
1.0 or Less	*	*		*	*	22
1.1 to 2.0	$40.00	$50.00		$54.05	$67.45	41
2.1 to 3.0	$47.72	$56.99		$60.44	$76.06	44
3.1 or More	$41.13	$53.75		$55.20	$66.50	42
Member Status						
Accredited Practice Member	$41.48	$52.50		$54.15	$67.10	53
Nonaccredited Member	*	*		*	*	24
Nonmember	$41.48	$56.24		$60.49	$75.00	66
Metropolitan Status						
Urban	*	*		*	*	15
Suburban	$43.27	$55.37		$58.85	$71.14	50
Second City	$42.63	$52.83		$57.12	$70.38	36
Town	$33.65	$53.00		$51.99	$64.93	33
Rural	*	*		*	*	14
Median Area Household Income						
$35,000 or Less	*	*		*	*	13
$35,000 to $49,999	$39.25	$53.62		$53.40	$64.70	52
$50,000 to $69,999	$41.85	$55.50		$58.06	$75.00	54
$70,000 or More	$39.63	$54.87		$60.38	$79.49	28

Note 1: Fee is for a 30-pound dog.
Note 2: An asterisk indicates that data was not reported due to an insufficient number of responses.
Note 3: 4% of respondents reported that they do not offer this service.

CHAPTER 13

TREATMENT PROCEDURES

Report cards, written home-care instructions, and discussions with your health-care team can help increase clients' perception of value for your treatment procedures. Learn about these tools and how to implement them in this chapter.

When Honey, a senior cat suffering from advanced liver disease, came to Cherokee Animal Clinic and Cherokee Cat Clinic in Overland Park, Kansas, the doctors didn't know the extent of the care that Honey would require. They knew that treating Honey would be costly and provided the owner with a written estimate. With the client's approval, Honey was hospitalized for 10 days and received fluid therapy, blood tests, an abdominal sonogram, feeding tubes, and various medications. The cat's owner stayed by its side, visiting every day until Honey was well enough to go home.

When the bill came, it was the highest that Practice Manager Kerry Meredith had ever seen for a feline patient. But the client didn't bat an eye. Instead of griping about the cost, the client was grateful to doctors and staff members who made it possible for Honey to go home. "We gave the client an estimate up front, and he knew the value he would receive," Meredith says. "He was able to make an educated decision because he knew what was involved in Honey's care."

Honey's owner didn't complain about the cost because he understood the value of the services that doctors and team members at Cherokee Animal Clinic and Cherokee Cat Clinic had provided. He felt confident that they knew what was best for his cat, and he trusted them to be honest with him. Client education, along with an open-door policy and adequate access to knowledgeable doctors and staff members, made this client's experience a meaningful one.

ADD VALUE TO YOUR TREATMENT PROCEDURES

To improve your clients' perception of value, you must first start by conveying value to your staff members. If team members understand procedures and their benefits and risks, they can and will effectively communicate those to clients. If everyone speaks the same language, clients will hear the same message consistently and be more apt to trust you, your hospital staff, and the quality of care you provide.

A practice's success stems from a group effort—every member of the team. Each person should walk the walk and talk the talk that is established for the whole team. Focus on the details and educate your staff by example. If they see you scrub and gown and place importance on preanesthetic blood work and pain medication, it will show in their communication with clients.

Once your team members speak the same language, follow these steps to demystify treatment procedures for clients:

1. Let clients see what you do "in the back." Take time to help clients see what you're doing, and compliance and appreciation will follow. For example, if a client brings a dog with an ear infection to you, recommend referencing a diagram of the ear while explaining why it's important to clean the ear thoroughly, performing cytology to determine the type of bacteria present, prescribing an anti-inflammatory medication, and scheduling a recheck exam. If possible, let the client look into the microscope so that she can actually see the culprit causing the infection.

2. Educate clients with hospital tours. Offer to take new clients on hospital tours, including visits to the surgery preparation area, surgery suites, exam rooms, boarding areas, etc. Explain a little about each area as you pass, and introduce staff members to clients. Speak with a staff member in the treatment area before starting a tour to make sure an emergency or life-saving procedure isn't in progress. If it's OK to proceed, tell clients ahead of time what to expect in the surgery and treatment areas. At Cherokee Animal Clinic and Cherokee Cat Clinic, team members don't set visiting hours on purpose. As long as the hospital is open, clients can see their pets.

3. Change your language. Vocabulary plays a big part in clients' perceptions of your hospital's caliber. A thousand little details make a practice successful. For instance, always call veterinarians "doctor," and use technical terms when appropriate. Call a "recheck" a "follow-up appointment" and use "day admission" rather than "drop-off." Say "comprehensive physical exam" instead of "office call," and use "immunizations" rather than "shots." Teach your staff to rephrase common questions. For example, instead of asking "Have we ever seen your pet before?" ask, "When was the last time we saw your pet?" Using professional language in a professional medical setting will go a long way toward increasing your team members' self-esteem, and that will come shining through.

ENSURE CONTINUITY OF CARE WITH EXCEPTIONAL HOME-CARE INSTRUCTIONS

A client's perception of value extends far beyond your hospital doors. Oftentimes, the health and recovery of an animal depends on the care it receives after leaving your hospital, placing the responsibility squarely on the client's shoulders. You can help make the recovery a positive experience by providing written and easy-to-understand home-care instructions.

- Keep directions simple so that it's easy for clients to see specifically what actions they need to take.

- Impress clients by scheduling discharge appointments so that a technician or veterinarian can orally review home-care instructions. Be sure to stagger release times so that you don't experience a late-afternoon rush and feel pressured to get clients out the door.

- Take clients to a quiet, private room to discuss instructions so that they can concentrate on what you're saying, and wait to bring the pet into the room until the bulk of the instructions have been covered. If necessary, demonstrate home-care procedures such as giving medication or changing bandages.

- Offer complimentary rechecks for the first two weeks after a procedure. You can build the cost of rechecks into the initial procedure. Clients won't be scared of incurring more fees for rechecks, and you'll have greater compliance with home-care instructions.

- Make a follow-up call one to three days after each patient is discharged. This gives clients a chance to ask questions and encourages compliance with home-care instructions.

Case Study

How We Educate Clients with Report Cards

To add value to comprehensive physical exams and show clients the extent of each physical, the doctors at Cherokee Animal Clinic and Cherokee Cat Clinic give clients report cards at the end of each visit.

"Every time a pet comes in, clients get a written report of everything the doctor checks," says Meredith. "Even if they just come in to get their pets' eyes checked, the doctors give a full-body exam, but clients may not always realize doctors are doing this in the exam room."

In addition to placing the pet's digital photo on the report card, staff members include basic information, such as the pet's name, weight, birthday and gender; the owner's name and contact information; and the exam date, technician's name, and doctor's name. Below this information, each body system is outlined with spaces for comments. Abnormal findings are highlighted in red. Categories include: appearance, weight, temperature; eyes; ears; mucous membranes; mouth and teeth; digestive system; lymph nodes; bones, joints and muscles; skin and hair coat; heart and circulation; respiratory and lungs; nervous system; and urinary and reproductive system.

A veterinarian or technician reviews the report card with the client, giving instructions for medications and recommendations for additional treatment, if necessary. A copy of the report card is placed in the patient's file, and one goes home with the client. Upon the client's return, the veterinarian can pull the report card out and address any concerns from the last visit.

"Report cards emphasize the thoroughness of our exams," Meredith says. "But they also summarize a lot of information for someone to absorb when the pet is sick. Having a personalized report to take home gives them something to refer back to in case they have questions."

Implementation Idea

Ways to Improve Your Home-Care Instructions

Here are a few ways to improve your home-care instructions that will improve compliance and the health of the pet:

- Personalize instructions with a picture of the pet, if possible.

- Use a yellow highlighter to call out specific instructions and important information.
- Provide hospital and emergency phone numbers (both day and night), enabling your clients to reach you in case they encounter problems after taking their pets home.
- List recheck appointments on the home-care instructions.
- Make the instructions easy to understand (see Figure 13.1).

HELPFUL RESOURCES

Home-Care Instructions Form, AAHA. American Animal Hospital Association Press, 2003.

Lifelearn Client Handouts on CD: Small Animal Series (800+ handouts), multiple authors. Lifelearn, Inc., 2004/2005. Available through the American Animal Hospital Association.

FIGURE 13.1
HOME-CARE INSTRUCTION FORM

(Adapted from a sample provided by Cherokee Animal Clinic and Cherokee Cat Clinic)

Dear Owners:

Thank you for ensuring that I receive proper care. My health is important to both of us, and I appreciate the time and money you've invested.

First: I had anesthesia recently. Staff members have been very careful when giving me food and water. They made sure that I did not get too much too soon. Now it's up to you. If you notice that I'm not eating normally, don't become alarmed. Sometimes it takes a few days to "bounce back" to my normal eating habits. Just like humans, each cat responds to anesthesia in his or her own unique way.

Second: Even though I really want to go home and romp around, I MUST be held back. You see, you were very kind to me when you chose the laser option for surgery. I didn't bleed hardly at all, and the incision feels just great! However, I HAVE NOT HEALED and I have a hard time remembering that. You need to remind me of this for the next few weeks by making sure I don't play too rough or jump down off of tall ledges, such as chairs, couches or beds. The sudden impact could cause my incision to pop open, and THAT will hurt and bleed. Please do your very best to minimize my activity as I will hurt myself if left to my own devices.

Also, if you see me chewing or rubbing my incision site excessively, please call my doctor so he can determine if an Elizabethan collar is needed to stop the destructive behavior.

Third: I doubt that this will happen, but I may appear a little groggy or not pass as many stools as usual. (How embarrassing!) This is OK for about 24 to 48 hours after anesthesia. However, if it persists, please call my doctor. He may want to see me again just to be sure that I'm OK.

Fourth: I have the following medications that must be given until they are gone. Please ask the doctor or technician the best way to give me these medications so that I will feel better soon.

Medication:
Dose:
Beginning on:
Ending on:

Fifth: Please make the following appointment(s) for me:

Recheck in _____ days, on _____. Thank you so much for taking such good care of me. If you need to call my doctors after regular office hours, here are their numbers. Please call my doctors in place of the emergency clinic.

Dr. John Smith xxx/xxx-xxxx
Dr. Jane Doe xxx/xxx-xxxx

Your loving cat,
Buggy

CHAPTER 13

DATA TABLES

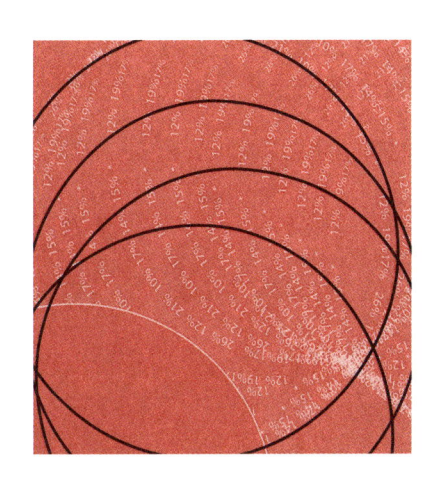

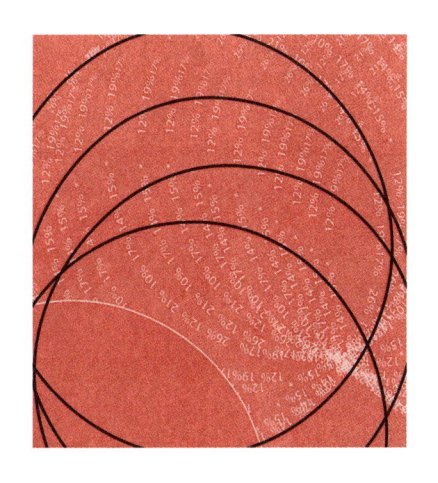

TABLE 13.1
ESTIMATE ROUTINELY PROVIDED FOR TREATMENT PLANS

	Yes	No	Number of Respondents
All Practices	**93%**	**7%**	**460**
Number of FTE Veterinarians			
1.0 or Less	91%	9%	142
1.1 to 2.0	93%	7%	139
2.1 to 3.0	98%	2%	86
3.1 or More	91%	9%	87
Member Status			
Accredited Practice Member	97%	3%	127
Nonaccredited Member	93%	8%	80
Nonmember	90%	10%	220
Metropolitan Status			
Urban	95%	5%	44
Suburban	92%	8%	131
Second City	95%	5%	95
Town	89%	11%	122
Rural	97%	3%	59
Median Area Household Income			
$35,000 or Less	92%	8%	53
$35,000 to $49,999	95%	5%	166
$50,000 to $69,999	92%	8%	154
$70,000 or More	90%	10%	77

Note 1: Some row totals do not equal 100% due to rounding.

TABLE 13.2
DEPOSIT ROUTINELY REQUIRED FOR TREATMENT

	Yes	No	Number of Respondents
All Practices	**51%**	**49%**	**459**
Number of FTE Veterinarians			
1.0 or Less	53%	47%	141
1.1 to 2.0	50%	50%	140
2.1 to 3.0	49%	51%	85
3.1 or More	53%	47%	87
Member Status			
Accredited Practice Member	52%	48%	127
Nonaccredited Member	53%	47%	79
Nonmember	51%	49%	220
Metropolitan Status			
Urban	55%	45%	44
Suburban	51%	49%	129
Second City	52%	48%	95
Town	50%	50%	125
Rural	53%	47%	57
Median Area Household Income			
$35,000 or Less	62%	38%	52
$35,000 to $49,999	61%	39%	167
$50,000 to $69,999	42%	58%	152
$70,000 or More	40%	60%	78

TABLE 13.3
DEPOSIT AMOUNT REQUIRED FOR TREATMENT

	25% of Estimate	50% of Estimate	More than 50% of Estimate	Number of Respondents
All Practices	**13%**	**77%**	**10%**	**237**
Number of FTE Veterinarians				
1.0 or Less	12%	76%	12%	75
1.1 to 2.0	13%	81%	6%	70
2.1 to 3.0	14%	76%	10%	42
3.1 or More	13%	75%	13%	48
Member Status				
Accredited Practice Member	11%	80%	9%	65
Nonaccredited Member	14%	80%	7%	44
Nonmember	13%	74%	13%	115
Metropolitan Status				
Urban	8%	88%	4%	25
Suburban	10%	75%	15%	67
Second City	8%	78%	14%	50
Town	17%	76%	6%	63
Rural	21%	72%	7%	29
Median Area Household Income				
$35,000 or Less	26%	71%	3%	31
$35,000 to $49,999	10%	77%	14%	103
$50,000 to $69,999	11%	83%	6%	64
$70,000 or More	12%	73%	15%	33

Note: Some row totals do not equal 100% due to rounding.

TABLE 13.4
TREATMENT CASE ONE: ADMITTING EXAMINATION

	25th Percentile	Median	Your Data	Average	75th Percentile	Number of Respondents
All Practices	**$35.61**	**$40.00**		**$40.76**	**$45.00**	**437**
Number of FTE Veterinarians						
1.0 or Less	$33.90	$38.00		$37.76	$42.00	134
1.1 to 2.0	$35.69	$40.00		$41.13	$45.25	134
2.1 to 3.0	$38.00	$41.00		$42.62	$45.53	85
3.1 or More	$38.00	$42.00		$42.93	$46.50	79
Member Status						
Accredited Practice Member	$39.00	$44.00		$43.99	$48.41	124
Nonaccredited Member	$36.00	$39.00		$40.34	$45.00	72
Nonmember	$34.00	$38.38		$38.79	$44.00	210
Metropolitan Status						
Urban	$38.00	$42.00		$41.86	$46.25	41
Suburban	$38.00	$42.00		$43.28	$48.00	120
Second City	$37.13	$40.96		$41.95	$45.00	88
Town	$34.19	$39.44		$39.54	$45.00	122
Rural	$30.50	$36.00		$35.71	$39.99	57
Median Area Household Income						
$35,000 or Less	$31.13	$37.23		$36.38	$40.75	52
$35,000 to $49,999	$34.50	$38.50		$38.64	$44.00	157
$50,000 to $69,999	$38.00	$40.55		$41.85	$45.02	146
$70,000 or More	$39.55	$45.00		$46.18	$52.25	73

Note 1: The patient is a dehydrated 60-pound golden retriever. You examine the dog, place an intravenous catheter, and administer 1,500 ml of lactated Ringer's solution over a 24-hour period.

TABLE 13.5
TREATMENT CASE ONE: CATHETERIZATION

	25th Percentile	Median	Your Data	Average	75th Percentile	Number of Respondents
All Practices	$29.81	$39.83	_____	$42.76	$50.00	428
Number of FTE Veterinarians						
1.0 or Less	$26.07	$36.00	_____	$39.40	$49.71	124
1.1 to 2.0	$28.25	$40.00	_____	$44.28	$54.35	129
2.1 to 3.0	$33.05	$40.96	_____	$42.75	$49.40	84
3.1 or More	$34.00	$40.95	_____	$45.30	$49.83	85
Member Status						
Accredited Practice Member	$34.00	$41.25	_____	$45.39	$54.50	119
Nonaccredited Member	$28.00	$37.00	_____	$40.66	$45.25	73
Nonmember	$27.88	$39.00	_____	$41.23	$50.00	206
Metropolitan Status						
Urban	$33.65	$46.42	_____	$49.20	$57.55	40
Suburban	$30.23	$40.00	_____	$44.52	$54.80	119
Second City	$31.88	$40.00	_____	$45.38	$52.50	89
Town	$28.25	$37.96	_____	$40.56	$46.38	116
Rural	$22.00	$34.00	_____	$34.80	$45.00	55
Median Area Household Income						
$35,000 or Less	$25.00	$34.00	_____	$36.66	$44.00	47
$35,000 to $49,999	$26.87	$37.00	_____	$40.59	$48.25	158
$50,000 to $69,999	$34.07	$42.00	_____	$44.39	$50.00	144
$70,000 or More	$35.68	$42.58	_____	$48.38	$58.25	70

Note 1: See the case description in the notes for Table 13.4.
Note 2: 3% of respondents reported that they perform this service as part of this case but do not charge for it separately.
Note 3: 3% of the respondents reported that they do not offer this service.

TABLE 13.6
TREATMENT CASE ONE: 1,500 ML LACTATED RINGER'S SOLUTION

	25th Percentile	Median	Your Data	Average	75th Percentile	Number of Respondents
All Practices	$27.75	$43.00		$45.44	$59.00	408
Number of FTE Veterinarians						
1.0 or Less	$25.00	$40.00		$45.35	$62.30	123
1.1 to 2.0	$27.00	$45.00		$45.73	$59.25	124
2.1 to 3.0	$26.00	$43.00		$45.50	$57.30	79
3.1 or More	$29.58	$44.21		$45.81	$58.27	76
Member Status						
Accredited Practice Member	$27.90	$44.00		$46.94	$59.50	111
Nonaccredited Member	$30.39	$46.00		$49.14	$67.50	69
Nonmember	$25.00	$40.00		$43.46	$55.00	201
Metropolitan Status						
Urban	$35.55	$50.00		$50.96	$62.30	35
Suburban	$35.00	$50.00		$52.95	$70.00	112
Second City	$30.00	$45.00		$46.57	$60.00	83
Town	$25.75	$37.60		$41.19	$51.98	114
Rural	$20.00	$30.00		$32.09	$43.20	56
Median Area Household Income						
$35,000 or Less	$23.50	$40.00		$41.34	$54.54	49
$35,000 to $49,999	$26.70	$39.50		$43.14	$57.50	149
$50,000 to $69,999	$28.00	$45.00		$47.14	$60.00	137
$70,000 or More	$30.00	$45.00		$47.68	$62.30	63

Note 1: See the case description in the notes for Table 13.4.
Note 2: 7% of respondents reported that they perform this service as part of this case but do not charge for it separately.
Note 3: 2% of the respondents reported that they do not offer this service.

TABLE 13.7
TREATMENT CASE ONE: FLUID INFUSION PUMP

	25th Percentile	Median	Your Data	Average	75th Percentile	Number of Respondents
All Practices	$12.30	$19.50		$20.39	$25.00	165
Number of FTE Veterinarians						
1.0 or Less	$13.28	$19.15		$20.58	$25.00	40
1.1 to 2.0	$12.00	$18.00		$20.06	$25.00	51
2.1 to 3.0	$16.38	$21.10		$21.85	$25.00	36
3.1 or More	$11.18	$17.50		$19.12	$23.92	36
Member Status						
Accredited Practice Member	$13.31	$20.00		$21.90	$28.63	52
Nonaccredited Member	*	*		*	*	23
Nonmember	$12.00	$18.50		$19.37	$23.00	75
Metropolitan Status						
Urban	*	*		*	*	21
Suburban	$13.76	$20.00		$20.88	$25.00	58
Second City	$13.25	$19.70		$21.46	$28.60	37
Town	$10.10	$15.50		$16.50	$21.88	32
Rural	*	*		*	*	12
Median Area Household Income						
$35,000 or Less	*	*		*	*	9
$35,000 to $49,999	$14.04	$19.10		$20.58	$24.69	58
$50,000 to $69,999	$11.35	$17.50		$18.60	$24.15	57
$70,000 or More	$12.50	$20.00		$21.63	$30.26	37

Note 1: See the case description in the notes for Table 13.4.
Note 2: An asterisk indicates that data was not reported due to an insufficient number of responses.
Note 3: 33% of respondents reported that they perform this service as part of this case but do not charge for it separately.
Note 4: 28% of the respondents reported that they do not offer this service.

TABLE 13.8
TREATMENT CASE ONE: HOSPITALIZATION

	25th Percentile	Median	Your Data	Average	75th Percentile	Number of Respondents
All Practices	**$28.00**	**$39.50**		**$42.73**	**$55.00**	**435**
Number of FTE Veterinarians						
1.0 or Less	$25.88	$34.50		$39.04	$52.61	134
1.1 to 2.0	$30.75	$40.80		$42.90	$55.00	130
2.1 to 3.0	$27.88	$39.50		$42.44	$52.95	82
3.1 or More	$32.38	$44.05		$49.04	$64.34	84
Member Status						
Accredited Practice Member	$34.75	$42.75		$46.27	$59.45	121
Nonaccredited Member	$30.00	$37.18		$43.86	$55.00	73
Nonmember	$25.00	$35.89		$40.54	$55.00	210
Metropolitan Status						
Urban	$39.00	$47.00		$51.79	$65.00	41
Suburban	$31.48	$45.00		$48.30	$64.00	119
Second City	$28.17	$39.30		$42.31	$53.50	90
Town	$25.00	$35.75		$38.32	$48.00	123
Rural	$24.84	$32.30		$34.26	$39.05	54
Median Area Household Income						
$35,000 or Less	$21.46	$32.63		$33.37	$43.22	52
$35,000 to $49,999	$26.00	$36.25		$41.88	$55.00	158
$50,000 to $69,999	$29.88	$40.60		$43.34	$52.25	145
$70,000 or More	$35.00	$48.50		$50.80	$67.25	70

Note 1: See the case description in the notes for Table 13.4.

TABLE 13.9
TREATMENT CASE ONE: VETERINARIAN/TECHNICIAN SUPERVISION

	25th Percentile	Median	Your Data	Average	75th Percentile	Number of Respondents
All Practices	$20.00	$28.50		$29.64	$34.99	121
Number of FTE Veterinarians						
1.0 or Less	$19.68	$29.15		$30.34	$35.00	33
1.1 to 2.0	$22.88	$30.00		$31.76	$36.50	41
2.1 to 3.0	*	*		*	*	23
3.1 or More	*	*		*	*	23
Member Status						
Accredited Practice Member	$20.23	$28.50		$28.97	$34.40	41
Nonaccredited Member	*	*		*	*	21
Nonmember	$23.00	$30.00		$32.57	$40.00	49
Metropolitan Status						
Urban	*	*		*	*	15
Suburban	$25.00	$32.00		$32.30	$36.65	33
Second City	$22.00	$29.00		$31.81	$36.50	27
Town	$19.35	$25.00		$27.45	$31.00	31
Rural	*	*		*	*	11
Median Area Household Income						
$35,000 or Less	*	*		*	*	16
$35,000 to $49,999	$18.74	$28.00		$28.45	$35.00	39
$50,000 to $69,999	$22.00	$29.75		$29.66	$34.15	40
$70,000 or More	*	*		*	*	20

Note 1: See the case description in the notes for Table 13.4.
Note 2: An asterisk indicates that data was not reported due to an insufficient number of responses.
Note 3: 56% of respondents reported that they perform this service as part of this case but do not charge for it separately.
Note 4: 15% of the respondents reported that they do not offer this service.

TABLE 13.10
TREATMENT CASE ONE: MEDICAL WASTE DISPOSAL

	25th Percentile	Median	Your Data	Average	75th Percentile	Number of Respondents
All Practices	$2.00	$3.30		$3.48	$4.66	121
Number of FTE Veterinarians						
1.0 or Less	$2.00	$3.91		$4.03	$5.50	39
1.1 to 2.0	$2.12	$3.50		$3.35	$4.68	35
2.1 to 3.0	*	*		*	*	23
3.1 or More	*	*		*	*	23
Member Status						
Accredited Practice Member	$2.00	$3.30		$3.15	$4.00	41
Nonaccredited Member	*	*		*	*	16
Nonmember	$2.00	$3.20		$3.60	$4.95	56
Metropolitan Status						
Urban	*	*		*	*	16
Suburban	$2.00	$3.40		$3.61	$4.87	45
Second City	$1.80	$2.74		$2.87	$3.63	25
Town	*	*		*	*	22
Rural	*	*		*	*	9
Median Area Household Income						
$35,000 or Less	*	*		*	*	8
$35,000 to $49,999	$1.73	$2.88		$3.20	$4.54	42
$50,000 to $69,999	$2.12	$3.47		$3.79	$5.00	43
$70,000 or More	$2.38	$3.30		$3.49	$4.38	25

Note 1: See the case description in the notes for Table 13.4.
Note 2: An asterisk indicates that data was not reported due to an insufficient number of responses.
Note 3: 44% of respondents reported that they perform this service as part of this case but do not charge for it separately.
Note 4: 26% of the respondents reported that they do not offer this service.

TABLE 13.11
TREATMENT CASE ONE: TOTAL

	25th Percentile	Median	Your Data	Average	75th Percentile	Number of Respondents
All Practices	$139.50	$174.00		$177.95	$214.00	453
Number of FTE Veterinarians						
1.0 or Less	$123.48	$165.00		$165.95	$206.66	137
1.1 to 2.0	$143.82	$177.00		$181.85	$215.88	137
2.1 to 3.0	$150.19	$178.41		$183.38	$216.06	86
3.1 or More	$148.96	$176.00		$186.73	$223.52	87
Member Status						
Accredited Practice Member	$157.08	$181.50		$191.43	$219.58	126
Nonaccredited Member	$144.10	$174.50		$178.09	$215.56	76
Nonmember	$128.00	$165.98		$170.36	$207.90	218
Metropolitan Status						
Urban	$167.38	$206.87		$205.16	$241.60	42
Suburban	$158.00	$196.10		$196.87	$232.35	125
Second City	$142.05	$178.41		$182.21	$217.50	94
Town	$134.78	$165.75		$163.15	$196.75	125
Rural	$109.69	$132.88		$139.46	$162.75	58
Median Area Household Income						
$35,000 or Less	$121.87	$153.55		$148.67	$182.69	54
$35,000 to $49,999	$132.10	$165.80		$169.48	$204.51	165
$50,000 to $69,999	$152.53	$177.29		$184.36	$215.75	150
$70,000 or More	$166.11	$203.00		$201.27	$232.30	74

Note 1: See the case description in the notes for Table 13.4.
Note 2: The total fee for treatment case one is the sum of the fees for the admitting exam, catheterization, 1,500 ml of lactated Ringer's solution, fluid pump, hospitalization, veterinarian/technician supervision, and medical waste disposal. Some respondents do not charge for some of the individual services provided in this case. In addition, some respondents reported that they do not offer some of the individual services. Therefore, the average total fee for the case may be significantly lower than the sum of the average fees for the individual services.

TABLE 13.12
TREATMENT CASE TWO: ADMITTING EXAMINATION

	25th Percentile	Median	Your Data	Average	75th Percentile	Number of Respondents
All Practices	$35.73	$40.00		$41.56	$45.00	457
Number of FTE Veterinarians						
1.0 or Less	$32.63	$38.00		$37.96	$42.15	140
1.1 to 2.0	$36.00	$40.50		$42.11	$46.50	139
2.1 to 3.0	$38.00	$41.00		$42.61	$46.25	86
3.1 or More	$38.00	$42.50		$45.20	$47.92	87
Member Status						
Accredited Practice Member	$39.00	$42.98		$43.83	$48.14	126
Nonaccredited Member	$36.00	$39.35		$41.77	$45.00	78
Nonmember	$34.25	$38.50		$40.10	$44.00	221
Metropolitan Status						
Urban	$38.00	$43.60		$43.74	$46.50	43
Suburban	$38.00	$42.02		$43.87	$49.00	128
Second City	$38.00	$41.95		$44.11	$46.00	96
Town	$34.06	$39.00		$39.33	$44.00	124
Rural	$30.00	$35.88		$35.77	$39.92	58
Median Area Household Income						
$35,000 or Less	$30.00	$37.00		$36.10	$41.00	54
$35,000 to $49,999	$34.60	$39.00		$40.46	$44.25	165
$50,000 to $69,999	$38.00	$41.09		$42.15	$45.25	153
$70,000 or More	$39.60	$45.00		$46.78	$53.00	75

Note: The patient is a 13-pound, neutered male cat with an abscess. You examine, sedate, treat, and hospitalize him overnight, and examine him in the morning. You dispense a one-week supply of antibiotics to the owner.

TABLE 13.13
TREATMENT CASE TWO: LIGHT SEDATION

	25th Percentile	Median	Your Data	Average	75th Percentile	Number of Respondents
All Practices	$28.00	$38.75		$42.77	$52.23	445
Number of FTE Veterinarians						
1.0 or Less	$24.41	$35.00		$38.34	$45.80	134
1.1 to 2.0	$29.31	$42.23		$46.70	$59.88	136
2.1 to 3.0	$27.50	$40.00		$42.18	$53.00	85
3.1 or More	$30.00	$38.63		$43.65	$50.00	86
Member Status						
Accredited Practice Member	$31.10	$41.50		$44.99	$55.14	121
Nonaccredited Member	$28.74	$38.80		$43.05	$52.15	76
Nonmember	$25.00	$36.00		$40.82	$50.06	216
Metropolitan Status						
Urban	$38.88	$53.10		$59.24	$80.14	40
Suburban	$32.75	$40.49		$44.73	$56.00	127
Second City	$30.00	$44.00		$46.35	$59.25	93
Town	$25.25	$34.00		$38.64	$45.20	121
Rural	$18.57	$28.97		$29.45	$38.75	56
Median Area Household Income						
$35,000 or Less	$22.91	$32.00		$34.71	$41.49	52
$35,000 to $49,999	$25.00	$37.00		$41.56	$52.10	161
$50,000 to $69,999	$30.00	$38.50		$43.07	$52.80	150
$70,000 or More	$36.56	$46.25		$50.63	$66.73	72

Note: See the case description in the notes for Table 13.12.

TREATMENT PROCEDURES AND SERVICES / 353

TABLE 13.14
TREATMENT CASE TWO: ABSCESS CURETTAGE, DEBRIDEMENT, AND FLUSH

	25th Percentile	Median	Your Data	Average	75th Percentile	Number of Respondents
All Practices	$30.00	$45.00		$53.14	$70.00	449
Number of FTE Veterinarians						
1.0 or Less	$29.62	$42.00		$50.52	$69.25	137
1.1 to 2.0	$29.25	$45.00		$52.82	$72.51	136
2.1 to 3.0	$33.00	$48.75		$56.01	$73.03	84
3.1 or More	$32.50	$49.50		$55.65	$66.56	87
Member Status						
Accredited Practice Member	$35.00	$50.00		$58.06	$75.00	127
Nonaccredited Member	$32.50	$50.00		$57.07	$73.00	77
Nonmember	$26.50	$41.71		$49.76	$65.00	215
Metropolitan Status						
Urban	$38.00	$59.50		$61.76	$78.55	43
Suburban	$39.63	$54.00		$62.40	$75.08	125
Second City	$33.75	$48.75		$55.37	$74.25	94
Town	$25.00	$40.00		$46.36	$61.00	123
Rural	$20.00	$32.00		$35.83	$44.15	57
Median Area Household Income						
$35,000 or Less	$25.00	$35.00		$40.52	$50.00	52
$35,000 to $49,999	$27.25	$42.00		$47.79	$61.41	160
$50,000 to $69,999	$32.75	$50.00		$56.82	$70.00	153
$70,000 or More	$40.00	$60.50		$65.57	$85.81	74

Note: See the case description in the notes for Table 13.12.

TABLE 13.15
TREATMENT CASE TWO: ANTIBIOTIC INJECTION

	25th Percentile	Median	Your Data	Average	75th Percentile	Number of Respondents
All Practices	$15.11	$20.10		$20.91	$26.00	442
Number of FTE Veterinarians						
1.0 or Less	$15.00	$19.98		$20.57	$25.50	134
1.1 to 2.0	$15.00	$20.00		$20.94	$26.00	134
2.1 to 3.0	$16.50	$21.90		$21.33	$26.40	83
3.1 or More	$15.75	$19.98		$21.18	$26.00	86
Member Status						
Accredited Practice Member	$16.00	$21.19		$21.46	$26.20	120
Nonaccredited Member	$16.00	$21.83		$22.23	$27.81	76
Nonmember	$14.93	$19.00		$19.74	$25.00	214
Metropolitan Status						
Urban	$21.00	$24.58		$25.93	$30.05	42
Suburban	$18.00	$23.00		$23.09	$28.45	124
Second City	$17.74	$22.00		$22.72	$27.13	93
Town	$13.09	$18.00		$17.93	$22.65	121
Rural	$11.15	$15.10		$15.93	$18.50	55
Median Area Household Income						
$35,000 or Less	$14.21	$18.00		$18.81	$24.94	52
$35,000 to $49,999	$15.00	$20.00		$20.25	$25.00	164
$50,000 to $69,999	$15.71	$21.00		$21.37	$26.00	145
$70,000 or More	$17.50	$23.00		$23.13	$28.00	71

Note: See the case description in the notes for Table 13.12.

TABLE 13.16
TREATMENT CASE TWO: PAIN MEDICATION

	25th Percentile	Median	Your Data	Average	75th Percentile	Number of Respondents
All Practices	**$14.85**	**$20.00**		**$21.04**	**$26.28**	**397**
Number of FTE Veterinarians						
1.0 or Less	$13.44	$19.97		$20.17	$24.23	112
1.1 to 2.0	$14.69	$20.68		$21.46	$27.37	126
2.1 to 3.0	$15.21	$20.00		$21.22	$26.77	76
3.1 or More	$15.00	$22.00		$21.64	$28.00	79
Member Status						
Accredited Practice Member	$15.75	$21.73		$22.18	$27.63	114
Nonaccredited Member	$15.00	$20.00		$22.33	$28.00	71
Nonmember	$12.96	$19.00		$19.56	$24.68	184
Metropolitan Status						
Urban	$19.50	$24.88		$24.62	$29.51	38
Suburban	$17.00	$23.00		$22.84	$28.00	111
Second City	$16.00	$22.50		$23.38	$30.00	87
Town	$13.00	$19.00		$18.48	$23.00	105
Rural	$10.00	$14.00		$15.52	$21.00	49
Median Area Household Income						
$35,000 or Less	$13.31	$17.19		$18.65	$23.38	44
$35,000 to $49,999	$12.99	$20.00		$20.08	$25.54	140
$50,000 to $69,999	$15.00	$20.00		$21.48	$26.88	141
$70,000 or More	$17.58	$24.50		$24.35	$31.86	62

Note 1: See the case description in the notes for Table 13.12.
Note 2: 5% of respondents reported that they perform this service as part of this case but do not charge for it separately.
Note 3: 6% of the respondents reported that they do not offer this service.

TABLE 13.17
TREATMENT CASE TWO: HOSPITALIZATION

	25th Percentile	Median	Your Data	Average	75th Percentile	Number of Respondents
All Practices	$22.00	$30.00		$34.55	$45.00	422
Number of FTE Veterinarians						
1.0 or Less	$20.00	$26.25		$30.93	$38.75	125
1.1 to 2.0	$21.70	$30.00		$33.53	$44.00	129
2.1 to 3.0	$22.50	$30.00		$34.60	$45.00	79
3.1 or More	$25.97	$35.75		$39.96	$50.00	85
Member Status						
Accredited Practice Member	$25.00	$33.00		$37.08	$45.33	117
Nonaccredited Member	$23.25	$29.85		$35.10	$43.91	76
Nonmember	$19.09	$26.87		$32.17	$42.00	198
Metropolitan Status						
Urban	$23.25	$32.61		$40.36	$49.75	41
Suburban	$25.64	$35.00		$39.71	$47.00	122
Second City	$22.00	$28.23		$33.77	$45.00	85
Town	$20.63	$29.90		$31.52	$39.74	116
Rural	$16.75	$23.00		$25.91	$33.00	51
Median Area Household Income						
$35,000 or Less	$14.93	$21.88		$26.90	$34.63	50
$35,000 to $49,999	$20.00	$27.00		$32.93	$42.75	149
$50,000 to $69,999	$23.63	$31.00		$35.09	$44.50	145
$70,000 or More	$27.63	$37.75		$42.14	$55.50	68

Note 1: See the case description in the notes for Table 13.12.
Note 2: 4% of respondents reported that they perform this service as part of this case but do not charge for it separately.
Note 3: 2% of the respondents reported that they do not offer this service.

TABLE 13.18
TREATMENT CASE TWO: MEDICAL PROGRESS EXAMINATION

	25th Percentile	Median	Your Data	Average	75th Percentile	Number of Respondents
All Practices	$21.00	$27.25		$27.25	$34.00	160
Number of FTE Veterinarians						
1.0 or Less	$19.08	$26.00		$25.24	$32.26	46
1.1 to 2.0	$21.23	$26.96		$27.16	$32.78	42
2.1 to 3.0	$24.50	$28.00		$28.84	$34.00	35
3.1 or More	$20.00	$28.00		$27.82	$36.25	35
Member Status						
Accredited Practice Member	$21.23	$28.33		$28.78	$35.31	54
Nonaccredited Member	$18.34	$27.25		$25.36	$32.26	30
Nonmember	$20.25	$25.00		$26.43	$33.75	60
Metropolitan Status						
Urban	*	*		*	*	16
Suburban	$21.00	$28.65		$28.80	$35.69	55
Second City	$24.00	$27.50		$27.72	$33.00	31
Town	$18.84	$24.50		$24.65	$30.25	42
Rural	*	*		*	*	13
Median Area Household Income						
$35,000 or Less	*	*		*	*	18
$35,000 to $49,999	$20.00	$25.75		$25.92	$30.25	58
$50,000 to $69,999	$21.00	$27.50		$26.92	$33.75	48
$70,000 or More	$23.50	$32.00		$31.20	$36.63	33

Note 1: See the case description in the notes for Table 13.12.
Note 2: An asterisk indicates that data was not reported due to an insufficient number of responses.
Note 3: 49% of respondents reported that they perform this service as part of this case but do not charge for it separately.
Note 4: 11% of the respondents reported that they do not offer this service.

TABLE 13.19
TREATMENT CASE TWO: ANTIBIOTIC, ORAL OR INJECTION

	25th Percentile	Median	Your Data	Average	75th Percentile	Number of Respondents
All Practices	$13.00	$18.26		$18.80	$24.00	314
Number of FTE Veterinarians						
1.0 or Less	$12.63	$18.00		$18.65	$23.30	93
1.1 to 2.0	$15.00	$20.00		$20.23	$25.33	97
2.1 to 3.0	$13.48	$18.00		$18.77	$24.25	57
3.1 or More	$10.91	$16.50		$16.93	$24.00	62
Member Status						
Accredited Practice Member	$14.00	$19.06		$19.24	$24.18	85
Nonaccredited Member	$13.13	$18.35		$19.79	$27.75	52
Nonmember	$12.00	$18.00		$18.10	$23.50	155
Metropolitan Status						
Urban	$14.03	$20.00		$21.09	$26.50	32
Suburban	$16.15	$20.74		$20.95	$26.75	88
Second City	$16.00	$20.00		$20.96	$26.78	65
Town	$11.00	$15.75		$16.10	$21.42	90
Rural	$10.00	$13.50		$14.34	$18.12	35
Median Area Household Income						
$35,000 or Less	$11.94	$17.00		$16.65	$19.56	37
$35,000 to $49,999	$12.00	$18.06		$18.22	$24.00	110
$50,000 to $69,999	$14.65	$19.33		$19.40	$24.31	106
$70,000 or More	$16.00	$19.00		$20.92	$26.98	53

Note 1: See the case description in the notes for Table 13.12.
Note 2: 17% of respondents reported that they perform this service as part of this case but do not charge for it separately.
Note 3: 9% of the respondents reported that they do not offer this service.

TABLE 13.20
TREATMENT CASE TWO: ANTIBIOTICS, ONE-WEEK SUPPLY

	25th Percentile	Median	Your Data	Average	75th Percentile	Number of Respondents
All Practices	**$15.00**	**$18.82**		**$19.35**	**$23.25**	**444**
Number of FTE Veterinarians						
1.0 or Less	$15.00	$18.99		$19.47	$23.88	137
1.1 to 2.0	$15.09	$19.25		$19.92	$24.63	133
2.1 to 3.0	$14.56	$18.95		$18.83	$23.00	83
3.1 or More	$14.42	$17.14		$18.81	$22.85	86
Member Status						
Accredited Practice Member	$15.00	$18.46		$19.32	$24.00	124
Nonaccredited Member	$15.00	$20.00		$20.38	$24.25	77
Nonmember	$14.00	$18.66		$18.83	$22.50	211
Metropolitan Status						
Urban	$15.00	$18.00		$20.15	$24.70	43
Suburban	$15.88	$20.00		$21.01	$25.00	126
Second City	$16.05	$20.00		$20.75	$25.00	91
Town	$14.00	$17.28		$17.93	$22.12	120
Rural	$12.45	$15.49		$15.68	$19.47	57
Median Area Household Income						
$35,000 or Less	$12.91	$17.38		$17.48	$21.10	54
$35,000 to $49,999	$15.00	$18.95		$19.54	$23.20	161
$50,000 to $69,999	$15.00	$18.75		$19.06	$23.00	147
$70,000 or More	$15.00	$21.62		$21.20	$25.24	72

Note 1: See the case description in the notes for Table 13.12.
Note 2: 2% of respondents reported that they perform this service as part of this case but do not charge for it separately.

TABLE 13.21
TREATMENT CASE TWO: MEDICAL WASTE DISPOSAL

	25th Percentile	Median	Your Data	Average	75th Percentile	Number of Respondents
All Practices	$2.00	$3.49		$3.63	$4.99	122
Number of FTE Veterinarians						
1.0 or Less	$2.19	$4.07		$4.05	$5.63	38
1.1 to 2.0	$2.41	$3.50		$3.73	$4.99	38
2.1 to 3.0	*	*		*	*	22
3.1 or More	*	*		*	*	23
Member Status						
Accredited Practice Member	$2.00	$3.45		$3.31	$4.21	41
Nonaccredited Member	*	*		*	*	16
Nonmember	$2.00	$3.50		$3.71	$5.00	57
Metropolitan Status						
Urban	*	*		*	*	17
Suburban	$2.35	$3.60		$3.97	$5.00	45
Second City	$1.78	$2.70		$3.06	$4.25	25
Town	*	*		*	*	23
Rural	*	*		*	*	8
Median Area Household Income						
$35,000 or Less	*	*		*	*	8
$35,000 to $49,999	$1.75	$3.45		$3.49	$4.69	43
$50,000 to $69,999	$2.55	$3.50		$3.92	$5.00	44
$70,000 or More	*	*		*	*	24

Note 1: See the case description in the notes for Table 13.12.
Note 2: An asterisk indicates that data was not reported due to an insufficient number of responses.
Note 3: 42% of respondents reported that they perform this service as part of this case but do not charge for it separately.
Note 4: 28% of the respondents reported that they do not offer this service.

TABLE 13.22
TREATMENT CASE TWO: TOTAL

	25th Percentile	Median	Your Data	Average	75th Percentile	Number of Respondents
All Practices	**$186.12**	**$240.51**		**$246.43**	**$301.53**	**460**
Number of FTE Veterinarians						
1.0 or Less	$171.43	$212.80		$226.73	$280.53	141
1.1 to 2.0	$185.14	$247.14		$250.83	$311.30	140
2.1 to 3.0	$201.73	$250.11		$253.26	$296.97	86
3.1 or More	$199.60	$261.35		$263.36	$313.33	88
Member Status						
Accredited Practice Member	$215.50	$264.00		$263.81	$310.70	127
Nonaccredited Member	$199.30	$241.50		$256.82	$314.06	79
Nonmember	$166.42	$218.35		$230.40	$287.47	222
Metropolitan Status						
Urban	$223.93	$304.50		$296.13	$363.60	43
Suburban	$225.94	$271.45		$273.80	$321.90	130
Second City	$204.33	$264.30		$259.95	$316.25	96
Town	$173.50	$216.50		$221.89	$269.95	125
Rural	$139.63	$172.63		$179.41	$209.29	58
Median Area Household Income						
$35,000 or Less	$164.41	$200.88		$204.45	$246.72	54
$35,000 to $49,999	$165.01	$225.25		$234.06	$284.06	166
$50,000 to $69,999	$195.00	$248.75		$252.53	$298.48	155
$70,000 or More	$233.04	$296.46		$290.30	$345.55	75

Note 1: See the case description in the notes for Table 13.12.
Note 2: The total fee for treatment case two is the sum of the fees for the admitting exam; light sedation; abscess curettage, debridement, and flush; antibiotic injection; pain medication; hospitalization; medical progress exam; antibiotic (oral or injection); one-week supply of antibiotics; and medical waste disposal. Some respondents do not charge for some of the individual services provided in this case. In addition, some respondents reported that they do not offer some of the individual services. Therefore, the average total fee for the case may be significantly lower than the sum of the average fees for the individual services.

TABLE 13.23
TREATMENT CASE THREE: ADMITTING EXAMINATION

	25th Percentile	Median	Your Data	Average	75th Percentile	Number of Respondents
All Practices	$35.13	$40.00		$40.78	$45.00	436
Number of FTE Veterinarians						
1.0 or Less	$33.30	$38.00		$37.70	$42.00	133
1.1 to 2.0	$35.63	$40.00		$41.08	$45.00	133
2.1 to 3.0	$38.00	$41.00		$42.64	$46.50	85
3.1 or More	$38.00	$42.25		$43.14	$46.88	80
Member Status						
Accredited Practice Member	$39.00	$44.00		$44.05	$48.78	125
Nonaccredited Member	$36.00	$39.00		$40.13	$45.00	71
Nonmember	$34.00	$38.00		$38.85	$44.00	208
Metropolitan Status						
Urban	$38.00	$42.00		$41.65	$46.25	41
Suburban	$38.00	$41.70		$42.98	$48.00	120
Second City	$38.00	$41.05		$42.22	$45.50	89
Town	$34.00	$39.00		$39.37	$45.00	123
Rural	$31.30	$37.50		$36.47	$40.00	55
Median Area Household Income						
$35,000 or Less	$31.00	$37.45		$36.69	$41.00	51
$35,000 to $49,999	$34.38	$38.50		$38.70	$44.00	157
$50,000 to $69,999	$38.00	$40.25		$41.79	$45.00	146
$70,000 or More	$39.53	$45.00		$46.15	$52.38	72

Note 1: The patient is a 30-pound, six-year-old male schnauzer with mild pancreatitis. You hospitalize him for three days, administer fluids, give him medical treatment and pain management medication each day, perform a chemistry panel on days two and three, and give him a special diet on the last day of hospitalization.
Note 2: 2% of the respondents reported that they do not offer this service.

TABLE 13.24
TREATMENT CASE THREE: IV CATHETER AND PLACEMENT

	25th Percentile	Median	Your Data	Average	75th Percentile	Number of Respondents
All Practices	**$30.00**	**$40.00**		**$42.78**	**$50.00**	**426**
Number of FTE Veterinarians						
1.0 or Less	$27.00	$35.00		$38.98	$48.00	127
1.1 to 2.0	$29.56	$40.00		$44.12	$54.05	126
2.1 to 3.0	$33.05	$42.23		$44.55	$50.00	84
3.1 or More	$35.13	$41.77		$45.21	$50.00	84
Member Status						
Accredited Practice Member	$34.00	$41.25		$45.23	$54.50	119
Nonaccredited Member	$28.00	$38.00		$41.37	$46.80	73
Nonmember	$28.75	$39.00		$41.48	$50.00	205
Metropolitan Status						
Urban	$32.68	$48.00		$49.52	$61.26	40
Suburban	$31.99	$41.00		$45.97	$56.24	121
Second City	$33.00	$40.00		$42.76	$48.50	87
Town	$29.00	$37.56		$39.96	$46.45	115
Rural	$25.00	$35.00		$37.32	$46.00	55
Median Area Household Income						
$35,000 or Less	$25.00	$35.00		$37.19	$45.50	47
$35,000 to $49,999	$28.00	$37.00		$39.22	$46.00	155
$50,000 to $69,999	$32.73	$42.25		$44.90	$51.91	144
$70,000 or More	$36.00	$45.00		$50.26	$60.00	71

Note 1: See the case description in the notes for Table 13.23.
Note 2: 2% of respondents reported that they perform this service as part of this case but do not charge for it separately.

TABLE 13.25
TREATMENT CASE THREE: LITER OF IV FLUIDS

	25th Percentile	Median	Your Data	Average	75th Percentile	Number of Respondents
All Practices	**$20.00**	**$30.00**		**$32.50**	**$42.35**	**379**
Number of FTE Veterinarians						
1.0 or Less	$20.00	$30.00		$33.07	$42.35	119
1.1 to 2.0	$22.00	$31.00		$32.30	$39.00	115
2.1 to 3.0	$20.00	$30.00		$31.08	$40.00	72
3.1 or More	$21.50	$32.53		$34.04	$45.75	68
Member Status						
Accredited Practice Member	$18.81	$30.50		$31.54	$43.38	100
Nonaccredited Member	$21.25	$30.00		$32.95	$42.18	69
Nonmember	$20.00	$30.00		$32.58	$42.00	187
Metropolitan Status						
Urban	$27.50	$39.00		$39.12	$49.00	37
Suburban	$25.00	$35.35		$36.40	$46.00	103
Second City	$20.00	$30.00		$33.00	$45.00	79
Town	$18.00	$26.25		$28.85	$38.06	110
Rural	$15.75	$20.75		$24.39	$34.40	42
Median Area Household Income						
$35,000 or Less	$21.50	$29.29		$30.79	$35.56	47
$35,000 to $49,999	$19.24	$28.00		$30.09	$40.00	140
$50,000 to $69,999	$23.25	$34.35		$33.89	$42.46	124
$70,000 or More	$22.00	$35.00		$34.61	$46.00	59

Note 1: See the case description in the notes for Table 13.23.
Note 2: 13% of respondents reported that they perform this service as part of this case but do not charge for it separately.

TABLE 13.26
TREATMENT CASE THREE: IV PUMP USE

	25th Percentile	Median	Your Data	Average	75th Percentile	Number of Respondents
All Practices	$12.00	$18.65		$19.96	$25.00	159
Number of FTE Veterinarians						
1.0 or Less	$14.10	$20.00		$20.52	$25.00	39
1.1 to 2.0	$11.25	$18.38		$19.81	$25.00	48
2.1 to 3.0	$18.00	$22.00		$22.40	$26.10	33
3.1 or More	$11.35	$15.00		$17.30	$20.25	37
Member Status						
Accredited Practice Member	$12.88	$20.00		$21.09	$27.36	50
Nonaccredited Member	*	*		*	*	21
Nonmember	$10.75	$18.50		$19.23	$24.48	74
Metropolitan Status						
Urban	*	*		*	*	19
Suburban	$14.43	$20.00		$20.96	$25.00	53
Second City	$11.25	$18.88		$19.89	$24.91	36
Town	$10.30	$16.28		$17.22	$22.80	34
Rural	*	*		*	*	13
Median Area Household Income						
$35,000 or Less	*	*		*	*	8
$35,000 to $49,999	$12.80	$18.25		$19.05	$23.35	57
$50,000 to $69,999	$10.92	$17.50		$18.40	$23.40	53
$70,000 or More	$14.63	$22.00		$22.78	$33.50	37

Note 1: See the case description in the notes for Table 13.23.
Note 2: An asterisk indicates that data was not reported due to an insufficient number of responses.
Note 3: 34% of respondents reported that they perform this service as part of this case but do not charge for it separately.
Note 4: 27% of the respondents reported that they do not offer this service.

TABLE 13.27
TREATMENT CASE THREE: TWO ANTIBIOTIC INJECTIONS

	25th Percentile	Median	Your Data	Average	75th Percentile	Number of Respondents
All Practices	**$28.00**	**$37.65**		**$39.08**	**$50.00**	**438**
Number of FTE Veterinarians						
1.0 or Less	$24.50	$37.10		$38.64	$50.00	135
1.1 to 2.0	$30.00	$38.50		$40.32	$50.00	131
2.1 to 3.0	$27.70	$36.00		$38.57	$50.00	82
3.1 or More	$28.48	$36.40		$38.51	$48.30	85
Member Status						
Accredited Practice Member	$30.00	$38.20		$39.24	$48.20	121
Nonaccredited Member	$27.40	$35.00		$38.93	$49.63	74
Nonmember	$25.00	$37.50		$38.26	$50.00	211
Metropolitan Status						
Urban	$32.25	$48.40		$47.15	$60.81	40
Suburban	$30.10	$42.00		$42.53	$56.00	119
Second City	$30.00	$42.95		$42.62	$52.33	90
Town	$25.00	$32.92		$33.91	$43.50	125
Rural	$20.00	$30.00		$31.09	$40.00	56
Median Area Household Income						
$35,000 or Less	$24.00	$33.00		$34.78	$43.18	52
$35,000 to $49,999	$26.40	$36.00		$37.29	$48.00	163
$50,000 to $69,999	$29.43	$39.00		$40.34	$49.88	145
$70,000 or More	$30.70	$43.50		$43.75	$56.00	68

Note 1: See the case description in the notes for Table 13.23.
Note 2: 2% of the respondents reported that they do not offer this service.

TABLE 13.28
TREATMENT CASE THREE: ANTIEMETICS

	25th Percentile	Median	Your Data	Average	75th Percentile	Number of Respondents
All Practices	$16.00	$22.00		$24.42	$30.00	392
Number of FTE Veterinarians						
1.0 or Less	$15.00	$20.00		$22.94	$29.00	116
1.1 to 2.0	$17.00	$22.00		$24.65	$30.00	121
2.1 to 3.0	$16.70	$22.00		$24.77	$29.94	72
3.1 or More	$15.13	$22.25		$25.65	$30.00	80
Member Status						
Accredited Practice Member	$18.00	$22.00		$24.45	$30.00	108
Nonaccredited Member	$16.00	$23.80		$24.27	$30.00	67
Nonmember	$15.00	$21.00		$24.00	$29.25	191
Metropolitan Status						
Urban	$17.70	$25.00		$27.02	$30.50	38
Suburban	$19.50	$24.02		$27.14	$31.50	111
Second City	$17.00	$25.00		$27.39	$33.00	82
Town	$14.95	$19.00		$20.77	$25.00	103
Rural	$12.04	$17.00		$19.89	$22.00	53
Median Area Household Income						
$35,000 or Less	$12.00	$18.60		$20.60	$25.00	49
$35,000 to $49,999	$15.38	$21.00		$23.54	$29.12	142
$50,000 to $69,999	$18.00	$22.75		$24.69	$30.00	128
$70,000 or More	$19.00	$25.98		$29.23	$37.50	63

Note 1: See the case description in the notes for Table 13.23.
Note 2: 2% of respondents reported that they perform this service as part of this case but do not charge for it separately.
Note 3: 5% of the respondents reported that they do not offer this service.

TABLE 13.29
TREATMENT CASE THREE: HOSPITALIZATION

	25th Percentile	Median	Your Data	Average	75th Percentile	Number of Respondents
All Practices	**$27.38**	**$38.00**		**$42.88**	**$53.66**	**434**
Number of FTE Veterinarians						
1.0 or Less	$25.30	$35.00		$39.42	$48.00	131
1.1 to 2.0	$29.13	$40.10		$44.59	$55.00	132
2.1 to 3.0	$26.00	$36.00		$39.17	$47.25	82
3.1 or More	$30.75	$43.00		$49.14	$62.00	85
Member Status						
Accredited Practice Member	$30.38	$41.98		$45.60	$55.38	122
Nonaccredited Member	$29.25	$36.00		$42.62	$52.30	73
Nonmember	$25.00	$35.50		$40.88	$54.30	207
Metropolitan Status						
Urban	$35.85	$47.00		$53.15	$66.88	41
Suburban	$30.00	$42.00		$45.70	$58.75	117
Second City	$29.75	$40.20		$44.47	$54.65	93
Town	$25.00	$35.75		$38.55	$47.35	121
Rural	$22.38	$30.00		$34.93	$40.65	54
Median Area Household Income						
$35,000 or Less	$21.36	$34.00		$37.68	$50.00	51
$35,000 to $49,999	$26.00	$38.50		$42.96	$53.25	161
$50,000 to $69,999	$28.00	$37.63		$40.54	$47.00	143
$70,000 or More	$32.00	$44.90		$50.31	$66.75	69

Note 1: See the case description in the notes for Table 13.23.
Note 2: 2% of the respondents reported that they do not offer this service.

TABLE 13.30
TREATMENT CASE THREE: TWO ABDOMINAL RADIOGRAPHS

	25th Percentile	Median	Your Data	Average	75th Percentile	Number of Respondents
All Practices	**$86.00**	**$105.00**		**$105.83**	**$125.30**	**437**
Number of FTE Veterinarians						
1.0 or Less	$75.00	$100.24		$98.34	$120.00	128
1.1 to 2.0	$85.25	$105.00		$107.12	$128.73	133
2.1 to 3.0	$90.00	$100.50		$107.21	$122.63	85
3.1 or More	$90.92	$110.00		$112.86	$131.13	86
Member Status						
Accredited Practice Member	$95.00	$108.00		$112.46	$132.00	123
Nonaccredited Member	$88.50	$106.50		$105.60	$125.15	74
Nonmember	$78.93	$100.00		$101.04	$120.00	209
Metropolitan Status						
Urban	$107.03	$122.50		$120.96	$140.75	40
Suburban	$95.23	$113.12		$113.42	$130.11	120
Second City	$90.00	$109.50		$110.45	$133.00	91
Town	$80.00	$95.55		$95.48	$110.00	123
Rural	$78.00	$90.00		$91.58	$106.85	55
Median Area Household Income						
$35,000 or Less	$75.00	$85.50		$94.96	$114.50	52
$35,000 to $49,999	$84.25	$100.00		$101.20	$119.95	160
$50,000 to $69,999	$90.00	$107.50		$108.91	$125.00	144
$70,000 or More	$97.03	$119.50		$118.84	$140.00	72

Note 1: See the case description in the notes for Table 13.23.
Note 2: 2% of the respondents reported that they do not offer this service.

TABLE 13.31
TREATMENT CASE THREE: PAIN MEDICATION, ADMINISTERED TWICE

	25th Percentile	Median	Your Data	Average	75th Percentile	Number of Respondents
All Practices	$23.20	$35.00		$36.08	$47.37	413
Number of FTE Veterinarians						
1.0 or Less	$20.00	$30.00		$33.53	$45.50	121
1.1 to 2.0	$24.00	$36.00		$36.64	$45.69	128
2.1 to 3.0	$24.25	$35.37		$37.20	$50.45	80
3.1 or More	$25.00	$38.20		$37.73	$48.60	79
Member Status						
Accredited Practice Member	$26.61	$38.74		$38.81	$47.56	120
Nonaccredited Member	$24.00	$34.00		$36.87	$48.95	70
Nonmember	$20.00	$30.00		$33.66	$45.00	194
Metropolitan Status						
Urban	$26.19	$39.95		$40.05	$52.50	38
Suburban	$29.59	$40.00		$41.50	$55.00	115
Second City	$25.00	$40.00		$38.92	$50.00	87
Town	$19.34	$30.00		$30.86	$42.24	114
Rural	$20.00	$25.86		$27.64	$36.00	52
Median Area Household Income						
$35,000 or Less	$18.00	$30.00		$31.03	$40.00	47
$35,000 to $49,999	$20.71	$32.00		$34.70	$46.07	150
$50,000 to $69,999	$24.00	$36.00		$36.57	$48.13	138
$70,000 or More	$25.25	$43.00		$42.07	$56.68	68

Note 1: See the case description in the notes for Table 13.23.
Note 2: 3% of respondents reported that they perform this service as part of this case but do not charge for it separately.
Note 3: 4% of the respondents reported that they do not offer this service.

TABLE 13.32
TREATMENT CASE THREE: MEDICAL WASTE DISPOSAL

	25th Percentile	Median	Your Data	Average	75th Percentile	Number of Respondents
All Practices	$2.00	$3.50		$3.75	$5.00	121
Number of FTE Veterinarians						
1.0 or Less	$2.00	$4.07		$4.10	$5.63	38
1.1 to 2.0	$2.09	$3.55		$3.94	$5.00	38
2.1 to 3.0	*	*		*	*	21
3.1 or More	*	*		*	*	23
Member Status						
Accredited Practice Member	$2.00	$3.47		$3.42	$4.63	41
Nonaccredited Member	*	*		*	*	15
Nonmember	$2.00	$3.30		$3.82	$5.00	57
Metropolitan Status						
Urban	*	*		*	*	15
Suburban	$2.35	$3.85		$4.07	$5.00	45
Second City	$2.00	$2.95		$3.38	$5.00	27
Town	*	*		*	*	22
Rural	*	*		*	*	9
Median Area Household Income						
$35,000 or Less	*	*		*	*	7
$35,000 to $49,999	$1.75	$3.13		$3.49	$4.73	44
$50,000 to $69,999	$2.41	$3.55		$4.14	$5.50	42
$70,000 or More	$2.80	$3.75		$3.84	$5.00	26

Note 1: See the case description in the notes for Table 13.23.
Note 2: An asterisk indicates that data was not reported due to an insufficient number of responses.
Note 3: 43% of respondents reported that they perform this service as part of this case but do not charge for it separately.
Note 4: 27% of the respondents reported that they do not offer this service.

TABLE 13.33
TREATMENT CASE THREE: TOTAL, DAY ONE

	25th Percentile	Median	Your Data	Average	75th Percentile	Number of Respondents
All Practices	$286.25	$354.48		$353.58	$416.19	448
Number of FTE Veterinarians						
1.0 or Less	$248.20	$325.50		$329.94	$404.98	135
1.1 to 2.0	$294.28	$365.31		$359.74	$420.10	136
2.1 to 3.0	$299.03	$359.20		$360.70	$409.25	85
3.1 or More	$305.00	$365.50		$373.25	$431.87	87
Member Status						
Accredited Practice Member	$315.99	$375.00		$370.47	$427.52	126
Nonaccredited Member	$292.00	$352.00		$355.34	$416.80	75
Nonmember	$265.15	$337.50		$339.95	$408.20	215
Metropolitan Status						
Urban	$350.00	$413.50		$406.18	$456.88	42
Suburban	$334.50	$380.10		$386.72	$444.20	123
Second City	$300.34	$363.23		$367.52	$441.06	94
Town	$262.54	$316.00		$316.65	$382.43	125
Rural	$239.88	$295.25		$295.03	$343.46	56
Median Area Household Income						
$35,000 or Less	$248.97	$310.75		$309.07	$363.23	53
$35,000 to $49,999	$273.94	$339.75		$337.46	$402.31	164
$50,000 to $69,999	$303.33	$358.70		$358.75	$407.45	149
$70,000 or More	$350.58	$411.25		$410.85	$466.03	72

Note 1: See the case description in the notes for Table 13.23.
Note 2: The total fee for day one of treatment case three is the sum of the fees for the admitting exam, IV catheter and placement, one liter of IV fluids, IV pump use, two antibiotic injections, antiemetics, hospitalization, two abdominal radiographs, pain medications (administered twice), and medical waste disposal.

TABLE 13.34
TREATMENT CASE THREE: INPATIENT EXAMINATION (VETERINARIAN SUPERVISION)

	25th Percentile	Median	Your Data	Average	75th Percentile	Number of Respondents
All Practices	**$21.00**	**$28.00**		**$28.41**	**$34.86**	**192**
Number of FTE Veterinarians						
1.0 or Less	$18.00	$25.00		$24.34	$30.00	59
1.1 to 2.0	$23.75	$29.00		$30.93	$35.00	51
2.1 to 3.0	$21.00	$29.00		$29.39	$36.06	42
3.1 or More	$20.25	$29.75		$30.16	$35.00	36
Member Status						
Accredited Practice Member	$21.00	$29.25		$31.03	$37.63	58
Nonaccredited Member	$19.50	$27.50		$26.39	$33.52	37
Nonmember	$20.25	$26.38		$27.28	$33.90	84
Metropolitan Status						
Urban	*	*		*	*	15
Suburban	$22.82	$28.75		$29.76	$35.09	58
Second City	$22.50	$29.50		$30.60	$37.14	44
Town	$18.95	$26.00		$26.42	$34.00	51
Rural	*	*		*	*	19
Median Area Household Income						
$35,000 or Less	*	*		*	*	22
$35,000 to $49,999	$20.00	$26.25		$26.97	$33.62	63
$50,000 to $69,999	$21.38	$28.25		$28.56	$34.61	62
$70,000 or More	$22.00	$31.50		$31.94	$37.46	40

Note 1: See the case description in the notes for Table 13.23.
Note 2: An asterisk indicates that data was not reported due to an insufficient number of responses.
Note 3: 44% of respondents reported that they perform this service as part of this case but do not charge for it separately.
Note 4: 9% of the respondents reported that they do not offer this service.

TABLE 13.35
TREATMENT CASE THREE: LITER OF FLUIDS

	25th Percentile	Median	Your Data	Average	75th Percentile	Number of Respondents
All Practices	$18.00	$25.00		$27.66	$36.00	432
Number of FTE Veterinarians						
1.0 or Less	$16.75	$25.00		$28.19	$38.27	129
1.1 to 2.0	$19.55	$26.50		$28.52	$36.63	129
2.1 to 3.0	$18.80	$25.00		$26.90	$34.82	85
3.1 or More	$16.29	$24.50		$26.56	$35.74	84
Member Status						
Accredited Practice Member	$17.50	$25.00		$27.21	$35.20	122
Nonaccredited Member	$18.00	$25.00		$27.81	$36.19	74
Nonmember	$18.30	$25.00		$27.27	$35.20	206
Metropolitan Status						
Urban	$25.00	$32.00		$33.32	$45.00	39
Suburban	$22.15	$31.00		$32.08	$40.57	117
Second City	$18.88	$26.10		$27.99	$35.00	93
Town	$16.06	$22.50		$24.84	$30.00	120
Rural	$12.00	$16.40		$19.16	$23.75	55
Median Area Household Income						
$35,000 or Less	$16.92	$25.00		$25.74	$31.38	50
$35,000 to $49,999	$16.00	$22.26		$25.67	$33.21	158
$50,000 to $69,999	$19.20	$25.00		$28.24	$36.25	145
$70,000 or More	$19.50	$31.04		$31.98	$43.40	69

Note 1: See the case description in the notes for Table 13.23.
Note 2: 2% of respondents reported that they perform this service as part of this case but do not charge for it separately.
Note 3: 2% of the respondents reported that they do not offer this service.

TABLE 13.36
TREATMENT CASE THREE: CBC WITH SIX CHEMISTRIES

	25th Percentile	Median	Your Data	Average	75th Percentile	Number of Respondents
All Practices	$59.00	$74.95		$77.13	$92.00	439
Number of FTE Veterinarians						
1.0 or Less	$55.00	$70.00		$73.27	$87.50	131
1.1 to 2.0	$59.89	$74.71		$77.87	$92.00	134
2.1 to 3.0	$65.00	$78.00		$80.50	$96.25	83
3.1 or More	$58.56	$74.46		$78.71	$94.25	86
Member Status						
Accredited Practice Member	$60.08	$77.00		$78.79	$92.00	124
Nonaccredited Member	$59.00	$77.96		$78.60	$94.00	75
Nonmember	$56.88	$70.00		$74.94	$90.00	210
Metropolitan Status						
Urban	$66.53	$85.30		$90.01	$108.00	40
Suburban	$61.95	$77.62		$80.68	$97.10	119
Second City	$60.50	$72.00		$76.73	$90.33	93
Town	$55.00	$71.50		$74.04	$90.85	123
Rural	$52.00	$62.00		$67.49	$77.90	56
Median Area Household Income						
$35,000 or Less	$57.60	$67.00		$69.92	$86.80	51
$35,000 to $49,999	$59.00	$75.00		$78.30	$93.94	160
$50,000 to $69,999	$55.00	$74.71		$76.55	$92.00	148
$70,000 or More	$61.99	$76.83		$80.77	$93.25	70

Note 1: See the case description in the notes for Table 13.23.
Note 2: 2% of the respondents reported that they do not offer this service.

TABLE 13.37
TREATMENT CASE THREE: TOTAL, DAY TWO

	25th Percentile	Median	Your Data	Average	75th Percentile	Number of Respondents
All Practices	$193.75	$249.93		$251.31	$305.95	445
Number of FTE Veterinarians						
1.0 or Less	$170.00	$231.37		$240.11	$296.39	134
1.1 to 2.0	$197.29	$252.93		$255.82	$307.90	134
2.1 to 3.0	$201.50	$258.00		$256.36	$297.81	85
3.1 or More	$189.00	$249.00		$256.05	$320.79	87
Member Status						
Accredited Practice Member	$217.98	$264.00		$264.57	$309.68	125
Nonaccredited Member	$193.50	$245.50		$256.48	$317.00	75
Nonmember	$174.78	$234.63		$238.77	$301.70	214
Metropolitan Status						
Urban	$257.00	$302.50		$298.20	$351.00	41
Suburban	$231.02	$282.00		$276.06	$320.62	121
Second City	$200.46	$261.09		$265.49	$327.21	94
Town	$175.65	$217.25		$222.29	$270.00	125
Rural	$159.17	$194.00		$201.48	$235.24	56
Median Area Household Income						
$35,000 or Less	$167.22	$225.50		$219.73	$263.20	52
$35,000 to $49,999	$180.30	$238.36		$244.59	$304.30	163
$50,000 to $69,999	$191.25	$256.00		$251.03	$296.45	149
$70,000 or More	$228.00	$282.50		$288.64	$345.24	71

Note 1: See the case description in the notes for Table 13.23.
Note 2: The total fee for day two of treatment case three is the sum of the fees for the inpatient exam (veterinarian supervision), one liter of fluids, continued use of fluid pump, two subcutaneous antibiotic injections, CBC with six chemistries, antiemetics, hospitalization, pain medications (administered twice), and medical waste disposal.

TABLE 13.38
TREATMENT CASE THREE: CBC WITH TWO CHEMISTRIES

	25th Percentile	Median	Your Data	Average	75th Percentile	Number of Respondents
All Practices	$50.00	$65.00		$65.16	$78.00	417
Number of FTE Veterinarians						
1.0 or Less	$45.00	$59.00		$61.23	$75.00	123
1.1 to 2.0	$51.50	$65.00		$66.85	$79.48	125
2.1 to 3.0	$52.00	$66.20		$67.50	$84.00	81
3.1 or More	$53.00	$65.00		$65.76	$77.56	83
Member Status						
Accredited Practice Member	$53.25	$66.28		$66.57	$78.68	121
Nonaccredited Member	$49.30	$67.50		$66.23	$82.63	69
Nonmember	$47.63	$60.50		$63.85	$75.52	200
Metropolitan Status						
Urban	$50.50	$66.30		$70.10	$86.25	33
Suburban	$54.19	$69.25		$69.37	$83.94	116
Second City	$53.75	$62.50		$63.98	$75.00	89
Town	$45.76	$59.80		$63.24	$77.30	115
Rural	$40.50	$56.55		$57.68	$69.01	56
Median Area Household Income						
$35,000 or Less	$49.00	$60.00		$61.05	$76.47	47
$35,000 to $49,999	$48.53	$63.00		$63.79	$78.10	150
$50,000 to $69,999	$49.00	$65.00		$65.10	$78.50	143
$70,000 or More	$55.00	$65.00		$71.12	$88.00	67

Note 1: See the case description in the notes for Table 13.23.
Note 2: 4% of the respondents reported that they do not offer this service.

TABLE 13.39
TREATMENT CASE THREE: SPECIAL DIET

	25th Percentile	Median	Your Data	Average	75th Percentile	Number of Respondents
All Practices	$2.00	$4.00		$5.75	$6.00	179
Number of FTE Veterinarians						
1.0 or Less	$1.92	$4.77		$6.41	$10.00	55
1.1 to 2.0	$1.95	$3.00		$5.45	$5.72	63
2.1 to 3.0	$2.00	$4.80		$6.39	$6.50	29
3.1 or More	$1.98	$3.20		$4.40	$5.00	29
Member Status						
Accredited Practice Member	$2.00	$4.30		$6.60	$8.50	49
Nonaccredited Member	$2.00	$3.00		$4.89	$5.00	33
Nonmember	$2.00	$4.10		$5.15	$6.15	84
Metropolitan Status						
Urban	*	*		*	*	15
Suburban	$2.00	$4.98		$6.18	$6.65	49
Second City	$1.82	$3.25		$5.18	$5.00	40
Town	$1.93	$3.20		$5.31	$5.50	53
Rural	*	*		*	*	18
Median Area Household Income						
$35,000 or Less	*	*		*	*	21
$35,000 to $49,999	$1.85	$4.00		$6.45	$8.50	61
$50,000 to $69,999	$1.96	$4.00		$6.38	$6.35	61
$70,000 or More	$2.00	$4.98		$5.08	$6.00	31

Note 1: See the case description in the notes for Table 13.23.
Note 2: An asterisk indicates that data was not reported due to an insufficient number of responses.
Note 3: 45% of respondents reported that they perform this service as part of this case but do not charge for it separately.
Note 4: 12% of the respondents reported that they do not offer this service.

TABLE 13.40
TREATMENT CASE THREE: TOTAL, DAY THREE

	25th Percentile	Median	Your Data	Average	75th Percentile	Number of Respondents
All Practices	**$166.08**	**$215.00**		**$214.52**	**$258.28**	**445**
Number of FTE Veterinarians						
1.0 or Less	$147.89	$202.12		$202.39	$248.50	134
1.1 to 2.0	$166.40	$221.41		$218.25	$263.92	134
2.1 to 3.0	$176.95	$227.00		$220.08	$252.72	85
3.1 or More	$172.50	$217.40		$220.49	$257.86	87
Member Status						
Accredited Practice Member	$188.81	$229.65		$227.16	$259.23	125
Nonaccredited Member	$172.50	$211.50		$216.94	$262.50	75
Nonmember	$148.00	$199.69		$204.88	$252.48	214
Metropolitan Status						
Urban	$183.50	$244.25		$242.08	$299.75	41
Suburban	$193.15	$240.85		$236.54	$279.86	121
Second City	$182.56	$231.25		$228.08	$267.34	94
Town	$151.00	$185.40		$190.99	$228.19	125
Rural	$125.63	$167.95		$172.54	$213.44	56
Median Area Household Income						
$35,000 or Less	$134.69	$195.00		$185.79	$220.19	52
$35,000 to $49,999	$153.70	$204.50		$207.11	$251.00	163
$50,000 to $69,999	$171.45	$214.00		$216.07	$259.43	149
$70,000 or More	$211.67	$244.32		$249.20	$294.75	71

Note 1: See the case description in the notes for Table 13.23.
Note 2: The total fee for day three of treatment case three is the sum of the fees for the inpatient exam (veterinarian supervision), one liter of fluids, continued use of fluid pump, two subcutaneous antibiotic injections, CBC with two chemistries, hospitalization, pain medications (administered twice), medical waste disposal, and a special diet in the hospital.

TABLE 13.41
TREATMENT CASE THREE: TOTAL

	25th Percentile	Median	Your Data	Average	75th Percentile	Number of Respondents
All Practices	$660.00	$814.55		$816.29	$974.95	448
Number of FTE Veterinarians						
1.0 or Less	$578.00	$754.10		$769.16	$937.00	135
1.1 to 2.0	$665.90	$840.33		$826.85	$980.68	136
2.1 to 3.0	$690.48	$837.67		$837.15	$935.64	85
3.1 or More	$669.50	$852.25		$849.78	$1,041.20	87
Member Status						
Accredited Practice Member	$741.58	$857.35		$858.29	$969.80	126
Nonaccredited Member	$669.50	$811.44		$828.75	$1,010.60	75
Nonmember	$589.50	$754.10		$781.54	$960.50	215
Metropolitan Status						
Urban	$806.55	$970.97		$933.60	$1,075.78	42
Suburban	$761.10	$901.33		$890.98	$1,049.50	123
Second City	$697.44	$858.97		$861.09	$1,004.10	94
Town	$599.45	$712.50		$729.93	$872.51	125
Rural	$543.86	$657.65		$669.05	$773.94	56
Median Area Household Income						
$35,000 or Less	$539.23	$742.95		$706.94	$833.14	53
$35,000 to $49,999	$608.44	$763.25		$786.40	$963.85	164
$50,000 to $69,999	$679.00	$818.00		$825.85	$955.51	149
$70,000 or More	$805.14	$915.00		$941.22	$1,078.07	72

Note 1: See the case description in the notes for Table 13.23.
Note 2: The total fee for treatment case three is the sum of the fees for all services provided on days one, two, and three (see Tables 13.33, 13.37, and 13.40). Some respondents do not charge for some of the individual services provided in this case. In addition, some respondents reported that they do not offer some of the individual services. Therefore, the average total fee for the case may be significantly lower than the sum of the average fees for the individual services.

TABLE 13.42
TREATMENT CASE FOUR: EXAMINATION

	25th Percentile	Median	Your Data	Average	75th Percentile	Number of Respondents
All Practices	$35.00	$40.00		$40.35	$45.00	437
Number of FTE Veterinarians						
1.0 or Less	$32.75	$38.00		$37.59	$42.00	133
1.1 to 2.0	$35.25	$40.00		$40.80	$45.00	133
2.1 to 3.0	$37.65	$40.00		$41.46	$45.00	85
3.1 or More	$37.75	$41.89		$42.56	$46.75	81
Member Status						
Accredited Practice Member	$38.93	$42.00		$43.45	$48.00	125
Nonaccredited Member	$36.00	$39.00		$40.36	$45.00	72
Nonmember	$33.98	$38.00		$38.23	$43.50	209
Metropolitan Status						
Urban	$38.00	$42.00		$41.82	$46.25	41
Suburban	$38.00	$41.70		$42.44	$48.00	122
Second City	$37.00	$41.00		$41.61	$45.00	89
Town	$34.19	$39.10		$39.26	$44.00	122
Rural	$30.00	$35.00		$35.24	$39.00	55
Median Area Household Income						
$35,000 or Less	$31.50	$37.00		$36.68	$41.00	51
$35,000 to $49,999	$34.00	$38.00		$37.91	$42.98	155
$50,000 to $69,999	$38.00	$40.00		$41.64	$45.00	148
$70,000 or More	$39.55	$45.00		$45.39	$50.88	73

Note: The patient is a 25-pound cocker spaniel with otitis externa in the left ear. You examine her and treat her as a day patient.

TABLE 13.43
TREATMENT CASE FOUR: PREANESTHETIC LAB TESTS

	25th Percentile	Median	Your Data	Average	75th Percentile	Number of Respondents
All Practices	$49.00	$62.00		$65.63	$78.53	421
Number of FTE Veterinarians						
1.0 or Less	$48.50	$62.00		$64.47	$75.75	125
1.1 to 2.0	$49.38	$63.61		$66.28	$79.14	128
2.1 to 3.0	$48.49	$65.00		$67.71	$84.00	82
3.1 or More	$48.25	$58.00		$64.10	$75.59	81
Member Status						
Accredited Practice Member	$50.25	$66.50		$68.56	$84.00	119
Nonaccredited Member	$48.00	$60.60		$65.60	$82.25	69
Nonmember	$48.75	$60.00		$63.72	$75.00	202
Metropolitan Status						
Urban	$52.00	$63.86		$69.45	$85.30	37
Suburban	$49.55	$63.35		$68.78	$84.38	117
Second City	$54.24	$65.00		$66.59	$78.50	89
Town	$44.88	$56.75		$62.52	$77.97	114
Rural	$46.31	$53.13		$59.17	$69.80	56
Median Area Household Income						
$35,000 or Less	$46.50	$58.65		$60.67	$73.89	50
$35,000 to $49,999	$47.13	$59.50		$63.13	$74.96	153
$50,000 to $69,999	$51.70	$65.00		$68.56	$84.38	141
$70,000 or More	$50.00	$65.00		$68.39	$80.00	67

Note 1: See the case description in the notes for Table 13.42.
Note 2: Lab tests include CBC, BUN, creatinine, sodium, and potassium.
Note 3: 5% of the respondents reported that they do not offer this service.

TABLE 13.44
TREATMENT CASE FOUR: ANESTHESIA, 30 MINUTES

	25th Percentile	Median	Your Data	Average	75th Percentile	Number of Respondents
All Practices	$48.00	$69.00		$73.33	$93.09	433
Number of FTE Veterinarians						
1.0 or Less	$40.00	$60.19		$63.55	$80.14	132
1.1 to 2.0	$50.00	$68.50		$74.83	$95.87	131
2.1 to 3.0	$52.00	$74.00		$78.66	$100.60	83
3.1 or More	$59.63	$76.50		$81.50	$96.58	82
Member Status						
Accredited Practice Member	$54.23	$72.00		$76.98	$95.00	121
Nonaccredited Member	$54.90	$66.99		$77.68	$95.15	73
Nonmember	$45.00	$65.00		$68.78	$91.70	208
Metropolitan Status						
Urban	$65.00	$95.87		$97.33	$120.00	39
Suburban	$56.00	$75.00		$79.13	$100.00	123
Second City	$54.85	$74.00		$76.42	$95.83	92
Town	$45.00	$61.50		$65.60	$80.41	116
Rural	$30.94	$50.00		$54.14	$71.50	55
Median Area Household Income						
$35,000 or Less	$35.00	$50.00		$59.09	$79.50	52
$35,000 to $49,999	$45.00	$64.00		$68.22	$85.93	158
$50,000 to $69,999	$52.00	$75.19		$77.05	$97.50	143
$70,000 or More	$64.99	$79.84		$87.71	$109.56	70

Note 1: See the case description in the notes for Table 13.42.
Note 2: 3% of the respondents reported that they do not offer this service.

TABLE 13.45
TREATMENT CASE FOUR: EAR SWAB/CYTOLOGY

	25th Percentile	Median	Your Data	Average	75th Percentile	Number of Respondents
All Practices	$16.99	$23.50		$25.21	$30.00	430
Number of FTE Veterinarians						
1.0 or Less	$16.95	$25.00		$26.30	$32.50	127
1.1 to 2.0	$16.00	$23.44		$24.54	$29.69	128
2.1 to 3.0	$16.63	$22.00		$25.26	$30.46	84
3.1 or More	$17.38	$24.03		$24.77	$32.00	86
Member Status						
Accredited Practice Member	$18.44	$24.80		$25.56	$30.66	122
Nonaccredited Member	$17.00	$23.09		$24.72	$28.44	72
Nonmember	$16.00	$23.00		$25.19	$30.00	208
Metropolitan Status						
Urban	$19.38	$25.00		$28.76	$36.98	41
Suburban	$19.75	$26.00		$27.88	$34.00	123
Second City	$16.50	$24.00		$24.04	$30.00	87
Town	$16.33	$21.00		$23.95	$27.68	117
Rural	$14.90	$18.00		$20.33	$25.00	54
Median Area Household Income						
$35,000 or Less	$13.13	$18.00		$20.92	$25.75	49
$35,000 to $49,999	$16.55	$23.25		$24.94	$30.00	157
$50,000 to $69,999	$17.97	$24.03		$25.72	$31.40	146
$70,000 or More	$19.07	$26.40		$27.23	$33.61	68

Note 1: See the case description in the notes for Table 13.42.
Note 2: 2% of respondents reported that they perform this service as part of this case but do not charge for it separately.
Note 3: 2% of the respondents reported that they do not offer this service.

TABLE 13.46
TREATMENT CASE FOUR: EAR CLEANING

	25th Percentile	Median	Your Data	Average	75th Percentile	Number of Respondents
All Practices	$15.50	$22.92		$25.55	$32.00	405
Number of FTE Veterinarians						
1.0 or Less	$15.00	$21.50		$23.42	$28.56	122
1.1 to 2.0	$15.20	$24.00		$27.04	$35.00	123
2.1 to 3.0	$16.13	$24.84		$27.12	$34.72	76
3.1 or More	$16.00	$22.15		$25.43	$33.00	79
Member Status						
Accredited Practice Member	$17.40	$23.78		$25.58	$31.98	112
Nonaccredited Member	$15.89	$21.63		$25.02	$30.13	70
Nonmember	$15.00	$22.90		$25.20	$31.63	198
Metropolitan Status						
Urban	$15.75	$26.00		$27.58	$34.95	39
Suburban	$18.44	$24.84		$27.81	$35.00	116
Second City	$17.75	$25.00		$27.32	$33.00	83
Town	$14.83	$20.00		$22.35	$28.20	110
Rural	$13.40	$18.00		$22.20	$25.78	50
Median Area Household Income						
$35,000 or Less	$14.44	$20.00		$22.97	$26.25	44
$35,000 to $49,999	$15.00	$20.00		$23.89	$30.00	146
$50,000 to $69,999	$17.93	$24.73		$26.50	$33.28	136
$70,000 or More	$19.00	$26.00		$29.37	$39.00	69

Note 1: See the case description in the notes for Table 13.42.
Note 2: 6% of respondents reported that they perform this service as part of this case but do not charge for it separately.

TABLE 13.47
TREATMENT CASE FOUR: ANTIBIOTIC INJECTION

	25th Percentile	Median	Your Data	Average	75th Percentile	Number of Respondents
All Practices	$15.78	$20.96		$21.29	$26.00	432
Number of FTE Veterinarians						
1.0 or Less	$15.00	$20.00		$20.56	$25.00	131
1.1 to 2.0	$16.10	$20.23		$21.35	$26.00	132
2.1 to 3.0	$17.50	$22.00		$22.71	$28.00	80
3.1 or More	$15.81	$20.25		$21.17	$25.97	84
Member Status						
Accredited Practice Member	$17.00	$22.00		$22.31	$26.85	123
Nonaccredited Member	$15.00	$20.82		$21.46	$26.00	71
Nonmember	$15.00	$20.00		$20.42	$25.61	209
Metropolitan Status						
Urban	$20.96	$25.00		$25.80	$30.47	39
Suburban	$18.00	$24.00		$24.01	$29.00	119
Second City	$17.03	$22.50		$22.55	$27.44	92
Town	$14.00	$19.00		$18.55	$23.47	119
Rural	$11.50	$17.00		$16.41	$20.00	55
Median Area Household Income						
$35,000 or Less	$12.38	$18.00		$18.36	$24.19	54
$35,000 to $49,999	$15.00	$20.00		$20.69	$25.45	155
$50,000 to $69,999	$16.00	$22.00		$21.99	$26.00	143
$70,000 or More	$18.00	$24.25		$23.80	$28.00	71

Note 1: See the case description in the notes for Table 13.42.
Note 2: 4% of the respondents reported that they do not offer this service.

TABLE 13.48
TREATMENT CASE FOUR: PAIN MEDICATION

	25th Percentile	Median	Your Data	Average	75th Percentile	Number of Respondents
All Practices	$15.00	$20.96		$21.24	$26.00	396
Number of FTE Veterinarians						
1.0 or Less	$14.35	$19.77		$19.92	$25.00	120
1.1 to 2.0	$14.94	$20.00		$20.68	$26.00	121
2.1 to 3.0	$16.24	$22.00		$22.53	$28.00	75
3.1 or More	$17.85	$23.00		$22.65	$27.12	75
Member Status						
Accredited Practice Member	$17.93	$22.88		$23.36	$28.00	113
Nonaccredited Member	$16.44	$20.50		$21.89	$26.38	66
Nonmember	$12.65	$20.00		$19.59	$25.00	191
Metropolitan Status						
Urban	$21.24	$25.63		$26.78	$33.00	36
Suburban	$18.00	$22.94		$23.29	$28.19	108
Second City	$18.00	$22.50		$23.29	$28.50	87
Town	$13.00	$18.00		$18.34	$23.48	108
Rural	$10.13	$15.00		$15.87	$21.35	52
Median Area Household Income						
$35,000 or Less	$11.63	$18.05		$18.46	$24.19	50
$35,000 to $49,999	$15.00	$20.00		$20.75	$26.81	144
$50,000 to $69,999	$15.23	$21.30		$21.92	$27.49	130
$70,000 or More	$17.50	$22.70		$23.26	$28.41	64

Note 1: See the case description in the notes for Table 13.42.
Note 2: 4% of respondents reported that they perform this service as part of this case but do not charge for it separately.
Note 3: 6% of the respondents reported that they do not offer this service.

TABLE 13.49
TREATMENT CASE FOUR: HOSPITALIZATION, DAY CHARGE

	25th Percentile	Median	Your Data	Average	75th Percentile	Number of Respondents
All Practices	$15.25	$20.73		$24.32	$30.00	324
Number of FTE Veterinarians						
1.0 or Less	$15.00	$20.00		$24.31	$30.00	94
1.1 to 2.0	$16.50	$20.00		$23.27	$28.00	95
2.1 to 3.0	$14.50	$20.00		$23.78	$31.26	67
3.1 or More	$16.00	$25.00		$26.17	$34.60	63
Member Status						
Accredited Practice Member	$16.50	$22.00		$24.49	$31.00	95
Nonaccredited Member	$15.45	$24.65		$26.25	$34.85	54
Nonmember	$15.00	$20.00		$23.45	$28.85	153
Metropolitan Status						
Urban	$18.64	$26.00		$29.40	$42.55	33
Suburban	$17.38	$22.80		$27.38	$35.00	93
Second City	$17.00	$22.05		$23.84	$30.00	71
Town	$12.81	$18.65		$22.10	$29.00	92
Rural	$12.00	$16.70		$17.84	$22.00	31
Median Area Household Income						
$35,000 or Less	$12.00	$17.50		$19.03	$20.00	35
$35,000 to $49,999	$15.00	$20.00		$23.29	$27.37	110
$50,000 to $69,999	$15.00	$21.04		$24.99	$33.00	111
$70,000 or More	$20.00	$25.60		$28.26	$33.50	59

Note 1: See the case description in the notes for Table 13.42.
Note 2: 15% of respondents reported that they perform this service as part of this case but do not charge for it separately.
Note 3: 8% of the respondents reported that they do not offer this service.

TABLE 13.50
TREATMENT CASE FOUR: ANTIBIOTIC, ONE-WEEK SUPPLY

	25th Percentile	Median	Your Data	Average	75th Percentile	Number of Respondents
All Practices	$15.78	$20.96		$21.29	$26.00	432
Number of FTE Veterinarians						
1.0 or Less	$16.00	$21.63		$22.07	$25.50	130
1.1 to 2.0	$16.35	$21.38		$22.84	$28.36	133
2.1 to 3.0	$15.08	$18.79		$20.55	$25.00	82
3.1 or More	$15.93	$21.86		$22.22	$26.18	80
Member Status						
Accredited Practice Member	$16.15	$21.05		$22.57	$27.74	121
Nonaccredited Member	$18.00	$22.50		$23.25	$28.00	71
Nonmember	$15.00	$20.00		$21.41	$25.50	208
Metropolitan Status						
Urban	$16.30	$21.94		$22.91	$27.19	40
Suburban	$19.00	$24.00		$25.05	$30.00	119
Second City	$15.60	$21.16		$21.62	$27.49	88
Town	$15.00	$18.98		$20.18	$25.00	118
Rural	$13.89	$18.44		$19.31	$25.00	56
Median Area Household Income						
$35,000 or Less	$12.50	$18.95		$19.68	$24.73	53
$35,000 to $49,999	$15.00	$20.00		$21.38	$26.97	153
$50,000 to $69,999	$16.02	$22.30		$22.77	$27.13	145
$70,000 or More	$19.26	$23.50		$24.50	$28.37	69

Note 1: See the case description in the notes for Table 13.42.
Note 2: 2% of the respondents reported that they do not offer this service.

TABLE 13.51
TREATMENT CASE FOUR: ANALGESICS, THREE-DAY SUPPLY

	25th Percentile	Median	Your Data	Average	75th Percentile	Number of Respondents
All Practices	$10.00	$14.38		$14.93	$19.11	376
Number of FTE Veterinarians						
1.0 or Less	$10.00	$14.00		$14.84	$19.04	110
1.1 to 2.0	$10.00	$15.00		$14.86	$18.59	121
2.1 to 3.0	$10.58	$15.00		$15.63	$19.18	69
3.1 or More	$9.00	$13.20		$13.86	$17.18	71
Member Status						
Accredited Practice Member	$10.00	$15.00		$15.87	$20.00	105
Nonaccredited Member	$10.18	$13.10		$13.77	$16.88	65
Nonmember	$10.00	$14.00		$14.70	$18.82	181
Metropolitan Status						
Urban	$11.91	$15.00		$15.79	$18.38	32
Suburban	$10.76	$15.00		$15.62	$20.00	102
Second City	$11.20	$15.00		$16.05	$20.00	81
Town	$8.47	$12.88		$13.82	$17.75	104
Rural	$8.00	$11.50		$12.63	$15.00	48
Median Area Household Income						
$35,000 or Less	$9.00	$13.20		$13.21	$16.60	51
$35,000 to $49,999	$10.81	$15.00		$15.13	$19.69	132
$50,000 to $69,999	$9.75	$14.00		$14.58	$19.33	125
$70,000 or More	$11.14	$15.00		$16.52	$20.00	59

Note 1: See the case description in the notes for Table 13.42.
Note 2: 4% of respondents reported that they perform this service as part of this case but do not charge for it separately.
Note 3: 8% of the respondents reported that they do not offer this service.

TABLE 13.52
TREATMENT CASE FOUR: MEDICAL WASTE DISPOSAL

	25th Percentile	Median	Your Data	Average	75th Percentile	Number of Respondents
All Practices	$2.00	$3.30		$3.48	$4.69	121
Number of FTE Veterinarians						
1.0 or Less	$2.00	$3.55		$3.81	$5.00	38
1.1 to 2.0	$2.12	$3.30		$3.48	$4.75	39
2.1 to 3.0	*	*		*	*	21
3.1 or More	*	*		*	*	22
Member Status						
Accredited Practice Member	$2.00	$3.47		$3.31	$4.21	41
Nonaccredited Member	*	*		*	*	15
Nonmember	$2.00	$3.00		$3.52	$4.99	57
Metropolitan Status						
Urban	*	*		*	*	16
Suburban	$2.13	$3.60		$3.68	$4.99	45
Second City	$1.58	$2.70		$2.86	$3.79	26
Town	*	*		*	*	22
Rural	*	*		*	*	8
Median Area Household Income						
$35,000 or Less	*	*		*	*	8
$35,000 to $49,999	$1.75	$2.98		$3.24	$4.61	44
$50,000 to $69,999	$2.50	$3.50		$3.94	$5.00	43
$70,000 or More	*	*		*	*	23

Note 1: See the case description in the notes for Table 13.42.
Note 2: An asterisk indicates that data was not reported due to an insufficient number of responses.
Note 3: 40% of respondents reported that they perform this service as part of this case but do not charge for it separately.
Note 4: 29% of the respondents reported that they do not offer this service.

TABLE 13.53
TREATMENT CASE FOUR: TOTAL

	25th Percentile	Median	Your Data	Average	75th Percentile	Number of Respondents
All Practices	**$249.75**	**$307.90**		**$308.70**	**$370.50**	**451**
Number of FTE Veterinarians						
1.0 or Less	$238.78	$296.47		$291.62	$354.57	136
1.1 to 2.0	$248.01	$304.26		$311.55	$372.97	137
2.1 to 3.0	$264.78	$319.91		$322.31	$384.34	86
3.1 or More	$252.00	$297.14		$315.20	$379.50	87
Member Status						
Accredited Practice Member	$272.88	$322.05		$326.43	$387.00	126
Nonaccredited Member	$254.66	$305.53		$310.80	$368.97	76
Nonmember	$237.38	$292.25		$295.01	$360.27	218
Metropolitan Status						
Urban	$308.84	$343.18		$351.35	$409.80	42
Suburban	$292.13	$351.18		$340.89	$393.82	124
Second City	$258.85	$316.71		$320.10	$385.86	94
Town	$230.31	$271.70		$281.44	$333.77	124
Rural	$214.81	$248.40		$248.85	$292.50	58
Median Area Household Income						
$35,000 or Less	$215.85	$260.75		$265.38	$313.31	54
$35,000 to $49,999	$233.38	$282.00		$293.77	$354.40	163
$50,000 to $69,999	$265.72	$318.00		$319.66	$379.00	151
$70,000 or More	$301.50	$353.36		$349.29	$402.54	73

Note 1: See the case description in the notes for Table 13.42.
Note 2: The total fee for treatment case four is the sum of the fees for the examination, preanesthetic lab tests, anesthesia, ear swab/cytology, ear cleaning, antibiotic injection, pain medication, hospitalization (day charge), antibiotics (one-week supply), analgesics (three-day supply), and medical waste disposal. Some respondents do not charge for some of the individual services provided in this case. In addition, some respondents reported that they do not offer some of the individual services. Therefore, the average total fee for the case may be significantly lower than the sum of the average fees for the individual services.

CHAPTER 14

SURGICAL PROCEDURES

How you educate clients about procedures, allay fears, and provide gentle, comforting support will differentiate your practice and enhance the value of your veterinary knowledge and expertise. In this chapter, you'll learn about the personal touches that make the difference.

Clients' emotions and anxieties run high when a pet needs surgery—whether it's for a routine spay or neuter procedure or an emergency surgery to remove an intestinal blockage. Progressive veterinarians have come up with unique ways to let their practice personalities show through.

BUILDING TRUST

At Seaside Animal Care in Calabash, North Carolina, practice owner Dr. Ernest Ward Jr. and his team strive to create a neighborly atmosphere where clients feel secure leaving their family pets in friendly, capable hands. Several years ago, Dr. Ward realized many clients were bringing along their children when dropping off pets for treatment or surgery. One day a young girl and her mother brought in their golden retriever puppy for surgery. The child was upset about leaving her friend alone. On a whim, Dr. Ward grabbed a stuffed dog someone had given him and handed it to her.

"It looked a lot like her dog," Dr. Ward recalls. "I told her, 'You take care of my friend tonight while I take care of yours, then tomorrow we can switch.'" It worked wonders, and soon Rusty the Dog began spending the night with different families. Over the years, the staff added a few more puppies and some cats as well. While it eases anxiety, Dr. Ward says, it also eases clients' anxiety about money. Clients feel they are getting more than they expect. It all comes back to the fact that clients' perception of value is tied to the emotional context of your veterinary services.

Dr. Ward readily admits this approach won't work in all practices. It depends on the demographics of the clients as well as how doctors and staff introduce the concept. "It's a gesture, sure, and some might say it's hokey, but it defuses some anxiety," Dr. Ward says. "We've found that it establishes a deeper level of trust between the child and me, and the parents know how much we care about their pets. Rusty is just one building block on the larger foundation of trust."

EASING THE PAIN

Some of clients' anxieties come from their perceptions of how their pets might suffer. A proactive approach to pain management will alleviate their fears. It wasn't all that long ago that many veterinarians didn't provide postsurgical pain relief for pets after spay or neuter procedures. Cathy Toft, a Mission, Kansas, pet owner, still remembers the shock she felt after taking her six-month-old kitten in to be declawed and neutered in 1990. When she asked the veterinarian about pain relief, her doctor responded, "Your little guy will be OK without it. Animals don't feel pain the same way people do."

Now veterinarians know better. The truth is that clients recognize how pain changes their pets' behavior and demeanor, and they expect you to provide a solution. Pain relief should not be an option. Today's forward-thinking doctors automatically include pain management as part of the total package for all surgical cases.

It wasn't that long ago that veterinarians regularly did not prescribe pain medications for post-operative patients. Today, practices more proactively make pain management a standard of care, and rarely do pet owners raise an eyebrow about the added expense. Make it an automatic charge that is part of surgery, and be sure to include it on your estimate.

The extra care and support you provide clients whose pets are undergoing elective and emergency surgery will go far in building trust. And the little touches, whether it be a Rusty, routine pain management, or a follow-up phone call, will further cement your caring compassion for your patients and their guardians.

Case study

Watch and Learn

When architects were designing Causeway Animal Hospital in Metairie, Louisiana, in 1999, they had to address the problem of where to put the heating, ventilation, and cooling (HVAC) system. The best place turned out to be on the second floor. The open loft area that accessed the HVAC closet also worked well as doctors' offices, where veterinarians could overlook surgery and treatment areas as well.

But it was during tours of the new facility that the loft area found its true calling. The safety glass wall allows viewers upstairs to look almost directly down at the surgical table. Impressed clients can't help but be wowed. And when clients see the amount of surgical equipment used in the suite, they understand the cost

Watching the doctors in action also enhances the perception of value of veterinary care. A group of licensed practitioner nursing students even came to observe surgery, saying it gave them an opportunity to see the action close up that they can't experience during human surgeries. The observation deck has turned out to be a real showstopper during tours because it helps clients understand that the practice is a true medical facility.

Implementation Idea

Start a Pain-Scoring Protocol at Your Hospital

No one wants to watch an animal endure pain. Clients want and expect you to keep their pets feeling as comfortable as possible. In his book, *Promoting the Human-Animal Bond in Veterinary Practice* (Iowa State University Press, 2001), Dr. Thomas E. Catanzaro, MHA, Dipl. FACHE, provides a simple way to assess pain using the scale below.

Pain Score	Description
P-0	No Pain
P-1	Maybe there was some pain.
P-2	There should be pain (minor wounds, preopertive dental, abrasions, etc.).
P-3	Postoperative soft-tissue surgery or more extensive wounds
P-4	Extractions or more extensive dental procedures
P-5	Multiple extractions, carnasal extractions
P-6	Declaw, postoperative cruciate trauma
P-7	Head-pressing attitude
P-8	Major soft-tissue wound, severe fracture, pancreatitis
P-9	Extensive burns, multiple fractures, spinal trauma, septic gut, eye injury
P-10	Patient SCREAMING!

Dr. Catanzaro notes that he endorses the pain-scoring system described in the *Compendium on Continuing Education for the Practicing Veterinarian* (February 1998, 140–153). "The eight factors, with the zero to three scoring for each factor, constitute the most thorough assessment tool available," he writes. "However, some practices like a simpler system. All pain scored two or higher should have intervention, and the practice protocol *cannot* override a pain score by a staff member."

HELPFUL RESOURCES

Home-Care Instructions Form, AAHA. American Animal Hospital Association Press, 2003.

Minor Surgical/Anesthetic Procedure Sticker for medical records, AAHA. American Animal Hospital Association Press, 2005.

Pain Management CD, William Tranquilli, DVM, MS, Dipl. ACVA and Leigh Lamont, DVM, MS. Lifelearn, Inc., 2005.

CHAPTER 14

DATA TABLES

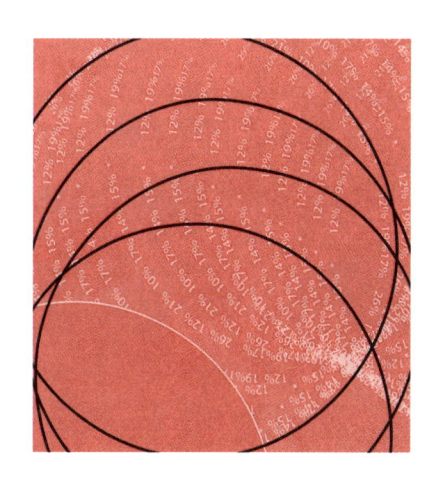

TABLE 14.1
SURGERY SETUP FEES

	25th Percentile	Median	Your Data	Average	75th Percentile	Number of Respondents
Total Surgery Setup Fee	$30.00	$44.02		$51.47	$64.00	70
Surgical Suite Use	$25.00	$38.10		$41.05	$50.25	48
Surgical Pack	$23.00	$32.00		$35.56	$46.50	97
Disposables	$8.50	$23.45		$21.13	$29.95	36
Medical Waste Disposal	$2.00	$3.50		$3.84	$4.99	65
Assistant Time (One Hour)	$26.40	$40.00		$57.21	$77.50	69
Suture Material	$9.00	$12.75		$15.07	$18.00	67
Staples	$9.99	$18.75		$19.67	$25.88	68
Use of Cold Tray Instruments	$15.05	$20.90		$23.38	$31.35	45

Note 1: Data was reported only for all practices due to an insufficient number of responses.
Note 2: 7% of the respondents reported that the surgery setup fee is included in the total surgery fee.
Note 3: 75% of respondents do not charge a setup fee for surgery.

TABLE 14.2
ELECTRONIC MONITORING FEE CHARGED

	Yes	No	Number of Respondents
All Practices	**40%**	**60%**	**453**
Number of FTE Veterinarians			
1.0 or Less	33%	67%	135
1.1 to 2.0	36%	64%	138
2.1 to 3.0	52%	48%	85
3.1 or More	46%	54%	89
Member Status			
Accredited Practice Member	55%	45%	125
Nonaccredited Member	43%	57%	76
Nonmember	31%	69%	220
Metropolitan Status			
Urban	45%	55%	42
Suburban	50%	50%	129
Second City	42%	58%	95
Town	31%	69%	124
Rural	33%	67%	55
Median Area Household Income			
$35,000 or Less	37%	63%	51
$35,000 to $49,999	38%	62%	164
$50,000 to $69,999	40%	60%	153
$70,000 or More	45%	55%	76

TABLE 14.3
ELECTRONIC MONITORING

	25th Percentile	Median	Your Data	Average	75th Percentile	Number of Respondents
All Practices	**$14.98**	**$21.00**		**$25.94**	**$31.75**	**169**
Number of FTE Veterinarians						
1.0 or Less	$14.13	$20.00		$22.78	$28.50	41
1.1 to 2.0	$10.63	$19.88		$23.35	$28.53	44
2.1 to 3.0	$16.00	$23.00		$26.80	$34.00	43
3.1 or More	$18.25	$23.50		$30.37	$38.50	38
Member Status						
Accredited Practice Member	$18.00	$21.62		$25.95	$34.00	63
Nonaccredited Member	$15.00	$21.75		$26.18	$33.00	31
Nonmember	$12.00	$20.00		$25.44	$28.85	65
Metropolitan Status						
Urban	*	*		*	*	18
Suburban	$19.00	$25.00		$30.32	$38.00	61
Second City	$12.63	$20.00		$26.31	$28.92	37
Town	$10.00	$15.00		$17.56	$21.51	36
Rural	*	*		*	*	16
Median Area Household Income						
$35,000 or Less	*	*		*	*	17
$35,000 to $49,999	$12.63	$21.32		$24.97	$34.88	60
$50,000 to $69,999	$15.50	$21.00		$26.62	$29.50	57
$70,000 or More	$17.00	$23.00		$28.43	$38.00	31

Note: An asterisk indicates that data was not reported due to an insufficient number of responses.

TABLE 14.4
ADDITIONAL FEE CHARGED FOR EMERGENCY SURGERY

	Yes	No	Number of Respondents
All Practices	**40%**	**60%**	**447**
Number of FTE Veterinarians			
1.0 or Less	46%	54%	132
1.1 to 2.0	37%	63%	135
2.1 to 3.0	43%	57%	86
3.1 or More	30%	70%	88
Member Status			
Accredited Practice Member	41%	59%	124
Nonaccredited Member	32%	68%	74
Nonmember	42%	58%	218
Metropolitan Status			
Urban	39%	61%	41
Suburban	36%	64%	129
Second City	36%	64%	90
Town	49%	51%	121
Rural	37%	63%	57
Median Area Household Income			
$35,000 or Less	42%	58%	50
$35,000 to $49,999	41%	59%	161
$50,000 to $69,999	37%	63%	153
$70,000 or More	38%	62%	73

TABLE 14.5
ADDITIONAL FEE FOR EMERGENCY SURGERY

	25th Percentile	Median	Your Data	Average	75th Percentile	Number of Respondents
All Practices	$48.00	$62.50		$70.74	$90.00	143
Number of FTE Veterinarians						
1.0 or Less	$40.00	$50.00		$64.00	$92.50	49
1.1 to 2.0	$49.08	$62.50		$64.00	$75.00	41
2.1 to 3.0	$50.00	$69.50		$77.79	$97.50	28
3.1 or More	*	*		*	*	21
Member Status						
Accredited Practice Member	$49.91	$64.25		$71.10	$80.63	42
Nonaccredited Member	*	*		*	*	21
Nonmember	$43.50	$55.00		$70.11	$85.99	71
Metropolitan Status						
Urban	*	*		*	*	12
Suburban	$40.00	$69.00		$72.95	$100.00	37
Second City	$49.25	$66.00		$70.25	$75.00	29
Town	$50.00	$70.75		$73.15	$93.75	44
Rural	*	*		*	*	18
Median Area Household Income						
$35,000 or Less	*	*		*	*	19
$35,000 to $49,999	$50.00	$60.50		$66.30	$75.00	51
$50,000 to $69,999	$49.83	$75.00		$83.75	$102.00	45
$70,000 or More	*	*		*	*	21

Note: An asterisk indicates that data was not reported due to an insufficient number of responses.

ELECTIVE PROCEDURES

TABLE 14.6
MINUTES ALLOTTED FOR ELECTIVE SURGICAL PROCEDURES

	25th Percentile	Median	Your Data	Average	75th Percentile	Number of Respondents
Canine Spay: < 25 Pounds	30	30		37	45	373
Canine Spay: 25–50 Pounds	30	40		40	50	366
Canine Spay: 51–75 Pounds	30	45		45	60	370
Canine Spay: > 75 Pounds	31	45		48	60	365
Canine Neuter: < 25 Pounds	20	30		29	30	365
Canine Neuter: 25–50 Pounds	20	30		30	30	361
Canine Neuter: 51–75 Pounds	20	30		32	40	359
Canine Neuter: > 75 Pounds	25	30		33	45	356
Feline Spay	20	30		32	40	357
Feline Neuter	10	15		16	20	352
Rabbit Spay	30	30		37	45	190
Rabbit Neuter	20	30		28	30	205
Ferret Spay	30	30		35	45	105
Ferret Neuter	15	30		26	30	106
Declaw Two Paws	20	30		30	35	362
Declaw Four Paws	30	40		42	53	248
Ear Cropping	45	60		56	60	78
Tail Docking (Neonatal)	5	10		13	15	261

Note: Data was reported only for all practices because there was no significant variation in the data based on practice size, member status, metropolitan status, or median area household income.

TABLE 14.7
CANINE SPAY: < 25 POUNDS

	25th Percentile	Median	Your Data	Average	75th Percentile	Number of Respondents
All Practices	$122.82	$170.00		$189.40	$241.75	425
Number of FTE Veterinarians						
1.0 or Less	$108.70	$145.00		$170.46	$211.65	130
1.1 to 2.0	$130.00	$185.00		$194.27	$246.00	131
2.1 to 3.0	$142.15	$171.80		$198.60	$250.50	81
3.1 or More	$143.75	$184.50		$199.95	$252.18	78
Member Status						
Accredited Practice Member	$148.38	$208.00		$221.20	$289.96	122
Nonaccredited Member	$124.05	$165.00		$191.94	$242.88	72
Nonmember	$110.00	$155.02		$168.09	$200.00	199
Metropolitan Status						
Urban	$163.25	$238.50		$240.55	$317.06	40
Suburban	$155.28	$199.25		$215.68	$258.05	116
Second City	$120.00	$169.00		$191.31	$238.00	87
Town	$111.38	$164.67		$170.15	$215.62	121
Rural	$100.00	$122.50		$129.09	$154.75	52
Median Area Household Income						
$35,000 or Less	$107.50	$135.00		$149.42	$180.00	51
$35,000 to $49,999	$108.90	$148.25		$168.07	$200.00	153
$50,000 to $69,999	$149.80	$187.00		$198.92	$252.95	145
$70,000 or More	$172.95	$234.50		$242.32	$320.09	66

Note 1: Respondents were asked to report the total fee for this procedure, including exam, anesthesia, pain management, surgery room setup, sutures, etc. The patient is six to seven months old and prepubertal.
Note 2: 5% of respondents reported that they do not offer this service.

TABLE 14.8
CANINE SPAY: 25–50 POUNDS

	25th Percentile	Median	Your Data	Average	75th Percentile	Number of Respondents
All Practices	$135.75	$184.00	_____	$204.13	$259.25	410
Number of FTE Veterinarians						
1.0 or Less	$120.25	$164.00	_____	$186.81	$235.69	128
1.1 to 2.0	$143.88	$197.62	_____	$212.04	$271.07	128
2.1 to 3.0	$148.56	$178.00	_____	$207.63	$257.50	76
3.1 or More	$156.75	$191.50	_____	$215.13	$273.92	74
Member Status						
Accredited Practice Member	$163.47	$217.35	_____	$236.09	$300.00	116
Nonaccredited Member	$135.00	$183.70	_____	$206.73	$260.28	71
Nonmember	$125.00	$167.50	_____	$183.39	$219.81	192
Metropolitan Status						
Urban	$175.00	$238.97	_____	$254.90	$342.75	37
Suburban	$166.95	$219.25	_____	$233.49	$279.00	111
Second City	$133.50	$178.50	_____	$205.17	$265.91	86
Town	$127.00	$171.20	_____	$185.24	$225.00	115
Rural	$111.00	$135.50	_____	$140.21	$165.00	52
Median Area Household Income						
$35,000 or Less	$115.00	$149.50	_____	$163.36	$190.00	51
$35,000 to $49,999	$125.00	$164.13	_____	$182.32	$215.35	147
$50,000 to $69,999	$156.00	$196.00	_____	$211.99	$271.07	140
$70,000 or More	$193.75	$248.00	_____	$264.13	$339.50	62

Note 1: Respondents were asked to report the total fee for this procedure, including exam, anesthesia, pain management, surgery room setup, sutures, etc. The patient is six to seven months old and prepubertal.
Note 2: 5% of respondents reported that they do not offer this service.

TABLE 14.9
CANINE SPAY: 51–75 POUNDS

	25th Percentile	Median	Your Data	Average	75th Percentile	Number of Respondents
All Practices	$155.00	$202.00		$226.28	$280.89	411
Number of FTE Veterinarians						
1.0 or Less	$136.55	$181.48		$207.83	$257.65	128
1.1 to 2.0	$159.63	$217.62		$233.46	$297.75	128
2.1 to 3.0	$168.23	$201.50		$233.40	$274.25	78
3.1 or More	$170.75	$213.00		$239.63	$292.32	74
Member Status						
Accredited Practice Member	$179.63	$239.75		$257.39	$327.51	116
Nonaccredited Member	$150.50	$208.00		$232.84	$285.00	71
Nonmember	$139.75	$187.00		$203.57	$243.20	193
Metropolitan Status						
Urban	$202.50	$291.52		$293.76	$384.31	40
Suburban	$189.66	$247.22		$255.63	$293.92	110
Second City	$151.00	$200.00		$227.36	$283.03	85
Town	$142.81	$190.00		$203.62	$243.40	115
Rural	$122.50	$154.25		$157.24	$179.75	52
Median Area Household Income						
$35,000 or Less	$121.50	$166.58		$179.23	$209.25	50
$35,000 to $49,999	$140.50	$180.03		$208.08	$251.34	148
$50,000 to $69,999	$175.38	$215.80		$233.53	$289.45	142
$70,000 or More	$220.20	$265.00		$285.91	$352.07	61

Note 1: Respondents were asked to report the total fee for this procedure, including exam, anesthesia, pain management, surgery room setup, sutures, etc. The patient is six to seven months old and prepubertal.

Note 2: 5% of respondents reported that they do not offer this service.

TABLE 14.10
CANINE SPAY: > 75 POUNDS

	25th Percentile	Median	Your Data	Average	75th Percentile	Number of Respondents
All Practices	**$170.00**	**$225.00**		**$250.74**	**$310.60**	**415**
Number of FTE Veterinarians						
1.0 or Less	$149.00	$193.00		$226.86	$287.50	125
1.1 to 2.0	$175.00	$243.35		$258.43	$322.03	128
2.1 to 3.0	$187.31	$224.00		$260.43	$311.05	80
3.1 or More	$184.50	$256.00		$262.01	$320.58	77
Member Status						
Accredited Practice Member	$191.00	$273.50		$290.24	$364.30	119
Nonaccredited Member	$166.53	$237.35		$252.27	$304.25	69
Nonmember	$159.89	$200.00		$223.93	$273.75	196
Metropolitan Status						
Urban	$247.25	$338.00		$329.69	$428.31	38
Suburban	$206.50	$274.22		$284.82	$337.50	112
Second City	$171.25	$217.25		$251.68	$315.13	86
Town	$163.75	$203.00		$224.69	$279.13	118
Rural	$134.25	$166.75		$172.83	$194.75	52
Median Area Household Income						
$35,000 or Less	$149.25	$184.75		$199.23	$236.88	46
$35,000 to $49,999	$159.31	$194.58		$229.94	$280.00	152
$50,000 to $69,999	$189.35	$239.00		$255.08	$311.03	141
$70,000 or More	$247.81	$309.30		$321.21	$397.00	66

Note 1: Respondents were asked to report the total fee for this procedure, including exam, anesthesia, pain management, surgery room setup, sutures, etc. The patient is six to seven months old and prepubertal.

Note 2: 5% of respondents reported that they do not offer this service.

TABLE 14.11
CANINE NEUTER: < 25 POUNDS

	25th Percentile	Median	Your Data	Average	75th Percentile	Number of Respondents
All Practices	$104.36	$149.92		$164.92	$208.99	425
Number of FTE Veterinarians						
1.0 or Less	$90.00	$130.00		$145.26	$187.10	131
1.1 to 2.0	$109.00	$160.00		$168.72	$211.00	131
2.1 to 3.0	$121.53	$152.96		$176.87	$233.84	82
3.1 or More	$120.00	$168.38		$176.48	$228.00	76
Member Status						
Accredited Practice Member	$135.00	$185.08		$195.77	$247.50	119
Nonaccredited Member	$100.38	$143.08		$163.57	$203.40	72
Nonmember	$98.38	$135.00		$146.14	$183.50	202
Metropolitan Status						
Urban	$150.00	$209.50		$207.12	$259.20	41
Suburban	$130.00	$180.50		$187.68	$236.56	116
Second City	$105.00	$152.00		$170.35	$213.00	87
Town	$98.80	$139.00		$147.49	$189.25	119
Rural	$81.50	$102.73		$109.19	$129.25	53
Median Area Household Income						
$35,000 or Less	$92.85	$120.00		$130.55	$160.25	50
$35,000 to $49,999	$94.50	$130.00		$147.70	$187.05	153
$50,000 to $69,999	$120.00	$156.00		$169.21	$213.00	143
$70,000 or More	$151.80	$210.95		$211.92	$262.88	69

Note 1: Respondents were asked to report the total fee for this procedure, including exam, anesthesia, pain management, surgery room setup, sutures, etc. The patient is six to seven months old and prepubertal.
Note 2: 5% of respondents reported that they do not offer this service.

TABLE 14.12
CANINE NEUTER: 25–50 POUNDS

	25th Percentile	Median	Your Data	Average	75th Percentile	Number of Respondents
All Practices	**$115.00**	**$158.23**		**$175.46**	**$223.68**	**414**
Number of FTE Veterinarians						
1.0 or Less	$97.63	$140.63		$154.60	$195.34	130
1.1 to 2.0	$121.88	$169.50		$181.68	$225.59	130
2.1 to 3.0	$130.00	$157.40		$186.00	$239.63	78
3.1 or More	$130.00	$174.10		$188.11	$244.00	71
Member Status						
Accredited Practice Member	$148.00	$197.00		$210.45	$267.00	115
Nonaccredited Member	$109.50	$149.00		$169.30	$218.09	69
Nonmember	$105.25	$144.50		$155.81	$192.08	198
Metropolitan Status						
Urban	$160.25	$235.16		$222.96	$284.05	40
Suburban	$147.54	$189.95		$199.13	$244.20	109
Second City	$117.00	$166.25		$182.10	$224.63	84
Town	$103.00	$147.20		$158.23	$200.00	119
Rural	$90.00	$113.90		$116.65	$138.75	53
Median Area Household Income						
$35,000 or Less	$96.50	$125.00		$138.20	$172.00	51
$35,000 to $49,999	$100.00	$140.00		$157.18	$194.63	148
$50,000 to $69,999	$126.03	$168.00		$182.73	$231.83	141
$70,000 or More	$162.38	$223.50		$224.22	$271.76	64

Note 1: Respondents were asked to report the total fee for this procedure, including exam, anesthesia, pain management, surgery room setup, sutures, etc. The patient is six to seven months old and prepubertal.
Note 2: 5% of respondents reported that they do not offer this service.

TABLE 14.13
CANINE NEUTER: 51–75 POUNDS

	25th Percentile	Median	Your Data	Average	75th Percentile	Number of Respondents
All Practices	**$130.00**	**$175.00**		**$189.19**	**$240.02**	**415**
Number of FTE Veterinarians						
1.0 or Less	$104.25	$150.50		$167.51	$216.50	128
1.1 to 2.0	$132.69	$183.50		$194.04	$244.48	130
2.1 to 3.0	$138.75	$178.00		$199.42	$254.80	78
3.1 or More	$138.38	$186.75		$203.92	$244.25	74
Member Status						
Accredited Practice Member	$154.33	$210.13		$224.04	$288.67	118
Nonaccredited Member	$121.31	$167.15		$186.22	$238.20	70
Nonmember	$116.25	$155.90		$167.76	$202.56	196
Metropolitan Status						
Urban	$164.90	$246.50		$234.33	$290.85	41
Suburban	$154.35	$212.50		$216.87	$258.84	108
Second City	$130.00	$177.50		$196.08	$244.20	84
Town	$120.00	$158.23		$169.41	$213.31	120
Rural	$98.50	$123.50		$128.73	$153.25	53
Median Area Household Income						
$35,000 or Less	$104.75	$137.50		$150.19	$187.15	50
$35,000 to $49,999	$115.13	$150.28		$170.70	$213.97	148
$50,000 to $69,999	$139.63	$182.98		$195.48	$242.47	144
$70,000 or More	$179.50	$240.02		$241.45	$296.23	63

Note 1: Respondents were asked to report the total fee for this procedure, including exam, anesthesia, pain management, surgery room setup, sutures, etc. The patient is six to seven months old and prepubertal.
Note 2: 5% of respondents reported that they do not offer this service.

TABLE 14.14
CANINE NEUTER: > 75 POUNDS

	25th Percentile	Median	Your Data	Average	75th Percentile	Number of Respondents
All Practices	$140.00	$186.00		$206.81	$253.70	417
Number of FTE Veterinarians						
1.0 or Less	$116.75	$163.51		$182.77	$238.50	125
1.1 to 2.0	$150.00	$194.88		$210.31	$265.75	128
2.1 to 3.0	$152.00	$190.00		$221.22	$275.00	81
3.1 or More	$145.00	$199.50		$221.04	$260.86	78
Member Status						
Accredited Practice Member	$160.00	$220.00		$239.81	$305.50	119
Nonaccredited Member	$130.00	$197.00		$212.45	$261.00	71
Nonmember	$133.50	$169.75		$183.02	$225.00	196
Metropolitan Status						
Urban	$190.00	$270.00		$267.24	$321.40	39
Suburban	$175.00	$230.00		$240.84	$286.32	113
Second City	$150.00	$186.00		$210.27	$267.75	85
Town	$130.00	$167.15		$180.09	$229.63	118
Rural	$105.50	$138.50		$140.87	$165.75	53
Median Area Household Income						
$35,000 or Less	$112.25	$156.63		$161.89	$190.96	46
$35,000 to $49,999	$131.60	$163.51		$188.48	$238.42	150
$50,000 to $69,999	$150.00	$195.00		$210.53	$253.45	144
$70,000 or More	$189.50	$250.00		$264.78	$323.35	67

Note 1: Respondents were asked to report the total fee for this procedure, including exam, anesthesia, pain management, surgery room setup, sutures, etc. The patient is six to seven months old and prepubertal.
Note 2: 5% of respondents reported that they do not offer this service.

TABLE 14.15
FELINE SPAY

	25th Percentile	Median	Your Data	Average	75th Percentile	Number of Respondents
All Practices	**$100.00**	**$141.25**		**$156.59**	**$199.00**	**428**
Number of FTE Veterinarians						
1.0 or Less	$87.88	$116.00		$139.97	$174.38	132
1.1 to 2.0	$102.00	$146.30		$155.15	$194.63	134
2.1 to 3.0	$117.75	$145.50		$173.24	$214.69	82
3.1 or More	$110.00	$159.00		$168.32	$222.40	75
Member Status						
Accredited Practice Member	$115.00	$173.00		$183.29	$235.00	119
Nonaccredited Member	$100.00	$147.50		$158.27	$212.90	72
Nonmember	$95.00	$129.00		$140.30	$174.00	207
Metropolitan Status						
Urban	$135.00	$177.50		$185.90	$238.38	40
Suburban	$126.00	$168.00		$185.25	$233.75	115
Second City	$98.75	$133.85		$156.96	$196.04	86
Town	$96.00	$133.00		$139.41	$173.00	123
Rural	$85.00	$100.00		$108.79	$130.00	55
Median Area Household Income						
$35,000 or Less	$90.00	$109.75		$119.68	$141.00	51
$35,000 to $49,999	$90.50	$119.77		$137.83	$165.00	155
$50,000 to $69,999	$112.00	$151.50		$165.70	$210.69	146
$70,000 or More	$146.72	$196.27		$199.87	$261.19	66

Note 1: Respondents were asked to report the total fee for this procedure, including exam, anesthesia, pain management, surgery room setup, sutures, etc. The patient is six to seven months old and prepubertal.
Note 2: 5% of respondents reported that they do not offer this service.

TABLE 14.16
FELINE NEUTER

	25th Percentile	Median	Your Data	Average	75th Percentile	Number of Respondents
All Practices	**$65.65**	**$90.00**		**$102.77**	**$128.00**	**431**
Number of FTE Veterinarians						
1.0 or Less	$60.00	$75.00		$95.60	$123.55	131
1.1 to 2.0	$65.08	$93.86		$102.79	$128.00	134
2.1 to 3.0	$70.00	$89.90		$105.97	$123.81	83
3.1 or More	$74.00	$95.50		$109.92	$142.19	78
Member Status						
Accredited Practice Member	$79.19	$105.10		$121.32	$163.50	121
Nonaccredited Member	$66.75	$91.50		$104.38	$144.42	74
Nonmember	$62.06	$83.08		$91.29	$115.00	206
Metropolitan Status						
Urban	$89.25	$111.90		$121.49	$162.00	39
Suburban	$80.75	$110.00		$121.68	$156.00	118
Second City	$65.65	$84.00		$102.32	$123.00	87
Town	$60.00	$84.00		$92.75	$121.00	123
Rural	$49.00	$67.00		$69.76	$80.00	55
Median Area Household Income						
$35,000 or Less	$56.00	$70.46		$82.30	$97.38	51
$35,000 to $49,999	$57.38	$80.00		$91.25	$110.47	154
$50,000 to $69,999	$68.00	$96.87		$108.69	$139.21	148
$70,000 or More	$83.62	$115.00		$124.89	$153.38	68

Note 1: Respondents were asked to report the total fee for this procedure, including exam, anesthesia, pain management, surgery room setup, sutures, etc. The patient is six to seven months old and prepubertal.
Note 2: 5% of respondents reported that they do not offer this service.

TABLE 14.17
RABBIT SPAY

	25th Percentile	Median	Your Data	Average	75th Percentile	Number of Respondents
All Practices	**$115.00**	**$150.00**		**$166.86**	**$200.00**	**223**
Number of FTE Veterinarians						
1.0 or Less	$96.13	$124.23		$160.10	$202.50	54
1.1 to 2.0	$117.43	$150.00		$164.54	$189.00	63
2.1 to 3.0	$136.04	$160.44		$178.91	$202.63	52
3.1 or More	$98.00	$158.00		$166.09	$227.00	51
Member Status						
Accredited Practice Member	$120.00	$169.97		$177.31	$222.25	68
Nonaccredited Member	$108.50	$147.74		$167.71	$228.63	34
Nonmember	$103.25	$145.35		$155.53	$183.75	104
Metropolitan Status						
Urban	*	*		*	*	21
Suburban	$125.00	$164.75		$185.66	$239.85	52
Second City	$111.38	$152.50		$171.25	$204.75	50
Town	$108.00	$150.00		$153.48	$183.74	69
Rural	$92.00	$120.00		$123.08	$139.15	27
Median Area Household Income						
$35,000 or Less	$99.25	$128.00		$131.25	$166.25	26
$35,000 to $49,999	$109.88	$135.50		$163.04	$186.25	86
$50,000 to $69,999	$120.00	$156.00		$167.56	$200.75	77
$70,000 or More	$139.13	$191.95		$198.33	$230.35	30

Note 1: Respondents were asked to report the total fee for this procedure, including exam, anesthesia, pain management, surgery room setup, sutures, etc. The patient is six to seven months old and prepubertal.
Note 2: An asterisk indicates that data was not reported due to an insufficient number of responses.
Note 3: 48% of respondents reported that they do not offer this service.

TABLE 14.18
RABBIT NEUTER

	25th Percentile	Median	Your Data	Average	75th Percentile	Number of Respondents
All Practices	**$80.00**	**$107.00**		**$126.80**	**$159.50**	**243**
Number of FTE Veterinarians						
1.0 or Less	$73.25	$95.00		$128.37	$163.90	61
1.1 to 2.0	$85.25	$116.50		$127.74	$158.83	68
2.1 to 3.0	$84.75	$107.00		$122.33	$153.88	59
3.1 or More	$80.00	$111.14		$131.04	$161.48	52
Member Status						
Accredited Practice Member	$87.75	$122.55		$142.34	$195.16	72
Nonaccredited Member	$79.00	$110.00		$130.86	$174.38	37
Nonmember	$80.00	$102.85		$117.23	$148.00	119
Metropolitan Status						
Urban	*	*		*	*	21
Suburban	$95.00	$139.00		$148.85	$185.00	61
Second City	$83.75	$105.37		$134.03	$175.44	50
Town	$78.50	$100.00		$111.88	$139.60	76
Rural	$70.46	$80.00		$85.99	$100.00	31
Median Area Household Income						
$35,000 or Less	$70.23	$93.25		$99.61	$118.50	25
$35,000 to $49,999	$76.50	$93.00		$119.58	$148.25	93
$50,000 to $69,999	$85.50	$120.50		$130.61	$165.60	84
$70,000 or More	$87.00	$140.50		$150.05	$195.56	36

Note 1: Respondents were asked to report the total fee for this procedure, including exam, anesthesia, pain management, surgery room setup, sutures, etc. The patient is six to seven months old and prepubertal.
Note 2: An asterisk indicates that data was not reported due to an insufficient number of responses.
Note 3: 44% of respondents reported that they do not offer this service.

TABLE 14.19
FERRET SPAY

	25th Percentile	Median	Your Data	Average	75th Percentile	Number of Respondents
All Practices	**$104.50**	**$146.50**		**$162.28**	**$199.34**	**121**
Number of FTE Veterinarians						
1.0 or Less	$75.00	$104.00		$144.44	$192.50	29
1.1 to 2.0	$121.00	$150.00		$181.95	$212.88	36
2.1 to 3.0	$112.63	$147.50		$157.67	$187.91	26
3.1 or More	$115.00	$140.00		$158.20	$186.43	29
Member Status						
Accredited Practice Member	$110.00	$159.00		$169.20	$214.00	43
Nonaccredited Member	*	*		*	*	18
Nonmember	$93.75	$139.26		$155.73	$185.00	54
Metropolitan Status						
Urban	*	*		*	*	6
Suburban	*	*		*	*	24
Second City	$117.63	$152.50		$183.70	$243.75	28
Town	$99.50	$150.00		$156.65	$191.93	45
Rural	*	*		*	*	15
Median Area Household Income						
$35,000 or Less	*	*		*	*	19
$35,000 to $49,999	$104.00	$136.00		$154.98	$168.45	45
$50,000 to $69,999	$110.88	$162.25		$170.13	$216.10	46
$70,000 or More	*	*		*	*	8

Note 1: Respondents were asked to report the total fee for this procedure, including exam, anesthesia, pain management, surgery room setup, sutures, etc. The patient is six to seven months old and prepubertal.
Note 2: An asterisk indicates that data was not reported due to an insufficient number of responses.
Note 3: 71% of respondents reported that they do not offer this service.

TABLE 14.20
FERRET NEUTER

	25th Percentile	Median	Your Data	Average	75th Percentile	Number of Respondents
All Practices	$75.00	$105.87		$125.66	$160.31	124
Number of FTE Veterinarians						
1.0 or Less	$51.50	$86.00		$115.33	$162.81	30
1.1 to 2.0	$91.50	$125.00		$142.96	$197.00	37
2.1 to 3.0	$75.00	$87.30		$109.98	$135.00	28
3.1 or More	$82.50	$112.25		$127.18	$152.75	28
Member Status						
Accredited Practice Member	$75.00	$97.00		$127.87	$169.20	43
Nonaccredited Member	*	*		*	*	18
Nonmember	$74.75	$107.00		$120.82	$158.78	57
Metropolitan Status						
Urban	*	*		*	*	6
Suburban	*	*		*	*	24
Second City	$86.63	$119.99		$142.68	$200.00	28
Town	$75.00	$95.50		$119.51	$161.75	47
Rural	*	*		*	*	16
Median Area Household Income						
$35,000 or Less	*	*		*	*	19
$35,000 to $49,999	$75.00	$94.10		$118.92	$140.00	47
$50,000 to $69,999	$75.00	$123.80		$135.58	$194.00	47
$70,000 or More	*	*		*	*	8

Note 1: Respondents were asked to report the total fee for this procedure, including exam, anesthesia, pain management, surgery room setup, sutures, etc. The patient is six to seven months old and prepubertal.
Note 2: An asterisk indicates that data was not reported due to an insufficient number of responses.
Note 3: 71% of respondents reported that they do not offer this service.

TABLE 14.21
DECLAW TWO PAWS

	25th Percentile	Median	Your Data	Average	75th Percentile	Number of Respondents
All Practices	**$126.48**	**$185.00**		**$207.62**	**$269.91**	**428**
Number of FTE Veterinarians						
1.0 or Less	$107.50	$150.00		$181.40	$217.00	131
1.1 to 2.0	$126.25	$187.50		$203.75	$265.11	132
2.1 to 3.0	$139.00	$230.00		$230.35	$302.00	83
3.1 or More	$162.50	$207.00		$228.64	$286.00	77
Member Status						
Accredited Practice Member	$153.25	$228.63		$241.02	$304.37	124
Nonaccredited Member	$133.75	$200.58		$208.28	$256.75	74
Nonmember	$110.00	$163.75		$186.00	$234.00	201
Metropolitan Status						
Urban	$175.25	$223.75		$252.21	$342.85	40
Suburban	$170.63	$236.00		$248.99	$316.75	122
Second City	$136.63	$191.88		$214.85	$272.00	82
Town	$108.49	$153.00		$175.13	$234.13	120
Rural	$100.00	$125.00		$141.03	$180.00	55
Median Area Household Income						
$35,000 or Less	$100.00	$139.38		$155.14	$185.44	52
$35,000 to $49,999	$107.50	$155.00		$184.75	$242.00	155
$50,000 to $69,999	$140.00	$208.00		$220.38	$280.00	147
$70,000 or More	$188.75	$250.74		$273.44	$336.29	66

Note 1: AAHA's position statement with regard to declawing is that declawing of domestic cats should be considered only after attempts have been made to prevent the cat from using his claws destructively or when clawing presents a significant health risk for people in the household. See www.aahanet.org for the full position statement.

Note 2: Respondents were asked to report the total fee for this procedure, including exam, anesthesia, pain management, surgery room setup, sutures, etc.

Note 3: 5% of respondents reported that they do not offer this service.

TABLE 14.22
DECLAW FOUR PAWS

	25th Percentile	Median	Your Data	Average	75th Percentile	Number of Respondents
All Practices	**$167.00**	**$250.60**		**$275.56**	**$359.00**	**295**
Number of FTE Veterinarians						
1.0 or Less	$137.75	$198.00		$231.22	$286.24	85
1.1 to 2.0	$175.00	$265.86		$281.49	$367.50	92
2.1 to 3.0	$193.98	$290.00		$311.11	$398.73	57
3.1 or More	$199.75	$261.38		$290.16	$363.25	58
Member Status						
Accredited Practice Member	$219.00	$300.00		$322.96	$400.00	91
Nonaccredited Member	$167.38	$267.21		$283.71	$363.50	60
Nonmember	$146.50	$198.50		$238.48	$290.00	122
Metropolitan Status						
Urban	$245.60	$305.63		$353.09	$513.71	28
Suburban	$233.00	$295.00		$329.89	$419.14	94
Second City	$169.00	$251.25		$269.52	$328.18	57
Town	$147.00	$189.00		$219.98	$295.00	79
Rural	$125.11	$164.75		$184.99	$234.61	30
Median Area Household Income						
$35,000 or Less	$131.76	$175.50		$206.53	$266.60	34
$35,000 to $49,999	$155.00	$230.00		$249.78	$304.00	107
$50,000 to $69,999	$193.86	$261.38		$291.26	$375.94	104
$70,000 or More	$250.93	$327.36		$351.24	$421.75	45

Note 1: AAHA's position statement with regard to declawing is that declawing of domestic cats should be considered only after attempts have been made to prevent the cat from using his claws destructively or when clawing presents a significant health risk for people in the household. See www.aahanet.org for the full position statement.
Note 2: Respondents were asked to report the total fee for this procedure, including exam, anesthesia, pain management, surgery room setup, sutures, etc.
Note 3: 34% of respondents reported that they do not offer this service.

TABLE 14.23
EAR CROPPING

	25th Percentile	Median	Your Data	Average	75th Percentile	Number of Respondents
All Practices	**$233.74**	**$293.79**		**$337.58**	**$412.26**	**90**
Number of FTE Veterinarians						
1.0 or Less	$172.50	$250.00		$280.13	$350.00	33
1.1 to 2.0	$242.88	$290.00		$333.13	$406.00	25
2.1 to 3.0	*	*		*	*	15
3.1 or More	*	*		*	*	17
Member Status						
Accredited Practice Member	*	*		*	*	21
Nonaccredited Member	*	*		*	*	15
Nonmember	$215.00	$275.00		$305.76	$350.00	49
Metropolitan Status						
Urban	*	*		*	*	6
Suburban	*	*		*	*	17
Second City	*	*		*	*	15
Town	$195.00	$260.50		$294.76	$352.50	37
Rural	*	*		*	*	13
Median Area Household Income						
$35,000 or Less	*	*		*	*	16
$35,000 to $49,999	$219.06	$292.50		$335.42	$400.00	40
$50,000 to $69,999	$270.75	$350.00		$374.70	$447.45	25
$70,000 or More	*	*		*	*	8

Note 1: AAHA's position statement with regard to ear cropping is that ear cropping for cosmetic reasons is not medically indicated nor of benefit to the patient. See www.aahanet.org for the full position statement.
Note 2: Respondents were asked to report the total fee for this procedure, including exam, anesthesia, pain management, surgery room setup, sutures, etc.
Note 3: An asterisk indicates that data was not reported due to an insufficient number of responses.
Note 4: 79% of respondents reported that they do not offer this service.

TABLE 14.24
TAIL DOCKING (NEONATAL)

	25th Percentile	Median	Your Data	Average	75th Percentile	Number of Respondents
All Practices	$15.00	$25.00		$31.09	$40.00	322
Number of FTE Veterinarians						
1.0 or Less	$11.86	$20.00		$28.42	$35.31	95
1.1 to 2.0	$15.00	$26.24		$32.29	$40.25	94
2.1 to 3.0	$15.00	$26.00		$34.40	$50.00	68
3.1 or More	$17.40	$25.00		$29.53	$38.00	61
Member Status						
Accredited Practice Member	$19.11	$27.05		$36.47	$44.50	97
Nonaccredited Member	$12.00	$23.88		$29.32	$42.53	54
Nonmember	$11.86	$21.00		$27.66	$35.00	147
Metropolitan Status						
Urban	$20.13	$35.31		$45.25	$57.34	25
Suburban	$20.25	$27.79		$37.46	$48.21	85
Second City	$16.13	$28.20		$32.65	$42.25	62
Town	$12.00	$20.00		$26.17	$36.00	96
Rural	$11.55	$17.20		$21.74	$28.00	48
Median Area Household Income						
$35,000 or Less	$10.25	$20.00		$21.98	$36.75	41
$35,000 to $49,999	$13.63	$24.00		$31.79	$39.50	128
$50,000 to $69,999	$15.00	$25.08		$31.93	$40.25	106
$70,000 or More	$22.00	$31.50		$36.63	$49.21	41

Note 1: AAHA's position statement with regard to tail docking is that tail docking for cosmetic reasons is not medically indicated nor of benefit to the patient. See www.aahanet.org for the full position statement.
Note 2: Respondents were asked to report the total fee for this procedure, including exam, anesthesia, pain management, surgery room setup, sutures, etc.
Note 3: 25% of respondents reported that they do not offer this service.

NONELECTIVE PROCEDURES

Note: For all nonelective procedures, respondents were asked to report the fee for the surgical procedure only. Fees do not include anesthesia, treatment, hospitalization, or supplies.

TABLE 14.25
MINUTES ALLOTTED FOR NONELECTIVE SURGICAL PROCEDURES

	25th Percentile	Median	Your Data	Average	75th Percentile	Number of Respondents
Abdominal Exploratory: Cat	40	60		56	60	370
Abdominal Exploratory: 30-Pound Dog	45	60		61	75	360
Abdominal Exploratory: 75-Pound Dog	60	60		70	90	359
Adrenalectomy: Ferret	39	60		54	60	70
Amputation: Pelvic Limb, Cat	60	60		69	90	321
Amputation: Tail, Cat	20	30		35	45	336
Amputation: Thoracic Limb, Cat	60	60		68	90	322
Amputation: Pelvic Limb, 30-Pound Dog	60	65		76	90	309
Amputation: Pelvic Limb, 75-Pound Dog	60	90		85	105	307
Amputation: Tail, 30-Pound Dog	30	30		38	60	331
Amputation: Tail, 75-Pound Dog	30	40		42	60	333
Amputation: Thoracic Limb, 30-Pound Dog	60	60		75	90	311
Amputation: Thoracic Limb, 75-Pound Dog	60	90		85	100	311
Anal Gland Resection: Bilateral, 30-Pound Dog	45	60		59	60	267
Aural Hematoma Repair	25	30		35	45	350
Broken Blood Feather Quill Removal	5	10		15	20	77
Cesarean Section: Cat with Three Kittens	40	60		56	60	349
Cesarean Section: 30-Pound Dog with Four Puppies	45	60		63	75	341
Cesarean Section: 75-Pound Dog with Four Puppies	60	60		71	90	340
Colonoscopy with Biopsy	30	45		49	60	62
Cruciate Repair: Anterior, 60-Pound Dog	60	75		79	90	221
Dewclaw Removal: Non-neonate Canine, Two Dewclaws, Front Limbs	20	30		29	40	320
Diaphragmatic Hernia: Cat	60	70		76	90	255
Diaphragmatic Hernia: 30-Pound Dog	60	70		76	90	255
Diaphragmatic Hernia: 75-Pound Dog	60	70		76	90	255
Ear Resection: Unilateral, 30-Pound Dog	45	60		61	60	185
Endoscopy: Upper GI, with Biopsy	37	55		50	60	81
Enucleation: One Eye, 30-Pound Dog	30	53		51	60	323
Femoral Head Removal: Cat	45	60		64	80	277
Gastrotomy: Foreign-Body Removal, Cat	45	60		62	75	350
Gastrotomy: Foreign-Body Removal, 30-Pound Dog	55	60		67	90	343
Gastrotomy: Foreign-Body Removal, 75-Pound Dog	60	65		74	90	334
Incisor Extraction: Rabbit	15	30		29	30	85
Intestinal Resection/Anastomosis: Cat	60	60		74	90	323
Intestinal Resection/Anastomosis: 30-Pound Dog	60	75		80	90	317
Intestinal Resection/Anastomosis: 75-Pound Dog	60	90		88	115	315
Laser Surgery In-house: Ear Surgery for Chronic Otitis	30	45		50	60	42
Laser Surgery In-house: Feline Declaw, Two Paws	30	30		34	45	72
Laser Surgery In-house: Oral Surgery	20	30		32	40	55
Laser Surgery In-house: Small Tumor Removal	15	20		27	30	72
Luxating Patella Repair: Miniature Poodle	53	60		67	90	171
Mastectomy: Unilateral, Cat	40	60		55	60	274
Mastectomy: Unilateral, 30-Pound Dog	45	60		62	75	269

TABLE 14.25 (CONTINUED)
MINUTES ALLOTTED FOR NONELECTIVE SURGICAL PROCEDURES

	25th Percentile	Median	Your Data	Average	75th Percentile	Number of Respondents
Mastectomy: Unilateral, 75-Pound Dog	60	60		71	90	267
Skin Tumor Removal (4 cm)	20	30		30	30	294
Splenectomy: 30-Pound Dog	60	60		72	90	261
Suture Laceration (4 cm)	15	20		25	30	287
Tibial Fracture Repair, Midshaft: Amazon Parrot	30	45		51	60	30
Tonsillectomy	25	30		35	45	71
Urethrostomy: Cat	53	60		67	90	196

Note: Data was reported only for all practices because there was no significant variation in the data based on practice size, member status, metropolitan status, or median area household income.

TABLE 14.26
ABDOMINAL EXPLORATORY: CAT

	25th Percentile	Median	Your Data	Average	75th Percentile	Number of Respondents
All Practices	**$170.00**	**$250.00**		**$275.61**	**$350.00**	**419**
Number of FTE Veterinarians						
1.0 or Less	$150.00	$225.00		$261.76	$350.00	123
1.1 to 2.0	$151.25	$250.00		$265.22	$350.00	132
2.1 to 3.0	$182.50	$262.50		$280.70	$342.54	81
3.1 or More	$200.00	$287.00		$300.16	$400.00	79
Member Status						
Accredited Practice Member	$200.00	$272.34		$299.96	$388.50	117
Nonaccredited Member	$152.50	$235.00		$267.62	$350.00	72
Nonmember	$150.00	$247.50		$260.53	$350.00	203
Metropolitan Status						
Urban	$162.53	$349.75		$356.89	$484.94	38
Suburban	$200.00	$300.00		$309.45	$400.00	119
Second City	$180.00	$250.00		$271.31	$350.00	82
Town	$150.00	$220.00		$247.81	$302.00	115
Rural	$108.75	$200.00		$216.16	$295.00	57
Median Area Household Income						
$35,000 or Less	$140.63	$200.00		$229.95	$295.38	52
$35,000 to $49,999	$150.00	$225.00		$246.54	$300.00	149
$50,000 to $69,999	$198.00	$296.00		$296.96	$387.00	143
$70,000 or More	$201.25	$317.25		$329.44	$400.00	68

Note: 2% of respondents reported that they do not offer this service.

TABLE 14.27
ABDOMINAL EXPLORATORY: 30-POUND DOG

	25th Percentile	Median	Your Data	Average	75th Percentile	Number of Respondents
All Practices	**$200.00**	**$277.25**		**$303.21**	**$399.58**	**406**
Number of FTE Veterinarians						
1.0 or Less	$158.00	$250.00		$280.92	$395.00	115
1.1 to 2.0	$194.00	$280.11		$296.53	$400.00	129
2.1 to 3.0	$202.15	$296.88		$310.19	$367.28	80
3.1 or More	$222.50	$298.00		$330.02	$409.50	78
Member Status						
Accredited Practice Member	$216.13	$300.00		$328.68	$400.00	116
Nonaccredited Member	$170.00	$256.00		$295.67	$400.00	67
Nonmember	$175.00	$260.72		$287.07	$375.00	196
Metropolitan Status						
Urban	$200.00	$380.00		$381.20	$520.85	37
Suburban	$219.75	$318.00		$330.19	$412.00	113
Second City	$217.50	$280.81		$306.94	$398.75	80
Town	$196.00	$250.00		$284.14	$360.00	114
Rural	$125.75	$225.00		$229.30	$301.27	54
Median Area Household Income						
$35,000 or Less	$150.00	$230.00		$257.31	$369.50	51
$35,000 to $49,999	$163.78	$250.00		$272.40	$352.25	146
$50,000 to $69,999	$200.00	$300.00		$322.08	$400.00	135
$70,000 or More	$250.00	$350.00		$364.81	$450.00	67

Note: 4% of respondents reported that they do not offer this service.

TABLE 14.28
ABDOMINAL EXPLORATORY: 75-POUND DOG

	25th Percentile	Median	Your Data	Average	75th Percentile	Number of Respondents
All Practices	**$210.00**	**$305.00**		**$346.54**	**$450.00**	**407**
Number of FTE Veterinarians						
1.0 or Less	$183.25	$300.00		$327.87	$450.00	116
1.1 to 2.0	$200.00	$300.00		$336.69	$462.50	129
2.1 to 3.0	$240.59	$302.00		$354.59	$414.25	79
3.1 or More	$250.00	$330.00		$372.81	$498.60	79
Member Status						
Accredited Practice Member	$250.00	$340.00		$365.74	$453.98	116
Nonaccredited Member	$185.00	$319.50		$344.50	$460.19	68
Nonmember	$200.00	$300.00		$330.51	$450.00	197
Metropolitan Status						
Urban	$243.75	$462.25		$443.15	$568.94	38
Suburban	$240.90	$369.94		$370.84	$476.25	114
Second City	$235.00	$320.00		$359.36	$450.00	80
Town	$200.00	$300.00		$323.27	$400.00	114
Rural	$160.95	$250.00		$260.92	$350.00	53
Median Area Household Income						
$35,000 or Less	$175.00	$276.00		$297.34	$431.21	51
$35,000 to $49,999	$190.00	$279.80		$318.69	$404.00	147
$50,000 to $69,999	$225.00	$334.92		$367.05	$480.00	135
$70,000 or More	$300.00	$400.00		$402.17	$500.00	67

Note: 4% of respondents reported that they do not offer this service.

TABLE 14.29
ADRENALECTOMY: FERRET

	25th Percentile	Median	Your Data	Average	75th Percentile	Number of Respondents
All Practices	$201.43	$300.00		$308.03	$400.00	76
Number of FTE Veterinarians						
1.0 or Less	*	*		*	*	14
1.1 to 2.0	*	*		*	*	17
2.1 to 3.0	*	*		*	*	19
3.1 or More	$174.32	$250.00		$308.92	$397.50	26
Member Status						
Accredited Practice Member	$222.25	$285.00		$328.27	$410.47	25
Nonaccredited Member	*	*		*	*	15
Nonmember	$225.75	$309.50		$314.21	$398.75	32
Metropolitan Status						
Urban	*	*		*	*	6
Suburban	*	*		*	*	20
Second City	*	*		*	*	21
Town	*	*		*	*	20
Rural	*	*		*	*	7
Median Area Household Income						
$35,000 or Less	*	*		*	*	8
$35,000 to $49,999	$166.50	$250.00		$297.44	$350.00	29
$50,000 to $69,999	$226.20	$312.50		$331.67	$420.00	28
$70,000 or More	*	*		*	*	10

Note 1: An asterisk indicates that data was not reported due to an insufficient number of responses.
Note 2: 82% of respondents reported that they do not offer this service.

TABLE 14.30
AMPUTATION: PELVIC LIMB, CAT

	25th Percentile	Median	Your Data	Average	75th Percentile	Number of Respondents
All Practices	$238.00	$338.00		$362.15	$469.50	369
Number of FTE Veterinarians						
1.0 or Less	$180.00	$300.00		$331.65	$443.45	109
1.1 to 2.0	$231.00	$329.50		$362.70	$477.25	113
2.1 to 3.0	$250.00	$342.50		$358.46	$426.80	68
3.1 or More	$275.00	$380.95		$403.04	$494.00	75
Member Status						
Accredited Practice Member	$270.00	$361.38		$393.23	$479.63	106
Nonaccredited Member	$250.00	$348.24		$377.83	$506.25	62
Nonmember	$214.25	$300.00		$336.40	$444.38	178
Metropolitan Status						
Urban	$323.75	$469.50		$487.06	$586.54	32
Suburban	$295.80	$360.00		$396.23	$498.00	95
Second City	$250.00	$350.00		$365.77	$480.00	75
Town	$226.05	$304.06		$353.12	$440.47	108
Rural	$135.00	$240.00		$241.13	$338.00	51
Median Area Household Income						
$35,000 or Less	$173.00	$275.00		$304.72	$400.00	49
$35,000 to $49,999	$201.25	$300.00		$318.33	$400.00	136
$50,000 to $69,999	$250.00	$350.00		$397.11	$500.00	123
$70,000 or More	$311.69	$416.00		$440.85	$531.89	54

Note: 11% of respondents reported that they do not offer this service.

TABLE 14.31
AMPUTATION: TAIL, CAT

	25th Percentile	Median	Your Data	Average	75th Percentile	Number of Respondents
All Practices	**$85.00**	**$125.00**		**$143.38**	**$175.00**	**396**
Number of FTE Veterinarians						
1.0 or Less	$75.00	$100.00		$129.50	$165.75	117
1.1 to 2.0	$85.25	$125.00		$145.91	$186.50	124
2.1 to 3.0	$99.85	$125.63		$146.26	$168.09	76
3.1 or More	$100.00	$135.00		$152.38	$199.00	75
Member Status						
Accredited Practice Member	$91.88	$128.88		$153.45	$199.13	110
Nonaccredited Member	$93.00	$115.00		$137.16	$159.70	69
Nonmember	$80.00	$120.00		$139.29	$175.00	192
Metropolitan Status						
Urban	$123.75	$177.50		$197.73	$231.75	30
Suburban	$100.00	$144.21		$152.41	$183.75	107
Second City	$81.75	$125.00		$143.56	$169.88	82
Town	$80.28	$120.00		$139.94	$169.27	113
Rural	$60.00	$83.00		$104.99	$145.58	56
Median Area Household Income						
$35,000 or Less	$60.00	$90.10		$110.19	$125.00	49
$35,000 to $49,999	$79.06	$101.50		$130.03	$153.75	144
$50,000 to $69,999	$100.00	$135.00		$153.89	$198.75	136
$70,000 or More	$120.00	$154.00		$178.97	$215.00	59

Note: 4% of respondents reported that they do not offer this service.

TABLE 14.32
AMPUTATION: THORACIC LIMB, CAT

	25th Percentile	Median	Your Data	Average	75th Percentile	Number of Respondents
All Practices	**$228.75**	**$315.00**		**$353.08**	**$450.00**	**370**
Number of FTE Veterinarians						
1.0 or Less	$180.00	$281.20		$322.46	$432.27	109
1.1 to 2.0	$216.00	$300.92		$347.22	$444.38	114
2.1 to 3.0	$255.00	$323.31		$360.55	$444.35	68
3.1 or More	$253.00	$350.00		$393.31	$487.00	75
Member Status						
Accredited Practice Member	$267.49	$356.55		$381.02	$450.00	106
Nonaccredited Member	$225.00	$308.42		$374.52	$507.50	62
Nonmember	$212.00	$300.00		$328.15	$420.00	179
Metropolitan Status						
Urban	$341.25	$451.00		$493.56	$593.48	30
Suburban	$267.65	$350.00		$387.01	$489.25	96
Second City	$250.00	$315.00		$354.11	$460.55	75
Town	$225.00	$298.75		$341.88	$408.94	109
Rural	$133.75	$245.00		$237.43	$305.21	52
Median Area Household Income						
$35,000 or Less	$173.00	$250.00		$292.47	$393.50	49
$35,000 to $49,999	$201.25	$293.24		$310.22	$375.00	136
$50,000 to $69,999	$250.00	$350.00		$389.24	$499.13	124
$70,000 or More	$299.73	$416.00		$426.22	$512.00	54

Note: 12% of respondents reported that they do not offer this service.

TABLE 14.33
AMPUTATION: PELVIC LIMB, 30-POUND DOG

	25th Percentile	Median	Your Data	Average	75th Percentile	Number of Respondents
All Practices	**$275.00**	**$370.00**		**$404.30**	**$516.00**	**357**
Number of FTE Veterinarians						
1.0 or Less	$200.00	$348.00		$357.10	$485.43	101
1.1 to 2.0	$275.00	$375.00		$408.10	$522.25	109
2.1 to 3.0	$300.00	$370.00		$410.86	$500.38	68
3.1 or More	$309.00	$400.00		$449.35	$550.00	75
Member Status						
Accredited Practice Member	$314.07	$393.20		$430.32	$505.25	104
Nonaccredited Member	$274.80	$400.00		$419.91	$546.00	55
Nonmember	$250.00	$350.00		$378.63	$488.14	176
Metropolitan Status						
Urban	$341.05	$508.00		$515.80	$665.91	28
Suburban	$304.75	$400.65		$441.36	$540.00	93
Second City	$300.00	$400.00		$419.65	$523.10	73
Town	$250.00	$350.00		$384.17	$480.00	107
Rural	$199.75	$275.00		$295.63	$367.50	49
Median Area Household Income						
$35,000 or Less	$195.00	$292.50		$326.55	$420.00	46
$35,000 to $49,999	$253.00	$350.00		$353.76	$439.25	133
$50,000 to $69,999	$300.00	$400.00		$452.71	$550.63	121
$70,000 or More	$357.50	$475.00		$482.11	$557.75	50

Note: 14% of respondents reported that they do not offer this service.

TABLE 14.34
AMPUTATION: PELVIC LIMB, 75-POUND DOG

	25th Percentile	Median	Your Data	Average	75th Percentile	Number of Respondents
All Practices	**$300.00**	**$427.98**		**$448.93**	**$540.00**	**354**
Number of FTE Veterinarians						
1.0 or Less	$230.00	$360.00		$388.66	$500.00	99
1.1 to 2.0	$298.88	$440.33		$453.51	$577.13	110
2.1 to 3.0	$310.00	$450.15		$466.11	$552.20	67
3.1 or More	$372.50	$451.50		$495.92	$596.25	74
Member Status						
Accredited Practice Member	$350.00	$450.15		$483.88	$540.00	103
Nonaccredited Member	$300.00	$447.50		$473.46	$629.88	56
Nonmember	$279.88	$400.00		$418.40	$525.00	174
Metropolitan Status						
Urban	$427.50	$523.57		$569.36	$698.75	28
Suburban	$329.88	$450.08		$486.40	$600.50	90
Second City	$350.00	$450.00		$462.36	$549.15	72
Town	$300.00	$400.00		$436.29	$529.13	108
Rural	$211.00	$325.00		$324.55	$391.10	49
Median Area Household Income						
$35,000 or Less	$207.50	$323.90		$355.56	$500.00	46
$35,000 to $49,999	$282.51	$360.00		$388.64	$492.50	133
$50,000 to $69,999	$350.00	$467.00		$506.54	$647.38	120
$70,000 or More	$450.00	$500.25		$548.38	$640.38	48

Note: 15% of respondents reported that they do not offer this service.

TABLE 14.35
AMPUTATION: TAIL, 30-POUND DOG

	25th Percentile	Median	Your Data	Average	75th Percentile	Number of Respondents
All Practices	$90.33	$144.11		$164.44	$208.23	396
Number of FTE Veterinarians						
1.0 or Less	$75.00	$121.50		$145.76	$194.18	112
1.1 to 2.0	$90.05	$150.00		$173.90	$222.75	125
2.1 to 3.0	$109.50	$150.00		$159.91	$187.00	75
3.1 or More	$101.25	$150.00		$175.13	$215.00	80
Member Status						
Accredited Practice Member	$103.18	$150.50		$178.86	$225.00	112
Nonaccredited Member	$88.50	$125.00		$142.88	$175.00	65
Nonmember	$85.00	$131.50		$160.07	$200.00	194
Metropolitan Status						
Urban	$135.00	$206.90		$215.66	$263.05	32
Suburban	$110.00	$150.00		$182.66	$225.00	107
Second City	$85.25	$150.00		$169.37	$208.60	80
Town	$90.00	$124.00		$161.03	$200.00	114
Rural	$63.00	$89.50		$102.18	$135.00	55
Median Area Household Income						
$35,000 or Less	$65.20	$100.00		$125.39	$150.75	50
$35,000 to $49,999	$82.43	$118.75		$146.77	$175.00	146
$50,000 to $69,999	$100.87	$150.00		$176.27	$221.78	132
$70,000 or More	$126.50	$192.50		$211.01	$250.00	60

Note: 5% of respondents reported that they do not offer this service.

TABLE 14.36
AMPUTATION: TAIL, 75-POUND DOG

	25th Percentile	Median	Your Data	Average	75th Percentile	Number of Respondents
All Practices	$100.00	$150.00		$178.89	$225.00	393
Number of FTE Veterinarians						
1.0 or Less	$82.00	$125.00		$163.23	$216.00	111
1.1 to 2.0	$95.00	$160.00		$180.07	$235.66	123
2.1 to 3.0	$120.00	$150.00		$177.69	$240.00	75
3.1 or More	$125.00	$157.75		$194.64	$237.11	80
Member Status						
Accredited Practice Member	$120.00	$175.00		$198.50	$255.00	113
Nonaccredited Member	$100.00	$138.00		$165.01	$210.00	67
Nonmember	$92.00	$149.30		$170.90	$225.00	189
Metropolitan Status						
Urban	$160.00	$219.38		$238.90	$309.38	32
Suburban	$125.00	$175.00		$190.42	$240.00	105
Second City	$87.00	$160.00		$180.77	$230.45	79
Town	$100.00	$140.88		$183.25	$226.05	114
Rural	$75.00	$90.10		$113.79	$150.00	55
Median Area Household Income						
$35,000 or Less	$71.74	$114.39		$145.18	$185.00	50
$35,000 to $49,999	$87.13	$131.29		$158.65	$199.75	144
$50,000 to $69,999	$120.00	$156.75		$185.66	$225.00	132
$70,000 or More	$150.00	$225.00		$236.92	$300.00	59

Note: 5% of respondents reported that they do not offer this service.

TABLE 14.37
AMPUTATION: THORACIC LIMB, 30-POUND DOG

	25th Percentile	Median	Your Data	Average	75th Percentile	Number of Respondents
All Practices	**$250.50**	**$360.00**		**$391.16**	**$500.00**	**356**
Number of FTE Veterinarians						
1.0 or Less	$200.00	$309.50		$344.93	$470.43	101
1.1 to 2.0	$250.00	$367.50		$390.52	$497.63	110
2.1 to 3.0	$300.00	$365.00		$406.44	$500.00	67
3.1 or More	$301.99	$388.90		$432.18	$545.20	74
Member Status						
Accredited Practice Member	$300.41	$377.38		$413.63	$500.00	104
Nonaccredited Member	$235.00	$361.50		$404.15	$547.50	56
Nonmember	$250.00	$350.00		$365.77	$450.00	175
Metropolitan Status						
Urban	$344.04	$500.00		$512.36	$660.58	29
Suburban	$300.00	$400.00		$418.99	$540.00	91
Second City	$299.63	$400.00		$409.13	$500.00	72
Town	$250.00	$350.00		$375.35	$451.31	108
Rural	$199.75	$270.00		$279.66	$350.00	49
Median Area Household Income						
$35,000 or Less	$200.00	$300.00		$329.24	$400.00	46
$35,000 to $49,999	$250.00	$327.50		$342.46	$420.00	133
$50,000 to $69,999	$298.50	$400.00		$437.62	$540.00	120
$70,000 or More	$326.22	$450.00		$451.25	$569.00	50

Note: 14% of respondents reported that they do not offer this service.

TABLE 14.38
AMPUTATION: THORACIC LIMB, 75-POUND DOG

	25th Percentile	Median	Your Data	Average	75th Percentile	Number of Respondents
All Practices	**$300.00**	**$405.00**		**$447.03**	**$538.00**	**359**
Number of FTE Veterinarians						
1.0 or Less	$227.10	$350.00		$395.17	$500.00	101
1.1 to 2.0	$275.00	$410.00		$429.81	$530.39	110
2.1 to 3.0	$328.97	$443.75		$459.44	$565.78	68
3.1 or More	$353.75	$450.00		$519.98	$594.08	76
Member Status						
Accredited Practice Member	$350.00	$442.35		$478.01	$533.50	106
Nonaccredited Member	$279.75	$443.75		$473.77	$630.00	56
Nonmember	$275.00	$373.00		$404.70	$500.13	174
Metropolitan Status						
Urban	$435.00	$550.00		$586.77	$813.75	29
Suburban	$325.41	$446.25		$477.00	$594.90	92
Second City	$300.00	$450.00		$460.70	$536.00	73
Town	$287.00	$375.00		$438.58	$518.95	109
Rural	$211.00	$322.80		$312.25	$391.10	49
Median Area Household Income						
$35,000 or Less	$215.00	$325.00		$369.69	$475.00	47
$35,000 to $49,999	$275.00	$360.00		$389.93	$450.44	134
$50,000 to $69,999	$326.49	$450.00		$485.45	$626.75	120
$70,000 or More	$411.00	$500.00		$529.15	$607.38	50

Note: 14% of respondents reported that they do not offer this service.

TABLE 14.39
ANAL GLAND RESECTION: BILATERAL, 30-POUND DOG

	25th Percentile	Median	Your Data	Average	75th Percentile	Number of Respondents
All Practices	**$200.00**	**$294.00**		**$303.23**	**$368.00**	**307**
Number of FTE Veterinarians						
1.0 or Less	$165.00	$250.00		$268.01	$348.00	83
1.1 to 2.0	$201.88	$290.38		$308.41	$368.88	90
2.1 to 3.0	$212.00	$300.00		$314.84	$400.00	63
3.1 or More	$207.92	$300.00		$323.66	$430.00	67
Member Status						
Accredited Practice Member	$255.00	$315.00		$337.91	$431.45	95
Nonaccredited Member	$177.00	$254.60		$300.45	$391.00	52
Nonmember	$184.00	$270.00		$282.58	$350.00	143
Metropolitan Status						
Urban	$235.49	$335.73		$367.33	$486.68	28
Suburban	$250.00	$306.00		$321.61	$375.00	79
Second City	$227.79	$300.00		$329.24	$417.20	64
Town	$180.00	$250.00		$284.24	$360.50	89
Rural	$150.00	$200.00		$229.76	$292.25	42
Median Area Household Income						
$35,000 or Less	$137.50	$230.00		$221.68	$301.47	33
$35,000 to $49,999	$167.25	$268.94		$285.76	$350.00	117
$50,000 to $69,999	$227.85	$300.00		$319.58	$375.00	108
$70,000 or More	$275.00	$350.00		$376.15	$469.50	43

Note: 27% of respondents reported that they do not offer this service.

TABLE 14.40
AURAL HEMATOMA REPAIR

	25th Percentile	Median	Your Data	Average	75th Percentile	Number of Respondents
All Practices	**$89.75**	**$125.00**		**$144.42**	**$180.00**	**421**
Number of FTE Veterinarians						
1.0 or Less	$80.00	$118.14		$143.44	$180.00	125
1.1 to 2.0	$95.00	$145.00		$151.60	$188.42	131
2.1 to 3.0	$81.95	$117.45		$128.37	$154.02	82
3.1 or More	$100.00	$125.00		$150.97	$200.00	80
Member Status						
Accredited Practice Member	$100.15	$132.00		$146.76	$176.90	117
Nonaccredited Member	$76.00	$120.00		$140.19	$180.00	75
Nonmember	$85.00	$120.00		$141.81	$180.00	201
Metropolitan Status						
Urban	$118.50	$152.50		$177.26	$238.75	36
Suburban	$95.25	$143.50		$158.29	$199.00	117
Second City	$99.00	$130.46		$151.62	$188.86	88
Town	$82.00	$115.00		$133.96	$175.00	115
Rural	$71.77	$95.00		$106.54	$135.00	57
Median Area Household Income						
$35,000 or Less	$73.53	$105.00		$119.52	$175.98	51
$35,000 to $49,999	$80.00	$116.00		$136.95	$175.00	152
$50,000 to $69,999	$99.80	$125.00		$142.17	$172.00	143
$70,000 or More	$115.00	$175.00		$180.23	$225.00	67

Note: 3% of respondents reported that they do not offer this service.

TABLE 14.41
BROKEN BLOOD FEATHER QUILL REMOVAL

	25th Percentile	Median	Your Data	Average	75th Percentile	Number of Respondents
All Practices	**$20.00**	**$34.48**		**$48.24**	**$50.00**	**90**
Number of FTE Veterinarians						
1.0 or Less	*	*		*	*	23
1.1 to 2.0	$16.39	$30.75		$41.83	$49.48	28
2.1 to 3.0	*	*		*	*	17
3.1 or More	*	*		*	*	20
Member Status						
Accredited Practice Member	$20.00	$36.00		$49.86	$48.38	29
Nonaccredited Member	*	*		*	*	13
Nonmember	$23.13	$34.48		$49.27	$60.01	42
Metropolitan Status						
Urban	*	*		*	*	10
Suburban	$20.00	$36.40		$49.29	$64.69	26
Second City	*	*		*	*	17
Town	$19.00	$30.75		$40.16	$50.00	26
Rural	*	*		*	*	9
Median Area Household Income						
$35,000 or Less	*	*		*	*	14
$35,000 to $49,999	$20.00	$33.40		$48.93	$50.00	31
$50,000 to $69,999	$20.00	$38.00		$51.88	$60.00	35
$70,000 or More	*	*		*	*	9

Note 1: An asterisk indicates that data was not reported due to an insufficient number of responses.
Note 2: 78% of respondents reported that they do not offer this service.

TABLE 14.42
CESAREAN SECTION: CAT WITH THREE KITTENS

	25th Percentile	Median	Your Data	Average	75th Percentile	Number of Respondents
All Practices	**$203.45**	**$300.00**		**$304.08**	**$395.00**	**413**
Number of FTE Veterinarians						
1.0 or Less	$200.00	$263.45		$285.27	$360.00	119
1.1 to 2.0	$200.00	$282.50		$301.88	$398.00	125
2.1 to 3.0	$200.00	$297.50		$307.34	$387.50	82
3.1 or More	$250.00	$320.00		$327.04	$402.06	84
Member Status						
Accredited Practice Member	$244.50	$300.00		$321.48	$397.00	114
Nonaccredited Member	$200.00	$327.50		$302.66	$398.75	72
Nonmember	$198.92	$281.25		$290.53	$379.63	198
Metropolitan Status						
Urban	$277.67	$350.00		$372.26	$441.00	35
Suburban	$231.60	$329.75		$332.54	$404.25	114
Second City	$248.25	$300.00		$312.15	$360.96	86
Town	$180.00	$262.00		$283.39	$382.50	113
Rural	$150.00	$227.00		$238.75	$313.22	58
Median Area Household Income						
$35,000 or Less	$160.17	$250.00		$268.05	$379.50	51
$35,000 to $49,999	$190.00	$260.75		$279.46	$346.00	154
$50,000 to $69,999	$241.50	$300.20		$320.74	$400.00	139
$70,000 or More	$235.65	$350.00		$347.28	$450.00	61

Note: 3% of respondents reported that they do not offer this service.

TABLE 14.43
CESAREAN SECTION: 30-POUND DOG WITH FOUR PUPPIES

	25th Percentile	Median	Your Data	Average	75th Percentile	Number of Respondents
All Practices	$250.00	$350.00		$368.53	$450.00	411
Number of FTE Veterinarians						
1.0 or Less	$240.00	$340.45		$340.93	$425.00	115
1.1 to 2.0	$250.00	$350.00		$375.54	$475.00	128
2.1 to 3.0	$255.25	$371.35		$376.28	$445.94	80
3.1 or More	$280.50	$366.35		$379.79	$463.50	84
Member Status						
Accredited Practice Member	$281.38	$361.88		$382.27	$475.00	114
Nonaccredited Member	$250.00	$356.47		$363.42	$449.25	70
Nonmember	$241.50	$350.00		$361.77	$450.00	199
Metropolitan Status						
Urban	$350.00	$421.80		$466.67	$574.50	35
Suburban	$301.74	$393.00		$393.02	$483.75	112
Second City	$300.00	$350.00		$374.08	$450.00	84
Town	$241.00	$337.50		$358.42	$450.00	116
Rural	$200.00	$279.50		$279.13	$343.00	57
Median Area Household Income						
$35,000 or Less	$200.00	$325.00		$315.31	$400.00	51
$35,000 to $49,999	$239.89	$309.50		$325.48	$400.00	151
$50,000 to $69,999	$300.00	$390.45		$392.28	$491.87	139
$70,000 or More	$301.91	$419.50		$455.74	$599.63	62

Note: 5% of respondents reported that they do not offer this service.

TABLE 14.44
CESAREAN SECTION: 75-POUND DOG WITH FOUR PUPPIES

	25th Percentile	Median	Your Data	Average	75th Percentile	Number of Respondents
All Practices	$292.00	$400.00		$418.03	$515.90	408
Number of FTE Veterinarians						
1.0 or Less	$275.00	$360.00		$399.01	$500.00	116
1.1 to 2.0	$279.63	$396.63		$415.22	$525.00	124
2.1 to 3.0	$300.00	$401.50		$436.06	$547.50	80
3.1 or More	$311.25	$419.10		$422.37	$500.00	84
Member Status						
Accredited Practice Member	$310.64	$409.50		$423.27	$515.60	113
Nonaccredited Member	$297.50	$380.00		$415.88	$516.25	70
Nonmember	$275.00	$397.25		$412.55	$500.00	197
Metropolitan Status						
Urban	$373.69	$520.10		$547.00	$661.25	36
Suburban	$349.25	$449.50		$449.15	$548.36	112
Second City	$329.50	$400.00		$432.07	$533.00	83
Town	$257.58	$360.00		$388.16	$478.75	113
Rural	$240.00	$325.00		$322.51	$404.50	57
Median Area Household Income						
$35,000 or Less	$225.00	$339.50		$352.86	$450.00	51
$35,000 to $49,999	$253.15	$350.00		$370.19	$450.00	151
$50,000 to $69,999	$350.00	$425.50		$453.13	$550.95	137
$70,000 or More	$355.58	$500.00		$503.89	$600.00	61

Note: 5% of respondents reported that they do not offer this service.

TABLE 14.45
COLONOSCOPY WITH BIOPSY

	25th Percentile	Median	Your Data	Average	75th Percentile	Number of Respondents
All Practices	$200.00	$254.70		$285.15	$383.00	71

Note 1: Data was reported only for all practices due to an insufficient number of responses.
Note 2: 83% of respondents reported that they do not offer this service.

TABLE 14.46
CRUCIATE REPAIR: ANTERIOR, 60-POUND DOG

	25th Percentile	Median	Your Data	Average	75th Percentile	Number of Respondents
All Practices	$398.50	$544.21		$610.99	$735.24	266
Number of FTE Veterinarians						
1.0 or Less	$354.75	$450.00		$605.63	$784.60	61
1.1 to 2.0	$372.30	$509.50		$580.97	$708.31	77
2.1 to 3.0	$400.00	$597.80		$569.19	$690.00	55
3.1 or More	$450.00	$603.50		$663.59	$800.00	70
Member Status						
Accredited Practice Member	$435.00	$600.00		$649.57	$747.73	87
Nonaccredited Member	$350.00	$500.00		$603.02	$717.50	41
Nonmember	$400.00	$500.00		$592.43	$730.00	119
Metropolitan Status						
Urban	*	*		*	*	22
Suburban	$474.28	$609.75		$639.23	$732.50	64
Second City	$385.00	$600.00		$643.94	$792.30	56
Town	$350.00	$467.25		$546.83	$734.48	84
Rural	$350.00	$422.00		$513.07	$600.00	34
Median Area Household Income						
$35,000 or Less	$367.65	$438.88		$497.71	$596.75	32
$35,000 to $49,999	$350.00	$500.00		$552.61	$649.00	93
$50,000 to $69,999	$404.75	$550.00		$626.40	$733.75	96
$70,000 or More	$500.00	$665.00		$780.44	$942.93	39

Note 1: An asterisk indicates that data was not reported due to an insufficient number of responses.
Note 2: 38% of respondents reported that they do not offer this service.

TABLE 14.47
DEWCLAW REMOVAL: NON-NEONATE CANINE, TWO DEWCLAWS, FRONT LIMBS

	25th Percentile	Median	Your Data	Average	75th Percentile	Number of Respondents
All Practices	**$60.00**	**$91.00**		**$106.14**	**$149.88**	**393**
Number of FTE Veterinarians						
1.0 or Less	$60.00	$80.00		$100.95	$148.00	115
1.1 to 2.0	$58.10	$90.00		$102.58	$134.05	122
2.1 to 3.0	$55.00	$94.13		$110.07	$150.00	76
3.1 or More	$60.00	$109.90		$113.21	$150.00	76
Member Status						
Accredited Practice Member	$60.00	$99.00		$109.38	$143.90	113
Nonaccredited Member	$60.00	$100.00		$109.04	$150.00	65
Nonmember	$55.00	$87.35		$103.05	$150.00	191
Metropolitan Status						
Urban	$60.00	$110.00		$120.74	$174.49	31
Suburban	$72.47	$116.28		$123.14	$160.00	106
Second City	$60.00	$90.00		$103.79	$145.00	79
Town	$50.00	$91.00		$102.76	$148.69	116
Rural	$37.50	$63.50		$73.12	$80.34	54
Median Area Household Income						
$35,000 or Less	$45.75	$67.50		$80.48	$112.00	48
$35,000 to $49,999	$50.00	$80.00		$89.93	$112.25	141
$50,000 to $69,999	$62.25	$110.00		$118.72	$159.75	133
$70,000 or More	$75.00	$125.00		$134.92	$190.00	63

Note: 7% of respondents reported that they do not offer this service.

TABLE 14.48
DIAPHRAGMATIC HERNIA: CAT

	25th Percentile	Median	Your Data	Average	75th Percentile	Number of Respondents
All Practices	**$250.00**	**$355.00**		**$410.27**	**$554.72**	**303**
Number of FTE Veterinarians						
1.0 or Less	$193.44	$350.00		$398.62	$564.86	84
1.1 to 2.0	$250.00	$360.00		$414.92	$540.00	91
2.1 to 3.0	$250.00	$351.50		$408.43	$571.50	59
3.1 or More	$300.00	$380.00		$418.13	$561.00	65
Member Status						
Accredited Practice Member	$303.33	$389.50		$442.46	$599.75	85
Nonaccredited Member	$233.00	$360.00		$407.46	$577.50	53
Nonmember	$225.00	$350.00		$385.30	$497.50	144
Metropolitan Status						
Urban	*	*		*	*	23
Suburban	$300.41	$410.00		$461.39	$600.00	76
Second City	$293.75	$389.75		$429.39	$600.00	66
Town	$247.50	$350.00		$384.11	$506.25	90
Rural	$150.00	$251.00		$288.60	$378.75	42
Median Area Household Income						
$35,000 or Less	$188.60	$290.54		$316.22	$422.50	40
$35,000 to $49,999	$250.00	$350.00		$371.08	$452.33	117
$50,000 to $69,999	$250.00	$397.50		$445.00	$600.00	98
$70,000 or More	$330.00	$490.00		$521.00	$672.59	43

Note 1: An asterisk indicates that data was not reported due to an insufficient number of responses.
Note 2: 27% of respondents reported that they do not offer this service.

TABLE 14.49
DIAPHRAGMATIC HERNIA: 30-POUND DOG

	25th Percentile	Median	Your Data	Average	75th Percentile	Number of Respondents
All Practices	$270.00	$400.00		$442.30	$584.00	299
Number of FTE Veterinarians						
1.0 or Less	$229.50	$360.00		$433.87	$600.00	83
1.1 to 2.0	$258.50	$420.00		$445.59	$599.75	89
2.1 to 3.0	$285.00	$357.50		$432.63	$574.67	58
3.1 or More	$323.70	$411.00		$453.31	$577.00	65
Member Status						
Accredited Practice Member	$333.63	$410.50		$463.05	$599.88	84
Nonaccredited Member	$257.50	$441.04		$428.11	$600.00	50
Nonmember	$250.00	$352.50		$426.27	$551.50	144
Metropolitan Status						
Urban	*	*		*	*	23
Suburban	$335.00	$449.50		$489.43	$639.00	73
Second City	$350.00	$445.00		$490.21	$615.59	64
Town	$250.00	$353.94		$416.68	$551.18	90
Rural	$175.00	$252.00		$308.59	$415.00	43
Median Area Household Income						
$35,000 or Less	$200.00	$307.82		$346.02	$499.25	40
$35,000 to $49,999	$300.00	$397.50		$407.10	$518.06	118
$50,000 to $69,999	$253.00	$430.00		$468.99	$650.00	95
$70,000 or More	$370.00	$561.15		$569.09	$694.75	41

Note 1: An asterisk indicates that data was not reported due to an insufficient number of responses.
Note 2: 28% of respondents reported that they do not offer this service.

TABLE 14.50
DIAPHRAGMATIC HERNIA: 75-POUND DOG

	25th Percentile	Median	Your Data	Average	75th Percentile	Number of Respondents
All Practices	$302.50	$450.00		$496.34	$650.00	297
Number of FTE Veterinarians						
1.0 or Less	$249.00	$450.00		$503.54	$692.40	82
1.1 to 2.0	$302.54	$450.00		$485.67	$650.00	89
2.1 to 3.0	$322.18	$450.00		$488.87	$625.05	57
3.1 or More	$366.33	$450.00		$504.95	$631.50	65
Member Status						
Accredited Practice Member	$351.90	$468.40		$510.67	$648.00	83
Nonaccredited Member	$270.00	$500.00		$506.82	$675.00	51
Nonmember	$282.50	$434.23		$475.91	$600.00	142
Metropolitan Status						
Urban	*	*		*	*	24
Suburban	$352.93	$517.50		$542.67	$687.25	72
Second City	$377.25	$525.00		$551.34	$692.90	64
Town	$281.90	$400.00		$461.06	$600.00	90
Rural	$205.00	$300.00		$349.89	$500.00	41
Median Area Household Income						
$35,000 or Less	$200.00	$342.02		$437.25	$600.00	39
$35,000 to $49,999	$315.00	$447.50		$456.27	$600.00	117
$50,000 to $69,999	$300.00	$450.00		$504.40	$689.00	95
$70,000 or More	$452.58	$625.09		$641.80	$790.28	41

Note 1: An asterisk indicates that data was not reported due to an insufficient number of responses.
Note 2: 27% of respondents reported that they do not offer this service.

TABLE 14.51
EAR RESECTION: UNILATERAL, 30-POUND DOG

	25th Percentile	Median	Your Data	Average	75th Percentile	Number of Respondents
All Practices	$210.56	$301.33		$332.04	$425.50	218
Number of FTE Veterinarians						
1.0 or Less	$175.00	$262.23		$296.78	$400.00	60
1.1 to 2.0	$216.00	$300.00		$323.93	$400.00	63
2.1 to 3.0	$240.00	$310.00		$355.60	$450.00	47
3.1 or More	$233.00	$348.84		$359.96	$439.20	47
Member Status						
Accredited Practice Member	$253.75	$323.50		$364.25	$447.25	70
Nonaccredited Member	$183.00	$289.80		$327.19	$401.49	42
Nonmember	$175.00	$285.00		$305.37	$418.00	95
Metropolitan Status						
Urban	*	*		*	*	22
Suburban	$200.00	$300.00		$333.58	$431.40	51
Second City	$275.00	$350.00		$393.76	$500.00	39
Town	$200.00	$299.38		$315.93	$395.63	70
Rural	$146.25	$227.50		$237.54	$340.21	30
Median Area Household Income						
$35,000 or Less	$132.42	$250.00		$280.81	$383.75	28
$35,000 to $49,999	$175.00	$274.73		$298.31	$369.25	78
$50,000 to $69,999	$230.75	$349.42		$357.27	$442.91	76
$70,000 or More	$300.00	$365.00		$395.58	$477.80	31

Note 1: An asterisk indicates that data was not reported due to an insufficient number of responses.
Note 2: 47% of respondents reported that they do not offer this service.

TABLE 14.52
ENDOSCOPY: UPPER GI, WITH BIOPSY

	25th Percentile	Median	Your Data	Average	75th Percentile	Number of Respondents
All Practices	**$173.50**	**$275.00**		**$274.96**	**$350.00**	**93**
Number of FTE Veterinarians						
1.0 or Less	*	*		*	*	16
1.1 to 2.0	$166.25	$296.10		$282.24	$354.50	25
2.1 to 3.0	*	*		*	*	18
3.1 or More	$199.63	$300.00		$292.90	$358.25	34
Member Status						
Accredited Practice Member	$213.34	$289.60		$279.05	$343.15	37
Nonaccredited Member	*	*		*	*	17
Nonmember	$161.00	$245.00		$277.84	$394.05	33
Metropolitan Status						
Urban	*	*		*	*	10
Suburban	*	*		*	*	22
Second City	$166.20	$250.00		$262.58	$347.50	29
Town	*	*		*	*	24
Rural	*	*		*	*	7
Median Area Household Income						
$35,000 or Less	*	*		*	*	14
$35,000 to $49,999	*	*		*	*	21
$50,000 to $69,999	$180.63	$275.23		$269.28	$347.38	40
$70,000 or More	*	*		*	*	15

Note 1: An asterisk indicates that data was not reported due to an insufficient number of responses.
Note 2: 77% of respondents reported that they do not offer this service.

TABLE 14.53
ENUCLEATION: ONE EYE, 30-POUND DOG

	25th Percentile	Median	Your Data	Average	75th Percentile	Number of Respondents
All Practices	**$180.75**	**$250.00**		**$262.48**	**$328.88**	**385**
Number of FTE Veterinarians						
1.0 or Less	$156.32	$239.50		$254.10	$328.86	106
1.1 to 2.0	$175.00	$250.00		$259.71	$345.66	121
2.1 to 3.0	$200.00	$250.00		$272.79	$333.75	73
3.1 or More	$200.00	$250.00		$269.59	$327.99	82
Member Status						
Accredited Practice Member	$200.00	$250.00		$273.27	$350.00	112
Nonaccredited Member	$180.00	$250.00		$263.89	$350.00	63
Nonmember	$165.00	$239.00		$252.14	$309.50	187
Metropolitan Status						
Urban	$277.50	$350.00		$356.05	$450.00	32
Suburban	$227.70	$280.00		$290.38	$368.50	101
Second City	$175.50	$250.00		$257.26	$326.45	81
Town	$163.38	$215.50		$239.93	$300.00	110
Rural	$146.25	$195.00		$199.87	$255.81	54
Median Area Household Income						
$35,000 or Less	$140.00	$200.00		$217.39	$273.00	52
$35,000 to $49,999	$163.38	$230.28		$236.03	$299.96	138
$50,000 to $69,999	$197.00	$274.38		$279.35	$359.98	129
$70,000 or More	$236.85	$300.00		$312.49	$384.00	58

Note: 9% of respondents reported that they do not offer this service.

TABLE 14.54
FEMORAL HEAD REMOVAL: CAT

	25th Percentile	Median	Your Data	Average	75th Percentile	Number of Respondents
All Practices	$250.00	$350.00		$370.41	$475.00	323
Number of FTE Veterinarians						
1.0 or Less	$212.16	$300.00		$328.73	$400.00	84
1.1 to 2.0	$250.00	$350.00		$368.93	$475.00	95
2.1 to 3.0	$250.00	$350.00		$369.44	$450.00	66
3.1 or More	$270.00	$384.60		$416.13	$534.00	75
Member Status						
Accredited Practice Member	$287.90	$400.00		$411.23	$500.00	100
Nonaccredited Member	$250.00	$360.00		$361.23	$494.50	50
Nonmember	$225.00	$317.50		$350.60	$450.00	152
Metropolitan Status						
Urban	*	*		*	*	21
Suburban	$295.00	$375.00		$404.74	$497.00	86
Second City	$268.50	$360.00		$391.53	$500.00	66
Town	$220.13	$290.77		$337.79	$450.00	98
Rural	$189.00	$300.00		$299.41	$400.00	47
Median Area Household Income						
$35,000 or Less	$180.00	$250.00		$270.93	$375.00	39
$35,000 to $49,999	$225.00	$300.00		$337.04	$450.00	115
$50,000 to $69,999	$270.00	$390.00		$404.49	$500.00	119
$70,000 or More	$329.56	$390.00		$440.30	$542.00	45

Note 1: An asterisk indicates that data was not reported due to an insufficient number of responses.
Note 2: 23% of respondents reported that they do not offer this service.

TABLE 14.55
GASTROTOMY: FOREIGN-BODY REMOVAL, CAT

	25th Percentile	Median	Your Data	Average	75th Percentile	Number of Respondents
All Practices	$225.00	$325.00		$338.27	$425.00	404
Number of FTE Veterinarians						
1.0 or Less	$200.00	$300.00		$326.56	$432.50	118
1.1 to 2.0	$237.00	$325.00		$341.64	$425.00	127
2.1 to 3.0	$225.00	$302.50		$328.64	$412.13	77
3.1 or More	$250.00	$350.00		$351.72	$441.25	78
Member Status						
Accredited Practice Member	$255.50	$345.00		$351.33	$440.50	113
Nonaccredited Member	$225.00	$350.00		$355.36	$420.00	71
Nonmember	$201.25	$300.00		$326.25	$430.00	197
Metropolitan Status						
Urban	$331.25	$456.03		$431.57	$501.50	32
Suburban	$284.98	$355.00		$372.13	$450.00	112
Second City	$247.50	$339.00		$346.79	$422.26	83
Town	$210.00	$300.00		$316.42	$410.10	115
Rural	$150.00	$243.60		$251.03	$330.00	55
Median Area Household Income						
$35,000 or Less	$195.00	$300.00		$296.89	$375.00	51
$35,000 to $49,999	$200.00	$285.00		$288.66	$360.00	144
$50,000 to $69,999	$271.17	$360.00		$376.78	$450.00	141
$70,000 or More	$285.00	$375.00		$396.27	$492.25	61

Note: 3% of respondents reported that they do not offer this service.

TABLE 14.56
GASTROTOMY: FOREIGN-BODY REMOVAL, 30-POUND DOG

	25th Percentile	Median	Your Data	Average	75th Percentile	Number of Respondents
All Practices	**$250.00**	**$350.00**		**$368.75**	**$450.00**	**399**
Number of FTE Veterinarians						
1.0 or Less	$225.00	$345.00		$362.03	$450.00	115
1.1 to 2.0	$264.38	$350.00		$368.40	$450.00	126
2.1 to 3.0	$250.00	$339.06		$356.61	$450.00	76
3.1 or More	$277.89	$370.00		$380.90	$475.76	78
Member Status						
Accredited Practice Member	$273.01	$354.50		$372.70	$450.00	112
Nonaccredited Member	$239.75	$360.00		$381.49	$464.00	69
Nonmember	$243.46	$350.00		$362.25	$450.00	196
Metropolitan Status						
Urban	$331.25	$456.03		$459.60	$570.88	32
Suburban	$300.00	$390.16		$398.37	$483.75	109
Second City	$272.95	$375.00		$388.41	$450.00	81
Town	$233.25	$308.70		$347.43	$438.74	116
Rural	$186.75	$258.71		$275.76	$350.00	54
Median Area Household Income						
$35,000 or Less	$195.25	$300.00		$318.78	$404.00	51
$35,000 to $49,999	$225.00	$300.00		$314.69	$400.00	144
$50,000 to $69,999	$300.00	$400.00		$410.32	$500.00	136
$70,000 or More	$300.00	$400.00		$436.91	$550.00	61

Note: 4% of respondents reported that they do not offer this service.

TABLE 14.57
GASTROTOMY: FOREIGN-BODY REMOVAL, 75-POUND DOG

	25th Percentile	Median	Your Data	Average	75th Percentile	Number of Respondents
All Practices	**$275.00**	**$385.00**		**$408.54**	**$500.00**	**399**
Number of FTE Veterinarians						
1.0 or Less	$243.88	$360.28		$399.58	$500.24	114
1.1 to 2.0	$286.88	$381.32		$415.41	$500.00	126
2.1 to 3.0	$291.50	$360.00		$396.63	$518.46	77
3.1 or More	$286.91	$402.04		$410.03	$500.00	78
Member Status						
Accredited Practice Member	$300.41	$399.78		$410.69	$500.00	112
Nonaccredited Member	$250.00	$396.00		$428.79	$600.00	69
Nonmember	$260.00	$375.00		$399.36	$500.00	195
Metropolitan Status						
Urban	$337.50	$497.50		$493.26	$580.20	32
Suburban	$307.15	$400.00		$445.43	$563.13	109
Second City	$300.00	$400.00		$430.21	$575.00	81
Town	$270.00	$350.00		$382.91	$468.00	115
Rural	$225.00	$300.00		$308.94	$400.00	55
Median Area Household Income						
$35,000 or Less	$195.25	$344.40		$352.64	$441.00	51
$35,000 to $49,999	$247.50	$324.00		$349.76	$450.00	143
$50,000 to $69,999	$302.58	$440.00		$453.43	$556.68	137
$70,000 or More	$349.06	$450.00		$482.04	$581.55	61

Note: 4% of respondents reported that they do not offer this service.

TABLE 14.58
INCISOR EXTRACTION: RABBIT

	25th Percentile	Median	Your Data	Average	75th Percentile	Number of Respondents
All Practices	$25.00	$49.17		$66.79	$93.25	106
Number of FTE Veterinarians						
1.0 or Less	$21.93	$40.00		$57.09	$65.00	26
1.1 to 2.0	$30.50	$50.00		$76.31	$122.50	29
2.1 to 3.0	*	*		*	*	24
3.1 or More	$30.00	$40.00		$67.52	$100.00	27
Member Status						
Accredited Practice Member	$28.38	$46.25		$54.95	$75.00	30
Nonaccredited Member	*	*		*	*	20
Nonmember	$30.00	$60.00		$73.39	$125.00	51
Metropolitan Status						
Urban	*	*		*	*	10
Suburban	$28.75	$57.50		$78.04	$127.25	26
Second City	*	*		*	*	21
Town	$20.00	$42.50		$57.43	$78.75	32
Rural	*	*		*	*	14
Median Area Household Income						
$35,000 or Less	*	*		*	*	15
$35,000 to $49,999	$24.16	$41.00		$57.15	$70.25	40
$50,000 to $69,999	$30.00	$72.85		$78.75	$125.00	31
$70,000 or More	*	*		*	*	18

Note 1: An asterisk indicates that data was not reported due to an insufficient number of responses.
Note 2: 75% of respondents reported that they do not offer this service.

TABLE 14.59
INTESTINAL RESECTION/ANASTOMOSIS: CAT

	25th Percentile	Median	Your Data	Average	75th Percentile	Number of Respondents
All Practices	$281.75	$400.00		$419.69	$529.13	368
Number of FTE Veterinarians						
1.0 or Less	$242.89	$400.00		$419.98	$540.00	102
1.1 to 2.0	$275.00	$400.00		$407.93	$500.00	115
2.1 to 3.0	$300.75	$400.00		$410.91	$493.25	74
3.1 or More	$324.75	$407.88		$446.55	$573.73	74
Member Status						
Accredited Practice Member	$325.75	$412.63		$445.48	$565.27	104
Nonaccredited Member	$295.00	$417.00		$419.48	$500.00	63
Nonmember	$250.00	$369.75		$398.60	$500.00	178
Metropolitan Status						
Urban	$405.00	$500.00		$549.92	$671.25	27
Suburban	$327.25	$445.00		$456.69	$560.81	98
Second City	$330.00	$450.00		$464.13	$575.00	75
Town	$246.00	$350.00		$377.40	$488.84	110
Rural	$210.00	$300.00		$303.44	$375.00	51
Median Area Household Income						
$35,000 or Less	$202.25	$300.00		$337.52	$443.75	46
$35,000 to $49,999	$237.00	$350.00		$367.18	$463.58	131
$50,000 to $69,999	$339.24	$450.00		$457.76	$579.88	126
$70,000 or More	$350.00	$486.37		$518.22	$624.38	58

Note: 11% of respondents reported that they do not offer this service.

TABLE 14.60
INTESTINAL RESECTION/ANASTOMOSIS: 30-POUND DOG

	25th Percentile	Median	Your Data	Average	75th Percentile	Number of Respondents
All Practices	**$300.00**	**$425.00**		**$454.57**	**$560.00**	**359**
Number of FTE Veterinarians						
1.0 or Less	$260.00	$425.00		$451.83	$550.63	97
1.1 to 2.0	$300.00	$422.50		$442.34	$550.00	110
2.1 to 3.0	$317.13	$403.50		$454.77	$589.15	74
3.1 or More	$328.00	$492.88		$474.82	$600.00	75
Member Status						
Accredited Practice Member	$339.00	$450.00		$478.86	$597.86	103
Nonaccredited Member	$300.00	$450.00		$454.79	$550.94	60
Nonmember	$293.00	$400.00		$433.63	$544.68	174
Metropolitan Status						
Urban	$443.75	$525.75		$556.69	$677.01	26
Suburban	$350.00	$475.00		$498.82	$600.00	95
Second City	$340.00	$507.00		$502.93	$600.00	73
Town	$271.25	$375.00		$415.86	$526.25	108
Rural	$230.25	$300.00		$328.62	$425.00	50
Median Area Household Income						
$35,000 or Less	$221.25	$311.00		$370.03	$516.00	46
$35,000 to $49,999	$276.25	$382.50		$399.62	$500.00	128
$50,000 to $69,999	$353.94	$475.00		$489.65	$600.00	121
$70,000 or More	$392.50	$539.50		$570.03	$681.03	57

Note: 12% of respondents reported that they do not offer this service.

TABLE 14.61
INTESTINAL RESECTION/ANASTOMOSIS: 75-POUND DOG

	25th Percentile	Median	Your Data	Average	75th Percentile	Number of Respondents
All Practices	**$350.00**	**$475.00**		**$493.88**	**$613.20**	**355**
Number of FTE Veterinarians						
1.0 or Less	$300.00	$450.00		$495.39	$600.00	95
1.1 to 2.0	$350.00	$450.00		$469.85	$599.88	108
2.1 to 3.0	$350.00	$467.50		$507.91	$617.88	74
3.1 or More	$357.87	$500.00		$507.89	$645.00	75
Member Status						
Accredited Practice Member	$389.00	$500.00		$530.29	$660.00	103
Nonaccredited Member	$344.00	$491.00		$498.36	$600.00	59
Nonmember	$311.25	$450.00		$467.47	$589.25	172
Metropolitan Status						
Urban	$486.65	$600.00		$602.94	$687.29	26
Suburban	$400.00	$500.00		$547.42	$679.00	95
Second City	$372.65	$550.00		$534.62	$671.59	71
Town	$311.25	$409.75		$455.40	$569.98	106
Rural	$250.00	$350.00		$357.51	$452.50	50
Median Area Household Income						
$35,000 or Less	$250.00	$369.60		$422.57	$567.00	45
$35,000 to $49,999	$303.75	$400.00		$442.39	$593.75	128
$50,000 to $69,999	$400.00	$500.00		$534.15	$650.00	121
$70,000 or More	$450.00	$539.75		$580.75	$679.25	54

Note: 12% of respondents reported that they do not offer this service.

TABLE 14.62
LASER SURGERY UNIT IN-HOUSE

	Yes	No	Number of Respondents
All Practices	**17%**	**83%**	**455**
Number of FTE Veterinarians			
1.0 or Less	10%	90%	139
1.1 to 2.0	17%	83%	139
2.1 to 3.0	15%	85%	86
3.1 or More	29%	71%	86
Member Status			
Accredited Practice Member	19%	81%	127
Nonaccredited Member	19%	81%	77
Nonmember	14%	86%	220
Metropolitan Status			
Urban	18%	83%	40
Suburban	21%	79%	130
Second City	17%	83%	94
Town	13%	87%	124
Rural	12%	88%	58
Median Area Household Income			
$35,000 or Less	9%	91%	53
$35,000 to $49,999	14%	86%	163
$50,000 to $69,999	23%	77%	154
$70,000 or More	15%	85%	75

Note: Some row totals do not equal 100% due to rounding.

TABLE 14.63
ACCESS TO LASER SURGERY UNIT VIA REFERRAL CENTER

	Yes	No	Number of Respondents
All Practices	**40%**	**60%**	**403**
Number of FTE Veterinarians			
1.0 or Less	35%	65%	124
1.1 to 2.0	45%	55%	128
2.1 to 3.0	42%	58%	79
3.1 or More	38%	62%	66
Member Status			
Accredited Practice Member	48%	52%	112
Nonaccredited Member	45%	55%	65
Nonmember	33%	67%	201
Metropolitan Status			
Urban	55%	45%	38
Suburban	38%	62%	112
Second City	38%	62%	79
Town	39%	61%	111
Rural	33%	67%	54
Median Area Household Income			
$35,000 or Less	26%	74%	50
$35,000 to $49,999	36%	64%	145
$50,000 to $69,999	42%	58%	131
$70,000 or More	51%	49%	69

TABLE 14.64
LASER SURGERY FEES

	25th Percentile	Median	Your Data	Average	75th Percentile	Number of Respondents
Ear Surgery for Chronic Otitis	$100.00	$250.00		$288.60	$410.00	43
Feline Declaw: Two Paws	$92.00	$160.05		$166.30	$215.33	76
Oral Surgery	$69.00	$110.00		$122.38	$150.00	57
Small Tumor Removal	$65.00	$98.50		$109.41	$144.13	78

Note: Data was reported only for all practices due to an insufficient number of responses. Between 76% and 87% of respondents reported that they do not offer these services.

TABLE 14.65
LUXATING PATELLA REPAIR: MINIATURE POODLE

	25th Percentile	Median	Your Data	Average	75th Percentile	Number of Respondents
All Practices	**$300.00**	**$416.00**		**$444.97**	**$557.10**	**199**
Number of FTE Veterinarians						
1.0 or Less	$253.20	$350.00		$406.89	$543.55	41
1.1 to 2.0	$285.00	$414.50		$428.49	$514.00	63
2.1 to 3.0	$300.00	$432.00		$447.63	$583.90	43
3.1 or More	$320.00	$450.00		$489.49	$583.00	51
Member Status						
Accredited Practice Member	$392.10	$491.75		$498.43	$631.16	70
Nonaccredited Member	*	*		*	*	24
Nonmember	$275.00	$360.00		$418.73	$530.00	95
Metropolitan Status						
Urban	*	*		*	*	14
Suburban	$375.23	$493.56		$503.10	$601.50	52
Second City	$300.00	$450.00		$464.81	$600.00	45
Town	$277.50	$375.00		$420.12	$511.25	60
Rural	$222.50	$280.00		$317.17	$450.00	25
Median Area Household Income						
$35,000 or Less	*	*		*	*	22
$35,000 to $49,999	$257.30	$351.00		$376.32	$473.63	68
$50,000 to $69,999	$355.00	$470.00		$497.83	$618.88	81
$70,000 or More	$420.00	$500.00		$553.25	$691.53	25

Note 1: An asterisk indicates that data was not reported due to an insufficient number of responses.
Note 2: 50% of respondents reported that they do not offer this service.

TABLE 14.66
MASTECTOMY: UNILATERAL, CAT

	25th Percentile	Median	Your Data	Average	75th Percentile	Number of Respondents
All Practices	$160.00	$233.00		$258.54	$328.00	303
Number of FTE Veterinarians						
1.0 or Less	$127.50	$215.00		$260.06	$363.75	81
1.1 to 2.0	$150.00	$215.00		$240.31	$300.00	99
2.1 to 3.0	$175.00	$230.50		$266.77	$316.55	60
3.1 or More	$198.50	$250.00		$276.06	$343.56	62
Member Status						
Accredited Practice Member	$176.25	$246.58		$265.95	$334.75	100
Nonaccredited Member	$154.50	$237.50		$263.89	$354.38	42
Nonmember	$150.00	$207.40		$255.08	$330.40	145
Metropolitan Status						
Urban	*	*		*	*	24
Suburban	$197.81	$273.00		$288.28	$361.88	90
Second City	$150.00	$200.00		$230.39	$300.00	61
Town	$177.50	$233.00		$262.31	$350.00	85
Rural	$100.00	$149.75		$176.62	$240.25	38
Median Area Household Income						
$35,000 or Less	$106.31	$197.50		$210.35	$282.00	34
$35,000 to $49,999	$145.00	$200.00		$217.88	$268.00	103
$50,000 to $69,999	$200.00	$275.00		$285.67	$367.50	107
$70,000 or More	$195.00	$300.00		$307.17	$400.00	55

Note 1: An asterisk indicates that data was not reported due to an insufficient number of responses.
Note 2: 25% of respondents reported that they do not offer this service.

TABLE 14.67
MASTECTOMY: UNILATERAL, 30-POUND DOG

	25th Percentile	Median	Your Data	Average	75th Percentile	Number of Respondents
All Practices	$200.00	$275.00		$310.21	$418.55	303
Number of FTE Veterinarians						
1.0 or Less	$186.56	$284.45		$326.27	$450.00	80
1.1 to 2.0	$178.75	$241.38		$281.29	$371.25	98
2.1 to 3.0	$200.00	$286.40		$324.62	$413.50	61
3.1 or More	$221.00	$300.00		$319.43	$415.00	63
Member Status						
Accredited Practice Member	$199.56	$272.50		$299.18	$392.50	98
Nonaccredited Member	$200.00	$269.00		$309.38	$450.00	40
Nonmember	$200.00	$277.50		$311.94	$415.16	148
Metropolitan Status						
Urban	*	*		*	*	24
Suburban	$230.55	$300.00		$344.51	$450.00	88
Second City	$188.00	$250.49		$307.96	$420.00	63
Town	$200.00	$252.50		$304.93	$403.75	86
Rural	$120.00	$180.00		$194.27	$233.50	37
Median Area Household Income						
$35,000 or Less	$129.35	$223.00		$238.84	$316.00	34
$35,000 to $49,999	$170.93	$238.50		$265.33	$311.25	106
$50,000 to $69,999	$217.50	$300.00		$346.28	$450.00	105
$70,000 or More	$229.35	$339.00		$359.46	$450.00	54

Note 1: An asterisk indicates that data was not reported due to an insufficient number of responses.
Note 2: 25% of respondents reported that they do not offer this service.

TABLE 14.68
MASTECTOMY: UNILATERAL, 75-POUND DOG

	25th Percentile	Median	Your Data	Average	75th Percentile	Number of Respondents
All Practices	**$225.00**	**$315.00**		**$361.86**	**$450.00**	**301**
Number of FTE Veterinarians						
1.0 or Less	$244.63	$350.00		$393.36	$500.00	80
1.1 to 2.0	$188.75	$275.00		$323.80	$449.75	97
2.1 to 3.0	$250.00	$325.40		$356.91	$477.80	59
3.1 or More	$250.00	$336.80		$375.48	$450.00	63
Member Status						
Accredited Practice Member	$230.00	$322.00		$362.36	$450.00	99
Nonaccredited Member	$250.00	$336.00		$376.61	$540.00	40
Nonmember	$201.65	$300.00		$347.76	$450.00	145
Metropolitan Status						
Urban	*	*		*	*	24
Suburban	$255.00	$350.00		$388.88	$478.20	87
Second City	$200.00	$299.50		$359.03	$450.00	63
Town	$229.00	$327.50		$366.36	$449.75	85
Rural	$149.75	$200.00		$227.08	$267.50	37
Median Area Household Income						
$35,000 or Less	$158.75	$262.50		$283.36	$359.00	34
$35,000 to $49,999	$183.75	$275.00		$304.57	$392.50	105
$50,000 to $69,999	$250.00	$375.00		$398.94	$500.00	103
$70,000 or More	$280.00	$427.00		$431.59	$500.00	55

Note 1: An asterisk indicates that data was not reported due to an insufficient number of responses.
Note 2: 25% of respondents reported that they do not offer this service.

TABLE 14.69
SKIN TUMOR REMOVAL (4 CM)

	25th Percentile	Median	Your Data	Average	75th Percentile	Number of Respondents
All Practices	**$75.00**	**$110.00**		**$128.00**	**$160.00**	**343**
Number of FTE Veterinarians						
1.0 or Less	$72.38	$106.00		$133.07	$180.00	100
1.1 to 2.0	$75.00	$102.13		$121.43	$157.25	108
2.1 to 3.0	$75.00	$109.52		$121.05	$150.00	63
3.1 or More	$79.75	$123.00		$132.11	$171.83	70
Member Status						
Accredited Practice Member	$82.00	$111.30		$125.74	$157.50	103
Nonaccredited Member	$75.00	$120.00		$130.21	$194.63	52
Nonmember	$75.00	$110.00		$130.14	$159.13	168
Metropolitan Status						
Urban	$75.00	$124.75		$140.14	$196.25	30
Suburban	$90.00	$125.00		$144.51	$183.06	104
Second City	$75.96	$107.12		$126.45	$163.63	64
Town	$75.00	$110.00		$125.71	$156.25	97
Rural	$54.00	$75.00		$87.53	$125.00	43
Median Area Household Income						
$35,000 or Less	$60.81	$94.50		$112.98	$147.00	37
$35,000 to $49,999	$75.00	$92.75		$113.92	$136.50	120
$50,000 to $69,999	$75.00	$119.50		$133.69	$181.00	119
$70,000 or More	$100.00	$130.00		$151.57	$200.00	63

Note: 16% of respondents reported that they do not offer this service.

TABLE 14.70
SPLENECTOMY: 30-POUND DOG

	25th Percentile	Median	Your Data	Average	75th Percentile	Number of Respondents
All Practices	$300.00	$400.00		$427.68	$550.00	302
Number of FTE Veterinarians						
1.0 or Less	$300.00	$400.00		$437.51	$598.12	77
1.1 to 2.0	$250.00	$375.00		$400.74	$492.50	98
2.1 to 3.0	$300.00	$379.00		$419.06	$540.00	59
3.1 or More	$322.50	$442.60		$455.59	$570.83	66
Member Status						
Accredited Practice Member	$300.83	$410.00		$426.08	$545.00	101
Nonaccredited Member	$256.50	$420.00		$435.48	$526.25	40
Nonmember	$300.00	$400.00		$433.78	$564.90	147
Metropolitan Status						
Urban	$441.25	$571.65		$575.39	$725.75	26
Suburban	$357.50	$450.00		$479.61	$591.56	86
Second City	$275.00	$375.00		$403.37	$490.00	59
Town	$250.00	$350.00		$395.44	$497.35	92
Rural	$200.00	$305.08		$313.42	$435.51	35
Median Area Household Income						
$35,000 or Less	$200.00	$310.55		$316.25	$430.00	35
$35,000 to $49,999	$250.00	$360.00		$379.95	$473.13	104
$50,000 to $69,999	$336.94	$440.00		$467.59	$594.00	107
$70,000 or More	$332.00	$465.44		$505.62	$615.49	53

Note: 25% of respondents reported that they do not offer this service.

TABLE 14.71
SUTURE LACERATION (4 CM)

	25th Percentile	Median	Your Data	Average	75th Percentile	Number of Respondents
All Practices	$60.00	$90.00		$98.42	$125.00	339
Number of FTE Veterinarians						
1.0 or Less	$50.00	$75.00		$94.56	$125.00	97
1.1 to 2.0	$65.00	$91.15		$99.06	$120.00	107
2.1 to 3.0	$57.40	$86.50		$96.04	$125.00	62
3.1 or More	$68.15	$93.14		$98.77	$125.00	70
Member Status						
Accredited Practice Member	$65.95	$92.00		$100.38	$125.00	103
Nonaccredited Member	$65.00	$85.00		$94.16	$120.00	51
Nonmember	$55.00	$85.00		$96.87	$125.00	165
Metropolitan Status						
Urban	$68.83	$100.00		$111.37	$149.75	29
Suburban	$70.00	$100.00		$109.69	$135.00	103
Second City	$62.50	$86.00		$90.45	$108.14	65
Town	$60.00	$80.00		$96.77	$125.00	96
Rural	$45.00	$60.00		$77.46	$100.00	41
Median Area Household Income						
$35,000 or Less	$57.40	$92.00		$93.07	$125.00	37
$35,000 to $49,999	$58.30	$78.00		$87.93	$109.00	119
$50,000 to $69,999	$60.00	$90.00		$97.28	$125.00	117
$70,000 or More	$74.00	$115.00		$120.36	$150.00	62

Note: 16% of respondents reported that they do not offer this service.

TABLE 14.72
TIBIAL FRACTURE REPAIR: MIDSHAFT, AMAZON PARROT

	25th Percentile	Median	Your Data	Average	75th Percentile	Number of Respondents
All Practices	$150.00	$245.00		$262.30	$387.50	29

Note 1: Data was reported only for all practices due to an insufficient number of responses.
Note 2: 93% of respondents reported that they do not offer this service.

TABLE 14.73
TONSILLECTOMY

	25th Percentile	Median	Your Data	Average	75th Percentile	Number of Respondents
All Practices	$95.50	$150.00		$187.95	$292.20	76
Number of FTE Veterinarians						
1.0 or Less	*	*		*	*	17
1.1 to 2.0	*	*		*	*	23
2.1 to 3.0	*	*		*	*	16
3.1 or More	*	*		*	*	20
Member Status						
Accredited Practice Member	$88.10	$177.75		$193.04	$270.95	29
Nonaccredited Member	*	*		*	*	13
Nonmember	$92.50	$150.00		$187.87	$300.00	29
Metropolitan Status						
Urban	*	*		*	*	8
Suburban	*	*		*	*	24
Second City	*	*		*	*	12
Town	*	*		*	*	21
Rural	*	*		*	*	10
Median Area Household Income						
$35,000 or Less	*	*		*	*	8
$35,000 to $49,999	$95.38	$135.00		$156.78	$209.25	30
$50,000 to $69,999	$95.00	$204.75		$211.08	$330.00	27
$70,000 or More	*	*		*	*	10

Note 1: 81% of respondents reported that they do not offer this service..

TABLE 14.74
TPLO

	25th Percentile	Median	Your Data	Average	75th Percentile	Number of Respondents
All Practices	$901.65	$1,400.00		$1,457.69	$2,262.50	32

Note 1: Data was reported only for all practices due to an insufficient number of responses.
Note 2: 92% of respondents reported that they do not offer this service.

TABLE 14.75
URETHROSTOMY: CAT

	25th Percentile	Median	Your Data	Average	75th Percentile	Number of Respondents
All Practices	$258.00	$360.00		$386.53	$482.72	241
Number of FTE Veterinarians						
1.0 or Less	$250.00	$350.00		$370.76	$450.00	60
1.1 to 2.0	$250.00	$361.75		$367.52	$449.63	74
2.1 to 3.0	$294.38	$358.47		$393.03	$500.00	46
3.1 or More	$270.00	$400.00		$410.14	$510.29	58
Member Status						
Accredited Practice Member	$292.50	$396.00		$391.59	$495.00	83
Nonaccredited Member	$275.75	$402.13		$398.93	$490.00	34
Nonmember	$251.50	$350.00		$393.37	$489.63	113
Metropolitan Status						
Urban	*	*		*	*	21
Suburban	$300.00	$396.00		$407.03	$495.00	67
Second City	$279.39	$400.00		$405.53	$532.28	46
Town	$249.00	$350.00		$367.19	$442.38	74
Rural	$130.00	$276.00		$289.78	$395.50	29
Median Area Household Income						
$35,000 or Less	$138.93	$276.00		$290.84	$369.50	25
$35,000 to $49,999	$250.00	$350.00		$356.92	$443.19	83
$50,000 to $69,999	$286.88	$400.00		$404.16	$500.00	92
$70,000 or More	$340.00	$454.75		$466.46	$588.75	38

Note 1: An asterisk indicates that data was not reported due to an insufficient number of responses.
Note 2: 39% of respondents reported that they do not offer this service.

TABLE 14.76
WING FRACTURE REPAIR: BUDGIE, SIMPLE

	25th Percentile	Median	Your Data	Average	75th Percentile	Number of Respondents
All Practices	$40.00	$62.50		$82.00	$100.00	40

Note 1: Data was reported only for all practices due to an insufficient number of responses.
Note 2: 90% of respondents reported that they do not offer this service.

TABLE 14.77
UNCLASSIFIED SURGERY: PER-MINUTE FEE, SURGICAL SUITE

	25th Percentile	Median	Your Data	Average	75th Percentile	Number of Respondents
All Practices	$3.58	$5.00		$5.18	$6.00	161
Number of FTE Veterinarians						
1.0 or Less	$2.00	$4.06		$4.63	$5.88	52
1.1 to 2.0	$4.00	$5.00		$5.10	$6.00	44
2.1 to 3.0	$3.90	$5.00		$5.09	$6.00	31
3.1 or More	$4.53	$5.25		$6.23	$8.10	33
Member Status						
Accredited Practice Member	$4.00	$5.38		$5.69	$6.58	48
Nonaccredited Member	$4.33	$5.00		$5.89	$7.00	25
Nonmember	$2.25	$4.38		$4.64	$5.38	80
Metropolitan Status						
Urban	*	*		*	*	9
Suburban	$4.07	$5.00		$5.52	$6.23	48
Second City	$4.75	$5.50		$5.65	$6.16	31
Town	$3.22	$4.65		$4.94	$6.00	44
Rural	$3.00	$4.00		$4.01	$4.50	27
Median Area Household Income						
$35,000 or Less	*	*		*	*	20
$35,000 to $49,999	$3.50	$5.00		$5.04	$6.00	56
$50,000 to $69,999	$4.00	$5.00		$5.60	$6.50	59
$70,000 or More	*	*		*	*	23

Note 1: An asterisk indicates that data was not reported due to an insufficient number of responses.
Note 2: 39% of respondents reported that they do not offer this service.

TABLE 14.78
MINIMUM FEE FOR UNCLASSIFIED SURGERY PERFORMED IN SURGICAL SUITE

	Yes	No	Number of Respondents
All Practices	**25%**	**75%**	**415**
Number of FTE Veterinarians			
1.0 or Less	33%	67%	123
1.1 to 2.0	22%	78%	130
2.1 to 3.0	25%	75%	76
3.1 or More	14%	86%	80
Member Status			
Accredited Practice Member	22%	78%	117
Nonaccredited Member	31%	69%	75
Nonmember	25%	75%	193
Metropolitan Status			
Urban	24%	76%	38
Suburban	29%	71%	119
Second City	19%	81%	86
Town	31%	69%	113
Rural	16%	84%	50
Median Area Household Income			
$35,000 or Less	28%	72%	46
$35,000 to $49,999	21%	79%	149
$50,000 to $69,999	24%	76%	145
$70,000 or More	31%	69%	64

TABLE 14.79
MINIMUM FEE FOR UNCLASSIFIED SURGERY PERFORMED IN SURGICAL SUITE

	25th Percentile	Median	Your Data	Average	75th Percentile	Number of Respondents
All Practices	**$45.00**	**$70.00**		**$83.63**	**$106.00**	**95**
Number of FTE Veterinarians						
1.0 or Less	$41.25	$60.00		$70.55	$100.00	36
1.1 to 2.0	$48.00	$75.00		$95.42	$119.00	29
2.1 to 3.0	*	*		*	*	18
3.1 or More	*	*		*	*	12
Member Status						
Accredited Practice Member	*	*		*	*	20
Nonaccredited Member	*	*		*	*	23
Nonmember	$43.53	$65.00		$73.85	$100.00	49
Metropolitan Status						
Urban	*	*		*	*	8
Suburban	$41.80	$85.00		$97.12	$150.00	26
Second City	*	*		*	*	17
Town	$45.00	$60.00		$84.28	$100.00	35
Rural	*	*		*	*	9
Median Area Household Income						
$35,000 or Less	*	*		*	*	13
$35,000 to $49,999	$45.00	$67.50		$88.11	$100.00	30
$50,000 to $69,999	$38.30	$72.00		$76.97	$108.00	33
$70,000 or More	*	*		*	*	17

Note 1: An asterisk indicates that data was not reported due to an insufficient number of responses.
Note 2: 31% of respondents reported that they do not offer this service.

TABLE 14.80
UNCLASSIFIED SURGERY: PER-MINUTE FEE, NONSURGICAL SUITE

	25th Percentile	Median	Your Data	Average	75th Percentile	Number of Respondents
All Practices	**$3.00**	**$4.40**		**$4.77**	**$6.00**	**148**
Number of FTE Veterinarians						
1.0 or Less	$2.25	$4.00		$4.47	$5.75	45
1.1 to 2.0	$3.00	$4.09		$4.49	$6.00	44
2.1 to 3.0	$3.00	$4.50		$4.73	$5.81	29
3.1 or More	$4.00	$5.00		$5.65	$7.06	30
Member Status						
Accredited Practice Member	$4.00	$5.00		$5.40	$6.40	45
Nonaccredited Member	$4.00	$5.00		$5.46	$7.00	30
Nonmember	$2.10	$3.76		$3.87	$5.00	65
Metropolitan Status						
Urban	*	*		*	*	8
Suburban	$3.04	$4.98		$4.91	$6.00	44
Second City	$4.00	$5.00		$5.37	$6.12	32
Town	$3.00	$4.05		$4.49	$6.06	38
Rural	*	*		*	*	24
Median Area Household Income						
$35,000 or Less	*	*		*	*	15
$35,000 to $49,999	$3.29	$4.35		$4.65	$6.00	54
$50,000 to $69,999	$3.00	$5.00		$4.96	$6.63	54
$70,000 or More	*	*		*	*	22

Note 1: An asterisk indicates that data was not reported due to an insufficient number of responses.
Note 2: 34% of respondents reported that they do not offer this service.

TABLE 14.81
MINIMUM FEE FOR UNCLASSIFIED SURGERY PERFORMED IN NONSURGICAL SUITE

	Yes	No	Number of Respondents
All Practices	**21%**	**79%**	**400**
Number of FTE Veterinarians			
1.0 or Less	26%	74%	115
1.1 to 2.0	20%	80%	125
2.1 to 3.0	17%	83%	72
3.1 or More	13%	88%	80
Member Status			
Accredited Practice Member	16%	84%	114
Nonaccredited Member	27%	73%	66
Nonmember	20%	80%	189
Metropolitan Status			
Urban	11%	89%	35
Suburban	23%	77%	116
Second City	20%	80%	84
Town	26%	74%	108
Rural	6%	94%	48
Median Area Household Income			
$35,000 or Less	21%	79%	43
$35,000 to $49,999	17%	83%	146
$50,000 to $69,999	21%	79%	136
$70,000 or More	24%	76%	62

Note: Some row totals do not equal 100% due to rounding.

TABLE 14.82
MINIMUM FEE FOR UNCLASSIFIED SURGERY PERFORMED IN NONSURGICAL SUITE

	25th Percentile	Median	Your Data	Average	75th Percentile	Number of Respondents
All Practices	$40.40	$60.00		$66.68	$89.97	72
Number of FTE Veterinarians						
1.0 or Less	$35.00	$60.00		$67.71	$90.00	27
1.1 to 2.0	*	*		*	*	22
2.1 to 3.0	*	*		*	*	12
3.1 or More	*	*		*	*	10
Member Status						
Accredited Practice Member	*	*		*	*	17
Nonaccredited Member	*	*		*	*	15
Nonmember	$36.25	$60.00		$68.19	$90.00	36
Metropolitan Status						
Urban	*	*		*	*	4
Suburban	*	*		*	*	23
Second City	*	*		*	*	17
Town	$40.80	$45.00		$61.77	$76.00	25
Rural	*	*		*	*	3
Median Area Household Income						
$35,000 or Less	*	*		*	*	9
$35,000 to $49,999	*	*		*	*	21
$50,000 to $69,999	$38.75	$52.50		$61.66	$84.68	26
$70,000 or More	*	*		*	*	14

Note 1: An asterisk indicates that data was not reported due to an insufficient number of responses.

Note 2: 36% of respondents reported that they do not offer this service.

SURGICAL CASES

TABLE 14.83
SURGICAL CASE ONE: PRESURGICAL EXAMINATION

	25th Percentile	Median	Your Data	Average	75th Percentile	Number of Respondents
All Practices	$33.90	$39.00		$39.24	$45.00	278
Number of FTE Veterinarians						
1.0 or Less	$33.70	$38.00		$37.40	$43.00	80
1.1 to 2.0	$31.45	$38.68		$39.05	$44.00	94
2.1 to 3.0	$34.38	$40.00		$37.27	$44.86	50
3.1 or More	$35.00	$41.50		$44.13	$48.50	51
Member Status						
Accredited Practice Member	$35.00	$42.00		$40.47	$45.30	70
Nonaccredited Member	$34.50	$39.00		$39.38	$44.50	49
Nonmember	$31.20	$37.00		$38.09	$42.50	135
Metropolitan Status						
Urban	$38.00	$42.00		$42.77	$46.38	32
Suburban	$34.50	$39.99		$39.64	$45.25	70
Second City	$34.85	$41.05		$41.62	$45.50	65
Town	$31.35	$38.50		$37.43	$43.25	73
Rural	$25.00	$35.00		$33.85	$39.43	36
Median Area Household Income						
$35,000 or Less	$29.18	$35.38		$35.11	$39.99	36
$35,000 to $49,999	$30.58	$38.00		$39.50	$45.00	103
$50,000 to $69,999	$34.00	$39.95		$38.98	$44.00	95
$70,000 or More	$38.00	$43.80		$42.78	$48.53	36

Note 1: The patient is a 25-pound female dog presented for removal of a single, large calculus from the bladder. Total surgery time is 45 minutes. Surgery is conducted in a single-use surgery room using full aseptic procedures.
Note 2: 31% of the respondents reported that the fee for this service is included in the total surgery fee and is not separately itemized.

TABLE 14.84
SURGICAL CASE ONE: CBC

	25th Percentile	Median	Your Data	Average	75th Percentile	Number of Respondents
All Practices	$32.00	$40.00		$44.13	$53.45	374
Number of FTE Veterinarians						
1.0 or Less	$28.00	$36.00		$40.32	$49.00	111
1.1 to 2.0	$35.00	$42.63		$46.97	$55.25	109
2.1 to 3.0	$33.18	$40.75		$44.93	$53.90	76
3.1 or More	$32.69	$41.25		$44.39	$53.48	74
Member Status						
Accredited Practice Member	$35.00	$44.00		$47.00	$55.00	105
Nonaccredited Member	$31.38	$41.55		$45.96	$54.63	64
Nonmember	$30.00	$36.88		$41.01	$48.75	180
Metropolitan Status						
Urban	$38.00	$49.00		$50.33	$65.00	31
Suburban	$35.00	$43.98		$45.72	$54.00	103
Second City	$30.15	$40.00		$43.05	$49.81	78
Town	$32.81	$39.75		$43.73	$50.38	104
Rural	$25.50	$33.92		$36.84	$46.00	51
Median Area Household Income						
$35,000 or Less	$30.00	$37.28		$42.43	$48.13	49
$35,000 to $49,999	$29.66	$36.93		$40.33	$47.44	134
$50,000 to $69,999	$34.78	$42.25		$46.40	$55.00	124
$70,000 or More	$37.33	$45.75		$49.71	$61.78	60

Note 1: See the case description in the notes for Table 14.83.
Note 2: 9% of the respondents reported that the fee for this service is included in the total surgery fee and is not separately itemized.

TABLE 14.85
SURGICAL CASE ONE: CHEMISTRY PANEL WITH SIX CHEMISTRIES

	25th Percentile	Median	Your Data	Average	75th Percentile	Number of Respondents
All Practices	$40.00	$50.00		$53.57	$65.00	325
Number of FTE Veterinarians						
1.0 or Less	$39.00	$46.80		$51.12	$63.75	97
1.1 to 2.0	$38.75	$53.00		$55.57	$70.02	98
2.1 to 3.0	$42.50	$50.00		$55.83	$68.68	65
3.1 or More	$40.00	$50.38		$51.70	$60.00	61
Member Status						
Accredited Practice Member	$43.00	$52.50		$55.52	$66.98	85
Nonaccredited Member	$40.00	$48.00		$53.45	$64.50	49
Nonmember	$38.00	$50.00		$52.18	$64.97	172
Metropolitan Status						
Urban	$40.00	$60.00		$62.07	$74.91	27
Suburban	$41.75	$50.13		$53.45	$62.38	90
Second City	$44.75	$57.95		$57.05	$70.70	69
Town	$37.00	$49.50		$52.41	$65.00	95
Rural	$32.63	$42.38		$43.88	$51.05	40
Median Area Household Income						
$35,000 or Less	$35.25	$45.35		$47.61	$53.31	40
$35,000 to $49,999	$38.63	$50.00		$53.03	$66.46	128
$50,000 to $69,999	$40.00	$52.00		$54.92	$66.25	107
$70,000 or More	$46.58	$55.00		$57.79	$71.00	45

Note 1: See the case description in the notes for Table 14.83.
Note 2: 22% of the respondents reported that the fee for this service is included in the total surgery fee and is not separately itemized.

TABLE 14.86
SURGICAL CASE ONE: SEDATION

	25th Percentile	Median	Your Data	Average	75th Percentile	Number of Respondents
All Practices	$24.56	$35.00		$43.41	$52.13	230
Number of FTE Veterinarians						
1.0 or Less	$20.00	$36.00		$44.17	$50.00	71
1.1 to 2.0	$25.25	$34.56		$44.20	$56.00	65
2.1 to 3.0	$23.05	$38.00		$45.37	$55.00	41
3.1 or More	$23.72	$30.35		$39.61	$45.38	50
Member Status						
Accredited Practice Member	$26.70	$38.00		$44.55	$56.00	65
Nonaccredited Member	$23.33	$29.00		$33.85	$42.03	39
Nonmember	$22.30	$35.00		$43.41	$49.88	109
Metropolitan Status						
Urban	$26.94	$44.25		$55.19	$74.03	28
Suburban	$27.30	$38.50		$47.99	$56.69	66
Second City	$24.87	$36.30		$44.06	$56.00	45
Town	$22.00	$30.35		$36.00	$41.00	66
Rural	*	*		*	*	21
Median Area Household Income						
$35,000 or Less	$20.36	$29.40		$39.63	$48.25	25
$35,000 to $49,999	$21.25	$32.95		$43.35	$54.94	84
$50,000 to $69,999	$25.60	$37.50		$42.98	$55.00	79
$70,000 or More	$26.20	$36.15		$46.83	$50.88	37

Note 1: See the case description in the notes for Table 14.83.
Note 2: An asterisk indicates that data was not reported due to an insufficient number of responses.
Note 3: 41% of the respondents reported that the fee for this service is included in the total surgery fee and is not separately itemized.

SURGICAL SERVICES / 451

TABLE 14.87
SURGICAL CASE ONE: FLUIDS

	25th Percentile	Median	Your Data	Average	75th Percentile	Number of Respondents
All Practices	$25.90	$43.30		$47.50	$65.50	393
Number of FTE Veterinarians						
1.0 or Less	$25.00	$35.50		$40.60	$55.00	111
1.1 to 2.0	$25.00	$41.00		$46.87	$62.75	118
2.1 to 3.0	$23.88	$43.15		$49.77	$72.00	78
3.1 or More	$36.31	$50.50		$55.25	$71.25	82
Member Status						
Accredited Practice Member	$34.65	$45.60		$52.73	$72.25	106
Nonaccredited Member	$32.79	$51.84		$51.73	$68.50	74
Nonmember	$23.00	$36.00		$41.53	$58.00	183
Metropolitan Status						
Urban	$30.13	$51.60		$53.32	$70.98	36
Suburban	$35.00	$48.00		$52.47	$69.89	105
Second City	$29.75	$48.34		$54.87	$76.38	82
Town	$23.00	$39.50		$41.94	$60.00	109
Rural	$19.50	$31.00		$34.84	$44.68	54
Median Area Household Income						
$35,000 or Less	$18.81	$31.95		$39.85	$65.00	46
$35,000 to $49,999	$25.00	$39.50		$46.49	$65.50	147
$50,000 to $69,999	$29.95	$42.63		$46.55	$60.00	130
$70,000 or More	$37.50	$54.00		$54.80	$71.03	61

Note 1: See the case description in the notes for Table 14.83.
Note 2: 7% of the respondents reported that the fee for this service is included in the total surgery fee and is not separately itemized.

TABLE 14.88
SURGICAL CASE ONE: MONITORING

	25th Percentile	Median	Your Data	Average	75th Percentile	Number of Respondents
All Practices	$15.00	$22.00		$25.25	$30.25	158
Number of FTE Veterinarians						
1.0 or Less	$14.96	$20.00		$21.69	$28.95	36
1.1 to 2.0	$13.46	$23.48		$27.72	$33.39	46
2.1 to 3.0	$13.13	$21.31		$23.87	$29.38	36
3.1 or More	$19.50	$23.80		$25.89	$32.50	37
Member Status						
Accredited Practice Member	$18.50	$24.19		$27.53	$32.75	59
Nonaccredited Member	$15.72	$21.88		$24.16	$28.20	34
Nonmember	$11.25	$19.06		$22.51	$28.42	57
Metropolitan Status						
Urban	*	*		*	*	19
Suburban	$18.99	$25.00		$28.47	$35.98	58
Second City	$13.50	$21.54		$23.69	$30.00	29
Town	$10.00	$16.00		$19.19	$24.81	38
Rural	*	*		*	*	14
Median Area Household Income						
$35,000 or Less	*	*		*	*	17
$35,000 to $49,999	$14.50	$21.54		$24.14	$31.00	55
$50,000 to $69,999	$15.25	$23.38		$26.40	$32.00	56
$70,000 or More	$18.13	$22.50		$26.76	$33.12	26

Note 1: See the case description in the notes for Table 14.83.
Note 2: An asterisk indicates that data was not reported due to an insufficient number of responses.
Note 3: 56% of the respondents reported that the fee for this service is included in the total surgery fee and is not separately itemized.

TABLE 14.89
SURGICAL CASE ONE: TOTAL SURGERY SETUP FEE

	25th Percentile	Median	Your Data	Average	75th Percentile	Number of Respondents
All Practices	$25.00	$44.02		$62.30	$68.34	62

Note 1: See the case description in the notes for Table 14.83.
Note 2: Data was reported only for all practices due to an insufficient number of responses.
Note 3: 72% of the respondents reported that the fee for this service is included in the total surgery fee and is not separately itemized.

TABLE 14.90
SURGICAL CASE ONE: SURGICAL SUITE USE

	25th Percentile	Median	Your Data	Average	75th Percentile	Number of Respondents
All Practices	$24.75	$39.69		$42.94	$50.96	50

Note 1: See the case description in the notes for Table 14.83.
Note 2: Data was reported only for all practices due to an insufficient number of responses.
Note 3: 74% of the respondents reported that the fee for this service is included in the total surgery fee and is not separately itemized.

TABLE 14.91
SURGICAL CASE ONE: SURGICAL PACK

	25th Percentile	Median	Your Data	Average	75th Percentile	Number of Respondents
All Practices	$21.96	$30.30		$35.67	$48.41	100
Number of FTE Veterinarians						
1.0 or Less	*	*		*	*	22
1.1 to 2.0	$17.00	$28.22		$31.35	$42.56	30
2.1 to 3.0	*	*		*	*	16
3.1 or More	$24.88	$35.00		$40.67	$58.43	29
Member Status						
Accredited Practice Member	$22.38	$31.75		$37.72	$51.00	30
Nonaccredited Member	*	*		*	*	21
Nonmember	$25.00	$31.50		$35.32	$46.00	39
Metropolitan Status						
Urban	*	*		*	*	7
Suburban	$22.75	$33.00		$38.52	$50.00	35
Second City	*	*		*	*	21
Town	*	*		*	*	23
Rural	*	*		*	*	11
Median Area Household Income						
$35,000 or Less	*	*		*	*	10
$35,000 to $49,999	$21.39	$34.70		$35.38	$45.00	36
$50,000 to $69,999	$20.88	$29.90		$34.47	$49.00	32
$70,000 or More	*	*		*	*	17

Note 1: See the case description in the notes for Table 14.83.
Note 2: An asterisk indicates that data was not reported due to an insufficient number of responses.
Note 3: 63% of the respondents reported that the fee for this service is included in the total surgery fee and is not separately itemized.

SURGICAL SERVICES / 453

TABLE 14.92
SURGICAL CASE ONE: DISPOSABLES

	25th Percentile	Median	Your Data	Average	75th Percentile	Number of Respondents
All Practices	$10.00	$17.29		$21.48	$30.00	46

Note 1: See the case description in the notes for Table 14.83.
Note 2: Data was reported only for all practices due to an insufficient number of responses.
Note 3: 73% of the respondents reported that the fee for this service is included in the total surgery fee and is not separately itemized.

TABLE 14.93
SURGICAL CASE ONE: MEDICAL WASTE DISPOSAL

	25th Percentile	Median	Your Data	Average	75th Percentile	Number of Respondents
All Practices	$2.50	$3.50		$4.16	$5.00	110
Number of FTE Veterinarians						
1.0 or Less	$2.50	$4.23		$4.29	$5.25	33
1.1 to 2.0	$2.58	$3.50		$4.25	$4.80	41
2.1 to 3.0	*	*		*	*	16
3.1 or More	*	*		*	*	19
Member Status						
Accredited Practice Member	$2.00	$3.68		$3.92	$5.00	34
Nonaccredited Member	*	*		*	*	15
Nonmember	$2.77	$3.50		$3.98	$4.95	52
Metropolitan Status						
Urban	*	*		*	*	14
Suburban	$2.31	$3.93		$4.28	$5.00	40
Second City	*	*		*	*	22
Town	$2.25	$3.50		$4.29	$5.25	25
Rural	*	*		*	*	6
Median Area Household Income						
$35,000 or Less	*	*		*	*	8
$35,000 to $49,999	$2.65	$3.96		$4.12	$4.71	38
$50,000 to $69,999	$2.70	$3.60		$4.63	$5.50	39
$70,000 or More	*	*		*	*	22

Note 1: See the case description in the notes for Table 14.83.
Note 2: An asterisk indicates that data was not reported due to an insufficient number of responses.
Note 3: 56% of the respondents reported that the fee for this service is included in the total surgery fee and is not separately itemized.

TABLE 14.94
SURGICAL CASE ONE: ONE HOUR ASSISTANT TIME

	25th Percentile	Median	Your Data	Average	75th Percentile	Number of Respondents
All Practices	**$28.68**	**$43.00**		**$54.98**	**$72.00**	**83**
Number of FTE Veterinarians						
1.0 or Less	*	*		*	*	19
1.1 to 2.0	$27.40	$43.00		$57.44	$69.00	33
2.1 to 3.0	*	*		*	*	13
3.1 or More	*	*		*	*	15
Member Status						
Accredited Practice Member	*	*		*	*	24
Nonaccredited Member	*	*		*	*	15
Nonmember	$24.50	$37.00		$50.16	$56.75	37
Metropolitan Status						
Urban	*	*		*	*	10
Suburban	*	*		*	*	20
Second City	*	*		*	*	19
Town	*	*		*	*	19
Rural	*	*		*	*	13
Median Area Household Income						
$35,000 or Less	*	*		*	*	10
$35,000 to $49,999	$26.00	$34.50		$55.61	$70.25	30
$50,000 to $69,999	$27.59	$50.00		$57.96	$87.50	25
$70,000 or More	*	*		*	*	14

Note 1: See the case description in the notes for Table 14.83.
Note 2: An asterisk indicates that data was not reported due to an insufficient number of responses.
Note 3: 67% of the respondents reported that the fee for this service is included in the total surgery fee and is not separately itemized.

TABLE 14.95
SURGICAL CASE ONE: SURGICAL REMOVAL OF BLADDER CALCULUS

	25th Percentile	Median	Your Data	Average	75th Percentile	Number of Respondents
All Practices	**$219.30**	**$300.00**		**$297.79**	**$371.25**	**403**
Number of FTE Veterinarians						
1.0 or Less	$200.00	$263.49		$276.48	$352.50	114
1.1 to 2.0	$225.00	$301.88		$314.81	$400.00	126
2.1 to 3.0	$210.00	$270.00		$284.77	$360.00	79
3.1 or More	$250.00	$300.00		$309.12	$392.63	80
Member Status						
Accredited Practice Member	$225.00	$309.50		$313.02	$390.79	113
Nonaccredited Member	$220.50	$300.00		$312.99	$400.00	69
Nonmember	$200.00	$270.00		$283.61	$359.06	194
Metropolitan Status						
Urban	$299.38	$375.00		$364.20	$422.00	35
Suburban	$250.00	$320.75		$313.85	$386.65	104
Second City	$230.38	$300.00		$322.94	$400.00	86
Town	$200.00	$250.00		$277.75	$344.50	115
Rural	$135.00	$236.00		$224.04	$300.00	55
Median Area Household Income						
$35,000 or Less	$183.75	$227.75		$260.21	$342.50	50
$35,000 to $49,999	$200.00	$274.25		$282.36	$359.81	152
$50,000 to $69,999	$231.15	$300.27		$309.54	$371.25	131
$70,000 or More	$250.00	$350.00		$343.04	$400.00	62

Note: See the case description in the notes for Table 14.83.

TABLE 14.96
SURGICAL CASE ONE: ANESTHESIA

	25th Percentile	Median	Your Data	Average	75th Percentile	Number of Respondents
All Practices	**$67.50**	**$92.50**		**$98.41**	**$125.00**	**393**
Number of FTE Veterinarians						
1.0 or Less	$49.50	$85.50		$89.36	$115.00	110
1.1 to 2.0	$63.00	$95.00		$95.58	$119.50	123
2.1 to 3.0	$75.19	$101.25		$109.52	$139.05	75
3.1 or More	$73.78	$95.00		$104.29	$128.00	81
Member Status						
Accredited Practice Member	$76.50	$100.00		$107.47	$133.75	105
Nonaccredited Member	$67.37	$100.63		$101.23	$128.23	72
Nonmember	$60.00	$90.00		$92.28	$119.50	187
Metropolitan Status						
Urban	$75.00	$117.60		$116.99	$149.50	39
Suburban	$79.25	$102.50		$109.58	$139.53	101
Second City	$63.00	$99.00		$98.51	$128.30	83
Town	$66.25	$82.00		$88.27	$102.18	109
Rural	$49.50	$73.45		$78.87	$99.88	54
Median Area Household Income						
$35,000 or Less	$55.00	$71.90		$82.07	$109.50	47
$35,000 to $49,999	$60.00	$83.13		$87.76	$114.33	144
$50,000 to $69,999	$75.16	$95.00		$103.36	$125.50	132
$70,000 or More	$91.25	$117.00		$122.88	$150.00	61

Note 1: Assume 45 minutes of anesthesia.
Note 2: See the case description in the notes for Table 14.83.
Note 3: 8% of the respondents reported that the fee for this service is included in the total surgery fee and is not separately itemized.

TABLE 14.97
SURGICAL CASE ONE: SUTURE MATERIAL

	25th Percentile	Median	Your Data	Average	75th Percentile	Number of Respondents
All Practices	**$12.00**	**$17.50**		**$20.86**	**$25.00**	**81**
Number of FTE Veterinarians						
1.0 or Less	*	*		*	*	17
1.1 to 2.0	*	*		*	*	24
2.1 to 3.0	*	*		*	*	12
3.1 or More	$9.75	$15.15		$21.63	$27.75	26
Member Status						
Accredited Practice Member	$10.00	$18.00		$21.96	$30.00	27
Nonaccredited Member	*	*		*	*	18
Nonmember	$11.25	$15.00		$17.91	$20.50	33
Metropolitan Status						
Urban	*	*		*	*	10
Suburban	*	*		*	*	24
Second City	*	*		*	*	17
Town	*	*		*	*	17
Rural	*	*		*	*	11
Median Area Household Income						
$35,000 or Less	*	*		*	*	11
$35,000 to $49,999	$12.44	$15.89		$17.81	$24.00	26
$50,000 to $69,999	$12.15	$18.00		$23.60	$32.25	28
$70,000 or More	*	*		*	*	14

Note 1: See the case description in the notes for Table 14.83.
Note 2: An asterisk indicates that data was not reported due to an insufficient number of responses.
Note 3: 73% of the respondents reported that the fee for this service is included in the total surgery fee and is not separately itemized.

TABLE 14.98
SURGICAL CASE ONE: STAPLES

	25th Percentile	Median	Your Data	Average	75th Percentile	Number of Respondents
All Practices	**$11.58**	**$20.41**		**$21.56**	**$27.02**	**70**

Note 1: See the case description in the notes for Table 14.83.
Note 2: Data was reported only for all practices due to an insufficient number of responses.
Note 3: 67% of the respondents reported that the fee for this service is included in the total surgery fee and is not separately itemized.

TABLE 14.99
SURGICAL CASE ONE: POST-OPERATIVE PAIN MEDICATION

	25th Percentile	Median	Your Data	Average	75th Percentile	Number of Respondents
All Practices	$16.21	$22.18		$25.27	$30.05	376
Number of FTE Veterinarians						
1.0 or Less	$16.00	$20.00		$23.08	$29.00	108
1.1 to 2.0	$15.97	$22.00		$26.45	$32.50	112
2.1 to 3.0	$17.34	$23.25		$25.60	$30.94	74
3.1 or More	$18.90	$25.00		$25.98	$31.81	78
Member Status						
Accredited Practice Member	$18.00	$25.00		$27.67	$36.90	107
Nonaccredited Member	$17.50	$24.00		$26.42	$32.00	71
Nonmember	$15.00	$20.08		$23.83	$27.60	174
Metropolitan Status						
Urban	$21.00	$27.00		$31.02	$35.00	35
Suburban	$18.22	$26.97		$28.96	$38.00	101
Second City	$18.00	$26.00		$27.46	$35.00	83
Town	$15.00	$20.00		$20.21	$23.67	96
Rural	$14.00	$17.35		$19.57	$24.15	53
Median Area Household Income						
$35,000 or Less	$15.71	$20.00		$22.17	$25.00	46
$35,000 to $49,999	$15.00	$21.00		$23.46	$28.27	137
$50,000 to $69,999	$18.00	$24.00		$27.38	$34.80	127
$70,000 or More	$18.33	$25.00		$27.63	$35.25	58

Note 1: See the case description in the notes for Table 14.83.
Note 2: 10% of the respondents reported that the fee for this service is included in the total surgery fee and is not separately itemized.

TABLE 14.100
SURGICAL CASE ONE: HOSPITALIZATION

	25th Percentile	Median	Your Data	Average	75th Percentile	Number of Respondents
All Practices	$21.28	$30.00		$34.13	$44.93	294
Number of FTE Veterinarians						
1.0 or Less	$19.50	$26.50		$31.01	$40.30	73
1.1 to 2.0	$20.00	$28.45		$32.56	$43.25	89
2.1 to 3.0	$23.00	$30.00		$31.80	$37.00	63
3.1 or More	$24.94	$42.25		$41.45	$53.80	66
Member Status						
Accredited Practice Member	$24.75	$33.50		$36.04	$45.00	83
Nonaccredited Member	$24.63	$30.00		$34.97	$45.50	54
Nonmember	$19.56	$26.50		$32.49	$43.87	137
Metropolitan Status						
Urban	$23.00	$43.63		$43.38	$56.20	31
Suburban	$24.76	$35.00		$36.44	$45.00	83
Second City	$20.00	$29.50		$31.51	$43.00	67
Town	$18.50	$28.50		$32.27	$43.00	75
Rural	$16.65	$24.00		$29.08	$35.38	34
Median Area Household Income						
$35,000 or Less	$17.50	$25.00		$30.21	$39.32	33
$35,000 to $49,999	$20.00	$28.34		$32.62	$44.83	98
$50,000 to $69,999	$21.50	$32.00		$34.76	$44.75	104
$70,000 or More	$25.50	$33.20		$38.82	$47.00	53

Note 1: See the case description in the notes for Table 14.83.
Note 2: 28% of the respondents reported that the fee for this service is included in the total surgery fee and is not separately itemized.

TABLE 14.101
SURGICAL CASE ONE: TOTAL

	25th Percentile	Median	Your Data	Average	75th Percentile	Number of Respondents
All Practices	$478.13	$618.50		$626.64	$758.36	438
Number of FTE Veterinarians						
1.0 or Less	$380.42	$561.38		$550.74	$716.00	128
1.1 to 2.0	$501.75	$657.00		$655.18	$811.39	133
2.1 to 3.0	$479.88	$613.00		$624.83	$732.28	86
3.1 or More	$551.63	$675.18		$687.26	$823.73	86
Member Status						
Accredited Practice Member	$520.00	$672.59		$667.23	$814.28	123
Nonaccredited Member	$515.75	$653.00		$659.64	$809.50	75
Nonmember	$456.97	$577.50		$589.34	$726.50	209
Metropolitan Status						
Urban	$552.65	$735.33		$724.41	$884.30	42
Suburban	$536.07	$672.59		$673.97	$850.75	117
Second City	$558.00	$673.65		$675.04	$830.00	91
Town	$436.59	$579.60		$570.17	$683.50	123
Rural	$364.25	$488.00		$492.49	$594.15	57
Median Area Household Income						
$35,000 or Less	$418.41	$523.78		$553.90	$669.38	52
$35,000 to $49,999	$435.35	$597.00		$593.87	$737.29	162
$50,000 to $69,999	$505.23	$642.50		$653.91	$807.38	144
$70,000 or More	$552.87	$683.50		$691.49	$864.35	71

Note 1: See the case description in the notes for Table 14.83.
Note 2: The total fee for surgical case one is the sum of the fees for the presurgical exam, CBC, chemistry panel with six chemistries, sedation, fluids, monitoring, total surgery setup fee, surgical suite use, surgical pack, disposables, medical waste disposal, one hour of assistant time, surgical removal of bladder calculus, anesthesia (45 minutes), suture material, staples, post-op pain medication, and hospitalization for one day. Some respondents do not charge for some of the individual services provided in this case. In addition, some respondents reported that they do not offer some of the individual services. Therefore, the average total fee for the case may be significantly lower than the sum of the average fees for the individual services.

TABLE 14.102
SURGICAL CASE TWO: PRESURGICAL EXAMINATION

	25th Percentile	Median	Your Data	Average	75th Percentile	Number of Respondents
All Practices	$33.90	$39.00		$38.81	$45.00	226
Number of FTE Veterinarians						
1.0 or Less	$32.00	$38.00		$36.93	$42.75	65
1.1 to 2.0	$30.94	$38.50		$37.68	$44.00	69
2.1 to 3.0	$35.00	$40.50		$40.23	$45.00	47
3.1 or More	$34.00	$39.84		$41.60	$47.44	42
Member Status						
Accredited Practice Member	$35.25	$42.00		$41.02	$45.00	61
Nonaccredited Member	$34.50	$39.50		$40.03	$45.00	35
Nonmember	$31.23	$36.25		$36.64	$42.03	108
Metropolitan Status						
Urban	*	*		*	*	24
Suburban	$34.99	$39.74		$40.16	$45.25	58
Second City	$36.75	$42.00		$40.46	$45.25	46
Town	$30.00	$36.00		$37.18	$43.75	69
Rural	$25.00	$35.00		$32.60	$38.00	27
Median Area Household Income						
$35,000 or Less	$29.18	$35.00		$34.56	$39.99	28
$35,000 to $49,999	$31.60	$38.00		$38.39	$44.90	85
$50,000 to $69,999	$33.90	$39.99		$38.97	$44.12	82
$70,000 or More	$39.38	$45.00		$44.65	$49.03	26

Note 1: The patient is a 45-pound male dog. After a presurgical examination, lab tests, and sedation, you repair a midshaft tibial fracture with one intramedullary pin. Total surgery time is 60 minutes.
Note 2: An asterisk indicates that data was not reported due to an insufficient number of responses.
Note 3: 24% of the respondents reported that the fee for this service is included in the total surgery fee and is not separately itemized.

TABLE 14.103
SURGICAL CASE TWO: CBC

	25th Percentile	Median	Your Data	Average	75th Percentile	Number of Respondents
All Practices	**$32.50**	**$40.00**		**$43.35**	**$50.50**	**280**
Number of FTE Veterinarians						
1.0 or Less	$30.50	$36.90		$40.44	$45.00	79
1.1 to 2.0	$35.00	$43.83		$46.91	$57.38	80
2.1 to 3.0	$32.69	$39.05		$43.04	$48.21	54
3.1 or More	$31.65	$40.48		$42.36	$52.08	64
Member Status						
Accredited Practice Member	$35.50	$43.98		$45.99	$54.50	85
Nonaccredited Member	$29.87	$38.50		$42.51	$47.50	37
Nonmember	$30.05	$37.28		$41.49	$49.25	137
Metropolitan Status						
Urban	*	*		*	*	19
Suburban	$34.50	$43.98		$44.61	$53.95	73
Second City	$30.00	$40.00		$42.41	$47.41	57
Town	$31.63	$38.50		$42.90	$49.88	88
Rural	$26.00	$35.00		$38.34	$46.63	37
Median Area Household Income						
$35,000 or Less	$30.00	$36.73		$39.93	$44.63	37
$35,000 to $49,999	$30.00	$38.00		$40.27	$45.25	102
$50,000 to $69,999	$34.65	$41.69		$45.25	$55.00	98
$70,000 or More	$38.50	$48.00		$50.13	$62.70	38

Note 1: See the case description in the notes for Table 14.102.
Note 2: An asterisk indicates that data was not reported due to an insufficient number of responses.
Note 3: 9% of the respondents reported that the fee for this service is included in the total surgery fee and is not separately itemized.

TABLE 14.104
SURGICAL CASE TWO: CHEMISTRY PANEL WITH SIX CHEMISTRIES

	25th Percentile	Median	Your Data	Average	75th Percentile	Number of Respondents
All Practices	**$40.00**	**$50.23**		**$54.29**	**$66.96**	**246**
Number of FTE Veterinarians						
1.0 or Less	$39.90	$46.90		$52.07	$65.25	70
1.1 to 2.0	$36.38	$53.00		$55.78	$70.02	66
2.1 to 3.0	$41.25	$52.25		$56.41	$71.73	52
3.1 or More	$40.00	$50.38		$53.24	$61.40	55
Member Status						
Accredited Practice Member	$44.75	$53.00		$56.95	$68.03	69
Nonaccredited Member	$39.50	$46.25		$50.56	$55.50	34
Nonmember	$39.13	$50.10		$53.24	$65.25	126
Metropolitan Status						
Urban	*	*		*	*	18
Suburban	$42.00	$50.00		$52.99	$60.93	67
Second City	$40.00	$51.13		$55.16	$70.28	50
Town	$38.00	$50.00		$55.26	$70.25	77
Rural	$32.25	$42.00		$44.47	$55.00	29
Median Area Household Income						
$35,000 or Less	$33.20	$45.00		$45.97	$50.88	29
$35,000 to $49,999	$38.25	$50.00		$52.79	$66.46	100
$50,000 to $69,999	$40.44	$55.75		$58.17	$73.23	86
$70,000 or More	$46.97	$55.60		$57.48	$69.05	28

Note 1: See the case description in the notes for Table 14.102.
Note 2: An asterisk indicates that data was not reported due to an insufficient number of responses.
Note 3: 19% of the respondents reported that the fee for this service is included in the total surgery fee and is not separately itemized.

TABLE 14.105
SURGICAL CASE TWO: SEDATION

	25th Percentile	Median	Your Data	Average	75th Percentile	Number of Respondents
All Practices	**$25.50**	**$40.00**		**$57.46**	**$75.00**	**193**
Number of FTE Veterinarians						
1.0 or Less	$29.32	$42.35		$53.89	$66.00	51
1.1 to 2.0	$26.00	$44.50		$59.64	$87.50	55
2.1 to 3.0	$24.30	$38.00		$61.09	$86.80	39
3.1 or More	$21.50	$38.65		$56.54	$77.00	45
Member Status						
Accredited Practice Member	$28.00	$46.50		$67.55	$90.50	61
Nonaccredited Member	$24.87	$29.40		$47.35	$48.00	27
Nonmember	$25.50	$40.00		$52.82	$74.78	92
Metropolitan Status						
Urban	*	*		*	*	21
Suburban	$33.00	$42.00		$61.54	$77.00	55
Second City	$24.73	$44.50		$59.83	$80.85	35
Town	$23.25	$36.13		$48.18	$65.00	62
Rural	*	*		*	*	17
Median Area Household Income						
$35,000 or Less	*	*		*	*	15
$35,000 to $49,999	$24.30	$37.00		$54.33	$75.00	75
$50,000 to $69,999	$30.70	$47.00		$66.46	$86.25	67
$70,000 or More	$27.50	$40.00		$51.70	$63.13	31

Note 1: See the case description in the notes for Table 14.102.
Note 2: An asterisk indicates that data was not reported due to an insufficient number of responses.
Note 3: 33% of the respondents reported that the fee for this service is included in the total surgery fee and is not separately itemized.

TABLE 14.106
SURGICAL CASE TWO: FLUIDS

	25th Percentile	Median	Your Data	Average	75th Percentile	Number of Respondents
All Practices	$25.00	$41.90		$46.59	$64.43	296
Number of FTE Veterinarians						
1.0 or Less	$25.00	$35.50		$39.96	$50.50	77
1.1 to 2.0	$25.00	$39.00		$46.54	$59.53	85
2.1 to 3.0	$24.50	$43.00		$49.99	$73.00	61
3.1 or More	$29.95	$49.00		$51.35	$70.00	70
Member Status						
Accredited Practice Member	$30.00	$45.00		$50.62	$66.58	87
Nonaccredited Member	$30.00	$39.00		$48.73	$70.00	46
Nonmember	$24.00	$35.00		$42.15	$55.00	139
Metropolitan Status						
Urban	*	*		*	*	21
Suburban	$35.00	$50.00		$53.31	$70.00	79
Second City	$26.91	$40.00		$51.57	$75.00	55
Town	$23.38	$39.75		$41.28	$55.50	94
Rural	$19.00	$32.00		$35.21	$44.26	41
Median Area Household Income						
$35,000 or Less	$16.33	$27.50		$34.65	$45.63	32
$35,000 to $49,999	$25.00	$39.50		$46.19	$66.00	113
$50,000 to $69,999	$29.65	$42.75		$46.52	$55.25	109
$70,000 or More	$28.50	$54.10		$54.47	$73.95	36

Note 1: See the case description in the notes for Table 14.102.
Note 2: An asterisk indicates that data was not reported due to an insufficient number of responses.
Note 3: 7% of the respondents reported that the fee for this service is included in the total surgery fee and is not separately itemized.

TABLE 14.107
SURGICAL CASE TWO: MONITORING

	25th Percentile	Median	Your Data	Average	75th Percentile	Number of Respondents
All Practices	$16.00	$23.50		$27.11	$31.25	113
Number of FTE Veterinarians						
1.0 or Less	*	*		*	*	24
1.1 to 2.0	$14.09	$25.00		$27.29	$32.73	32
2.1 to 3.0	$14.25	$20.50		$27.23	$31.00	26
3.1 or More	$19.75	$23.50		$27.67	$35.75	30
Member Status						
Accredited Practice Member	$18.50	$22.00		$25.35	$30.00	43
Nonaccredited Member	*	*		*	*	22
Nonmember	$13.78	$21.00		$29.48	$34.00	43
Metropolitan Status						
Urban	*	*		*	*	13
Suburban	$19.00	$25.00		$27.48	$34.50	43
Second City	*	*		*	*	21
Town	$9.00	$17.00		$18.90	$25.00	27
Rural	*	*		*	*	8
Median Area Household Income						
$35,000 or Less	*	*		*	*	7
$35,000 to $49,999	$14.96	$25.00		$26.18	$31.75	44
$50,000 to $69,999	$19.00	$25.00		$29.78	$32.88	45
$70,000 or More	*	*		*	*	14

Note 1: See the case description in the notes for Table 14.102.
Note 2: An asterisk indicates that data was not reported due to an insufficient number of responses.
Note 3: 52% of the respondents reported that the fee for this service is included in the total surgery fee and is not separately itemized.

TABLE 14.108
SURGICAL CASE TWO: TOTAL SURGERY SETUP FEE

	25th Percentile	Median	Your Data	Average	75th Percentile	Number of Respondents
All Practices	$25.75	$45.38		$80.07	$73.50	50

Note 1: See the case description in the notes for Table 14.102.
Note 2: Data was reported only for all practices due to an insufficient number of responses.
Note 3: 62% of the respondents reported that the fee for this service is included in the total surgery fee and is not separately itemized.

TABLE 14.109
SURGICAL CASE TWO: SURGICAL SUITE USE

	25th Percentile	Median	Your Data	Average	75th Percentile	Number of Respondents
All Practices	$25.00	$40.50		$48.06	$55.00	44

Note 1: See the case description in the notes for Table 14.102.
Note 2: Data was reported only for all practices due to an insufficient number of responses.
Note 3: 66% of the respondents reported that the fee for this service is included in the total surgery fee and is not separately itemized.

TABLE 14.110
SURGICAL CASE TWO: SURGICAL PACK

	25th Percentile	Median	Your Data	Average	75th Percentile	Number of Respondents
All Practices	$25.00	$34.00		$41.20	$50.50	77
Number of FTE Veterinarians						
1.0 or Less	*	*		*	*	17
1.1 to 2.0	*	*		*	*	22
2.1 to 3.0	*	*		*	*	12
3.1 or More	*	*		*	*	24
Member Status						
Accredited Practice Member	*	*		*	*	22
Nonaccredited Member	*	*		*	*	14
Nonmember	$23.75	$37.19		$40.66	$50.75	34
Metropolitan Status						
Urban	*	*		*	*	4
Suburban	$26.00	$35.00		$42.11	$54.85	27
Second City	*	*		*	*	14
Town	*	*		*	*	19
Rural	*	*		*	*	11
Median Area Household Income						
$35,000 or Less	*	*		*	*	7
$35,000 to $49,999	$24.25	$39.85		$41.77	$59.55	30
$50,000 to $69,999	$23.88	$30.05		$40.23	$48.50	26
$70,000 or More	*	*		*	*	11

Note 1: See the case description in the notes for Table 14.102.
Note 2: An asterisk indicates that data was not reported due to an insufficient number of responses.
Note 3: 58% of the respondents reported that the fee for this service is included in the total surgery fee and is not separately itemized.

TABLE 14.111
SURGICAL CASE TWO: DISPOSABLES

	25th Percentile	Median	Your Data	Average	75th Percentile	Number of Respondents
All Practices	$11.44	$20.00		$23.18	$28.93	36

Note 1: See the case description in the notes for Table 14.102.
Note 2: Data was reported only for all practices due to an insufficient number of responses.
Note 3: 65% of the respondents reported that the fee for this service is included in the total surgery fee and is not separately itemized.

TABLE 14.112
SURGICAL CASE TWO: MEDICAL WASTE DISPOSAL

	25th Percentile	Median	Your Data	Average	75th Percentile	Number of Respondents
All Practices	$2.58	$3.50		$4.17	$5.00	81
Number of FTE Veterinarians						
1.0 or Less	*	*		*	*	23
1.1 to 2.0	$2.50	$3.50		$3.91	$4.68	27
2.1 to 3.0	*	*		*	*	12
3.1 or More	*	*		*	*	18
Member Status						
Accredited Practice Member	$2.00	$3.49		$3.69	$4.42	26
Nonaccredited Member	*	*		*	*	10
Nonmember	$2.93	$3.81		$4.25	$5.00	38
Metropolitan Status						
Urban	*	*		*	*	12
Suburban	$2.19	$4.24		$4.46	$5.18	30
Second City	*	*		*	*	14
Town	*	*		*	*	20
Rural	*	*		*	*	3
Median Area Household Income						
$35,000 or Less	*	*		*	*	4
$35,000 to $49,999	$2.00	$3.50		$3.69	$4.45	25
$50,000 to $69,999	$3.00	$4.00		$4.89	$5.71	35
$70,000 or More	*	*		*	*	14

Note 1: See the case description in the notes for Table 14.102.
Note 2: An asterisk indicates that data was not reported due to an insufficient number of responses.
Note 3: 50% of the respondents reported that the fee for this service is included in the total surgery fee and is not separately itemized.

TABLE 14.113
SURGICAL CASE TWO: ONE HOUR ASSISTANT TIME

	25th Percentile	Median	Your Data	Average	75th Percentile	Number of Respondents
All Practices	**$30.52**	**$50.00**	_____	**$65.75**	**$90.00**	**67**

Note 1: See the case description in the notes for Table 14.102.
Note 2: Data was reported only for all practices due to an insufficient number of responses.
Note 3: 60% of the respondents reported that the fee for this service is included in the total surgery fee and is not separately itemized.

TABLE 14.114
SURGICAL CASE TWO: SURGICAL FRACTURE REPAIR WITH IM PIN

	25th Percentile	Median	Your Data	Average	75th Percentile	Number of Respondents
All Practices	**$300.00**	**$400.00**	_____	**$432.42**	**$540.50**	**294**
Number of FTE Veterinarians						
1.0 or Less	$238.43	$350.00	_____	$395.66	$527.50	81
1.1 to 2.0	$301.88	$425.00	_____	$436.97	$536.00	85
2.1 to 3.0	$313.75	$399.00	_____	$398.96	$500.00	58
3.1 or More	$320.00	$450.00	_____	$493.87	$620.00	67
Member Status						
Accredited Practice Member	$311.50	$442.58	_____	$454.83	$542.28	84
Nonaccredited Member	$338.70	$439.29	_____	$460.37	$550.25	46
Nonmember	$271.25	$350.00	_____	$404.44	$512.25	144
Metropolitan Status						
Urban	*	*	_____	*	*	22
Suburban	$343.31	$450.00	_____	$487.40	$631.04	74
Second City	$350.00	$450.00	_____	$456.93	$550.50	61
Town	$270.00	$350.00	_____	$400.28	$500.00	87
Rural	$180.00	$302.50	_____	$321.33	$400.00	43
Median Area Household Income						
$35,000 or Less	$178.75	$312.50	_____	$315.02	$387.85	38
$35,000 to $49,999	$275.00	$359.50	_____	$387.41	$480.00	113
$50,000 to $69,999	$302.13	$429.00	_____	$456.15	$550.00	104
$70,000 or More	$457.13	$583.15	_____	$637.57	$804.66	34

Note 1: See the case description in the notes for Table 14.102.
Note 2: An asterisk indicates that data was not reported due to an insufficient number of responses.
Note 3: 2% of the respondents reported that the fee for this service is included in the total surgery fee and is not separately itemized.

TABLE 14.115
SURGICAL CASE TWO: SUTURE MATERIAL

	25th Percentile	Median	Your Data	Average	75th Percentile	Number of Respondents
All Practices	$12.00	$18.00		$21.64	$27.75	62

Note 1: See the case description in the notes for Table 14.102.
Note 2: Data was reported only for all practices due to an insufficient number of responses.
Note 3: 66% of the respondents reported that the fee for this service is included in the total surgery fee and is not separately itemized.

TABLE 14.116
SURGICAL CASE TWO: STAPLES

	25th Percentile	Median	Your Data	Average	75th Percentile	Number of Respondents
All Practices	$12.00	$20.04		$23.35	$29.75	56

Note 1: See the case description in the notes for Table 14.102.
Note 2: Data was reported only for all practices due to an insufficient number of responses.
Note 3: 59% of the respondents reported that the fee for this service is included in the total surgery fee and is not separately itemized.

TABLE 14.117
SURGICAL CASE TWO: POST-OPERATIVE PAIN MEDICATION

	25th Percentile	Median	Your Data	Average	75th Percentile	Number of Respondents
All Practices	$18.00	$25.00		$30.97	$36.00	278
Number of FTE Veterinarians						
1.0 or Less	$18.00	$21.69		$27.71	$32.44	72
1.1 to 2.0	$16.82	$25.19		$30.78	$38.88	84
2.1 to 3.0	$18.85	$24.00		$30.11	$35.10	53
3.1 or More	$19.69	$28.00		$35.96	$47.19	66
Member Status						
Accredited Practice Member	$18.00	$28.00		$32.33	$40.00	83
Nonaccredited Member	$20.00	$28.00		$34.91	$42.33	44
Nonmember	$17.95	$23.81		$29.07	$32.13	134
Metropolitan Status						
Urban	*	*		*	*	22
Suburban	$21.38	$28.00		$34.47	$40.00	67
Second City	$19.75	$30.00		$34.22	$40.00	59
Town	$16.00	$22.00		$28.30	$30.00	83
Rural	$14.63	$20.00		$24.06	$26.35	41
Median Area Household Income						
$35,000 or Less	$15.72	$19.98		$29.52	$34.25	34
$35,000 to $49,999	$17.84	$22.00		$26.79	$30.00	105
$50,000 to $69,999	$20.00	$28.75		$34.05	$41.08	102
$70,000 or More	$21.38	$28.55		$37.08	$47.00	31

Note 1: See the case description in the notes for Table 14.102.
Note 2: An asterisk indicates that data was not reported due to an insufficient number of responses.
Note 3: 10% of the respondents reported that the fee for this service is included in the total surgery fee and is not separately itemized.

TABLE 14.118
SURGICAL CASE TWO: HOSPITALIZATION

	25th Percentile	Median	Your Data	Average	75th Percentile	Number of Respondents
All Practices	**$21.68**	**$30.30**		**$34.52**	**$44.50**	**225**
Number of FTE Veterinarians						
1.0 or Less	$20.04	$28.00		$31.94	$40.45	56
1.1 to 2.0	$20.71	$32.25		$34.58	$43.53	64
2.1 to 3.0	$21.00	$30.00		$30.31	$36.00	47
3.1 or More	$24.19	$38.66		$40.11	$47.00	56
Member Status						
Accredited Practice Member	$24.59	$32.75		$35.28	$43.48	66
Nonaccredited Member	$26.00	$36.00		$37.00	$45.00	35
Nonmember	$19.56	$28.23		$33.37	$44.50	109
Metropolitan Status						
Urban	*	*		*	*	20
Suburban	$24.81	$36.59		$37.42	$45.00	60
Second City	$21.00	$30.00		$31.89	$42.82	53
Town	$19.96	$30.00		$32.97	$40.15	62
Rural	$16.93	$24.00		$26.36	$33.50	26
Median Area Household Income						
$35,000 or Less	$18.60	$24.00		$29.82	$35.00	27
$35,000 to $49,999	$20.00	$30.00		$32.73	$43.53	80
$50,000 to $69,999	$24.19	$34.25		$35.88	$45.00	80
$70,000 or More	$25.28	$34.75		$40.50	$49.00	32

Note 1: See the case description in the notes for Table 14.102.
Note 2: An asterisk indicates that data was not reported due to an insufficient number of responses.
Note 3: 24% of the respondents reported that the fee for this service is included in the total surgery fee and is not separately itemized.

TABLE 14.119
SURGICAL CASE TWO: TOTAL

	25th Percentile	Median	Your Data	Average	75th Percentile	Number of Respondents
All Practices	$459.14	$663.30		$676.46	$844.05	330
Number of FTE Veterinarians						
1.0 or Less	$382.75	$581.48		$597.96	$749.90	92
1.1 to 2.0	$470.13	$669.75		$677.88	$879.38	96
2.1 to 3.0	$515.59	$645.00		$651.50	$784.41	65
3.1 or More	$574.26	$748.50		$787.58	$1,032.25	73
Member Status						
Accredited Practice Member	$518.84	$728.50		$732.88	$921.59	96
Nonaccredited Member	$564.75	$727.00		$753.05	$905.00	47
Nonmember	$425.38	$584.56		$619.75	$760.23	162
Metropolitan Status						
Urban	$687.13	$906.29		$854.26	$1,059.92	26
Suburban	$557.87	$720.45		$765.00	$976.10	85
Second City	$528.13	$725.50		$718.49	$915.29	66
Town	$431.00	$601.50		$606.79	$743.08	101
Rural	$374.13	$500.70		$521.51	$618.00	44
Median Area Household Income						
$35,000 or Less	$366.94	$550.00		$545.47	$704.79	38
$35,000 to $49,999	$444.75	$623.50		$625.19	$816.59	126
$50,000 to $69,999	$535.50	$677.50		$727.39	$896.15	115
$70,000 or More	$612.75	$795.15		$808.17	$1,108.89	44

Note 1: See the case description in the notes for Table 14.102.
Note 2: The total fee for surgical case two is the sum of the fees for the presurgical exam, CBC, chemistry panel with six chemistries, sedation, fluids, monitoring, total surgery setup fee, surgical suite usage, surgical pack, disposables, medical waste disposal, assistant time (one hour), surgical fracture repair with one IM pin, suture material, staples, post-op pain medication, and hospitalization for one day. Some respondents do not charge for some of the individual services provided in this case. In addition, some respondents reported that they do not offer some of the individual services. Therefore, the average total fee for the case may be significantly lower than the sum of the average fees for the individual services.

CHAPTER 15

END-OF-LIFE SERVICES

One of the toughest parts about being a member of the veterinary health-care team is dealing with the short life spans of the patients that come through the doors. But part of every team member's job is to take care of the clients too, especially during the difficult decision to euthanize. This chapter will address how to help a client decide when it is time to euthanize, guide the client through the experience, and address costs.

THE STRENGTH OF THE HUMAN-ANIMAL BOND

The 2005-2006 National Pet Owners Survey from the American Pet Products Manufacturers Association reported that 75% of dog owners consider their dogs as part of the family and more than half of cat owners feel the same about their pets. The same study showed that nearly all pet owners reported companionship, love, company, and affection as the top benefits of having a pet.

Fifty-nine percent of respondents said the pets were good for their health and helped them relax. And 49% said that having a dog was a major influence on their own exercise routines. "Pets not only provide unconditional love and affection, but research now shows they also provide significant health benefits," stated Bob Vetere, president of the American Pet Products Manufacturers Association. "The steady increase in pet ownership confirms that a growing number of us are realizing pets truly enhance our lives."

Laurel Lagoni, MS, describes the human-animal bond in her book *The Practical Guide to Client Grief* (AAHA Press, 1997) as, "a popular way of referring to the types of relationships and attachments people form with animals, particularly companion animals like dogs, cats, birds, horses, goats, rabbits, snakes, ferrets, and guinea pigs. This bond can be immensely rewarding as long as the relationship continues. But, when the bond is broken by the animal's death, it can exact a high price."

CASE STUDY

Punkin's Passing

Punkin, a beloved Saint Bernard, came into Windsor Veterinary Clinic in Windsor, Colorado, with her owner, Kim Albert. Albert struggled with the decision, but took one long look at her sweet dog and knew it was time. "Let me preface this by saying that until Sharon [referring to the practice manager, Sharon DeNayer] and Doc [referring to clinic owner, Dr. Robin Downing] came to Windsor, I couldn't stay with our pets when they were put down," said Kim in an interview with *Veterinary Economics* for the December 2006 cover story article. "But when it came time for Punkin, I said, 'OK, Doc, I want to give her one last kiss on her big fat nose,' and I stayed with her and let her go."

"Our staff focused on being there for Kim during that time," says DeNayer, remembering that day. "She was heartbroken, but I sat with her as she sobbed uncontrollably that morning." Obviously the team at Windsor Veterinary Clinic works very hard to honor the human-animal bond and makes a difference to clients because they provide such compassionate care.

WOMB-TO-TOMB CARE

Veterinary medicine is different than any other medical discipline because veterinarians see their patients grow and develop from birth to death. In *The Practical Guide to Client Grief*, the author reports that veterinarians experience the deaths of their patients five times more frequently than their human-medicine counterparts. "That is why we work with our whole staff to bond with clients on every visit over the course of the pet's life," says Sharon DeNayer, Practice Manager and Bereavement Companion for Windsor Veterinary Clinic in Windsor, Colorado. "This helps us reach the goal of making the deceased pet's family feel as comfortable as possible under the most difficult circumstances."

The best way a veterinary team can help their client handle this difficult decision is to guide them through the process. "Support their decision, give them a quiet comfortable place to be, and even cry with them," says DeNayer. She stresses that there are several preparatory steps that will guide the clients through the decision.

1. Discuss the pet's health. Set up a time to discuss the pet's health with the whole family, including the children. Do this as early as possible, and as frequently as appropriate, to help ease the time when it comes.

2. Work out the details. Start reviewing the details with the client, such as does the client want to be with the pet during euthanasia and would the client prefer private or general cremation. Establishing this early on eliminates the need to discuss this when the client will be emotional.

3. Decide who will be present. According to *The Loss of Your Pet* brochure from AAHA Press, avoiding your pet's death doesn't make the situation less painful, but may actually make it more difficult to accept. Many people also feel that a child seeing a pet die could be detrimental, when actually it can help them say goodbye and learn about death. Both adults and children should choose to be with their pet during euthanasia because they want to:

- Say goodbye
- Hold and comfort their pets
- Realize that their pets are actually dead
- Know that their pets died peacefully and with dignity

- Be there for their pets the way their pets were always there for them

4. Determine the right time. Give the client a tool to evaluate her pet's quality of life, such as the "HHHHHMM" evaluation tool established by Alice Villalobos, DVM, coauthor of *Canine and Feline Geriatric Oncology: Honoring the Human-Animal Bond* (Blackwell Publishing, 2007). This list of items, which the client can score from one to 10, includes: hurt, hunger, hydration, hygiene, happiness, mobility, more good days than bad. If the client scores a majority of these items at a five or above, Villalobos says that it supports the decision for euthanasia.

CASE STUDY

Trooper's Passing

Leslie Hauck called her daughter one Saturday morning, crying. "We just got back from the vet and Trooper has cancer throughout his abdomen," she wailed. "The vet suggests doing exploratory surgery to see if the cancer is treatable, but I don't know if I want to do that. If I don't, the vet estimates Trooper has only a few days to live."

Her daughter, a former veterinary clinic receptionist, went through the HHHHHMM list with her (see Step 4 in the list under "Womb-to-Tomb Care" in this chapter). Trooper had not been eating for several days and had stopped drinking the day before. He had become lethargic over the previous couple of weeks and spent most of his days sleeping at the top of the steps.

Her daughter asked her, "If you do the surgery and the cancer is treatable, how long could you expect Trooper to be around?"

"A few months to a year," Leslie replied.

"Mom, will those few months give Trooper with the quality of life you want for him?"

"There is no guarantee, and the vet said he may not be strong enough to ever fully recover from surgery."

"I think you have your answer."

Trooper was euthanized that Monday. Leslie painted an oil painting of him to commemorate his companionship that hangs on the kitchen wall to this day.

"Euthanasia is not an easy situation to experience," says Dr. Robin Downing, a diplomate of the American Academy of Pain Management and Owner of Windsor Veterinary Clinic in Windsor, Colorado. "But not being sensitive to the situation is a sure way to upset your client further or potentially lose that person as a client."

Dr. Downing lists some of the most common reasons clients are NOT happy with the way a veterinarian handled a pet's euthanasia:

- You don't allow the client or family, and specifically the children, to be present
- You don't explain the sequence of events or what will happen.
- You use inappropriate drugs that may cause a difficult death.
- You demonstrate a lack of compassion or indifference.
- You have the client walk through a filled reception area posteuthanasia.
- The client can hear ambient noise, such as inappropriate laughter, outside of the room.

HOW TO HANDLE BILLING

Discussing the cost of euthanizing a pet is the last thing a client will want to talk about either before or after the euthanasia. There are some options for handling the billing that will ensure that you get paid for your services and inventory while still showing compassion for the client's feelings.

- Discuss payment in the preplanning meeting with the client and the family.
- Provide options for payment: Send an invoice after the euthanasia, accept a prepayment, or give the client an estimate prior to the euthanasia and have the client bring in a check at the time of euthanasia.
- Be flexible in changing the handling of the remains (it will change the fees).
- Do not surprise the client—give an estimate prior to the euthanasia.

HELPING CLIENTS GRIEVE FOR THEIR PETS

Grief is a normal part of loss. It is also a healthy step in the healing process. According to Downing and DeNayer, the best thing to do is handle each scenario individually because each person's relationship with the pet and each person's personality are different.

They recommend the following steps to help grieving clients:

- If possible, create an appropriate euthanasia site outside of the hospital.
- Whenever possible, schedule euthanasia when there are no other appointments (e.g., after appointments have finished at the end of the day).
- Offer choices, such as euthanasia at home.
- Have a designated team member who is trained and responsible for bereavement care in the practice.
- Never make the client feel rushed.
- Make a donation in the pet's name to an organization supporting the welfare of animals, such as the Morris Animal Foundation or the AAHA Helping Pets Fund.
- Don't use a standard condolence letter; each death is an individualized event so send the client a personalized card that is signed by all staff members.
- Say the right things. (See Implementation Idea in this chapter.)

In addition, it is very important to establish a comfort room in your practice, preferably away from the routine activities of the practice. "Euthanasia is one of the biggest reasons clinics lose clients," says DeNayer. "If you don't have a comfort room and use an exam room for the euthanasia, for your client, that room will always be associated with her pet's death, and it may make the client uncomfortable and unwilling to come back."

Taking the steps you've read about in this chapter will increase the level of service in your practice and bond clients to you. Though you can't make the experience painless, you can certainly do your best to ease clients' minds and let them know you care.

Implementation Idea

Make a List of Right and Wrong Words to Use

It's difficult to know just what to say when a client's pet passes away. Help your team understand the right and wrong words to use by making a list like the one below and discussing it in a staff meeting dedicated to the topic. Ask for additional ideas, and expand your list as necessary.

What to Say

"May I proceed?"

"I'm sorry for your loss."

"I heard about your loss. I don't know what to say."

"I'm thinking of you."

"I'm here for you."

"I remember the story you told me about…"

"I've been thinking about you and your family."

"You're very important to me."

"What can I do to help you?"

"You're in my thoughts and prayers."

Expressions to Avoid

"Are you ready?" (No one is ever ready for euthanasia.)

"I know just how you feel." (Each loss is unique.)

"Put to sleep" (especially around children who may be "put to sleep" for surgery some day)

"Time heals all wounds."

"You must get on with your life."

"It was God's will."

"God never gives us more than we can bear."

"Everything happens for a reason."

"You're holding up so well."

"Think of all you still have to be thankful for."

"You'll get over it."

"Just be happy he's out of his pain."

"At least she's not suffering."

"Now you have an angel in heaven."

"Aren't you over it yet?"

"You can always have/get another (dog, cat, etc.)."

"It's a blessing in disguise."

HELPFUL RESOURCES

For Clients

Cat Heaven, Cynthia Rylant. Blue Sky Press, 1997.

Dog Heaven, Cynthia Rylant. Blue Sky Press, 1995

A Final Act of Caring: Ending the Life of an Animal Friend, by Mary and Herb Montgomery. Montgomery Press, 1993. Available through the American Animal Hospital Association.

Forever in My Heart: Remembering My Pet's Life, by Mary and Herb Montgomery. Montgomery Press, 2000. Available through the American Animal Hospital Association.

Good-bye My Friend: Grieving the Loss of a Pet, by Mary and Herb Montgomery. Montgomery Press, 1991. Available through the American Animal Hospital Association.

The Loss of Your Pet Brochure, AAHA. American Animal Hospital Association Press, 2002.

A Special Place for Charlee, by Debby Morehead. Partners in Publishing, LLC, 1996. Available through the American Animal Hospital Association.

When Your Pet Dies: A Guide to Mourning, Remembering, and Healing, Alan D. Wolfelt, PhD. Companion Press, 2004.

For Veterinarians

Canine and Feline Geriatric Oncology: Honoring the Human-Animal Bond, Alice Villalobos, DVM and Laurie Kaplan, MSC. Blackwell Publishing, 2007.

Euthanasia Sticker for medical records, AAHA. American Animal Hospital Association Press.

Connecting with Clients, Laurel Lagoni, MS, and Dana Durrance, MA. AAHA Press, 1998.

The Human-Animal Bond, Laurel Lagoni, MS, et al. Elsevier, 1994.

Pet Loss & Client Grief, Drs. Cindy Adams and Susan Cohen. Lifelearn, 1999.

The Practical Guide to Client Grief, Laurel Lagoni, MS. AAHA Press, 1997.

CHAPTER 15

DATA TABLES

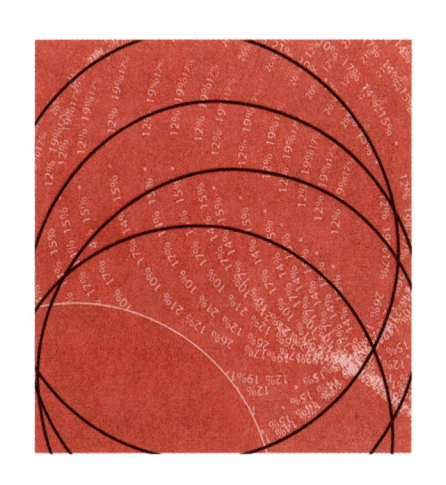

EUTHANASIA

TABLE 15.1
EUTHANASIA CASE ONE: IV CATHETER AND PLACEMENT

	25th Percentile	Median	Your Data	Average	75th Percentile	Number of Respondents
All Practices	**$23.75**	**$32.00**		**$33.30**	**$40.00**	**167**
Number of FTE Veterinarians						
1.0 or Less	$20.00	$29.25		$31.01	$37.88	44
1.1 to 2.0	$29.50	$34.75		$35.31	$45.25	54
2.1 to 3.0	$24.88	$30.00		$32.39	$38.50	33
3.1 or More	$22.00	$34.13		$33.28	$40.09	34
Member Status						
Accredited Practice Member	$24.94	$31.10		$34.00	$38.00	46
Nonaccredited Member	*	*		*	*	23
Nonmember	$23.63	$34.50		$33.99	$40.73	78
Metropolitan Status						
Urban	*	*		*	*	19
Suburban	$23.75	$32.50		$32.44	$43.75	47
Second City	$22.26	$30.00		$31.44	$38.00	27
Town	$24.75	$32.00		$36.40	$45.00	51
Rural	*	*		*	*	19
Median Area Household Income						
$35,000 or Less	$20.56	$31.50		$34.40	$44.59	26
$35,000 to $49,999	$21.25	$30.00		$29.50	$36.32	56
$50,000 to $69,999	$25.00	$34.50		$34.85	$42.90	57
$70,000 or More	*	*		*	*	23

Note 1: This patient is a 30-pound dog presented for euthanasia; exam not included. Owner is not present.
Note 2: An asterisk indicates that data was not reported due to an insufficient number of responses.
Note 3: 24% of respondents reported that they do not charge an additional fee for this service.
Note 4: 44% of respondents reported that they do not offer this service.

TABLE 15.2
EUTHANASIA CASE ONE: PREANESTHETIC SEDATIVE

	25th Percentile	Median	Your Data	Average	75th Percentile	Number of Respondents
All Practices	$15.50	$23.00		$25.21	$30.00	151
Number of FTE Veterinarians						
1.0 or Less	$15.00	$22.00		$24.39	$31.25	45
1.1 to 2.0	$15.88	$23.00		$25.90	$29.45	41
2.1 to 3.0	$20.00	$25.00		$26.21	$32.00	35
3.1 or More	$17.25	$25.00		$25.30	$30.50	25
Member Status						
Accredited Practice Member	$17.50	$28.70		$28.93	$35.00	43
Nonaccredited Member	*	*		*	*	21
Nonmember	$15.00	$22.00		$23.45	$30.00	69
Metropolitan Status						
Urban	*	*		*	*	14
Suburban	$15.63	$25.00		$26.82	$31.21	37
Second City	$18.16	$22.50		$25.64	$32.95	26
Town	$15.00	$23.00		$24.39	$30.00	51
Rural	*	*		*	*	21
Median Area Household Income						
$35,000 or Less	*	*		*	*	20
$35,000 to $49,999	$15.00	$20.00		$23.65	$31.30	54
$50,000 to $69,999	$20.00	$25.00		$27.48	$31.50	53
$70,000 or More	*	*		*	*	19

Note 1: See the case description in the notes for Table 15.1.
Note 2: An asterisk indicates that data was not reported due to an insufficient number of responses.
Note 3: 24% of respondents reported that they do not charge an additional fee for this service.

TABLE 15.3
EUTHANASIA CASE ONE: EUTHANASIA

	25th Percentile	Median	Your Data	Average	75th Percentile	Number of Respondents
All Practices	$35.00	$46.00		$48.00	$58.00	512
Number of FTE Veterinarians						
1.0 or Less	$32.13	$45.00		$46.27	$56.86	136
1.1 to 2.0	$35.00	$46.30		$46.46	$56.00	154
2.1 to 3.0	$37.00	$50.00		$52.48	$63.13	105
3.1 or More	$35.75	$45.05		$47.98	$60.00	102
Member Status						
Accredited Practice Member	$40.00	$48.02		$51.13	$60.73	134
Nonaccredited Member	$36.75	$49.00		$50.74	$61.00	107
Nonmember	$33.50	$45.00		$45.68	$56.00	227
Metropolitan Status						
Urban	$40.00	$48.14		$51.77	$59.00	35
Suburban	$40.00	$51.25		$53.07	$63.00	139
Second City	$37.77	$48.02		$49.86	$60.00	108
Town	$32.00	$42.00		$44.34	$52.00	159
Rural	$30.00	$40.00		$41.54	$54.50	63
Median Area Household Income						
$35,000 or Less	$27.13	$35.00		$38.93	$50.75	64
$35,000 to $49,999	$35.00	$45.00		$46.51	$57.00	185
$50,000 to $69,999	$39.63	$47.25		$49.77	$59.39	168
$70,000 or More	$44.05	$53.38		$54.74	$64.46	82

Note 1: See the case description in the notes for Table 15.1.
Note 2: 3% of respondents reported that they do not offer this service.

TABLE 15.4
EUTHANASIA CASE ONE: TOTAL

	25th Percentile	Median	Your Data	Average	75th Percentile	Number of Respondents
All Practices	**$43.47**	**$57.93**		**$64.53**	**$80.00**	**526**
Number of FTE Veterinarians						
1.0 or Less	$38.63	$55.43		$62.53	$78.75	140
1.1 to 2.0	$41.14	$57.10		$64.07	$80.38	158
2.1 to 3.0	$46.13	$61.38		$70.72	$89.41	106
3.1 or More	$43.75	$59.56		$62.81	$75.44	106
Member Status						
Accredited Practice Member	$45.00	$61.93		$69.49	$88.15	139
Nonaccredited Member	$44.39	$57.93		$60.73	$73.00	108
Nonmember	$40.00	$56.20		$62.83	$79.25	233
Metropolitan Status						
Urban	$45.00	$79.44		$79.64	$111.04	36
Suburban	$46.00	$62.00		$68.24	$82.75	145
Second City	$44.00	$57.60		$63.31	$75.74	109
Town	$40.00	$52.73		$61.89	$75.00	164
Rural	$35.00	$50.00		$57.00	$72.80	63
Median Area Household Income						
$35,000 or Less	$32.13	$49.28		$56.13	$73.76	68
$35,000 to $49,999	$40.00	$55.43		$61.35	$76.88	188
$50,000 to $69,999	$45.00	$60.25		$68.63	$82.25	172
$70,000 or More	$46.00	$60.00		$70.07	$90.00	84

Note 1: See the case description in the notes for Table 15.1.
Note 2: The total fee for euthanasia case one is the sum of the fees for the IV catheter and placement, preanesthetic sedative, and euthanasia procedure. Some respondents do not charge for some of the individual services provided in this case. In addition, some respondents reported that they do not offer some of the individual services. Therefore, the average total fee for the case may be significantly lower than the sum of the average fees for the individual services.

TABLE 15.5
EUTHANASIA CASE TWO: IV CATHETER AND PLACEMENT

	25th Percentile	Median	Your Data	Average	75th Percentile	Number of Respondents
All Practices	**$23.44**	**$32.00**		**$33.25**	**$40.00**	**200**
Number of FTE Veterinarians						
1.0 or Less	$20.00	$26.04		$31.31	$38.00	51
1.1 to 2.0	$29.00	$34.50		$34.99	$45.00	65
2.1 to 3.0	$24.81	$30.11		$32.47	$37.75	44
3.1 or More	$21.50	$33.71		$34.03	$41.02	37
Member Status						
Accredited Practice Member	$25.00	$31.63		$34.36	$37.75	52
Nonaccredited Member	$20.75	$30.00		$31.09	$39.00	35
Nonmember	$24.50	$34.50		$34.19	$42.00	91
Metropolitan Status						
Urban	*	*		*	*	20
Suburban	$24.50	$33.35		$34.15	$45.00	59
Second City	$21.44	$29.99		$30.00	$35.80	35
Town	$24.75	$32.00		$36.22	$45.00	63
Rural	*	*		*	*	19
Median Area Household Income						
$35,000 or Less	$20.19	$31.50		$35.10	$46.64	28
$35,000 to $49,999	$21.00	$30.00		$29.68	$36.57	70
$50,000 to $69,999	$25.00	$33.50		$34.59	$42.90	72
$70,000 or More	$24.88	$35.00		$39.60	$48.61	25

Note 1: This patient is a 30-pound dog presented for euthanasia; exam not included. Owner is present.
Note 2: An asterisk indicates that data was not reported due to an insufficient number of responses.
Note 3: 27% of respondents reported that they do not charge an additional fee for this service.
Note 4: 34% of respondents reported that they do not offer this service.

TABLE 15.6
EUTHANASIA CASE TWO: PREANESTHETIC SEDATIVE

	25th Percentile	Median	Your Data	Average	75th Percentile	Number of Respondents
All Practices	$16.00	$25.00		$26.95	$31.00	175
Number of FTE Veterinarians						
1.0 or Less	$15.00	$22.50		$27.22	$32.88	52
1.1 to 2.0	$17.56	$25.00		$27.36	$30.00	46
2.1 to 3.0	$19.00	$25.00		$26.78	$32.50	45
3.1 or More	$17.25	$25.00		$25.36	$30.50	25
Member Status						
Accredited Practice Member	$18.00	$28.70		$29.59	$35.00	47
Nonaccredited Member	$15.40	$20.00		$23.55	$27.73	29
Nonmember	$15.00	$23.00		$25.47	$30.00	78
Metropolitan Status						
Urban	*	*		*	*	16
Suburban	$16.44	$28.00		$29.93	$34.80	44
Second City	$16.40	$23.00		$26.65	$30.90	31
Town	$15.38	$24.93		$25.16	$30.25	62
Rural	*	*		*	*	21
Median Area Household Income						
$35,000 or Less	*	*		*	*	23
$35,000 to $49,999	$15.00	$20.00		$24.19	$31.30	62
$50,000 to $69,999	$20.00	$26.74		$28.99	$33.47	60
$70,000 or More	*	*		*	*	24

Note 1: See the case description in the notes for Table 15.5.
Note 2: An asterisk indicates that data was not reported due to an insufficient number of responses.
Note 3: 41% of respondents reported that they do not charge an additional fee for this service.
Note 4: 25% of respondents reported that they do not offer this service.

TABLE 15.7
EUTHANASIA CASE TWO: EUTHANASIA

	25th Percentile	Median	Your Data	Average	75th Percentile	Number of Respondents
All Practices	$36.00	$48.00		$50.12	$60.00	515
Number of FTE Veterinarians						
1.0 or Less	$33.84	$45.00		$48.52	$59.87	136
1.1 to 2.0	$35.00	$48.00		$47.95	$57.10	154
2.1 to 3.0	$38.64	$51.48		$53.63	$64.56	108
3.1 or More	$39.25	$49.00		$50.62	$64.50	101
Member Status						
Accredited Practice Member	$40.00	$51.35		$53.19	$64.21	134
Nonaccredited Member	$36.48	$48.96		$50.63	$60.75	108
Nonmember	$34.89	$45.00		$48.37	$58.88	228
Metropolitan Status						
Urban	$46.00	$56.38		$58.87	$70.00	35
Suburban	$45.00	$55.00		$55.54	$67.00	139
Second City	$37.44	$48.52		$50.51	$60.00	106
Town	$34.00	$42.46		$46.01	$54.50	163
Rural	$30.00	$43.00		$42.66	$55.00	63
Median Area Household Income						
$35,000 or Less	$29.38	$35.50		$41.07	$52.08	66
$35,000 to $49,999	$35.00	$45.60		$48.93	$59.75	190
$50,000 to $69,999	$40.00	$50.00		$51.94	$60.50	163
$70,000 or More	$44.80	$55.00		$55.23	$66.39	82

Note 1: See the case description in the notes for Table 15.5.
Note 2: 2% of respondents reported that they do not offer this service.

TABLE 15.8
EUTHANASIA CASE TWO: TOTAL

	25th Percentile	Median	Your Data	Average	75th Percentile	Number of Respondents
All Practices	**$45.00**	**$64.75**		**$69.76**	**$86.53**	**533**
Number of FTE Veterinarians						
1.0 or Less	$41.75	$62.75		$67.68	$85.00	142
1.1 to 2.0	$45.00	$61.22		$68.24	$86.00	160
2.1 to 3.0	$50.00	$68.13		$76.60	$98.01	110
3.1 or More	$45.03	$64.45		$67.36	$79.00	104
Member Status						
Accredited Practice Member	$48.14	$65.74		$73.61	$94.51	140
Nonaccredited Member	$47.00	$64.00		$65.21	$79.40	111
Nonmember	$44.20	$61.25		$68.33	$85.75	236
Metropolitan Status						
Urban	$58.50	$88.80		$91.12	$113.69	36
Suburban	$50.00	$72.50		$74.17	$90.00	149
Second City	$45.00	$62.00		$66.33	$78.50	109
Town	$42.00	$60.00		$67.92	$89.50	167
Rural	$35.00	$53.00		$58.13	$73.00	63
Median Area Household Income						
$35,000 or Less	$35.00	$58.00		$61.45	$79.00	69
$35,000 to $49,999	$44.50	$62.00		$66.71	$81.75	193
$50,000 to $69,999	$48.57	$65.82		$74.68	$99.21	170
$70,000 or More	$50.00	$68.00		$72.40	$90.00	87

Note 1: See the case description in the notes for Table 15.5.
Note 2: The total fee for euthanasia case two is the sum of the fees for the IV catheter and placement, preanesthetic sedative, and euthanasia procedure. Some respondents do not charge for some of the individual services provided in this case. In addition, some respondents reported that they do not offer some of the individual services. Therefore, the average total fee for the case may be significantly lower than the sum of the average fees for the individual services.

TABLE 15.9
EUTHANASIA CASE THREE: IV CATHETER AND PLACEMENT

	25th Percentile	Median	Your Data	Average	75th Percentile	Number of Respondents
All Practices	**$23.46**	**$32.00**		**$33.10**	**$39.08**	**197**
Number of FTE Veterinarians						
1.0 or Less	$20.00	$27.27		$30.60	$36.11	50
1.1 to 2.0	$29.00	$34.50		$35.23	$45.00	65
2.1 to 3.0	$24.94	$30.58		$33.58	$39.25	46
3.1 or More	$21.25	$31.63		$32.26	$38.56	32
Member Status						
Accredited Practice Member	$25.00	$32.00		$34.70	$38.00	55
Nonaccredited Member	$20.81	$30.50		$31.95	$42.56	32
Nonmember	$24.25	$34.50		$33.47	$38.63	89
Metropolitan Status						
Urban	*	*		*	*	22
Suburban	$24.38	$33.18		$33.64	$45.00	58
Second City	$21.44	$29.99		$29.97	$35.55	31
Town	$24.75	$32.00		$35.48	$39.00	63
Rural	*	*		*	*	19
Median Area Household Income						
$35,000 or Less	$20.19	$33.25		$35.61	$46.50	28
$35,000 to $49,999	$22.00	$30.00		$30.40	$36.42	71
$50,000 to $69,999	$24.94	$32.00		$33.87	$41.60	70
$70,000 or More	*	*		*	*	23

Note 1: This patient is a cat presented for euthanasia.
Note 2: An asterisk indicates that data was not reported due to an insufficient number of responses.
Note 3: 27% of respondents reported that they do not charge an additional fee for this service.
Note 4: 34% of respondents reported that they do not offer this service.

TABLE 15.10
EUTHANASIA CASE THREE: PREANESTHETIC SEDATIVE

	25th Percentile	Median	Your Data	Average	75th Percentile	Number of Respondents
All Practices	$15.04	$23.00		$25.71	$30.93	170
Number of FTE Veterinarians						
1.0 or Less	$11.19	$20.75		$24.05	$30.63	50
1.1 to 2.0	$17.19	$23.00		$26.99	$30.00	44
2.1 to 3.0	$20.00	$25.00		$26.78	$32.00	43
3.1 or More	$17.13	$25.00		$24.51	$31.00	26
Member Status						
Accredited Practice Member	$19.33	$28.70		$29.43	$35.00	49
Nonaccredited Member	$15.50	$20.00		$20.96	$25.43	27
Nonmember	$15.00	$21.50		$24.36	$30.00	74
Metropolitan Status						
Urban	*	*		*	*	14
Suburban	$16.06	$26.72		$28.87	$34.66	44
Second City	$12.00	$23.00		$24.56	$32.95	26
Town	$15.00	$23.00		$24.44	$30.50	63
Rural	*	*		*	*	21
Median Area Household Income						
$35,000 or Less	*	*		*	*	20
$35,000 to $49,999	$15.00	$20.50		$23.73	$30.13	62
$50,000 to $69,999	$18.75	$25.00		$28.05	$33.31	57
$70,000 or More	*	*		*	*	24

Note 1: See the case description in the notes for Table 15.9.
Note 2: An asterisk indicates that data was not reported due to an insufficient number of responses.
Note 3: 41% of respondents reported that they do not offer this service.
Note 4: 24% of respondents reported that they do not charge an additional fee for this service.

TABLE 15.11
EUTHANASIA CASE THREE: EUTHANASIA

	25th Percentile	Median	Your Data	Average	75th Percentile	Number of Respondents
All Practices	**$34.00**	**$45.00**		**$45.46**	**$55.00**	**523**
Number of FTE Veterinarians						
1.0 or Less	$30.73	$43.25		$43.92	$55.00	138
1.1 to 2.0	$32.00	$44.50		$43.39	$52.13	158
2.1 to 3.0	$36.40	$45.00		$49.37	$58.26	109
3.1 or More	$34.81	$45.00		$46.26	$56.79	102
Member Status						
Accredited Practice Member	$38.85	$49.13		$50.23	$60.63	138
Nonaccredited Member	$35.00	$45.00		$45.37	$55.00	110
Nonmember	$30.00	$41.00		$43.22	$52.63	230
Metropolitan Status						
Urban	$37.24	$49.00		$50.85	$63.00	35
Suburban	$40.00	$50.00		$51.06	$60.00	145
Second City	$34.25	$44.05		$46.63	$57.00	108
Town	$30.00	$40.00		$41.76	$50.00	165
Rural	$25.04	$35.50		$38.41	$50.00	62
Median Area Household Income						
$35,000 or Less	$25.01	$35.00		$37.89	$46.57	64
$35,000 to $49,999	$30.30	$41.50		$43.53	$51.90	193
$50,000 to $69,999	$36.81	$46.50		$47.72	$56.00	168
$70,000 or More	$41.23	$50.04		$51.80	$60.00	85

Note: See the case description in the notes for Table 15.9.

TABLE 15.12
EUTHANASIA CASE THREE: TOTAL

	25th Percentile	Median	Your Data	Average	75th Percentile	Number of Respondents
All Practices	**$41.00**	**$59.00**		**$64.32**	**$81.00**	**539**
Number of FTE Veterinarians						
1.0 or Less	$35.25	$53.96		$61.06	$79.75	144
1.1 to 2.0	$40.38	$59.50		$63.01	$80.00	164
2.1 to 3.0	$48.00	$62.00		$72.77	$97.93	111
3.1 or More	$43.91	$59.00		$62.63	$76.31	102
Member Status						
Accredited Practice Member	$47.00	$63.00		$70.91	$90.00	145
Nonaccredited Member	$44.00	$56.00		$59.27	$73.00	111
Nonmember	$37.35	$55.00		$62.38	$79.00	236
Metropolitan Status						
Urban	$55.71	$73.00		$81.23	$110.50	37
Suburban	$48.39	$64.00		$69.44	$88.68	153
Second City	$40.83	$55.00		$61.15	$72.98	108
Town	$40.00	$55.00		$63.11	$80.00	169
Rural	$30.00	$50.00		$52.79	$65.00	63
Median Area Household Income						
$35,000 or Less	$30.50	$52.50		$55.60	$73.75	69
$35,000 to $49,999	$40.00	$55.00		$61.38	$76.40	196
$50,000 to $69,999	$45.00	$61.77		$68.89	$84.85	174
$70,000 or More	$47.75	$60.00		$69.80	$89.72	86

Note 1: See the case description in the notes for Table 15.9.
Note 2: The total fee for euthanasia case three is the sum of the fees for the IV catheter and placement, preanesthetic sedative, and euthanasia procedure. Some respondents do not charge for some of the individual services provided in this case. In addition, some respondents reported that they do not offer some of the individual services. Therefore, the average total fee for the case may be significantly lower than the sum of the average fees for the individual services.

TABLE 15.13
EUTHANASIA CASE FOUR: IV CATHETER AND PLACEMENT

	25th Percentile	Median	Your Data	Average	75th Percentile	Number of Respondents
All Practices	$21.75	$30.00		$32.21	$39.25	78
Number of FTE Veterinarians						
1.0 or Less	*	*		*	*	22
1.1 to 2.0	$25.00	$31.00		$34.85	$45.50	30
2.1 to 3.0	*	*		*	*	15
3.1 or More	*	*		*	*	10
Member Status						
Accredited Practice Member	*	*		*	*	19
Nonaccredited Member	*	*		*	*	11
Nonmember	$22.13	$30.00		$31.50	$38.25	40
Metropolitan Status						
Urban	*	*		*	*	8
Suburban	*	*		*	*	19
Second City	*	*		*	*	10
Town	$24.75	$32.00		$35.95	$45.00	31
Rural	*	*		*	*	8
Median Area Household Income						
$35,000 or Less	*	*		*	*	12
$35,000 to $49,999	$20.50	$30.00		$28.37	$32.15	33
$50,000 to $69,999	*	*		*	*	19
$70,000 or More	*	*		*	*	11

Note 1: This patient is a small mammal presented for euthanasia.
Note 2: An asterisk indicates that data was not reported due to an insufficient number of responses.
Note 3: 10% of respondents reported that they do not charge an additional fee for this service.
Note 4: 74% of respondents reported that they do not offer this service.

TABLE 15.14
EUTHANASIA CASE FOUR: PREANESTHETIC SEDATIVE

	25th Percentile	Median	Your Data	Average	75th Percentile	Number of Respondents
All Practices	**$15.00**	**$22.00**		**$26.49**	**$31.00**	**88**
Number of FTE Veterinarians						
1.0 or Less	$10.00	$15.00		$25.00	$35.50	25
1.1 to 2.0	*	*		*	*	24
2.1 to 3.0	*	*		*	*	19
3.1 or More	*	*		*	*	16
Member Status						
Accredited Practice Member	$20.08	$30.00		$31.46	$40.00	29
Nonaccredited Member	*	*		*	*	13
Nonmember	$12.88	$20.00		$23.70	$26.50	37
Metropolitan Status						
Urban	*	*		*	*	6
Suburban	*	*		*	*	18
Second City	*	*		*	*	13
Town	$17.55	$24.63		$27.01	$31.75	36
Rural	*	*		*	*	13
Median Area Household Income						
$35,000 or Less	*	*		*	*	10
$35,000 to $49,999	$14.69	$18.50		$23.64	$31.75	34
$50,000 to $69,999	$16.88	$25.00		$27.56	$31.50	25
$70,000 or More	*	*		*	*	13

Note 1: See the case description in the notes for Table 15.13.
Note 2: An asterisk indicates that data was not reported due to an insufficient number of responses.
Note 3: 25% of respondents reported that they do not charge an additional fee for this service.
Note 4: 56% of respondents reported that they do not offer this service.

TABLE 15.15
EUTHANASIA CASE FOUR: EUTHANASIA

	25th Percentile	Median	Your Data	Average	75th Percentile	Number of Respondents
All Practices	**$20.95**	**$30.50**		**$34.67**	**$47.00**	**383**
Number of FTE Veterinarians						
1.0 or Less	$25.00	$30.59		$34.34	$46.75	84
1.1 to 2.0	$25.00	$35.00		$36.69	$49.50	117
2.1 to 3.0	$20.00	$26.25		$32.35	$41.00	81
3.1 or More	$19.75	$31.00		$33.29	$42.00	90
Member Status						
Accredited Practice Member	$25.00	$34.25		$36.74	$50.00	109
Nonaccredited Member	$24.00	$32.00		$35.89	$48.70	81
Nonmember	$20.00	$30.00		$33.00	$42.50	167
Metropolitan Status						
Urban	*	*		*	*	22
Suburban	$24.50	$33.00		$36.62	$50.50	109
Second City	$18.88	$30.00		$35.01	$50.50	73
Town	$20.50	$30.00		$33.64	$45.00	129
Rural	$20.00	$30.00		$32.71	$42.38	44
Median Area Household Income						
$35,000 or Less	$18.19	$30.00		$31.40	$43.75	40
$35,000 to $49,999	$20.75	$30.30		$34.46	$45.20	142
$50,000 to $69,999	$20.00	$30.88		$34.63	$47.35	126
$70,000 or More	$25.00	$36.00		$37.73	$50.71	64

Note 1: See the case description in the notes for Table 15.13.
Note 2: An asterisk indicates that data was not reported due to an insufficient number of responses.
Note 3: 25% of respondents reported that they do not offer this service.

TABLE 15.16
EUTHANASIA CASE FOUR: TOTAL

	25th Percentile	Median	Your Data	Average	75th Percentile	Number of Respondents
All Practices	$25.00	$39.00		$45.88	$62.00	395
Number of FTE Veterinarians						
1.0 or Less	$25.00	$39.75		$45.71	$60.00	90
1.1 to 2.0	$26.38	$45.00		$49.84	$70.75	120
2.1 to 3.0	$20.00	$36.41		$44.88	$61.25	82
3.1 or More	$21.00	$37.00		$40.72	$56.00	91
Member Status						
Accredited Practice Member	$25.00	$40.00		$48.62	$65.61	114
Nonaccredited Member	$25.00	$39.44		$44.64	$62.00	82
Nonmember	$22.00	$37.50		$44.46	$60.00	172
Metropolitan Status						
Urban	*	*		*	*	23
Suburban	$25.00	$37.44		$44.08	$60.00	116
Second City	$21.75	$33.00		$44.26	$57.76	73
Town	$25.00	$40.00		$49.06	$64.50	131
Rural	$24.00	$37.50		$42.43	$60.50	45
Median Area Household Income						
$35,000 or Less	$25.00	$38.00		$43.67	$60.00	43
$35,000 to $49,999	$25.00	$40.00		$46.06	$59.94	144
$50,000 to $69,999	$22.00	$35.50		$44.54	$63.00	129
$70,000 or More	$25.00	$40.00		$49.82	$65.00	67

Note 1: See the case description in the notes for Table 15.13.
Note 2: An asterisk indicates that data was not reported due to an insufficient number of responses.
Note 3: The total fee for euthanasia case four is the sum of the fees for the IV catheter and placement, preanesthetic sedative, and euthanasia procedure. Some respondents do not charge for some of the individual services provided in this case. In addition, some respondents reported that they do not offer some of the individual services. Therefore, the average total fee for the case may be significantly lower than the sum of the average fees for the individual services.

TABLE 15.17
EUTHANASIA CASE FIVE: PREANESTHETIC SEDATIVE

	25th Percentile	Median	Your Data	Average	75th Percentile	Number of Respondents
All Practices	$16.25	$26.00		$29.22	$37.30	51

Note 1: This patient is a bird or reptile presented for euthanasia.
Note 2: Data was reported only for all practices due to an insufficient number of responses.
Note 3: 14% of respondents reported that they do not charge an additional fee for this service.
Note 4: 75% of respondents reported that they do not offer this service.

TABLE 15.18
EUTHANASIA CASE FIVE: EUTHANASIA

	25th Percentile	Median	Your Data	Average	75th Percentile	Number of Respondents
All Practices	$20.00	$30.00		$32.35	$43.76	254
Number of FTE Veterinarians						
1.0 or Less	$24.25	$30.00		$34.16	$47.83	45
1.1 to 2.0	$20.00	$29.00		$31.45	$42.00	75
2.1 to 3.0	$20.00	$26.25		$31.30	$40.00	59
3.1 or More	$20.00	$32.50		$32.45	$43.24	68
Member Status						
Accredited Practice Member	$21.25	$32.00		$35.15	$50.00	81
Nonaccredited Member	$24.00	$30.84		$34.14	$45.00	54
Nonmember	$16.90	$25.00		$29.22	$38.50	106
Metropolitan Status						
Urban	*	*		*	*	15
Suburban	$24.25	$30.00		$32.80	$42.00	72
Second City	$15.50	$26.00		$32.46	$47.50	53
Town	$20.00	$30.00		$31.19	$43.20	83
Rural	$20.00	$33.13		$33.68	$45.00	26
Median Area Household Income						
$35,000 or Less	$19.50	$25.00		$28.26	$37.24	27
$35,000 to $49,999	$20.00	$29.85		$32.43	$44.66	92
$50,000 to $69,999	$20.00	$30.00		$33.44	$43.88	84
$70,000 or More	$20.00	$27.13		$32.70	$48.43	42

Note 1: See the case description in the notes for Table 15.17.
Note 2: An asterisk indicates that data was not reported due to an insufficient number of responses.
Note 3: 48% of respondents reported that they do not offer this service.

TABLE 15.19
EUTHANASIA CASE FIVE: TOTAL

	25th Percentile	Median	Your Data	Average	75th Percentile	Number of Respondents
All Practices	$20.00	$30.34		$37.34	$50.00	260
Number of FTE Veterinarians						
1.0 or Less	$25.00	$30.00		$39.99	$53.75	48
1.1 to 2.0	$20.00	$30.00		$35.81	$49.60	76
2.1 to 3.0	$20.00	$29.13		$36.71	$48.26	60
3.1 or More	$20.00	$35.00		$36.02	$50.47	69
Member Status						
Accredited Practice Member	$22.00	$38.00		$42.71	$55.40	83
Nonaccredited Member	$25.00	$35.00		$38.57	$45.50	55
Nonmember	$19.50	$27.53		$32.61	$41.00	109
Metropolitan Status						
Urban	*	*		*	*	15
Suburban	$25.00	$30.00		$36.23	$45.98	75
Second City	$18.10	$27.53		$39.24	$53.90	53
Town	$20.00	$35.00		$36.53	$49.68	85
Rural	$20.00	$34.25		$37.02	$50.00	27
Median Area Household Income						
$35,000 or Less	$19.75	$25.00		$31.41	$37.62	29
$35,000 to $49,999	$21.15	$33.00		$37.95	$50.00	93
$50,000 to $69,999	$20.00	$30.50		$37.76	$54.63	86
$70,000 or More	$20.00	$30.00		$37.96	$50.00	43

Note 1: See the case description in the notes for Table 15.17.
Note 2: An asterisk indicates that data was not reported due to an insufficient number of responses.
Note 3: The total fee for euthanasia case five is the sum of the fees for the preanesthetic sedative and euthanasia procedure. Some respondents do not charge for some of the individual services provided in this case. In addition, some respondents reported that they do not offer some of the individual services. Therefore, the average total fee for the case may be significantly lower than the sum of the average fees for the individual services.

NECROPSY

TABLE 15.20
NECROPSY: 30-POUND DOG

	25th Percentile	Median	Your Data	Average	75th Percentile	Number of Respondents
All Practices	$75.05	$119.50		$138.74	$181.79	377
Number of FTE Veterinarians						
1.0 or Less	$77.81	$120.00		$141.21	$195.87	93
1.1 to 2.0	$71.45	$114.50		$129.35	$158.00	107
2.1 to 3.0	$85.00	$115.80		$138.65	$183.75	80
3.1 or More	$80.00	$120.00		$145.88	$200.00	87
Member Status						
Accredited Practice Member	$93.34	$132.50		$155.11	$200.00	114
Nonaccredited Member	$72.20	$96.50		$135.63	$182.25	66
Nonmember	$75.00	$114.44		$133.17	$167.95	162
Metropolitan Status						
Urban	*	*		*	*	20
Suburban	$100.00	$138.00		$163.64	$213.85	99
Second City	$82.45	$121.00		$148.94	$197.89	85
Town	$75.00	$109.94		$124.99	$151.50	120
Rural	$55.00	$80.50		$95.73	$128.00	45
Median Area Household Income						
$35,000 or Less	$61.25	$92.00		$110.36	$150.00	48
$35,000 to $49,999	$75.00	$100.00		$126.07	$154.00	137
$50,000 to $69,999	$83.13	$125.59		$153.84	$211.06	120
$70,000 or More	$100.00	$141.75		$162.39	$201.00	62

Note 1: The fee is for a complete necropsy (excluding brain and spinal column) and related diagnostics (e.g., histopathology).
Note 2: An asterisk indicates that data was not reported due to an insufficient number of responses.
Note 3: 28% of respondents reported that they do not offer this service.

TABLE 15.21
NECROPSY: 75-POUND DOG

	25th Percentile	Median	Your Data	Average	75th Percentile	Number of Respondents
All Practices	**$80.45**	**$126.40**		**$155.74**	**$200.00**	**381**
Number of FTE Veterinarians						
1.0 or Less	$79.63	$142.50		$164.14	$210.00	98
1.1 to 2.0	$75.00	$125.00		$144.59	$175.00	108
2.1 to 3.0	$91.25	$120.00		$153.20	$200.00	77
3.1 or More	$84.75	$135.90		$154.16	$200.00	86
Member Status						
Accredited Practice Member	$95.28	$150.00		$170.56	$202.96	116
Nonaccredited Member	$75.00	$124.28		$151.60	$200.00	67
Nonmember	$80.00	$125.00		$150.95	$184.05	162
Metropolitan Status						
Urban	*	*		*	*	22
Suburban	$100.00	$150.00		$189.34	$250.00	101
Second City	$95.44	$139.10		$162.52	$207.50	84
Town	$75.63	$118.00		$137.98	$165.00	120
Rural	$60.00	$93.93		$113.81	$146.32	46
Median Area Household Income						
$35,000 or Less	$60.00	$111.00		$122.23	$157.00	50
$35,000 to $49,999	$75.00	$110.13		$144.52	$172.35	136
$50,000 to $69,999	$85.00	$150.00		$171.00	$236.00	123
$70,000 or More	$120.00	$155.50		$181.82	$217.39	62

Note 1: The fee is for a complete necropsy (excluding brain and spinal column) and related diagnostics (e.g., histopathology).
Note 2: An asterisk indicates that data was not reported due to an insufficient number of responses.
Note 3: 27% of respondents reported that they do not offer this service.

TABLE 15.22
NECROPSY: CAT

	25th Percentile	Median	Your Data	Average	75th Percentile	Number of Respondents
All Practices	$75.00	$113.65		$134.27	$181.13	384
Number of FTE Veterinarians						
1.0 or Less	$77.52	$115.60		$134.77	$195.56	98
1.1 to 2.0	$67.04	$105.00		$128.68	$169.85	112
2.1 to 3.0	$78.63	$110.00		$130.95	$179.48	77
3.1 or More	$75.08	$120.00		$142.57	$198.50	86
Member Status						
Accredited Practice Member	$90.00	$126.18		$147.79	$193.00	115
Nonaccredited Member	$71.45	$85.00		$130.94	$178.00	67
Nonmember	$75.00	$111.63		$131.12	$175.50	166
Metropolitan Status						
Urban	$69.75	$150.00		$149.21	$198.25	25
Suburban	$87.50	$135.00		$162.66	$212.83	101
Second City	$78.63	$118.58		$142.00	$197.89	85
Town	$68.00	$100.00		$116.61	$150.00	119
Rural	$50.00	$80.00		$94.11	$125.75	46
Median Area Household Income						
$35,000 or Less	$58.75	$92.00		$109.24	$150.00	50
$35,000 to $49,999	$75.00	$100.00		$126.37	$158.47	140
$50,000 to $69,999	$76.81	$120.00		$145.29	$200.00	122
$70,000 or More	$100.00	$141.75		$153.28	$200.00	62

Note 1: The fee is for a complete necropsy (excluding brain and spinal column) and related diagnostics (e.g., histopathology).
Note 2: 26% of respondents reported that they do not offer this service.

TABLE 15.23
NECROPSY: SMALL MAMMAL

	25th Percentile	Median	Your Data	Average	75th Percentile	Number of Respondents
All Practices	$64.35	$100.00		$122.28	$160.00	210
Number of FTE Veterinarians						
1.0 or Less	$75.00	$100.00		$125.14	$150.00	48
1.1 to 2.0	$50.00	$75.50		$106.33	$144.51	64
2.1 to 3.0	$61.25	$102.29		$128.37	$183.49	36
3.1 or More	$66.76	$100.00		$128.19	$176.63	58
Member Status						
Accredited Practice Member	$71.50	$119.50		$142.51	$200.00	69
Nonaccredited Member	$63.70	$80.00		$120.22	$150.00	37
Nonmember	$60.00	$100.00		$111.50	$150.00	93
Metropolitan Status						
Urban	*	*		*	*	12
Suburban	$75.00	$120.00		$155.25	$211.80	51
Second City	$69.69	$100.00		$135.61	$175.75	38
Town	$56.63	$91.75		$106.89	$136.50	78
Rural	$40.00	$68.03		$84.53	$101.75	26
Median Area Household Income						
$35,000 or Less	$47.50	$99.00		$103.03	$150.00	25
$35,000 to $49,999	$60.00	$92.05		$113.48	$146.25	78
$50,000 to $69,999	$69.38	$100.00		$136.97	$195.00	65
$70,000 or More	$75.00	$125.00		$135.57	$185.00	35

Note 1: The fee is for a complete necropsy (excluding brain and spinal column) and related diagnostics (e.g., histopathology).
Note 2: An asterisk indicates that data was not reported due to an insufficient number of responses.
Note 3: 58% of respondents reported that they do not offer this service.

TABLE 15.24
NECROPSY: FERRET

	25th Percentile	Median	Your Data	Average	75th Percentile	Number of Respondents
All Practices	$65.00	$100.00		$123.58	$161.00	213
Number of FTE Veterinarians						
1.0 or Less	$75.00	$110.00		$136.93	$190.12	49
1.1 to 2.0	$50.00	$75.00		$105.91	$143.34	65
2.1 to 3.0	$62.50	$100.05		$118.82	$168.79	37
3.1 or More	$68.38	$100.00		$130.17	$175.00	57
Member Status						
Accredited Practice Member	$75.00	$109.87		$137.39	$185.75	71
Nonaccredited Member	$63.70	$80.00		$120.81	$150.00	37
Nonmember	$62.50	$100.00		$117.20	$150.00	93
Metropolitan Status						
Urban	*	*		*	*	12
Suburban	$75.00	$127.50		$158.35	$214.01	54
Second City	$75.00	$100.00		$134.71	$172.75	40
Town	$52.88	$96.84		$105.98	$139.50	76
Rural	$38.75	$68.03		$83.76	$101.75	26
Median Area Household Income						
$35,000 or Less	*	*		*	*	23
$35,000 to $49,999	$64.35	$98.38		$117.78	$150.00	82
$50,000 to $69,999	$70.00	$100.00		$134.77	$190.00	63
$70,000 or More	$76.25	$122.50		$135.87	$189.75	38

Note 1: The fee is for a complete necropsy (excluding brain and spinal column) and related diagnostics (e.g., histopathology).
Note 2: An asterisk indicates that data was not reported due to an insufficient number of responses.
Note 3: 57% of respondents reported that they do not offer this service.

TABLE 15.25
NECROPSY: BIRD

	25th Percentile	Median	Your Data	Average	75th Percentile	Number of Respondents
All Practices	**$52.25**	**$92.50**		**$108.82**	**$150.00**	**149**
Number of FTE Veterinarians						
1.0 or Less	$65.00	$107.50		$117.92	$150.00	31
1.1 to 2.0	$41.14	$65.00		$86.49	$127.50	42
2.1 to 3.0	$60.00	$92.50		$112.92	$156.98	29
3.1 or More	$53.75	$97.75		$117.88	$168.25	42
Member Status						
Accredited Practice Member	$58.75	$99.70		$124.82	$182.38	58
Nonaccredited Member	$53.31	$75.50		$97.86	$149.09	26
Nonmember	$50.00	$100.00		$103.78	$150.00	58
Metropolitan Status						
Urban	*	*		*	*	10
Suburban	$70.36	$112.25		$137.78	$200.50	32
Second City	$54.69	$100.00		$121.65	$168.25	26
Town	$46.00	$76.00		$91.09	$126.40	59
Rural	*	*		*	*	18
Median Area Household Income						
$35,000 or Less	*	*		*	*	19
$35,000 to $49,999	$50.00	$75.00		$96.58	$125.00	51
$50,000 to $69,999	$57.75	$107.00		$129.77	$182.07	47
$70,000 or More	*	*		*	*	24

Note 1: The fee is for a complete necropsy (excluding brain and spinal column) and related diagnostics (e.g., histopathology).
Note 2: An asterisk indicates that data was not reported due to an insufficient number of responses.
Note 3: 70% of respondents reported that they do not offer this service.

TABLE 15.26
NECROPSY: REPTILE

	25th Percentile	Median	Your Data	Average	75th Percentile	Number of Respondents
All Practices	$57.75	$97.58		$117.71	$165.00	131
Number of FTE Veterinarians						
1.0 or Less	*	*		*	*	24
1.1 to 2.0	$50.00	$65.00		$91.52	$135.00	35
2.1 to 3.0	$60.00	$92.50		$113.58	$178.95	27
3.1 or More	$51.25	$97.75		$125.02	$175.00	40
Member Status						
Accredited Practice Member	$62.00	$119.50		$138.38	$185.75	51
Nonaccredited Member	*	*		*	*	24
Nonmember	$58.88	$100.00		$107.07	$150.00	50
Metropolitan Status						
Urban	*	*		*	*	8
Suburban	$75.00	$130.00		$153.26	$212.83	29
Second City	*	*		*	*	24
Town	$50.00	$85.00		$95.91	$126.40	51
Rural	*	*		*	*	16
Median Area Household Income						
$35,000 or Less	*	*		*	*	16
$35,000 to $49,999	$52.50	$76.00		$101.09	$125.00	43
$50,000 to $69,999	$75.00	$125.00		$148.20	$208.00	43
$70,000 or More	*	*		*	*	22

Note 1: The fee is for a complete necropsy (excluding brain and spinal column) and related diagnostics (e.g., histopathology).
Note 2: An asterisk indicates that data was not reported due to an insufficient number of responses.
Note 3: 74% of respondents reported that they do not offer this service.

CREMATION AND DISPOSAL

TABLE 15.27
COMMUNAL CREMATION: 30-POUND DOG

	25th Percentile	Median	Your Data	Average	75th Percentile	Number of Respondents
All Practices	$38.50	$50.00		$54.18	$65.00	482
Number of FTE Veterinarians						
1.0 or Less	$35.00	$47.00		$50.06	$59.50	117
1.1 to 2.0	$40.00	$50.00		$52.35	$63.75	145
2.1 to 3.0	$40.00	$56.10		$58.51	$71.95	105
3.1 or More	$40.00	$52.00		$55.63	$66.58	101
Member Status						
Accredited Practice Member	$40.00	$52.00		$56.90	$68.00	135
Nonaccredited Member	$39.00	$50.00		$53.82	$65.50	98
Nonmember	$35.00	$50.00		$51.87	$62.94	208
Metropolitan Status						
Urban	$35.75	$49.50		$54.55	$69.25	38
Suburban	$45.00	$56.00		$58.78	$68.25	138
Second City	$38.79	$50.00		$54.45	$65.55	106
Town	$35.28	$49.80		$50.99	$60.00	141
Rural	$34.88	$44.26		$49.81	$65.00	50
Median Area Household Income						
$35,000 or Less	$34.25	$44.00		$47.66	$59.00	52
$35,000 to $49,999	$35.78	$50.00		$51.70	$62.13	177
$50,000 to $69,999	$40.00	$54.00		$55.70	$66.79	164
$70,000 or More	$46.19	$56.00		$59.72	$70.00	77

Note: 11% of respondents reported that they do not offer this service.

TABLE 15.28
COMMUNAL CREMATION: CAT

	25th Percentile	Median	Your Data	Average	75th Percentile	Number of Respondents
All Practices	$35.00	$45.00		$47.16	$59.00	483
Number of FTE Veterinarians						
1.0 or Less	$30.00	$42.00		$44.60	$52.00	119
1.1 to 2.0	$33.50	$44.00		$46.08	$55.50	145
2.1 to 3.0	$35.53	$47.25		$49.55	$61.11	106
3.1 or More	$35.00	$48.00		$49.09	$62.45	99
Member Status						
Accredited Practice Member	$36.00	$47.00		$49.49	$61.50	135
Nonaccredited Member	$33.00	$45.00		$46.20	$59.00	99
Nonmember	$32.18	$43.25		$46.24	$58.75	208
Metropolitan Status						
Urban	$35.00	$41.85		$49.02	$60.00	40
Suburban	$40.00	$50.00		$51.95	$64.00	139
Second City	$35.00	$45.77		$48.74	$60.00	108
Town	$33.00	$40.05		$43.02	$52.00	139
Rural	$28.33	$36.10		$41.40	$53.50	49
Median Area Household Income						
$35,000 or Less	$30.00	$38.19		$41.19	$49.75	52
$35,000 to $49,999	$32.00	$40.00		$44.93	$55.55	177
$50,000 to $69,999	$35.00	$48.00		$49.78	$60.00	167
$70,000 or More	$40.60	$50.00		$51.43	$62.50	77

Note: 9% of respondents reported that they do not offer this service.

TABLE 15.29
INDIVIDUAL CREMATION AND RETURN OF ASHES TO OWNER: 30-POUND DOG

	25th Percentile	Median	Your Data	Average	75th Percentile	Number of Respondents
All Practices	**$111.21**	**$139.13**		**$143.23**	**$167.95**	**472**
Number of FTE Veterinarians						
1.0 or Less	$108.53	$130.00		$141.34	$170.00	114
1.1 to 2.0	$109.00	$139.50		$139.62	$165.00	147
2.1 to 3.0	$120.00	$142.00		$145.79	$169.75	105
3.1 or More	$115.75	$137.88		$146.25	$165.23	94
Member Status						
Accredited Practice Member	$124.38	$149.75		$154.12	$176.00	130
Nonaccredited Member	$110.75	$140.00		$141.17	$165.00	94
Nonmember	$104.40	$130.00		$137.01	$159.63	206
Metropolitan Status						
Urban	$124.25	$150.00		$155.21	$165.00	32
Suburban	$124.30	$145.89		$152.94	$179.75	133
Second City	$109.28	$139.65		$141.26	$165.70	102
Town	$105.00	$131.25		$138.43	$164.18	145
Rural	$90.63	$120.00		$127.85	$156.00	53
Median Area Household Income						
$35,000 or Less	$100.00	$130.00		$128.90	$150.00	53
$35,000 to $49,999	$108.50	$130.50		$136.76	$158.40	177
$50,000 to $69,999	$117.90	$140.00		$145.19	$170.00	154
$70,000 or More	$129.50	$153.00		$162.36	$196.25	77

Note: 13% of respondents reported that they do not offer this service.

TABLE 15.30
INDIVIDUAL CREMATION AND RETURN OF ASHES TO OWNER: CAT

	25th Percentile	Median	Your Data	Average	75th Percentile	Number of Respondents
All Practices	**$100.00**	**$125.00**		**$130.00**	**$155.00**	**478**
Number of FTE Veterinarians						
1.0 or Less	$98.50	$124.50		$128.97	$155.00	117
1.1 to 2.0	$96.25	$122.00		$125.96	$151.00	149
2.1 to 3.0	$105.00	$125.00		$130.86	$155.00	106
3.1 or More	$105.00	$130.00		$134.62	$155.00	93
Member Status						
Accredited Practice Member	$115.00	$135.00		$141.21	$156.90	132
Nonaccredited Member	$98.70	$120.00		$124.49	$146.75	97
Nonmember	$95.00	$121.00		$125.51	$150.56	206
Metropolitan Status						
Urban	$123.00	$139.00		$140.87	$162.50	33
Suburban	$109.00	$130.00		$139.46	$162.50	137
Second City	$99.75	$129.00		$128.86	$150.00	102
Town	$95.00	$120.00		$124.11	$150.00	147
Rural	$79.17	$104.30		$115.01	$142.50	52
Median Area Household Income						
$35,000 or Less	$87.50	$115.00		$116.03	$145.00	53
$35,000 to $49,999	$97.00	$120.00		$122.44	$145.29	179
$50,000 to $69,999	$100.63	$130.00		$133.64	$155.00	156
$70,000 or More	$116.00	$140.00		$147.85	$175.00	79

Note: 11% of respondents reported that they do not offer this service.

TABLE 15.31
DISPOSAL OF SMALL MAMMAL

	25th Percentile	Median	Your Data	Average	75th Percentile	Number of Respondents
All Practices	$20.00	$30.00		$32.14	$42.75	268
Number of FTE Veterinarians						
1.0 or Less	$20.00	$30.00		$33.09	$45.00	55
1.1 to 2.0	$20.00	$27.00		$30.20	$39.75	88
2.1 to 3.0	$20.00	$30.00		$32.04	$41.00	53
3.1 or More	$20.19	$31.30		$33.34	$43.19	64
Member Status						
Accredited Practice Member	$20.00	$35.00		$34.95	$50.00	69
Nonaccredited Member	$20.00	$28.00		$31.37	$39.00	56
Nonmember	$20.00	$28.00		$29.99	$38.13	126
Metropolitan Status						
Urban	*	*		*	*	17
Suburban	$20.00	$30.00		$32.07	$45.00	81
Second City	$19.00	$28.00		$31.27	$38.69	49
Town	$20.50	$30.00		$31.59	$40.00	81
Rural	$18.00	$25.00		$31.23	$46.00	35
Median Area Household Income						
$35,000 or Less	$21.25	$31.50		$31.14	$38.97	28
$35,000 to $49,999	$20.00	$29.50		$31.60	$40.00	100
$50,000 to $69,999	$17.12	$28.00		$30.80	$40.00	89
$70,000 or More	$20.00	$35.00		$36.10	$50.00	47

Note: 41% of respondents reported that they do not offer this service.

TABLE 15.32
DISPOSAL OF REPTILE OR BIRD

	25th Percentile	Median	Your Data	Average	75th Percentile	Number of Respondents
All Practices	$20.00	$28.21		$31.84	$43.00	218
Number of FTE Veterinarians						
1.0 or Less	$16.00	$25.10		$32.23	$45.50	41
1.1 to 2.0	$17.55	$25.00		$29.11	$38.00	72
2.1 to 3.0	$20.00	$29.50		$33.06	$43.94	44
3.1 or More	$20.00	$30.00		$33.08	$45.00	55
Member Status						
Accredited Practice Member	$20.00	$32.80		$34.48	$50.00	64
Nonaccredited Member	$20.91	$28.40		$32.21	$39.00	40
Nonmember	$17.55	$25.00		$29.09	$37.15	104
Metropolitan Status						
Urban	*	*		*	*	16
Suburban	$18.75	$30.00		$31.21	$42.75	66
Second City	$15.25	$26.50		$31.63	$41.38	45
Town	$20.00	$28.42		$31.07	$40.00	63
Rural	*	*		*	*	24
Median Area Household Income						
$35,000 or Less	*	*		*	*	22
$35,000 to $49,999	$20.00	$26.50		$30.55	$39.00	83
$50,000 to $69,999	$17.00	$28.00		$30.31	$40.00	71
$70,000 or More	$18.75	$35.50		$37.05	$50.50	38

Note 1: An asterisk indicates that data was not reported due to an insufficient number of responses.
Note 2: 52% of respondents reported that they do not offer this service.

SURVEYS

2006 AAHA Veterinary Fee Survey

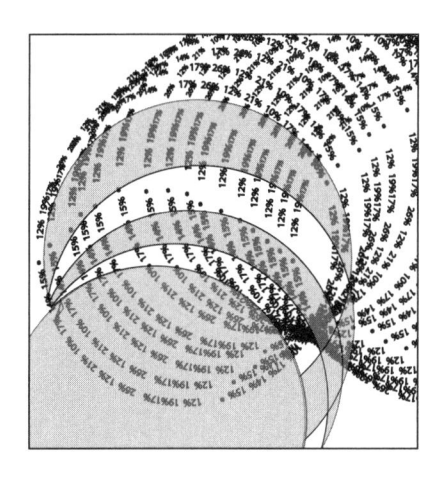

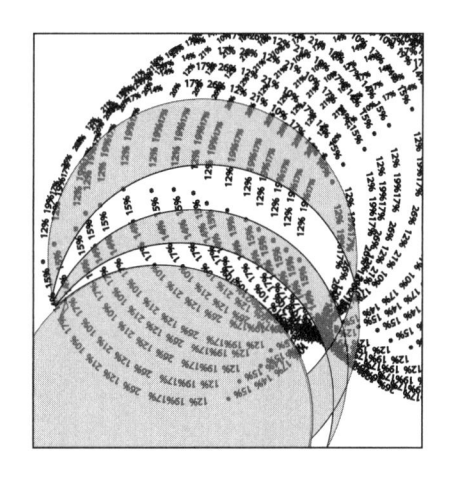

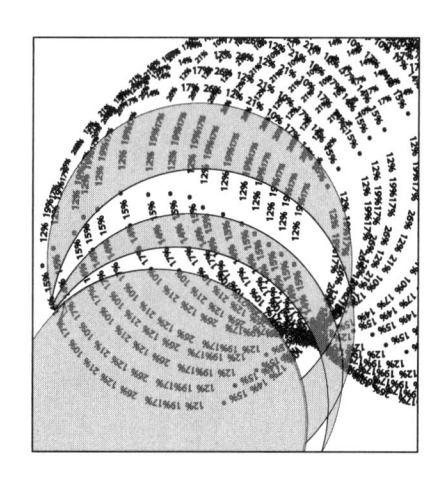

Survey Due:

November 8, 2006

AAHA
AMERICAN ANIMAL HOSPITAL ASSOCIATION

*Healthy Practices.
Healthier Pets.*

Survey A

INSTRUCTIONS

Thank you for participating in AAHA's fee survey. We realize that practices establish fees in different ways, so we designed this form to accommodate the most common formats used. We described various scenarios thought to be typical for most practices so that we could provide additional insight regarding fees related to case management.

Please follow the instructions for each section, and respond as completely as you can. If you are part of a multi-practice corporation, please complete the survey for your individual practice. <u>Please report fees that were in effect as of October 1, 2006.</u> Keep a copy of your survey for your records.

Your information is completely confidential. Your practice's name will not be attached to the survey. All results will be reported in aggregate, and individual responses will not be released.

SECTION I: EXAMINATIONS & CONSULTATIONS

1. Please report the fee charged and the time scheduled for the following examinations and consultations. If you do not offer the service, please write "NA" in the blank.

	Fee	Allotted Time (Minutes)
a. Examination—wellness	_____	_____
b. Examination—senior pet	_____	_____
c. Examination—sick pet	_____	_____
d. Examination—single health problem	_____	_____
e. Do you include the fee for a medical progress exam in the examination fee in question 1d above?	☐ Yes (go to 1g)	☐ No (go to 1f)
f. Examination—medical progress exam of previously identified problem	_____	_____
g. Examination—health certificate exam	_____	_____
h. Do you include the fee for the certificate preparation in the examination fee in question 1g above?	☐ Yes (go to 1j)	☐ No (go to 1i)
i. Certificate preparation fee	_____	_____
j. Examination—inpatient (veterinarian supervision of a hospitalized patient; excludes hospitalization fee)	_____	_____
k. Examination—behavior consultation	_____	_____
l. Examination—second opinion at owner's request	_____	_____
m. Examination—referral from another veterinarian	_____	_____
n. Examination—emergency during routine office hours	_____	_____
o. Examination—emergency after hours (6:00 p.m.–12:00 a.m.)	_____	_____
p. Examination—emergency after hours (12:00 a.m.–7:00 a.m.)	_____	_____
q. Examination—avian	_____	_____
r. Examination—reptile	_____	_____
s. Examination—rabbit	_____	_____

	Fee	Allotted Time (Minutes)
t. Examination—small mammal	_____	_____
u. Examination—ferret	_____	_____
v. Consultation only—in office (pet not present)	_____	_____
w. Consultation only—by telephone (10 minutes)	_____	10 minutes

SECTION II: WELLNESS VISITS

2. Next to each type of wellness visit, check the "Yes" or "No" box to indicate whether you offer that type of wellness visit in your practice. Report fees for the wellness services you offer, assuming each pet is presented separately and that a routine prevaccination physical is performed. If you do not charge a separate fee for a service listed, but you do include it as part of the wellness visit, write "Included" in the blank. If you do not offer a service, write "NA" in the blank.

   There are other services included in a wellness visit than the ones we've listed. We will add together all of the applicable fees, some of which you've reported in other sections of this survey, in order to obtain the total fee for the wellness visit.

   a. Pediatric Canine (Puppy) Wellness Visits

       i. Offered? ☐ Yes ☐ No   **Fee**

       ii. DHPP vaccine   _____

       iii. One-year rabies vaccine   _____

       iv. Corona vaccine   _____

       v. Bordetella vaccine   _____

       vi. Parvovirus vaccine   _____

       vii. Leptospirosis vaccine   _____

       viii. Prophylactic deworming treatment (report fee for one visit)_____

   b. Adult Canine Wellness Visit

       i. Offered? ☐ Yes ☐ No   **Fee**

       ii. Rabies vaccine (three-year)   _____

       iii. Lyme disease vaccine   _____

       iv. Of the heartworm tests listed below, which one do you most commonly use during an adult canine wellness visit? (check one)

           ☐ Occult/antigen only

           ☐ Occult/antigen plus Lyme and *E. canis* (three-way screen)

           ☐ Do not routinely perform a heartworm test

           ☐ Other _____

c. Pediatric Feline (Kitten) Wellness Visit

    i. Offered? ☐ Yes ☐ No **Fee**

    ii. FVRCP vaccine _____

    iii. FeLV and FIV tests _____

    iv. FeLV vaccine _____

    v. Bartonella test _____

d. Adult Feline Wellness Visit

    i. Offered? ☐ Yes ☐ No **Fee**

    ii. FIP vaccine _____

    iii. Of the heartworm tests listed, which one do you most commonly perform during an adult feline wellness visit? (check one)

        ☐ Occult/antibody

        ☐ Occult/antigen

        ☐ Combined occult/antibody and occult/antigen

e. Avian Wellness Visit

    i. Offered? ☐ Yes ☐ No **Fee**

    ii. Fecal microscopic examination _____

    iii. Choanal slit culture/sensitivity _____

    iv. Cloacal culture/sensitivity _____

    v. Polyoma vaccine _____

    vi. Pacheco's vaccine _____

    vii. Complete blood count _____

f. Reptile Wellness Visit

    i. Offered? ☐ Yes ☐ No **Fee**

    ii. Fecal examination _____

    iii. Fecal culture _____

    iv. Complete blood count _____

    v. Gender determination (sexing) _____

    vi. Nail trim _____

g. Small Mammal Wellness Visit (Not Ferret)

    i. Offered? ☐ Yes ☐ No **Fee**

    ii. Ear mite examination _____

    iii. Tooth trim _____

h. Ferret Wellness Visit

    i. Offered? ☐ Yes ☐ No **Fee**

    ii. Distemper vaccine _____

	Fee
iii. Heartworm antigen test	_____
iv. Plasma chemistry profile	_____

SECTION III: VACCINATION PROTOCOLS

3. Do you evaluate each canine and feline patient to determine vaccination protocols based on risk assessment?   ☐ Yes   ☐ No

4. Do you use the Core (Recommended)/Non-core (Optional) vaccination protocols in your practice?
    ☐ Yes   ☐ No

    If yes:

    a. Do you follow the AAHA Canine Vaccine Guidelines?   ☐ Yes   ☐ No   ☐ Don't know

    b. Do you follow the AAFP's feline vaccination recommendations?   ☐ Yes   ☐ No   ☐ Don't know

5. How frequently do you recommend a dog or cat be vaccinated for the core vaccines (DHPP or FVRCP, respectively)? (check one)

    ☐ Annually

    ☐ Every two years

    ☐ Every three years

    ☐ Other (please specify) _____

6. If you recommend that a dog or cat be vaccinated for the core vaccines every two years or less, do you stagger the core vaccines with other vaccines (e.g., give DHPP in 2006, rabies in 2007)?
    ☐ Yes   ☐ No

    a. If you do <u>not</u> stagger the vaccines, how do you encourage the pet owner to bring the animal in for examinations during the "off" years (years when no vaccine would be scheduled)? (check all that apply)

    ☐ Recommend annual or biannual physical examinations

    ☐ Heartworm testing

    ☐ Other (please specify) _____

7. Do you administer Leptospirosis vaccinations?   ☐ Yes   ☐ No

    a. If yes, what is your recommendation for how frequently a dog should be vaccinated?

    ☐ Annually

    ☐ Biannually

    ☐ Other (specify frequency) _____

8. Are you checking immunity titers rather than vaccinating?   ☐ Yes   ☐ No

    a. If yes, please report fee _____

SECTION IV: DISCOUNTS

9. Do you reduce or discount wellness examination fees in any way if your client presents three or more pets for physical and booster vaccinations in the same office visit?   ☐ Yes   ☐ No

a. If yes, report the discount amount in dollars if client receives flat discount (report one or the other, not both).

　　i. _____ or the percentage discount

　　ii. _____ %

10. Do you offer discounts on medical services for pets belonging to senior citizens?
    ☐ Yes   ☐ No

    a. If yes, report the discount and describe if necessary _____

11. Do you offer discounts on prescription medications for pets belonging to senior citizens?
    ☐ Yes   ☐ No

    a. If yes, report the discount and describe if necessary _____

12. Do you offer discounts on over-the-counter (OTC) products for pets belonging to senior citizens?
    ☐ Yes   ☐ No

    a. If yes, report the discount and describe if necessary _____

SECTION V: LABORATORY FEES

13. Please report the lab and blood collection fee for each of the following species. Write "Included" in the blank if this fee is included in the fee for the laboratory service. If you do not perform blood work for a particular species, write "NA" in the blank.

	Fee		Fee
a. Canine	_____	d. Reptile	_____
b. Feline	_____	e. Small mammal (other than ferret)	_____
c. Avian	_____	f. Ferret	_____

14. Please report the lab fee charged to the client for each of the following tests and mark whether that test is <u>most commonly</u> performed in-house or by an outside lab (check one box only).

    If it is the outside lab that most commonly performs the test, please also indicate the fee you pay to the outside lab. If you check "In-house," do not report a "Fee Paid to Outside Lab."

    Unless another species is specified, report the fee for a canine test if you see dogs in your practice. If you do not see dogs, report the fee for the most common species you see (e.g., a feline-only practice should report fees for cats). If you are reporting for a species other than dog, please indicate the species here: a. _____

    Write "NA" in the blank if you do not offer the service.

	Fee Charged to Client	In-house	Outside Lab	Fee Paid to Outside Lab
b. ACTH stimulation	_____	☐	☐	_____
c. Arterial/Venous blood gases	_____	☐	☐	_____
d. Avian *Chlamydia*	_____	☐	☐	_____
e. Avian chromosomal sexing	_____	☐	☐	_____

	Fee Charged to Client	In–house	Outside Lab	Fee Paid to Outside Lab
f. Avian ELISA allergy testing	_____	☐	☐	_____
g. Avian PBFDV	_____	☐	☐	_____
h. Avian Polyomavirus test	_____	☐	☐	_____
i. Bacterial culture and sensitivity	_____	☐	☐	_____
j. Bladder stone analysis	_____	☐	☐	_____
k. Blood parasite testing				
i. *Ehrlichia*	_____	☐	☐	_____
ii. *Haemobartonella*	_____	☐	☐	_____
iii. *Babesia*	_____	☐	☐	_____
l. CBC (no differential)	_____	☐	☐	_____
m. CBC (automated)	_____	☐	☐	_____
n. CBC (with manual differential)	_____	☐	☐	_____
o. Chemistries				
i. Chemistry setup fee	_____	☐	☐	_____
ii. Single chemistry	_____	☐	☐	_____
iii. 2 chemistries	_____	☐	☐	_____
iv. 3 chemistries	_____	☐	☐	_____
v. 4 chemistries	_____	☐	☐	_____
vi. 5–7 chemistries	_____	☐	☐	_____
vii. 8–12 chemistries	_____	☐	☐	_____
viii. 16–24 chemistries	_____	☐	☐	_____
p. Cytology				
i. Fine-needle aspirate	_____	☐	☐	_____
ii. Vaginal	_____	☐	☐	_____
iii. Ear swab	_____	☐	☐	_____
iv. Skin swab	_____	☐	☐	_____
q. Dexamethasone suppression	_____	☐	☐	_____
r. Electrolytes (assume sodium, chloride, and potassium)	_____	☐	☐	_____
s. Fecal Diff-Quik stain (Clostridium)	_____	☐	☐	_____
t. Fecal examination				
i. Direct smear	_____	☐	☐	_____

		Fee Charged to Client	In–house	Outside Lab	Fee Paid to Outside Lab
ii.	Flotation (gravitational)	_____	☐	☐	_____
iii.	Flotation (centrifugation; zinc sulfate)	_____	☐	☐	_____
iv.	Sedimentation (Baermann)	_____	☐	☐	_____
v.	Wet mount for *Giardia*	_____	☐	☐	_____
u.	Fecal Gram's stain	_____	☐	☐	_____
v.	Feline leukemia (FeLV) test	_____	☐	☐	_____
w.	FeLV and feline immuno-deficiency virus (FIV) test	_____	☐	☐	_____
x.	FIV test	_____	☐	☐	_____
y.	Fructosamine test	_____	☐	☐	_____
z.	Fungal culture	_____	☐	☐	_____
aa.	Giardia antigen test	_____	☐	☐	_____
bb.	Glucose curve (6)	_____	☐	☐	_____
cc.	Glucose single	_____	☐	☐	_____
dd.	Health check—CBC with 16–24 chemistries and T4	_____	☐	☐	_____
ee.	Health check—CBC with 8–12 chemistries	_____	☐	☐	_____
ff.	Heartworm test, feline—occult/antibody	_____	☐	☐	_____
gg.	Heartworm test, feline—occult/antigen	_____	☐	☐	_____
hh.	Heartworm tests, feline—occult/antibody and occult/antigen	_____	☐	☐	_____
ii.	Heartworm test, canine—occult/antigen only	_____	☐	☐	_____
jj.	Heartworm test, canine—occult/antigen plus Lyme and *E. canis* (three-way screen)	_____	☐	☐	_____
kk.	Histopathology—multiple tissues	_____	☐	☐	_____
ll.	Histopathology—single tissue	_____	☐	☐	_____
mm.	Lyme testing	_____	☐	☐	_____
nn.	Pancreatic evaluation	_____	☐	☐	_____

		Fee Charged to Client	In–house	Outside Lab	Fee Paid to Outside Lab
	i. Trypsin-like immuno-reactivity (TLI)	_____	☐	☐	_____
	ii. Pancreatic lipase immuno-reactivity (PLI)	_____	☐	☐	_____
	iii. Canine pancreatic lipase (CPL)	_____	☐	☐	_____
	iv. Specific canine pancreatic lipase (Spec cPL)	_____	☐	☐	_____
oo.	Parvovirus test	_____	☐	☐	_____
pp.	PTH assay	_____	☐	☐	_____
qq.	Reticulocyte count	_____	☐	☐	_____
rr.	Serum testing for allergen-specific IgE	_____	☐	☐	_____
ss.	T4	_____	☐	☐	_____
tt.	T4, T3, free T4, and free T4ED	_____	☐	☐	_____
uu.	TSH level	_____	☐	☐	_____
vv.	Uric acid				
	i. Avian	_____	☐	☐	_____
	ii. Reptiles	_____	☐	☐	_____
ww.	Urinalysis (complete; assume specific gravity, dipstick, sediment)	_____	☐	☐	_____

Report the fees for each of the following tests individually:

	i. Urine specific gravity	_____	☐	☐	_____
	ii. Dipstick	_____	☐	☐	_____
	iii. Sediment	_____	☐	☐	_____
	iv. Microalbuminaria	_____	☐	☐	_____
	v. Urine protein:creatinine (UP:C) ratio	_____	☐	☐	_____

SECTION VI: MISCELLANEOUS DIAGNOSTIC SERVICES

15. For the following diagnostic services, please report the total fee charged to the client, including procedure, outside lab or diagnostic service charges, and interpretation, but excluding the office call/exam fee. If you do not offer a service, please write "NA" in the space provided.

a. Routine electrocardiogram (ECG) (assume evaluation of arrhythmia in a nine-year-old dachhund)   **Fee**

    In-house-six—lead ECG   _____

    In-house—lead II only   _____

    Outside service   _____

b. Schirmer tear test   _____

c. Corneal stain   _____

d. Tonometry   _____

e. Blood pressure evaluation   _____

f. Ear swab exam/stain (otitis externa)   _____

g. Wood's lamp examination   _____

SECTION VII: DIAGNOSTIC IMAGING

16. What is your radiographic setup fee (include fee for use of X-ray machine, initial setup of machine, and recording in appropriate logs)? _____

17. How does your practice charge for radiographs? (check one)

    ☐ By the number of films taken

    ☐ By the body part (e.g., chest, spine)

    ☐ By the number of films taken and by the body part

18. Please report the total fee charged to the client for routine radiographs, including the fee for the film, procedure (taking and developing radiographs), and interpretation, but <u>excluding</u> the office call/exam fee. If you do not offer a service, please write "NA" in the space provided.

   Fee

a. 1 view—8 x 10 cassette for dog or cat   _____

b. 1 view—8 x 10 additional   _____

c. 1 view—14 x 17 cassette for dog or cat   _____

d. 1 view—14 x 17 additional   _____

e. 2 views—chest, 60-lb dog   _____

f. 3 views—chest, 60-lb dog   _____

g. 2 views—pelvis, 60-lb dog   _____

h. 2 views—abdomen, 60-lb dog   _____

i. 2 views—spine, dachshund   _____

j. 2 views—abdomen, cat   _____

k. 2 views—chest, cat   _____

l. 2 views—forearm, cat   _____

m. 1 view—8 x 10 cassette, bird/reptile/small mammal   _____

	Fee
n. 1 view—14 x 17 cassette, bird/reptile/small mammal	_____
o. 2 views—8 x 10 cassette, bird/reptile/small mammal	_____
p. 2 views—14 x 17 cassette, bird/reptile/small mammal	_____
q. Dental radiography (rostral and caudal mandible and rostral and caudal maxilla)	_____

19. Please report the total fee charged to the client for myelograms, including fee for film, procedure, and interpretation, but <u>excluding</u> the office call/exam fee. If you do not offer a service, please write "NA" in the space provided.

	Fee
a. Spinal puncture	_____
b. Dye (assume Omnipaque™)	_____
c. 2 views—cervical spine	_____
d. CSF examination	_____

20. Please report the total fee charged to the client for ultrasound services, including procedure and interpretation, but <u>excluding</u> the office call/exam fee. If you do not offer a service, please write "NA" in the space provided.

	Fee
a. Does your practice have an ultrasound unit in-house?	☐ Yes ☐ No
b. Chest and abdomen	_____
c. Chest only	_____
d. Abdomen only	_____
e. Guided biopsy collection, liver	_____

21. Please report the total fee charged to the client for a CAT scan, including procedure and interpretation, but excluding the office call/exam fee. If you do not offer a service, please write "NA" in the space provided.

a. Does your practice have a CAT scan unit in-house?	☐ Yes ☐ No
b. Chest and abdomen	_____
c. Chest only	_____
d. Abdomen only	_____
e. Head	_____
f. Spine	_____
g. Thoracic limb	_____
h. Pelvic limb	_____

22. Please report the total fee charged to the client for an MRI, including procedure and interpretation, but <u>excluding</u> the office call/exam fee. If you do not offer a service, please write "NA" in the space provided.

a. Does your practice have a MRI unit in-house? ☐ Yes ☐ No

	Fee
b. Chest and abdomen	_____
c. Chest only	_____
d. Abdomen only	_____
e. Head	_____
f. Spine	_____
g. Thoracic limb	_____
h. Pelvic limb	_____

23. Please report the fees you charge for the following radiograph and ultrasound studies. Do not include charges for office call/examination, sedation, anesthesia, or hospitalization. For the radiographs, assume use of one cassette per view, using 10 x 12 film. Please report the fee you would normally charge if you perform this service, or indicate the fee you normally charge if you process the films and have a specialist interpret them. If you do not offer a service, please write "NA" in the space provided.

CASE ONE

Two-year-old, 25-lb male, mixed-breed dog presented with lameness of the right forelimb; pain and crepitation were noted over the radius and ulna during physical examination; two radiographic views were obtained of the right radius and ulna.

	Fee
a. Two films and your interpretation	_____

OR

| b. Films | _____ |
| c. AND specialist interpretation fee (as charged to client) | _____ |

CASE TWO

Nine-year-old, neutered, male miniature poodle with chronic hematuria; two scout films followed by ultrasound evaluation of bladder and urethra and then a double contrast cystogram using two films (total of four radiographic films).

	Fee
d. Cystogram procedure	_____
e. Contrast materials	_____
f. Bladder catheterization	_____
g. Double contrast cystogram (procedure)	_____
h. All films and your interpretation	_____

OR

i. Films	_____
j. **AND** specialist interpretation fee (as charged to client)	_____
k. **PLUS** ultrasound evaluation of bladder and urethra	_____

SECTION VIII: MEDICATIONS

If you do not offer a service, please write "NA" in the space provided.

24. What is the prescription fee for medications that are dispensed from your hospital? _____

25. Do you charge a fee to write a prescription to a client who wants to have the prescription filled elsewhere? ☐ Yes ☐ No

    a. If yes, please report fee _____

26. On average, what percentage markup do you apply to the following inventory items? Percentage markup is defined as [(Selling Price − Cost of Product)/Cost of Product] x 100.

    a. Prescription medications _____%
    b. Heartworm preventives _____%
    c. Food _____%
    d. Flea and tick products _____%
    e. Over-the-counter products _____%

SECTION IX: MISCELLANEOUS FEES

Grooming

27. For each of the following services, report the fee only if you routinely offer the service. Write "NA" in the blank if you do not routinely offer the service.

	Fee
a. Nail trim, dog	_____
b. Nail trim, cat	_____
c. Nail trim, bird	_____
d. Nail trim, small mammal	_____
e. Anal gland expression	_____
f. Bath and brush	_____
g. Wing trim	_____
h. Beak trim	_____

Boarding (report fee per night)

28. For each of the following services, report the fee only if you routinely offer the service. Write "NA" in the blank if you do not routinely offer the service.

	Fee
a. <30-lb dog in small cage	_____
b. <30-lb dog in small run	_____
c. 30–60-lb dog in medium run	_____
d. 61–90-lb dog in large run	_____

	Fee
e. >90-lb dog in large run	_____
f. Cat	_____
g. Bird	_____
h. Reptile	_____
i. Small mammal (other than ferret)	_____
j. Ferret	_____

Euthanasia

29. Write "Included" in the blank if you would not assess an additional fee. Write "NA" in the blank if you do not provide this service.

   a. 30-lb dog, admitted/owner not present (not including exam) **Fee**

       i. IV catheter and placement _____

       ii. Preanesthetic sedative _____

       iii. Euthanasia _____

   b. 30-lb dog, with owner present (not including exam)

       i. IV catheter and placement _____

       ii. Preanesthetic sedative _____

       iii. Euthanasia _____

   c. Cat

       i. IV catheter and placement _____

       ii. Preanesthetic sedative _____

       iii. Euthanasia _____

   d. Small mammal

       i. IV catheter and placement _____

       ii. Preanesthetic sedative _____

       iii. Euthanasia _____

   e. Bird/Reptile

       i. Preanesthetic sedative _____

       ii. Euthanasia _____

Necropsy

30. Please provide the fee you would charge a client for a complete necropsy (excluding brain and spinal column) and related diagnostics (e.g., histopathology) for the following animals. Write "NA" in the blank if you do not provide this service.

       Fee

   a. 30-lb dog _____

	Fee
b. 75-lb dog	_____
c. Cat	_____
d. Small mammal (not ferret)	_____
e. Ferret	_____
g. Bird	_____
h. Reptile	_____

Cremation

33. For each of the following services, report the fee only if you routinely offer the service. Write "NA" in the blank if you do not routinely offer the service.

	Fee
a. Communal (mass) cremation, 30-lb dog	_____
b. Communal (mass) cremation, cat	_____
c. Individual cremation, return of ashes to owner, 30-lb dog	_____
d. Individual cremation of cat, return of ashes to owner	_____
e. Disposal of small mammal	_____
f. Disposal of reptile or bird	_____

SECTION X: ABOUT YOUR PRACTICE

34. Report the following information about your practice.

    a. Please report the number of full-time-equivalent* (FTE) veterinarians who work in this practice: _____

    Full-time-equivalent = 40 hours worked per week. Example: A veterinarian who works 20 hours per week should be reported as a 0.5 FTE.

    b. Are you a member of AAHA?   ☐ Yes   ☐ No   ☐ Don't Know

      i. If yes, is your practice an AAHA–accredited practice?   ☐ Yes   ☐ No   ☐ Don't Know

    c. Is your practice small animal or mixed animal?

      ☐ Small animal

      ☐ Mixed animal

      ☐ Other (please specify) _____

    d. What type of practice do you have? (check all that apply) ☐ General practice

      ☐ Emergency only

      ☐ Referral/specialty

      ☐ Emergency and referral/specialty

      ☐ Feline only

      ☐ Exotic/avian only

- ☐ Exotic only
- ☐ Avian only
- ☐ Other (please specify) _____

e. How frequently do you adjust your fees?
   - ☐ Monthly
   - ☐ Every six months
   - ☐ Annually
   - ☐ Other (please specify frequency) _____

f. Do you adjust all fees at the same time, or do you adjust certain targeted fees only?   ☐ All at same time   ☐ Targeted fees

   i. If you increase fees all at the same time, what percentage do you increase them by? _____

g. Are the practice owners:
   - ☐ Females only
   - ☐ Males only
   - ☐ Females and males

SURVEY DUE BY NOVEMBER 8, 2006.

Thank you for completing this important survey. Please return it to AAHA in the postage-paid envelope provided. If you've lost your envelope, please mail to the address below. Feel free to call or email us if you have questions!

AAHA
AMERICAN ANIMAL HOSPITAL ASSOCIATION

Healthy Practices.
Healthier Pets.

American Animal Hospital Association
12575 W. Bayaud Avenue
Lakewood, CO 80228
800/252-2242 or 303/986-2800
AAHAPress@AAHAnet.org
© 2006 American Animal Hospital Association

2006
AAHA
Veterinary Fee Survey

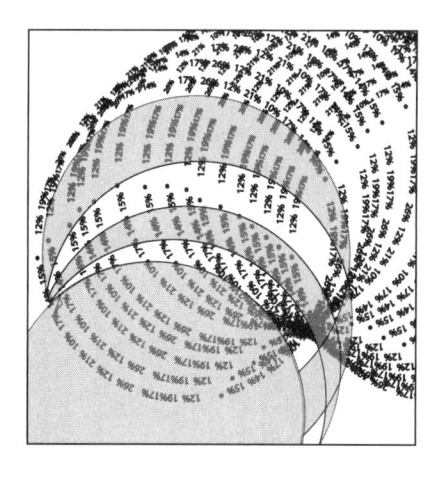

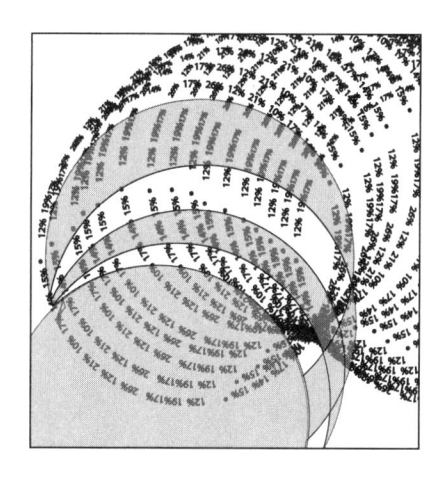

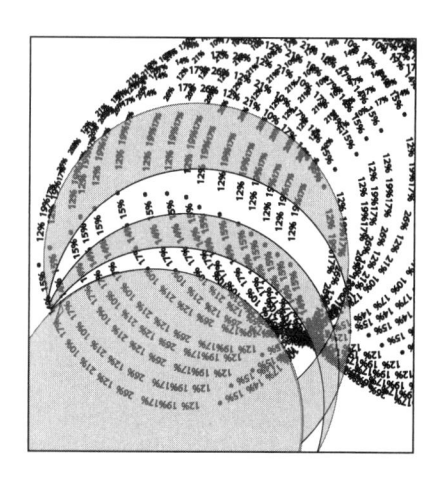

Survey Due:

November 8, 2006

AAHA
AMERICAN
ANIMAL
HOSPITAL
ASSOCIATION

Healthy Practices.
Healthier Pets.

Survey B #

INSTRUCTIONS

Thank you for participating in AAHA's fee survey. We realize that practices establish fees in different ways, so we designed this form to accommodate the most common formats used. We described various scenarios thought to be typical for most practices so that we could provide additional insight regarding fees related to case management.

Please follow the instructions for each section, and respond as completely as you can. If you are part of a multi-practice corporation, please complete the survey for your individual practice. Please report fees that were in effect as of October 1, 2006. Keep a copy of your survey for your records.

Your information is completely confidential. Your practice's name will not be attached to the survey. All results will be reported in aggregate, and individual responses will not be released.

SECTION I: HOSPITALIZATION

1. Report the fee assessed for hospitalization (not including examination, veterinarian supervision, or treatment) of a routine medical case. Please include the fees for IV catheter and placement in addition to the IV fluids in the fees you report for hospitalization with IV. If you do not base fees on the weight of the pet, please repeat the same fee for different weights. If you do not offer the service, please write "NA" in the blank.

	Hospitalization Fee with IV (Per Day)	Hospitalization Fee without IV (Per Day)
No Overnight Stay		
a. 10-lb cat	_____	_____
b. 25-lb dog	_____	_____
c. 60-lb dog	_____	_____
d. 100-lb dog	_____	_____
e. Bird	_____	_____
f. Reptile	_____	_____
g. Small mammal (other than ferret)	_____	_____
h. Ferret	_____	_____
With Overnight Stay		
i. 10-lb cat	_____	_____
j. 25-lb dog	_____	_____
k. 60-lb dog	_____	_____
l. 100-lb dog	_____	_____
m. Bird	_____ (w/ tube feeding)	_____ (w/o tube feeding)
n. Reptile	_____	_____
o. Small mammal (other than ferret)	_____	_____
p. Ferret	_____	_____

2. Does your practice have someone physically present in the hospital 24 hours a day?
   ☐ Yes   ☐ No

   a. If yes, who is present in the hospital 24 hours a day?

   i. ☐ Full medical staff (24-hour hospital)

   ii. ☐ Full medical staff (daytime practice plus overnight/weekend emergency)

   iii. ☐ Full medical staff daytime and assigned person overnight/weekends

      1. ☐ Assigned person is a technician

      2. ☐ Assigned person is a veterinarian

      3. ☐ Assigned person is other (indicate title) _____

   iv. ☐ Full medical staff night time/weekends and assigned person daytime

      1. ☐ Assigned person is a technician

      2. ☐ Assigned person is a veterinarian

      3. ☐ Assigned person is other (indicate title) _____

   b. If no, is practice an outpatient clinic only?   ☐ Yes   ☐ No

SECTION II: FLUID THERAPY

3. Please report the fee you charge a client for providing fluid therapy to a patient. Please provide fees for each type of catheter. If a fee is included in the fee you charge for the fluid (see question 4 below), write "Included" in the blank. If you do not offer a service, please write "NA" in the blank.

IV Fluids Setup

	Fee
a. Catheter	
i. Butterfly catheter	_____
ii. IV indwelling catheter	_____
iii. Jugular catheter (catheter thru needle)	_____
b. Catheter placement	
i. Butterfly catheter	_____
ii. IV indwelling catheter	_____
iii. Jugular catheter (catheter thru needle)	_____
c. Initial bag of fluids (assume 1,000 ml lactated Ringer's)	_____
d. Fluid infusion pump setup	_____
e. Fluid infusion pump use (assume 8 hours)	_____
f. Fluid warmer	_____

4. Please report the fee you charge a client for administering 1,000 ml of the following fluids to a patient. If you do not offer a particular fluid, please write "NA" in the space provided.

        Fee

a. Lactated Ringer's solution   _____

b. Ringer's solution   _____

c. Normosol-R   _____

d. 0.9% NaCl   _____

e. D5W (5% dextrose)   _____

f. Dextran   _____

g. Hetastarch   _____

h. Other (please specify) _____   _____

SECTION III: ANESTHESIA

5. Assume you examined a patient a week ago (and charged an exam fee) and recommended that the patient be admitted today for a minor surgical procedure. You conduct a preanesthetic exam. Do you charge for the preanesthetic exam?   ☐ Yes   ☐ No

   a. If yes, how much do you charge? _____

6. Assume you are anesthetizing a 40-pound dog for a surgical procedure. Please indicate your fee for anesthesia, and itemize the fees for those services you charge a separate fee for. If the fee for a service is included in the total fee for administering anesthesia (items d through f below), write "Included" in the blank. If you do not routinely perform the service, write "NA" in the blank.

Sedation

        Fee

a. IV sedative for 30-minute radiology procedure (no intubation/inhalant, 30-lb dog)   _____

b. IM sedative for fractious cat for abscess treatment   _____

General Anesthesia

c. Preanesthetic sedation   _____

d. IV induction   _____

e. Intubation   _____

f. Inhalant, 30 minutes (assume isoflurane for 40-lb dog)   _____

g. Inhalant, 60 minutes (assume isoflurane for 40-lb dog)   _____

h. Additional hour inhalant (assume isoflurane for 40-lb dog)   _____

i. Anesthetic monitoring, electronic   _____

j. Anesthetic monitoring, manual   _____

Other Anesthesia   Fee

    k. Mask inhalation anesthesia for small mammal   _____

    l. Mask inhalation anesthesia for bird   _____

Pain Management

    m. Does your practice include pain management in the fees for elective procedures?

        ☐ Yes, for no additional fee

        ☐ Yes, for an additional fee

        ☐ Not routinely provided

    n. Does your practice include pain management in the fees for nonelective procedures?

        ☐ Yes, for no additional fee

        ☐ Yes, for an additional fee

        ☐ Not routinely provided

    o. Does your practice use fentanyl patches?   ☐ Yes   ☐ No

        i. If yes, please report the fee for fentanyl patch for a 30-lb dog _____

SECTION IV: TREATMENT PROCEDURES

7. Do you routinely provide an estimate for treatment plans?   ☐ Yes   ☐ No

8. Do you routinely require a deposit for treatment?   ☐ Yes   ☐ No

    a. If yes, what type of deposit do you require? (check one)

        ☐ 25% of estimate

        ☐ 50% of estimate

        ☐ More than 50% of estimate

Treatment Cases

For each of the following cases, report the fees for nonsurgical treatment of the patient. Write "Included" in the blank if the service is performed as part of the case but is not charged for separately. Write "NA" in the blank if a service is not provided.

CASE ONE

9. A 60-lb golden retriever that is dehydrated has an intravenous catheter placed and has 1,500 ml of lactated Ringer's solution administered over a 24-hour period.

     Fee

    a. Admitting examination   _____

    b. Catheterization fee   _____

    c. 1,500 ml lactated Ringer's solution   _____

    d. Fluid infusion pump   _____

	Fee
e. Hospitalization	_____
f. Veterinarian/technician supervision	_____
g. Medical waste disposal	_____

CASE TWO

10. A 13-lb, neutered male cat is sedated for treatment of a draining abscess. He will be treated today, hospitalized overnight, and examined in the morning. A one-week supply of antibiotics will be dispensed to go home.

	Fee
a. Admitting examination	_____
b. Light sedation	_____
c. Abscess curettage, debridement, and flush	_____
d. Antibiotic injection	_____
e. Pain management medication post-treatment	_____
f. Hospitalization	_____
g. Medical progress exam	_____
h. Antibiotic (oral or injection)	_____
i. One-week supply of antibiotics	_____
j. Medical waste disposal	_____

CASE THREE

11. A 30-lb, six-year-old male schnauzer has moderate pancreatitis, which you suspected and confirmed with diagnostic testing. The schnauzer is hospitalized for three days. You give fluids, medical treatment and pain management medication each day, perform a chemistry panel on days 2 and 3, and give the dog a special diet on the last day of hospitalization.

	Fee
Day One	
a. Admitting examination	_____
b. IV catheter and placement	_____
c. First liter of IV fluids	_____
d. Fluid infusion pump use	_____
e. Two antibiotic injections	_____
f. Antiemetics	_____
g. Hospitalization	_____
h. Two views, abdominal radiographs	_____
i. Pain management medication (fee for administering twice)	_____
j. Medical waste disposal	_____

                                                             Fee

Day Two

   k. Inpatient examination (veterinarian supervision)   _____

   l. Second liter of fluids late on second day   _____

   m. Fluid infusion pump, continuing use   _____

   n. Two subcutaneous antibiotic injections   _____

   o. CBC with six chemistries   _____

   p. Antiemetics   _____

   q. Hospitalization   _____

   r. Pain management medication (fee for administering twice)   _____

   s. Medical waste disposal   _____

                                                             Fee

Day Three

   t. Inpatient examination (veterinarian supervision)   _____

   u. Third liter of fluids late on third day   _____

   v. Fluid infusion pump, continuing use   _____

   w. Two subcutaneous antibiotic injections   _____

   x. CBC with two chemistries   _____

   y. Hospitalization   _____

   z. Pain management medication (fee for administering twice)   _____

   aa. Medical waste disposal   _____

   bb. Special diet—in hospital   _____

CASE FOUR

12. A 25-pound cocker spaniel presents with otitis externa in the left ear and is treated as a day patient.

                                                             Fee

   a. Examination   _____

   b. Preanesthetic lab work
(CBC, BUN, creatinine, sodium, and potassium)   _____

   c. Anesthesia, 30 minutes   _____

   d. Ear swab/cytology   _____

   e. Ear cleaning   _____

   f. Antibiotic injection   _____

   g. Pain management medication post-treatment   _____

   h. Hospitalization, day charge   _____

	Fee
i. Antibiotic, one-week supply	_____
j. Analgesics, three-day supply	_____
k. Medical waste disposal	_____

SECTION V: SURGERY

Surgery Setup

13. Please report your fees for the following services. If you do not separate fees, but instead charge one setup fee, enter a fee in the blank for "Total surgery setup fee" only. If you do not charge for setup at all, enter "NA" in the blank for "Total surgery setup fee" only.

	Fee
a. Total surgery setup fee	_____
b. Surgical suite usage	_____
c. Surgical pack fee	_____
d. Disposables	_____
e. Medical waste disposal	_____
f. Assistant time (assume 1 hour)	_____
g. Suture material	_____
h. Staples	_____
i. Use of cold tray instruments	_____

14. Do you charge an electronic monitoring fee?   ☐ Yes   ☐ No

    a. If yes, how much do you charge?   _____

15. Do you add an additional fee for a surgery procedure performed as an emergency?   ☐ Yes   ☐ No

    a. If yes, what is the additional fee?   _____

Elective Procedures

Spay/Neuter

16. Please report the total fee (including exam, anesthesia, pain management, surgery room setup, sutures, etc.) you charge for these services and the time allotted for each. In each case, assume the patient is six-to-seven months of age but prepubertal (i.e., not an early or pediatric spay/neuter). Write "NA" in the blank if you do not provide this service.

	Total Fee	Time Allotted (Minutes)
CANINE SPAY		
a. <25 lbs	_____	_____
b. 25–50 lbs	_____	_____
c. 51–75 lbs	_____	_____
d. >75 lbs	_____	_____

	Total Fee	Time Allotted (Minutes)
CANINE NEUTER		
e. <25 lbs	_____	_____
f. 25–50 lbs	_____	_____
g. 51–75 lbs	_____	_____
h. >75 lbs	_____	_____
OTHER SPECIES		
i. Feline spay	_____	_____
j. Feline neuter	_____	_____
k. Rabbit spay	_____	_____
l. Rabbit neuter	_____	_____
m. Ferret spay	_____	_____
n. Ferret neuter	_____	_____

17. Do you offer a discounted early spay/neuter (i.e., pediatric spay/neuter) price for puppies and kittens? ☐ Yes ☐ No

    If yes, please report the fees below.

	Total Fee
a. Pediatric/puppy spay	_____
b. Pediatric/puppy neuter	_____
c. Pediatric/kitten spay	_____
d. Pediatric/kitten neuter	_____

Declawing

18. AAHA's position statement with regard to declawing is as follows: Declawing of domestic cats should be considered only after attempts have been made to prevent the cat from using its claws destructively or when clawing presents a significant health risk for people within the household. See www.AAHAnet.org for the full position statement.

    Please report the total fee (including exam, anesthesia, pain management, surgery room setup, sutures, etc.) you charge for these services and the time allotted for each. Write "NA" in the blank if you do not provide this service.

	Total Fee	Time Allotted (Minutes)
a. Declaw two paws	_____	_____
b. Declaw four paws	_____	_____

Ear Cropping and Tail Docking

19. AAHA's position statement with regard to ear cropping and tail docking is as follows: Ear cropping and/or tail docking in pets for cosmetic reasons are not medically indicated nor of benefit to the

patient. These procedures cause pain and distress, and, as with all surgical procedures, are accompanied by inherent risks of anesthetic complications, hemorrhage, and infection. Therefore, the American Animal Hospital Association opposes both the cropping of ears and the docking of tails when done solely for cosmetic reasons. See www.AAHAnet.org for the full position statement.

Please report the total fee (including exam, anesthesia, pain management, surgery room setup, sutures, etc.) you charge for these services and the time allotted for each. Write "NA" in the blank if you do not provide this service.

	Total Fee	Time Allotted (Minutes)
a. Ear cropping	_____	_____
b. Tail docking (neonatal)	_____	_____

Nonelective Procedures

20. Please report the fee for the <u>surgical procedure only</u>, along with the average number of minutes scheduled for each of the following procedures. (Do not include anesthesia, treatment, hospitalization, or supplies.) If you do not offer a service, write "NA" in the blank.

	Fee	Time Allotted (Minutes)
a. Abdominal exploratory, cat	_____	_____
b. Abdominal exploratory, 30-lb dog	_____	_____
c. Abdominal exploratory, 75-lb dog	_____	_____
d. Adrenalectomy, ferret	_____	_____
e. Amputation		
i. Cat, thoracic limb	_____	_____
ii. Cat, pelvic limb	_____	_____
iii. Cat, tail	_____	_____
iv. Dog, 30-lb, thoracic limb	_____	_____
v. Dog, 75-lb, thoracic limb	_____	_____
vi. Dog, 30-lb, pelvic limb	_____	_____
vii. Dog, 75-lb, pelvic limb	_____	_____
viii. Dog, 30-lb, tail	_____	_____
ix. Dog, 75-lb, tail	_____	_____
f. Anal gland resection, bilateral, 30-lb dog	_____	_____
g. Aural hematoma repair	_____	_____
h. Broken blood feather quill removal	_____	_____
i. C-section, cat with three kittens	_____	_____
j. C-section, 30-lb dog with four puppies	_____	_____
k. C-section, 75-lb dog with four puppies	_____	_____

		Fee	Time Allotted (Minutes)
l.	Cataract surgery, one lens	_____	_____
m.	Colonoscopy with biopsy	_____	_____
n.	Cruciate repair, anterior, 60-lb dog	_____	_____
o.	Dewclaw removal, non-neonate canine, two dewclaws on front limbs	_____	_____
p.	Diaphragmatic hernia, cat	_____	_____
q.	Diaphragmatic hernia, 30-lb dog	_____	_____
r.	Diaphragmatic hernia, 75-lb dog	_____	_____
s.	Ear resection, unilateral, 30-lb dog	_____	_____
t.	Endoscopy, upper GI, with biopsy	_____	_____
u.	Enucleation, one eye, 30-lb dog	_____	_____
v.	Femoral head removal, cat	_____	_____
w.	Gastrotomy, foreign-body removal, cat	_____	_____
x.	Gastrotomy, foreign-body removal, 30-lb dog	_____	_____
y.	Gastrotomy, foreign-body removal, 75-lb dog	_____	_____
z.	Hemilaminectomy, L1-L2, 15-lb dog	_____	_____
aa.	Incisor extraction, rabbit	_____	_____
bb.	Intestinal resection anastomosis, cat	_____	_____
cc.	Intestinal resection anastomosis, 30-lb dog	_____	_____
dd.	Intestinal resection anastomosis, 75-lb dog	_____	_____
ee.	Laser surgery		
	i. Does your practice have a laser surgery unit in house?	☐ Yes	☐ No
	ii. Does your practice have access to a laser surgery unit through a referral center in your area?	☐ Yes	☐ No

Please provide the following fees for surgery performed with laser surgery in house. Do not include charges for office call/examination, sedation, anesthesia, or hospitalization. If you do not have an in-house laser surgery unit, please write "NA."

		Fee	Time Allotted (Minutes)
	iii. Small tumor removal	_____	_____
	iv. Feline declaw (two paws)	_____	_____
	v. Ear surgery (for chronic otitis)	_____	_____
	vi. Oral surgery	_____	_____
ff.	Luxating patella repair, miniature poodle	_____	_____
gg.	Mastectomy, unilateral, cat	_____	_____
hh.	Mastectomy, unilateral, 30-lb dog	_____	_____
ii.	Mastectomy, unilateral, 75-lb dog	_____	_____

	Fee	Time Allotted (Minutes)

jj. Skin tumor removal, 4 cm _____ _____

kk. Splenectomy, 30-lb dog _____ _____

ll. Suture laceration, 4 cm _____ _____

mm. Tibial fracture repair, midshaft, Amazon parrot _____ _____

nn. Tonsillectomy _____ _____

oo. TPLO _____ _____

pp. Urethrostomy, cat _____ _____

qq. Unclassified surgery, per-minute fee, surgical suite _____ _____

  i. Do you charge a minimum surgical fee for surgery performed in the surgical suite?
   ☐ Yes ☐ No

  ii. If yes, what is the minimum surgical fee? _____

rr. Unclassified surgery, per-minute fee, nonsurgical suite _____ _____

  i. Do you charge a minimum surgical fee for surgery performed in the clean surgery area?
   ☐ Yes ☐ No

  ii. If yes, what is the minimum surgical fee? _____

ss. Wing fracture repair on budgie, simple _____ _____

Surgical Cases

In the following two cases, report the fees for the services listed. Assume that the surgery is conducted in a single-use surgery room using full aseptic procedures. Write "Included" in the blank if the service is included in the total fee (not separately itemized). Write "NA" in the blank if you do not offer this service.

CASE ONE

21. A 25-lb female dog is presented for removal of a single, large calculus from the bladder. Total surgical time: 45 minutes. If you do not charge a separate fee for a service but include it as part of the surgery fee, write "Included" in the blank. If you do not provide this service, write "NA" here: _____

	Fee

a. Presurgical examination _____

b. CBC _____

c. Chemistry panel with six chemistries _____

d. Sedation _____

e. Fluids _____

f. Monitoring _____

g. Total surgery setup fee _____

h. Surgical suite usage _____

i. Surgical pack fee _____

	Fee
j. Disposables	_____
k. Medical waste disposal	_____
l. Assistant time (assume one hour)	_____
m. Surgical removal of bladder calculus	_____
n. Anesthesia, 45 minutes	_____

(NOTE: If you charge one lump sum for the entire surgical procedure enter "Included" in the blanks for a–f, h, and i, and enter total fee here.)

o. Suture material	_____
p. Staples	_____
q. Pain management medication post-op	_____
r. Hospitalization (day of surgery only)	_____

CASE TWO

22. A 45-lb male dog; repair of midshaft tibial fracture with one intramedullary pin; total surgical time: 60 minutes

	Fee
a. Presurgical examination	_____
b. CBC	_____
c. Chemistry panel with six chemistries	_____
d. Sedation	_____
e. Fluids	_____
f. Monitoring	_____
g. Total surgery setup fee	_____
h. Surgical suite usage	_____
i. Surgical pack fee	_____
j. Disposables	_____
k. Medical waste disposal	_____
l. Assistant time (assume one hour)	_____
m. Surgical fracture repair (IM pin)	_____

(NOTE: If you charge one lump sum for the entire surgical procedure enter "Included" in the blanks for a–f, h, and i, and enter total fee here.)

n. Suture material	_____
o. Staples	_____
p. Pain management medication post-op	_____
q. Hospitalization (day of surgery only)	_____

SECTION VIII: DENTISTRY

Please report your fees for the following case. If a service is included in the case but not charged for separately, write "Included" in the blank. If you do not offer a service, write "NA" in the blank.

CASE ONE

23. A six-year-old, 35-lb, spayed female cocker spaniel was examined one week ago, vaccinated, and found to have moderate dental calculus and slight gingivitis. The patient is admitted this morning, anesthetized for routine dental scaling, subgingival curettage, and tooth polishing; no extractions are performed.

		Fee
a.	Preanesthetic exam	___
b.	CBC with differential	___
c.	Chemistry panel with eight chemistries	___
d.	Anesthesia (30 minutes)	___
e.	IV catheter and placement	___
f.	IV fluids (assume 1,000-ml bag of lactated Ringer's solution)	___
g.	Dental scaling and polishing	___
h.	Subgingival curettage	___
i.	Fluoride application	___
j.	Electronic monitoring	___
k.	Pain medication post-procedure	___
l.	Injectable antibiotics post-procedure	___
m.	Hospitalization	___
n.	Antibiotics, one-week supply	___

24. Please report the fees for each of the following additional dental procedures, assuming the same facts as in Case One above. Write "Included" in the blank if you would not assess an additional fee. Write "NA" in the blank if you do not provide this service.

		Fee
a.	Extraction of moderately loose premolar tooth	___
b.	Extraction of firmly implanted upper fourth premolar tooth	___
c.	Endodontic treatment of upper fourth premolar tooth	___
d.	Endodontic treatment of lower first molar	___
e.	Endodontic treatment of upper canine tooth	___
f.	Post-op injectable antibiotic	___
g.	Post-op antibiotics, one-week supply	___

SECTION X: ABOUT YOUR PRACTICE

25. Please report the number of full-time-equivalent* (FTE) veterinarians who work in this practice: _____

    (*Full-time-equivalent = 40 hours worked per week. Example: A veterinarian who works 20 hours per week should be reported as a 0.5 FTE.)

26. Are you a member of AAHA?   ☐ Yes   ☐ No   ☐ Don't Know

    a. If yes, is your practice an AAHA-accredited practice?   ☐ Yes   ☐ No   ☐ Don't Know

27. Is your practice small animal or mixed animal?
    - ☐ Small animal
    - ☐ Mixed animal
    - ☐ Other (please specify) _____

28. What type of practice do you have? (check all that apply)
    - ☐ General practice
    - ☐ Emergency only
    - ☐ Referral/specialty
    - ☐ Emergency and referral/specialty
    - ☐ Feline only
    - ☐ Exotic/avian only
    - ☐ Exotic only
    - ☐ Avian only
    - ☐ Other (please specify) _____

29. How frequently do you adjust your fees?
    - ☐ Monthly
    - ☐ Every six months
    - ☐ Annually
    - ☐ Other (please specify frequency) _____

30. Do you adjust all fees at the same time, or do you adjust certain targeted fees only?
    ☐ All at same time   ☐ Targeted fees

    a. If you increase fees all at the same time, what percentage do you increase them by? _____

31. Are the practice owners:
    - ☐ Females only
    - ☐ Males only
    - ☐ Females and males

SURVEY DUE BY NOVEMBER 8, 2006.

Thank you for completing this important survey. Please return it to AAHA in the postage-paid envelope provided. If you've lost your envelope, please mail to the address below. Feel free to call or email us if you have questions!

AAHA
AMERICAN ANIMAL HOSPITAL ASSOCIATION

Healthy Practices.
Healthier Pets.

American Animal Hospital Association
12575 W. Bayaud Avenue
Lakewood, CO 80228
800/252-2242 or 303/986-2800
AAHAPress@AAHAnet.org
© 2006 American Animal Hospital Association